THEORY AND APPROACH OF INFORMATION RETRIEVALS FROM ELECTROMAGNETIC SCATTERING AND REMOTE SENSING

Theory and Approach of Information Retrievals from Electromagnetic Scattering and Remote Sensing

by

YA-QIU JIN
Fudan University,
Shanghai, China

Springer

A C.I.P. Catalogue record for this book is available from the Library of Congress.

ISBN-10 1-4020-4029-6 (HB)
ISBN 13 978-1-4020-4029-0 (HB)
ISBN 10 1-4020-4030-X (e-book)
ISBN 13 978-1-4020-4030-6 (e-book)

Published by Springer,
P.O. Box 17, 3300 AA Dordrecht, The Netherlands.

www.springeronline.com

Printed on acid-free paper

Printed in the Netherlands.

Contents

Chapter 3
SAR Imagery and Applications 91

Chapter 4
Remote Sensing Data Validation and Applications 151

Chapter 5

Numerical Forward-Backward Method (FBM) for Scattering from Randomly Rough Surface and Target **218**

Chapter 6

Numerical Finite Element Method (FEM) for Scattering from Randomly Rough Surface and Target **260**

Chapter 7
Inverse Scattering and Parameter Retrievals 310

Chapter 8
Electromagnetics of Complex Particulate Media 341

Index 383

Preface

Advances in space-borne remote sensing have significantly changed the mankind viewpoint how to observe our own Earth planet. Great amount of remote sensing data and images presents new resources to quantitatively describe and monitor our Earth environment, atmosphere, oceanic and land surfaces. In remote sensing, electromagnetic (EM) scattering, emission and wave propagation, as interaction with the Earth environment, lay the physical basis for understanding and extracting geoscientific information. Study of electromagnetic waves with remote sensing application has become an active and interdisciplinary area.

This book presents some new progress on the theoretical and numerical approaches for information retrieval of the remote sensing via EM scattering and emission.

We begin in Chapter 1 with the vector radiative transfer (VRT) theory for inhomogeneous scatter media. The VRT takes account of multiple scattering, emission and propagation of random scatter media, and quantitatively leads to insights of elucidating and understanding EM wave-terrain surface interaction. Meanwhile, it is extensively applicable to carrying out data interpretation and validation, and to solving the inverse problem, e.g. iteratively, physically or statistically. In Chapter 1, iterative solutions of multiple scattering and emission from inhomogeneous dense scatter media, and inhomogeneous non-spherical scatter media are discussed. Three-dimensional VRT equation (3D-VRT) for spatially inhomogeneous random scatter media for high resolution observation is also investigated.

The polarimetric imagery of synthetic aperture radar (SAR) technology is one of most important advances in space-borne microwave remote sensing during recent decades. Chapter 2 deals with fully polarimetric scattering theory. The parameterized model of random non-spherical scatter media is developed. The theories of polarimetric Mueller matrix with coherency matrix, entropy, copolarized and cross-polarized indexes, and the probability density functions of multi-look four Stokes parameters are discussed. When a pulse wave is propagating through the media, its echo shows temporal variation due to time delay, and is able to identify different volumetric and surface scatterings. In this chapter, a theoretical model of layered media of random non-spherical scatterers under a polarized pulse incidence, and its Mueller matrix solution are developed, and applied to estimation of the canopy height and surface geophysical properties.

Chapter 3 devotes to applications of the SAR imagery. Making use of the Mueller matrix solution and morphological processing, the digital elevation mapping of the earth surface is inverted from a single-pass polarimetric SAR image data. Two-thresholds expectation maximum and Markov random field algorithms are studied to make automatic change detection from multi-temporal SAR images over the urban area. To make unsupervised surface classification, the deorientation theory of polarimetric scattering targets is developed and applied to SAR images over inhomogeneous terrain surfaces. Some other algorithms such as for ship detection and moving target refocusing in SAR images are also discussed.

Great amount of remote sensing data and images in temporal and spatial scales presents a challenge how to quantitatively extract the information from data-image and make new knowledge based on those information. Data validation from remote sensing is always required to meet various operational demands. It might be as a final objective and process that encompass the entire remote sensing system. This is discussed in Chapter 4 along with several applications, for example, the terrain surface moisture mapping based on multi-year passive microwave remote sensing, monitoring the sandstorm using the SSM/I (special sensor microwave/imager) data and Getis statistics, a genetic algorithm to retrieve the land surface roughness and soil wetness, etc. Correlations of active SAR and passive SSM/I observations over snowpack and oceanic surfaces are also demonstrated. Data fusions of Radarsat SAR and DMSP SSM/I for monitoring sea ice, and Landsat ETM+ and ERS-2 SAR for classification of urban terrain surfaces are studied.

Chapter 5 and 6 are devoted to fast computational methods in bistatic scattering simulation of a composite model of randomly rough surface with a target presence when a tapered EM wave is incident upon at low grazing angle. Chapter 5 deals with the generalized forward-backward method with the spectral accelerate algorithm (GFBM-SAA) for numerical scattering simulations of a ship-like target on wind-driven oceanic rough surface. Bistatic scattering from a fractal dielectric rough surface is also studied. Parameterization of the tapered incident wave for numerical scattering simulation is discussed.

Chapter 6 deals with the finite element methods (FEM) in bistatic scattering of a target on or above rough surface under an EM wave incidence at low grazing angle. The methods of FEM-DDM (domain decomposition method) for large scale rough surface, FEM-TLQSA (two level quasi-stationary algorithm) for the Doppler spectrum of a flying target above dynamic oceanic surface are developed. The angular correlation function of scattering from an object over rough surface in the FEM approach is also illustrated.

Inverse problems, such as inverse scattering, dielectric reconstruction and scatter characteristic recovery and so on, require further comprehensive studies in both the theoretical approach and operational data validation. In Chapter 7, we discuss some inversion issues. Using the parabolic approximation, roughness profile of the fractal surfaces is reconstructed from scattering measurements at low grazing incidence. Iterative inversions of the scalar radiative transfer and vector radiative transfer equations for a layer of random spheroids are developed. Iterative inversion is applied to retrievals of inhomogeneous vegetation canopy biomass and land surface moisture in SAR observation.

Chapter 8 presents some discussions on electromagnetics of complex media, such as polarimetric scattering from a layer of spatially oriented chiral spheroids, or metamaterial spheroids. Time-frequency analysis of EM pulse response from a large spherical target is demonstrated. Monte Carlo simulation for clustered scattering from very densely (fractional volume is large than 0.4) random spherical particles is also studied.

This book, based on my works published during most recent years, is intended as a textbook for graduate students and a reference book for scientists to see the recent progress in my research laboratory.

I would like to acknowledge the collaboration with my colleagues and graduate students. Research supports from the State Ministry of Science and Technology (2001CB309400), National Natural Science Foundation of China, and Shanghai Science and Technology Committee are gratefully acknowledged.

Chapter 1

Vector Radiative Transfer (VRT) Theory of Inhomogeneous Random Media 1

1-1 Iterative Solution of Multiple Scattering and Emission From Inhomogeneous Dense Scatter Media

The vector radiative transfer (VRT) equation is an integral-deferential equation to describe multiple scattering, absorption and transmission of four Stokes parameters in random scatter media. From the integral formal solution of VRT equation, the lower order solutions, such as the first-order scattering for a layer medium or the second order scattering for a half space, can be obtained [Jin 1994, Tsang et al. 1985]. The lower order solutions are usually good at low frequency when high-order scattering is negligible. It is not feasible to continue iteration for obtaining a high order scattering solution because too many folds of integration would be involved.

For example, in the space-borne microwave remote sensing the Defense Meteorological Satellite Program (DMSP) Special Sensor Microwave/Imager (SSM/I) employed seven channels of 19, 37 and 85*GHz* with dual polarization and 22 *GHz* with only vertical polarization. Multiple scattering from the terrain surfaces such as snowpack cannot be neglected at these channels. The discrete ordinate and eigen-analysis method has been studied to take into account multiple scattering and apply it to remote sensing of atmospheric precipitation, snowpack, etc. [Jin 1994]. A snowpack was modeled as a layer of random and dense spherical particles, and the VRT for a layer of uniformly dense spherical particles has been numerically studied by the discrete ordinate method [Jin 1994, Tsang et al. 1985].

However, due to surface melting and refrozen crust presence, the snowpack undergoes stratifying and might have inhomegeneous profiles of the ice grain size, fractional volume and physical temperature etc. It becomes necessary to study multiple scattering and emission from stratified snowpack of dense particles. However, the discrete ordinate and eigen-analysis method cannot be simply applied to multi-layers model, because numerically solving a set of the VRT equations is difficult.

By stratifying the inhomogeneous media into multiple thin slabs and employing the first order Mueller matrix of each thin slab, an iterative method is developed to derive high orders scattering solutions of the whole scatter media. High-order scattering and emission from inhomogeneous stratifying media of dense spherical particles are numerically obtained [Jin and Liang 2003]. The brightness temperatures at low frequency, such as 5.3 *GHz*, without high-order scattering and at SSM/I channels with high order scattering are discussed. This approach is also compared with the conventional discrete ordinate method for a uniform layer model. Numerical simulation for an inhomogeneous snowpack is also compared with the measurements of microwave remote sensing.

1-1.1 Iterative Solution of VRT Equation for Dense Scatterers Media

The effective propagation constant K is defined to take into account the scattering coherence of random dense particles [Jin 1994, Tsang et al. 1985]. It has been derived as

$$K^2 = k_0^2 + fB[1 + i\frac{2}{9}Ka^3 B\frac{(1-f)^4}{(1+2f)^2}] \tag{1-1.1}$$

where

$$B = \frac{k_s^2 - k_0^2}{1 + (k_s^2 - k_0^2)(1-f)/(3K^2)} \tag{1-1.2}$$

The radius of spherical particles is a, the fractional volume is f, and k_0, k_s are the wave numbers of the free space and particles, respectively. Eq.(1-1.2) yields the effective permittivity of the medium as

$$\varepsilon_f = K^2/k_0^2 \tag{1-1.3}$$

The VRT equation of the stratified media of random spherical particles is written as [Jin 1994]

$$\cos\theta \frac{d}{dz}\bar{I}(\theta,z) = -\bar{\bar{\kappa}}_e(\theta,z)\cdot\bar{I}(\theta,z) + \int_0^\pi \bar{\bar{P}}(\theta,\theta',z)\cdot\bar{I}(\theta',z)d\theta'\sin\theta' + \bar{\bar{\kappa}}_a(\theta,z)\cdot C\bar{T}_0(z) \tag{1-1.4}$$

where $\bar{\bar{\kappa}}_e, \bar{\bar{\kappa}}_a$ are the extinction and absorption matrices, respectively. $\bar{\bar{P}}$ is the phase matrix for multiple scattering, and $\bar{T}_0(z)$ is the physical temperature profile of the scatter media.

The boundary conditions at $z=0$ and $z=-d$ are written as

$$\bar{I}(\pi-\theta, z=0) = \bar{\bar{R}}_{f0}(\theta)\cdot\bar{I}(\theta,z=0) \tag{1-1.5a}$$

$$\bar{I}(\theta, z=-d) = \bar{\bar{R}}_{fu}(\theta)\cdot\bar{I}(\pi-\theta,z=-d) + \bar{\bar{T}}_{fu}(\theta)\cdot C\bar{T}_{0u} \tag{1-1.5b}$$

where $\bar{\bar{R}}_{f0}(\theta)$, $\bar{\bar{R}}_{fu}(\theta)$ are the reflectivity matrices at the top and bottom (underlying, *u*) boundaries, respectively. $\bar{\bar{T}}_{fu}(\theta)$ is the transmittivity matrix of the underlying surface at *z=-d*, and $T_0(z)$ and T_{0u} are the physical temperatures of the scatter media and underlying surface, respectively.

In the multi-layered snowpack, it is assumed that all ice particles within each layer (separated by the solid lines in Fig.1-1.1) are identical and that fractional volume is constant. The reflection from and transmission through those interfaces between the layers are calculated by the effective permittivities of the layers.

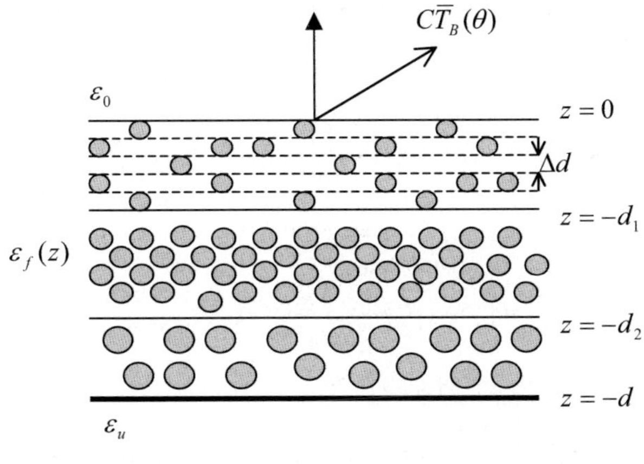

Fig.1-1.1. Geometry of inhomogeneous snowpack

High-order scattering, compared with low-order scattering, is related to the albedo and optical depth of the layer. In other words, the scattering can be approximated by low order if the optical depth and albedo of the layer are small enough. Thus, if each layer of the whole scatter media is stratified into many thin slabs whose optical depths are small (see the separation by the dotted lines in Fig.1-1.1), scattering in each isolated slab can be well approximated by its low-order solution. Taking *L* slabs from $z=0$ to $z=-d$ indicated by the number $\ell=1,2,\cdots,L$ from the top to bottom, the zero-th order emission and low order Mueller matrix can be employed for each isolated slab [Jin and Liang 2003].

If $(\Delta\tau)^2 = (\kappa_{e\ell}\Delta d)^2 << 1$ in each thin slab, the emission of the slab is approximated as

$$\bar{I}_{\ell s}^{(n=0)}(\theta) \approx C\bar{T}_{\ell}^{(0)}(\theta) = \frac{\kappa_{a\ell}}{\kappa_{e\ell}} C\bar{T}_{0\ell}[1-\exp(-\kappa_{e\ell}\Delta d\,|\sec\theta|)], \qquad 0^\circ \le \theta < 180^\circ \qquad (1\text{-}1.6)$$

where the subscript s denotes the emission coming from the isolated slab self.

Also, the first order Mueller matrix of this slab is written as

$$\bar{\bar{M}}_{\ell}(\theta,\theta_i) \approx \bar{\bar{M}}_{\ell}^{(1)}(\theta,\theta_i)$$

$$= \frac{\bar{\bar{P}}_{\ell}(\theta,\pi-\theta_i)\sec\theta}{\kappa_{e\ell}(\sec\theta_i+\sec\theta)} \begin{cases} 1-\exp[-\kappa_{e\ell}\Delta d(\sec\theta_i+\sec\theta)] \\ \qquad 0^\circ < \theta < 90^\circ, 0^\circ < \theta_i < 90^\circ \\ \exp(-\kappa_{e\ell}\Delta d\sec\theta_i) - \exp(\kappa_{e\ell}\Delta d\sec\theta) \\ \qquad 90^\circ < \theta < 180^\circ, 0^\circ < \theta_i < 90^\circ \\ \exp(\kappa_{e\ell}\Delta d\sec\theta_i) - \exp(-\kappa_{e\ell}\Delta d\sec\theta) \\ \qquad 0^\circ < \theta < 90^\circ, 90^\circ < \theta_i < 180^\circ \\ 1-\exp[\kappa_{e\ell}\Delta d(\sec\theta_i+\sec\theta)] \\ \qquad 90^\circ < \theta < 180^\circ, 90^\circ < \theta_i < 180^\circ \end{cases} \qquad (1\text{-}1.7)$$

If the optical depth is large, yielding too many thin slabs according to of (1-1.6,7), the thickness of each slab Δd can be chosen such that $(\Delta\tau)^3 = (\kappa_{e\ell}\Delta d)^3 << 1$, and the zero-th order emission is now approximated as

$$\bar{I}_{\ell s}^{(n=0)}(\theta) \approx C\bar{T}_{\ell}^{(0)}(\theta) + C\bar{T}_{\ell}^{(1)}(\theta) \qquad (1\text{-}1.8)$$

where

$$C\bar{T}_{\ell}^{(1)}(\theta<90^\circ) = \kappa_{a\ell} C\bar{T}_{0\ell}\sec\theta \int_0^{\pi/2} [\bar{\bar{P}}_{\ell}(\theta,\theta')[D_3(\theta) - D_1(\theta,\theta')\exp(-\kappa_{e\ell}\Delta d\sec\theta')]$$
$$+ \bar{\bar{P}}_{\ell}(\theta,\pi-\theta')[D_3(\theta) - D_2(\theta,\theta')]\sin\theta' d\theta' \qquad (1\text{-}1.9a)$$

$$C\bar{T}_{\ell}^{(1)}(\theta>90^\circ) = \kappa_{a\ell} C\bar{T}_{0\ell}\sec\theta \int_0^{\pi/2} \{\bar{\bar{P}}_{\ell}(\theta,\theta')[D_3(\theta) - D_2(\pi-\theta,\theta')]$$
$$+ \bar{\bar{P}}_{\ell}(\theta,\pi-\theta')[D_3(\theta) - D_1(\pi-\theta,\theta')\exp(-\kappa_{e\ell}\Delta d\sec\theta')]\}\sin\theta' d\theta' \qquad (1\text{-}1.9b)$$

here

$$D_1(\theta,\theta') = \frac{1-\exp[-\kappa_{e\ell}\Delta d(\sec\theta - \sec\theta')]}{\kappa_{e\ell}(\sec\theta - \sec\theta')} \qquad (1\text{-}1.10a)$$

$$D_2(\theta,\theta') = \frac{1-\exp[-\kappa_{e\ell}\Delta d(\sec\theta + \sec\theta')]}{\kappa_{e\ell}(\sec\theta + \sec\theta')} \qquad (1\text{-}1.10b)$$

$$D_3(\theta) = \frac{1-\exp(-\kappa_{e\ell}\Delta d\sec\theta)}{\kappa_{e\ell}\sec\theta} \qquad (1\text{-}1.10c)$$

Correspondingly, the Mueller matrix of each slab is first approximated by

$$\bar{\bar{M}}_{\ell}(\theta,\theta_i) \approx \bar{\bar{M}}_{\ell}^{(1)}(\theta,\theta_i) + \bar{\bar{M}}_{\ell}^{(2)}(\theta,\theta_i) \qquad (1\text{-}1.11)$$

where

$$\overline{\overline{M}}_\ell^{(2)}(\theta,\theta_i<90^\circ)$$
$$=\frac{1}{|\cos\theta|}\begin{cases}\int_0^{\pi/2}[\overline{\overline{P}}_\ell(\theta,\pi-\theta')\cdot\overline{\overline{P}}_\ell(\pi-\theta',\pi-\theta_i)A_2(\theta,\theta',\theta_i)\\+\overline{\overline{P}}_\ell(\theta,\theta')\cdot\overline{\overline{P}}_\ell(\theta',\pi-\theta_i)A_1(\theta,\theta',\theta_i)]\sin\theta' d\theta' \quad 0^\circ<\theta<90^\circ\\ \\ \int_0^{\pi/2}[\overline{\overline{P}}_\ell(\theta,\pi-\theta')\cdot\overline{\overline{P}}_\ell(\pi-\theta',\pi-\theta_i)A_4(\pi-\theta,\theta',\theta_i)\\+\overline{\overline{P}}_\ell(\theta,\theta')\cdot\overline{\overline{P}}_\ell(\theta',\pi-\theta_i)A_3(\pi-\theta,\theta',\theta_i)]\sin\theta' d\theta' \quad 90^\circ<\theta<180^\circ\end{cases} \qquad (1\text{-}1.12a)$$

$$\overline{\overline{M}}_\ell^{(2)}(\theta,\theta_i>90^\circ)$$
$$=\frac{1}{|\cos\theta|}\begin{cases}\int_0^{\pi/2}[\overline{\overline{P}}_\ell(\theta,\pi-\theta')\cdot\overline{\overline{P}}_\ell(\pi-\theta',\pi-\theta_i)A_4(\theta,\theta',\pi-\theta_i)\\+\overline{\overline{P}}_\ell(\theta,\theta')\cdot\overline{\overline{P}}_\ell(\theta',\pi-\theta_i)A_3(\theta,\theta',\pi-\theta_i)]\sin\theta' d\theta' \quad 0^\circ<\theta<90^\circ\\ \\ \int_0^{\pi/2}[\overline{\overline{P}}_\ell(\theta,\pi-\theta')\cdot\overline{\overline{P}}_\ell(\pi-\theta',\pi-\theta_i)A_2(\pi-\theta,\theta',\pi-\theta_i)\\+\overline{\overline{P}}_\ell(\theta,\theta')\cdot\overline{\overline{P}}_\ell(\theta',\pi-\theta_i)A_1(\pi-\theta,\theta',\pi-\theta_i)]\sin\theta' d\theta' \quad 90^\circ<\theta<180^\circ\end{cases} \qquad (1\text{-}1.12b)$$

where

$$A_1(\theta,\theta',\theta_i)=\frac{\sec\theta'\{D_2(\theta,\theta_i)-D_1(\theta,\theta')\exp[-\kappa_{e\ell}\Delta d(\sec\theta'+\sec\theta_i)]\}}{\kappa_{e\ell}(\sec\theta'+\sec\theta_i)} \qquad (1\text{-}1.13a)$$

$$A_2(\theta,\theta',\theta_i)=\frac{\sec\theta'[D_2(\theta,\theta')-D_2(\theta,\theta_i)]}{\kappa_{e\ell}(\sec\theta'-\sec\theta_i)} \qquad (1\text{-}1.13b)$$

$$A_3(\theta,\theta',\theta_i)=\frac{\sec\theta'\exp(-\kappa_{e\ell}\Delta d\sec\theta_i)[D_1(\theta,\theta_i)-D_2(\theta,\theta')]}{\kappa_{e\ell}(\sec\theta'+\sec\theta_i)} \qquad (1\text{-}1.13c)$$

$$A_4(\theta,\theta',\theta_i)=\frac{\sec\theta'[D_1(\theta,\theta_i)\exp(-\kappa_{e\ell}\Delta d\sec\theta_i)-D_1(\theta,\theta')\exp(-\kappa_{e\ell}\Delta d\sec\theta')]}{\kappa_{e\ell}(\sec\theta'-\sec\theta_i)} \qquad (1\text{-}1.13d)$$

The emission and Mueller matrix in Eqs.(1-1.6,8) and Eq.(1-1.11) are, respectively, employed for calculation of scattering intensity in each isolated slab.

To obtain scattering and emission from whole media, it should take into account all interactions of all slabs for which scattering and emission have been known. The transmission and reflection are taken into account only through the boundaries of the sublayers (e.g. in Fig.1-1.1, the solid lines at $z=-d_1$ and $z=-d_2$), and $\overline{\overline{T}}_{\ell-1,\ell}(\theta)=1$ and $\overline{\overline{R}}_{\ell-1,\ell}(\theta)=0$ at the interfaces of those thin slabs indicated by the dotted lines.

The downward emission in the ℓ-slab from the ($\ell-1$)-slab is written as

$$\bar{I}_\ell^{(n)}(\pi-\theta)=\bar{I}_{\ell s}^{(n)}(\pi-\theta)+[\overline{\overline{R}}_{\ell,\ell-1}(\theta)\cdot\bar{I}_\ell^{(n-1)}(\theta)+\overline{\overline{T}}_{\ell-1,\ell}(\theta)\cdot\bar{I}_{\ell-1}^{(n)}(\pi-\theta)]\exp(-\kappa_{e\ell}\Delta d\sec\theta) \qquad (1\text{-}1.14a)$$

The upward emission in the ℓ-slab from the ($\ell+1$)-slab is written as

$$\bar{I}_\ell^{(n)}(\theta)=\bar{I}_{\ell s}^{(n)}(\theta)+[\overline{\overline{R}}_{\ell,\ell+1}(\theta)\cdot\bar{I}_\ell^{(n-1)}(\pi-\theta)+\overline{\overline{T}}_{\ell+1,\ell}(\theta)\cdot\bar{I}_{\ell+1}^{(n)}(\theta)]\exp(-\kappa_{e\ell}\Delta d\sec\theta) \qquad (1\text{-}1.14b)$$

where $\theta\in(0^\circ,90^\circ)$, the superscript n denotes the iteration number, $\bar{I}_{\ell s}^{(n)}(\theta)$ is first calculated by Eq. (1-1.6) or Eq. (1-1.8) ($n=0$) and then by Eq. (1-1.15) ($n>0$) as follows.

The emission from $z=0$ and $z=-d$, $\bar{I}_{\ell=0}^{(n)}(\pi-\theta)$ and $\bar{I}_{\ell=L+1}^{(n)}(\theta)$, are defined by Eqs. (1-1.5a,b) as

$$\bar{I}^{(n)}_{\ell=0}(\pi-\theta)=\begin{cases}0 & n=0\\ \overline{\overline{R}}_{f0}(\theta)\cdot\bar{I}^{(n-1)}_{\ell=1}(\theta) & n>0\end{cases} \tag{1-1.15a}$$

$$\bar{I}^{(n)}_{\ell=L+1}(\theta)=\begin{cases}\overline{\overline{R}}_{fu}(\theta)\cdot\bar{I}^{(n)}_{\ell=L}(\pi-\theta)+\overline{\overline{T}}_{fu}(\theta)\cdot C\bar{T}_u & n=0\\ \overline{\overline{R}}_{fu}(\theta)\cdot\bar{I}^{(n)}_{\ell=L}(\pi-\theta) & n>0\end{cases} \tag{1-1.15b}$$

From Eqs.(1-1.13,14), the n-th order emissions, upwards and downwards, to take into account the interactions of the slabs, $\bar{I}^{(n)}_{\ell}(\theta)$ and $\bar{I}^{(n)}_{\ell}(\pi-\theta)$, can be obtained. Further, the (n+1)-th order emission of the ℓ-slab-self can be calculated by its Mueller matrix as

$$\begin{aligned}\bar{I}^{(n+1)}_{\ell s}(\theta)=&\int_{\pi/2}^{\pi}\overline{\overline{M}}_{\ell}(\theta,\theta')\cdot\overline{\overline{T}}_{\ell-1,\ell}(\theta')\cdot\bar{I}^{(n)}_{\ell-1}(\theta')\sin\theta' d\theta'\\ &+\int_{0}^{\pi/2}\overline{\overline{M}}_{\ell}(\theta,\theta')\cdot\overline{\overline{T}}_{\ell+1,\ell}(\theta')\cdot\bar{I}^{(n)}_{\ell+1}(\theta')\sin\theta' d\theta'\end{aligned} \tag{1-1.16}$$

Thus, we obtain a high-order iterative solution. Iteration process is summarized as follows

(1) Stratifying the media into multi-slabs, calculate $\bar{I}^{(n=0)}_{\ell s}(\theta)$ and $\overline{\overline{M}}_{\ell}(\theta,\theta_i)$ based on Eqs. (1-1.6,7) or (1-1.8,11), according to $\Delta\tau$.

(2) Employing Eqs.(1-1.14,15), calculate $\bar{I}^{(n)}_{\ell}(\theta)$. The upward emission coming from whole scatter media is the *n-th* order solution $\bar{I}^{(n)}_{\ell=1}(\theta)$.

(3) Using Eq.(1-1.16), the (n+1)-*th* emission of the ℓ-slab, $\bar{I}^{(n+1)}_{\ell s}(\theta)$, is calculated by $\bar{I}^{(n)}_{\ell}(\theta)$.

(4) Repeating the steps (2) and (3) to a final convergence, calculation is finished. The high-order scattered emission is obtained. .

Taking $\bar{I}_{\ell=1}(\theta)$ as the sum of all orders $\bar{I}^{(n)}_{\ell=1}(\theta)$, the brightness temperature observed at the region 0 is obtained as

$$\bar{T}_B(\theta_0)=\frac{1}{C}\overline{\overline{T}}_{f0}(\theta)\cdot\bar{I}_{\ell=1}(\theta) \tag{1-1.17}$$

It should mention that in each iteration, only one reflection between the interfaces of the stratified layers is taken into account. Multi-reflections between those layers interfaces are calculated during multi-order iterations. Thus, $\bar{I}^{(n)}_{\ell=1}(\theta)$ is different from the n-th order integral formal solution.

Emission of each slab-self takes account of the zeroth order solution (as $(\Delta\tau)^2<<1$) or the zeroth and first-order solutions (as $(\Delta\tau)^3<<1$). Correspondingly, the first-order or the first- and second-order Mueller matrices are employed to take account of interactions of the slabs. Total emission from the scatter media, $\bar{I}^{(n)}_{\ell=1}(\theta)$, includes the emissions of all slabs and n times interactions among them.

1-1.2 Numerical Results

A. Uniform snowpack

Consider an uniform snowpack of the depth $d=100cm$. The ice particles are identical spheres with radius $a=0.06cm$, fractional volume $f=0.3$, and dielectric constant $\varepsilon_s=(3.2+i0.0032)\varepsilon_0$. The underlying medium has $\varepsilon_u=(6+i)\varepsilon_0$. The physical temperatures of both the snowpack and underlying surface are simply assumed as $263K$. Based on Eq.(1-1.2), the effective propagation constant of the snowpack is calculated as

$\varepsilon_f = (1.49 + i0.0009)\varepsilon_0$.

Taking $\Delta d = 1cm$ of each slab in our layering method and uniformly discretizing the scattering angle θ by each 6° equally, Eqs. (1-1.6,7) and the Mueller matrix are calculated. Iteration is carried out to the fourth order. Fig.1-1.2 presents the brightness temperature of $19GHz$ and comparison with the result of the discrete ordinate and eigen-analysis method [Jin 1994] (Gaussian quadrature number is $N = 8$). The results of two methods are almost exactly identical.

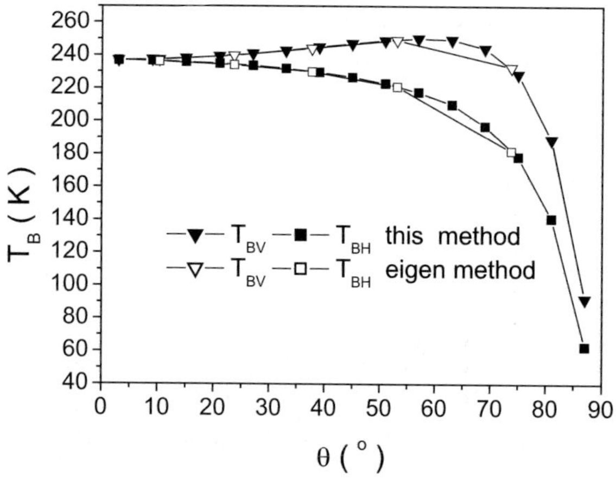

Fig.1-1.2. Brightness temperature of an uniform snowpack at $19GHz$

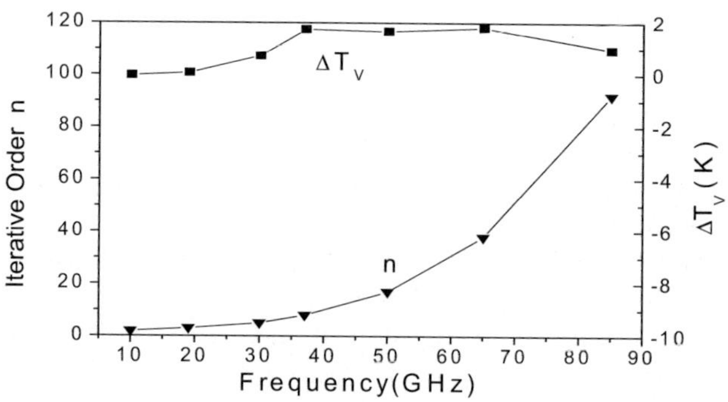

Fig.1-1.3. Iteration number n and comparison with the discrete ordinate eigen-analysis

Fig.1-1.3 shows the iteration number n for different frequency, and comparison with the discrete ordinate and eigen-analysis $\Delta T_v = T_{iter} - T_{eigen}$. The observation angle is $\theta = 53^\circ$, $d = 50cm$, $a = 0.05cm$, and other parameters are the same as Fig.1-1.2.

It can be seen that as the frequency increases, high-order iteration is needed due to larger albedo andstronger scattering. For example,the albedo at $10GHz$ is $\omega_0 = 0.042$, while it becomes $\omega_0 = 0.965$ at $85GHz$. It is obvious the iteration orders at such high frequency increases if one can see $0.965^{90} \approx 0.04$.

B. *Inhomogeneous Snowpack*

Calculate the brightness temperature of inhomogeneous snowpack of $d = 10m$ at $5.3GHz$. The ice particle radius is $a = 0.06cm$, and the particle density is assumed to have the profile as (in Fig.1-1.4)

$$\rho(z) = 0.25 + 0.16[1 - \exp(0.05z)] + 0.16(R - 0.5) \qquad (z < 0) \qquad (1\text{-}1.19)$$

where $R \in (0,1)$ is a random number uniformly $\in(0.1)$. The profile of fractional volume is obtained from $f(z) = \rho(z)/0.91$. And the physical temperature profile (in *K*) is also assumed as

$$T_0(z) = 234 + 24\exp(0.81z) \qquad (K) \qquad (z < 0) \qquad (1\text{-}1.20)$$

It is reasonable to have higher temperature at the top layers due to the sunshine.

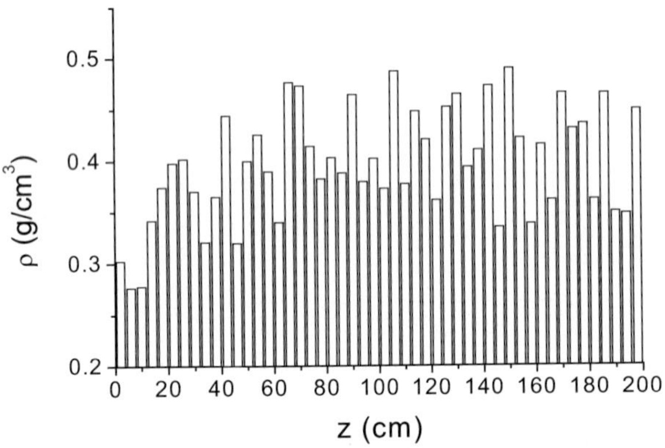

Fig.1-1.4. Density profile of an inhomogeneous snowpack

The inhomegeneous snowpack of $10m$ is seen as the stratified layers with each layer $4cm$. Every layer has different density (as shown in Fig.1-1.4), physical temperature, and the Mueller matrix. Each layer is now divided into thin multi-slabs of thickness $\Delta d = 1cm$ for iteration calculation. Due to random noise R in Eq.(1-1.18), calculations of five realizations are averaged to obtain final result.

Fig.1-1.5(a) shows the brightness temperature at $5.3GHz$. Discrete points are the measurements of Tsang et al. (2001). The temperature of uniform snowpack is obtained from Eq.(1-1.19) and the density is $0.41g/cm^3$. It can be seen that the result of the inhomogeneous snowpack model is better. Because the frequency of $5.3GHz$ is low, multiple scattering is actually negligible.

To take into account of multiple scattering, Fig.1-1.5(b) presents the brightness temperature at $19GHz$ and $37GHz$ under the same situation. It becomes significant to take into account multiple scattering at these high frequencies.

Particle scattering can darken or shadow the emission from underlying media. As the frequency becomes higher or the layer depth becomes thicker, this darkening becomes stronger and yields a lower brightness temperature. Thus, it usually yields $T_{B19} > T_{B37} > T_{B85}$. However, it might become anomalous for a inhomogeneous snowpack [Andrew and Mark 2000]. For example, Fig.1-1.6 gives vertically polarized brightness temperature of the summer 2001 at the location ($80^\circ S, 30^\circ E$) observed by the SSM/I AS F15 at 19-, 37- and 85 *GHz* channels. It can be seen that $T_{B85v} < T_{B19v}$ during the days 210~300, and anomalous $T_{B85v} > T_{B19v}$ during the days 300~365. This is mainly due to inhomogeneous structures of snowpack, such as ice crusts or coarser grains formed during melting, refrozen stratigraphy. The radiometers at different frequencies seem to be sensitive to the snowpack with different effective depths, that is, thick depth for the low-frequency sensor and thin depth for high-frequency one.

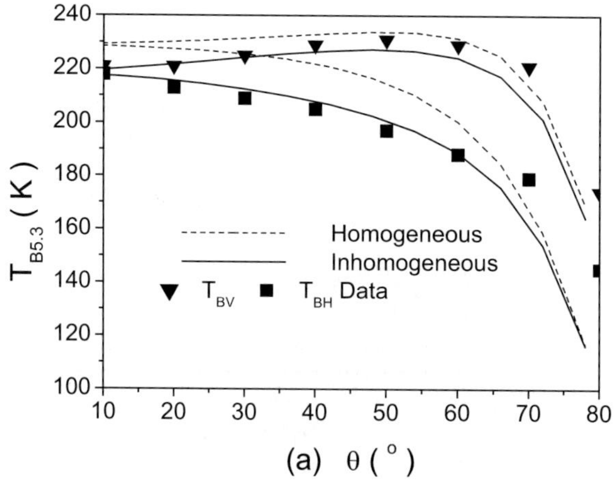

Fig.1-1.5 (a). Brightness temperature of homogeneous and inhomogeneous snowpacks

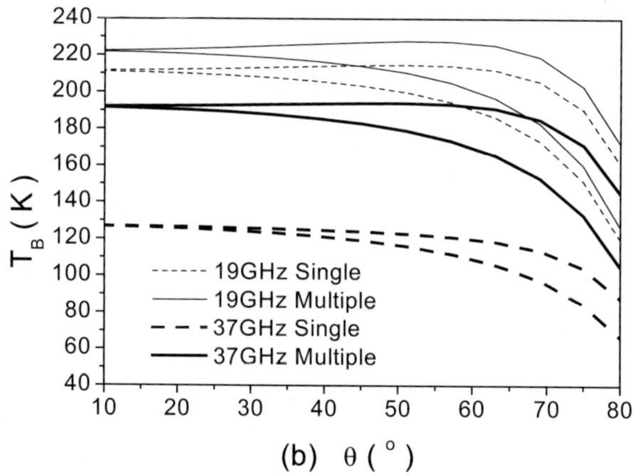

Fig.1-1.5(b). Brightness temperature of single and multiple scattering

Fig.1-1.7 shows the brightness temperature of 1-, 3- and 5-layers snowpack. The 1-layer model takes $a = 0.03cm$, $f = 0.3$, $d = 50cm$. The ice crust in the 3- and 5-layers models has a thickness $1cm$. The top layer is fine snow with the thickness $2.0cm$ and $a = 0.02cm$. The bottom layer is coarser snow with the thickness $200cm$ and $a = 0.15cm$. Other snow layers have $f = 0.3$, and ice radius $a = 0.03cm$. The middle layer of the 5-layer model has thickness $10cm$, $a = 0.03cm$. The underlying bottom medium is pure ice. The snowpack has temperature $213K$. The observation angle is $\theta = 53°$.

Numerical results of inhomogeneous snowpack show that anomalous brightness temperature $T_{B19} < T_{B37} < T_{B85}$.

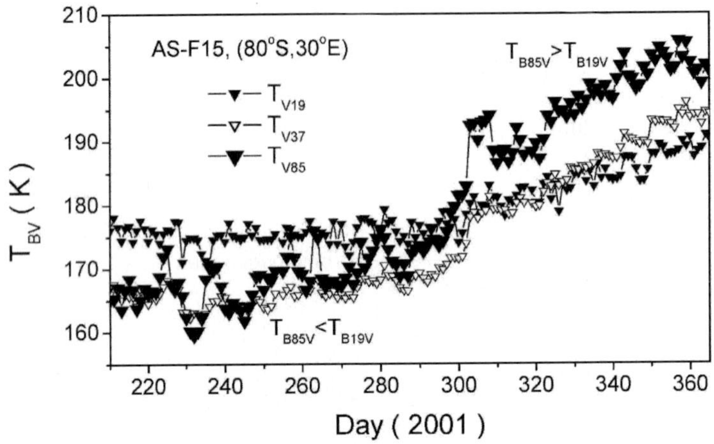

Fig.1-1.6. Vertically polarized brightness temperature observed by the SSM/I.

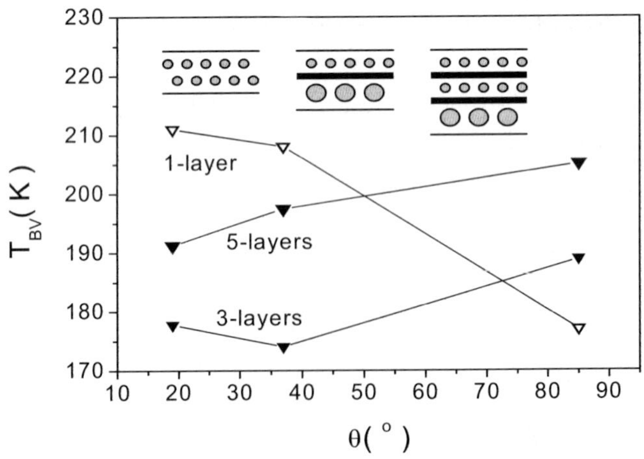

Fig.1-1.7. Brightness temperature of inhomogeneous snowpack vs. frequency

1-2 Iterative Solution of High-Order Scattering Solution for Inhomogeneous Non-Spherical Scatter Media

The VRT takes into account multiple scattering, absorption and transmitting of four Stokes parameters in random media. Iterative approach of the VRT can demonstrate scattering mechanism in different orders. The first-order solution of the Mueller matrix for a single layer of random medium, and second solution of the Mueller matrix for a half-space have been derived [Jin 1994, Tsang et al. 1985]. However, it seems very difficult to continue iterations to find higher order scattering solution because multi-fold integrations over the scattering angles have to be involved.

In numerical approach, the discrete ordinate and eigen-analysis method [Jin 1994, Tsang et al. 1985, Ishimaru et al. 1982], invariant imbedding method [Tsang et al. 1985], the Monte Carlo method etc. [Jin 1994, Tsang et al. 1990] have been studied. These numerical approaches usually need more computations. Moreover, the singularity appeared when the incident angle happens be equal to the discrete Gaussian angle [Siewert 2000, Barichello et al. 2000] in active VRT problem has not been well discussed. The invariant imbedding method was only studied for passive VRT problem [Tsang et al. 1985].

As discussed in the last session 1-1, by stratifying a random scatter media into multiple thin-layers in the z direction, the first-order scattering solution of each thin layer is employed to derive high-order scattering solution of whole random media. An iterative approach to solve high-order scattering solution from the Mueller matrices in discrete ordinates is developed. Numerical results demonstrate that this newly iterative approach as feasible, effective and especially applicable to high-order solution of VRT for both bistatic scattering and thermal emission of inhomogeneous non-spherical scatter media.

1-2.1 Stratifying the Random Media and First-Order Iterative Solution

As an electromagnetic wave, $\bar{I}_0\delta(\cos\theta-\cos\theta_0)\delta(\varphi)$, is incident upon the scatter media, the VRT equation for a layer of inhomogeneous scattering media is written as

$$\cos\theta\frac{d}{dz}\bar{I}(\theta,\varphi,z)=-\bar{\bar{\kappa}}_e(\theta,\varphi,z)\cdot\bar{I}(\theta,\varphi,z)+\int_0^{\pi}d\theta'\sin\theta'\int_0^{2\pi}d\varphi'\bar{\bar{P}}(\theta,\varphi;\theta',\varphi',z)\cdot\bar{I}(\theta',\varphi',z) \qquad (1\text{-}2.1)$$

The boundary conditions are written as

$$\begin{cases}\bar{I}(\pi-\theta,\varphi,z=0)=\bar{I}_0\delta(\cos\theta-\cos\theta_0)\delta(\varphi)\\ \bar{I}(\theta,\varphi,z=-d)=\bar{\bar{R}}_{12}(\theta)\cdot\bar{I}(\pi-\theta,\varphi,z=-d)\end{cases} \qquad (1\text{-}2.2)$$

where $\bar{I}_0$ is incident Stokes vector, $(\pi-\theta_0,\varphi_0=0)$ are the incident angles, $\bar{\bar{R}}_{12}(\theta)$ is the reflectivity matrix of the underlying surface, $\bar{\bar{P}}$ is the phase matrix, $\bar{\bar{\kappa}}_e$ is the extinction matrix and is written as

$$\bar{\bar{\kappa}}_e(\theta,\varphi,z)=\frac{2\pi}{k}n_0\begin{bmatrix}2\,\mathrm{Im}<f_{vv}^0> & 0 & \mathrm{Im}<f_{vh}^0> & -\mathrm{Re}<f_{vh}^0>\\ 0 & 2\,\mathrm{Im}<f_{hh}^0> & \mathrm{Im}<f_{hv}^0> & \mathrm{Re}<f_{hv}^0>\\ 2\,\mathrm{Im}<f_{hv}^0> & 2\,\mathrm{Im}<f_{vh}^0> & \mathrm{Im}<f_{vv}^0+f_{hh}^0> & \mathrm{Re}<f_{vv}^0-f_{hh}^0>\\ \mathrm{Re}<f_{hv}^0> & -\mathrm{Re}<f_{vh}^0> & \mathrm{Re}<f_{hh}^0-f_{vv}^0> & \mathrm{Im}<f_{vv}^0+f_{hh}^0>\end{bmatrix} \qquad (1\text{-}2.3)$$

where f_{pq}^0 $(p,q=v,h)$ are the forward scattering amplitude functions (where the superscript 0 indicates $f_{pq}^0=f_{pq}(\theta,\varphi;\theta,\varphi)$), the angular brackets $< >$ denote the ensemble averaging over spatial orientation of the Euler angles. In following calculations of each thin layer, the location z is not specially indicated in $\bar{\bar{\kappa}}_e$ and $\bar{\bar{P}}$.

Generally, $\bar{\bar{\kappa}}_e$ is non-diagonal. However, it is desirable to find the matrix $\bar{\bar{E}}(\theta,\varphi)$ and its inverse $\bar{\bar{E}}^{-1}(\theta,\varphi)$ to diagonize $\bar{\bar{\kappa}}_e$ as follows

$$\bar{\bar{\beta}}(\theta,\varphi)=\bar{\bar{E}}^{-1}(\theta,\varphi)\cdot\bar{\bar{\kappa}}_e(\theta,\varphi)\cdot\bar{\bar{E}}(\theta,\varphi) \tag{1-2.4}$$

where the ii element of $\bar{\bar{\beta}}$ is denoted as β_i ($i=1,2,3,4$). It has been known that β_i is eigen-value of $\bar{\bar{\kappa}}_e(\theta,\varphi)$, and $\bar{\bar{E}}(\theta,\varphi)$ is composed by the eigen-vector [Jin 1994, Tsang et al. 1985]. The analytic formulations can be found in Jin (1994) and Tsang et al. (1985). When $<f_{vh}^0>=<f_{hv}^0>=0$, $\bar{\bar{E}}$ and $\bar{\bar{E}}^{-1}$ simply become constant matrixes:

$$\bar{\bar{E}}=\begin{bmatrix}1&0&0&0\\0&1&0&0\\0&0&1&1\\0&0&i&-i\end{bmatrix},\quad \bar{\bar{E}}^{-1}=\begin{bmatrix}1&0&0&0\\0&1&0&0\\0&0&1/2&-i/2\\0&0&1/2&i/2\end{bmatrix} \tag{1-2.5}$$

Now multiplying $\bar{\bar{E}}^{-1}$ to the LHS of Eqs. (1-2.1,2.2), it yields

$$\cos\theta\frac{d}{dz}\bar{I}^E(\theta,\varphi,z)=-\bar{\bar{\beta}}(\theta,\varphi)\cdot\bar{I}^E(\theta,\varphi,z)+\int_0^\pi d\theta'\sin\theta'\int_0^{2\pi}d\varphi'\bar{\bar{P}}^E(\theta,\varphi;\theta',\varphi')\cdot\bar{I}^E(\theta',\varphi',z) \tag{1-2.6}$$

$$\begin{cases}\bar{I}^E(\pi-\theta,\varphi,z=0)=\bar{I}_0^E\delta(\cos\theta-\cos\theta_0)\delta(\varphi)\\ \bar{I}^E(\theta,\varphi,z=-d)=\bar{\bar{R}}_{12}^E(\theta)\cdot\bar{I}^E(\pi-\theta,\varphi,z=-d)\end{cases} \tag{1-2.7}$$

where

$$\bar{I}^E(\theta,\varphi,z)=\bar{\bar{E}}^{-1}\cdot\bar{I}(\theta,\varphi,z) \tag{1-2.8}$$

$$\bar{\bar{P}}^E(\theta,\varphi;\theta',\varphi')=\bar{\bar{E}}^{-1}\cdot\bar{\bar{P}}(\theta,\varphi;\theta',\varphi')\cdot\bar{\bar{E}} \tag{1-2.9a}$$

$$\bar{\bar{R}}_{12}^E(\theta)=\bar{\bar{E}}^{-1}\cdot\bar{\bar{R}}_{12}(\theta)\cdot\bar{\bar{E}} \tag{1-2.9b}$$

It can be seen that as $\bar{I}^E$ is solved from Eqs. (1-2.6,1-2.7), $\bar{I}$ of Eq. (1-2.1) can be obtained.

Let us consider a thin layer with the thickness Δd without solid boundary. As an electromagnetic plane wave at $(\theta_0,0)$ is incident upon this layer, the first-order scattering intensities from this layer obtained as [Jin 1994]:

$$I_{si}^{E(1)}(\theta,\varphi,0)=\sum_{j=1}^4 P_{ij}^E(\theta,\varphi;\pi-\theta_0,0)\frac{\sec\theta}{\beta_j\sec\theta_0+\beta_i\sec\theta}\times\{1-\exp[-\Delta d(\beta_j\sec\theta_0+\beta_i\sec\theta)]\}\cdot I_{0j}^E \tag{1-2.10a}$$

$$I_{si}^{E(1)}(\pi-\theta,\varphi,-\Delta d)=\sum_{j=1}^4 P_{ij}^E(\pi-\theta,\varphi;\pi-\theta_0,0)\frac{\sec\theta}{\beta_j\sec\theta_0-\beta_i\sec\theta}\times[\exp(-\beta_i\Delta d\sec\theta)-\exp(-\beta_j\Delta d\sec\theta_0)]\cdot I_{0j}^E \tag{1-2.10b}$$

where $i,j=1,2,3,4$ denote four Stokes parameters, respectively.

When the optical depth $\Delta\tau_i=\beta_i\Delta d<<1$ is assumed and scattering angle θ is not close to 90°, higher-order terms of the $(\Delta\tau)^2$ order are neglected. It yields Eqs.(1-2.10a,2.10b) as

$$I_{si}^{E(1)}(\theta,\varphi,0)\approx\Delta d\sec\theta\sum_{j=1}^4 P_{ij}^E(\theta,\varphi;\pi-\theta_0,0)\cdot I_{0j}^E \tag{1-2.11a}$$

$$I_{si}^{E(1)}(\pi-\theta,\varphi,-\Delta d)\approx\Delta d\sec\theta\sum_{j=1}^4 P_{ij}^E(\pi-\theta,\varphi;\pi-\theta_0,0)\cdot I_{0j}^E \tag{1-2.11b}$$

Thus total scattered power per unit area becomes

$$\Delta P_{si} = \int_0^{2\pi} d\varphi \int_0^{\pi/2} [\bar{I}_{si}^{E(1)}(\theta,\varphi,0) + \bar{I}_{si}^{E(1)}(\pi-\theta,\varphi,-d)]\cos\theta\sin\theta d\theta$$
$$\approx \Delta d \sum_{j=1}^{4} I_{0j}^E \int_0^{2\pi} d\varphi \int_0^{\pi/2} [P_{ij}^E(\theta,\varphi;\pi-\theta_0,0) + P_{ij}^E(\pi-\theta,\varphi;\pi-\theta_0,0)]\sin\theta d\theta \quad (1\text{-}2.12)$$

where

$$\kappa_{sij}^E(\theta_0) = \int_0^{2\pi} d\varphi \int_0^{\pi/2} [P_{ij}^E(\theta,\varphi;\pi-\theta_0,0) + P_{ij}^E(\pi-\theta,\varphi;\pi-\theta_0,0)]\sin\theta d\theta \quad (1\text{-}2.13)$$

is defined as the scattering coefficient.

Eq.(1-2.12) yields

$$\Delta \bar{P}_s \approx \Delta d \bar{\bar{\kappa}}_s^E(\theta_0) \cdot \bar{I}_0^E \quad (1\text{-}2.14)$$

It has been known from the definition that the total scattered power propagating through $\Delta d \sec\theta_0$ shall be

$$\Delta \bar{P}_s' = \int \bar{\bar{\kappa}}_s^E(\theta_0)\Delta d \sec\theta_0 \cdot \bar{I}_0^E \delta(\theta-\theta_0)\delta(\varphi)\cos\theta d\Omega = \Delta d \bar{\bar{\kappa}}_s^E(\theta_0) \cdot \bar{I}_0^E \quad (1\text{-}2.15)$$

It concludes that as $\Delta\tau << 1$, the first-order scattering is approximately equal to the total scattering. The error is the order of $(\Delta\tau)^2$.

Define

$$M_{ij}^E(\theta,\varphi;\theta_0,0) = \frac{P_{ij}^E(\theta,\varphi;\pi-\theta_0,0)\sec\theta}{\beta_j\sec\theta_0 + \beta_i\sec\theta} \begin{cases} 1-\exp[-\Delta d(\beta_j\sec\theta_0 + \beta_i\sec\theta)] \\ \quad as\ 0^\circ < \theta < 90^\circ, 0^\circ < \theta_0 < 90^\circ \\ \exp(-\beta_j\Delta d\sec\theta_0) - \exp(\beta_i\Delta d\sec\theta) \\ \quad as\ 90^\circ < \theta < 180^\circ, 0^\circ < \theta_0 < 90^\circ \\ \exp(\beta_j\Delta d\sec\theta_0) - \exp(-\beta_i\Delta d\sec\theta) \\ \quad as\ 0^\circ < \theta < 90^\circ, 90^\circ < \theta_0 < 180^\circ \\ 1-\exp[\Delta d(\beta_j\sec\theta_0 + \beta_i\sec\theta)] \\ \quad as\ 90^\circ < \theta < 180^\circ, 90^\circ < \theta_0 < 180^\circ \end{cases} \quad (1\text{-}2.16)$$

where $0^\circ < \theta_0 < 90^\circ$ means down-going propagation and $90^\circ < \theta_0 < 180^\circ$ up-going propagation. From Eqs. (1-2.10,2.16), it yields the Mueller matrix form as

$$\bar{I}_s^E = \bar{\bar{M}}^E \cdot \bar{I}_0^E \quad (1\text{-}2.17)$$

In passive VRT equation for thermal emission, the brightness temperature of a thin layer Δd with $\Delta\tau << 1$ is

$$T_i^E(\theta) = \frac{1-\exp(-\beta_i\Delta d\,|\sec\theta|)}{\beta_i} \sum_{j=1}^{4} \kappa_{aij}^E \cdot T_{0j}^E \quad (1\text{-}2.18)$$

where $\bar{\bar{\kappa}}_a^E = \bar{\bar{E}}^{-1} \cdot \bar{\bar{\kappa}}_a \cdot \bar{\bar{E}}$, $\bar{\bar{\kappa}}_a$ is the absorption matrix. $\bar{T}_0^E = \bar{\bar{E}}^{-1} \cdot [T,T,0,0]^t$, where T is physical temperature of the layer.

1-2.2 High-Order Scattering Solution in Iterative Discrete Ordinates

When the scatterers are uniformly distributed in the azimuth direction, $\bar{\bar{\kappa}}_e^E$ has no φ dependence. Expanding $\bar{I}^E(\theta,\varphi,z)$ and $\bar{\bar{P}}^E(\theta,\varphi;\theta',\varphi')$ to Fourier series as following

$$\bar{I}^E(\theta,\varphi,z) = \sum_{m=0}^{\infty} [\bar{I}^{mcE}(\theta,z)\cos(m\varphi) + \bar{I}^{msE}(\theta,z)\sin(m\varphi)] \quad (1\text{-}2.19)$$

$$\bar{\bar{P}}^E(\theta,\varphi;\theta',\varphi') = \sum_{m=0}^{\infty} [\bar{\bar{P}}^{mcE}(\theta,\theta')\cos m(\varphi-\varphi') + \bar{\bar{P}}^{msE}(\theta,\theta')\sin m(\varphi-\varphi')] \quad (1\text{-}2.20)$$

where the superscripts *c* and *s* denote the functions of $\cos m\varphi$ and $\sin m\varphi$, respectively.

Using the Fourier transform, Eq.(1-2.6) of VRT becomes

$$\cos\theta \frac{d}{dz} \bar{I}^{mcE}(\theta,z) = -\bar{\bar{\beta}}(\theta)\cdot \bar{I}^{mcE}(\theta,z) + (1+\delta_m)\pi \int_0^\pi d\theta' \sin\theta' \times [\bar{\bar{P}}^{mcE}(\theta,\theta')\cdot \bar{I}^{mcE}(\theta',z) - \bar{\bar{P}}^{msE}(\theta,\theta')\cdot \bar{I}^{msE}(\theta',z)] \tag{1-2.21a}$$

$$\cos\theta \frac{d}{dz} \bar{I}^{msE}(\theta,z) = -\bar{\bar{\beta}}(\theta)\cdot \bar{I}^{msE}(\theta,z) + (1+\delta_m)\pi \int_0^\pi d\theta' \sin\theta' \times [\bar{\bar{P}}^{msE}(\theta,\theta')\cdot \bar{I}^{mcE}(\theta',z) + \bar{\bar{P}}^{mcE}(\theta,\theta')\cdot \bar{I}^{msE}(\theta',z)] \tag{1-2.21b}$$

Suppose to stratify the scatter media into H layers with each thin depth Δd in the z direction, and discretize the scattering angle θ to N sub-angles (N is an even number), i.e. $\theta_k = \pi(k-0.5)/N$, $k = 1,2,\cdots,N$. Eqs.(1-2.21a,2.21b) in discrete ordinates yield

$$\bar{I}_{h,\ell}^{(n)m\alpha E} = \bar{I}_{h,\ell}^{(n)m\alpha E} + \bar{I}_{h+1,\ell}^{(n)m\alpha E} \cdot \bar{\bar{U}}_{\beta\ell} \qquad \ell = 1,2,\cdots N/2 \tag{1-2.22a}$$

$$\bar{I}_{h,\ell}^{(n)m\alpha E} = \bar{I}_{h,\ell}^{(n)m\alpha E} + \bar{I}_{h-1,\ell}^{(n)m\alpha E} \cdot \bar{\bar{U}}_{\beta\ell} \qquad \ell = N/2+1,\cdots N \tag{1-2.22b}$$

$$\bar{I}_{h,k}^{(n+1)mcE} = \sum_{\ell=N/2+1}^{N} \bar{\bar{M}}_{k\ell}^{mcE} \cdot \bar{I}_{h-1,\ell}^{(n)mcE} + \sum_{\ell=1}^{N/2} \bar{\bar{M}}_{k\ell}^{mcE} \cdot \bar{I}_{h+1,\ell}^{(n)mcE} - \sum_{\ell=N/2+1}^{N} \bar{\bar{M}}_{k\ell}^{msE} \cdot \bar{I}_{h-1,\ell}^{(n)msE} - \sum_{\ell=1}^{N/2} \bar{\bar{M}}_{k\ell}^{msE} \cdot \bar{I}_{h+1,\ell}^{(n)msE} \tag{1-2.23a}$$

$$\bar{I}_{h,k}^{(n+1)msE} = \sum_{\ell=N/2+1}^{N} \bar{\bar{M}}_{k\ell}^{msE} \cdot \bar{I}_{h-1,\ell}^{(n)mcE} + \sum_{\ell=1}^{N/2} \bar{\bar{M}}_{k\ell}^{msE} \cdot \bar{I}_{h+1,\ell}^{(n)mcE} + \sum_{\ell=N/2+1}^{N} \bar{\bar{M}}_{k\ell}^{mcE} \cdot \bar{I}_{h-1,\ell}^{(n)msE} + \sum_{\ell=1}^{N/2} \bar{\bar{M}}_{k\ell}^{mcE} \cdot \bar{I}_{h+1,\ell}^{(n)msE} \tag{1-2.23b}$$

where the superscript n is iterative number, $\alpha = c$ or s, $\bar{\bar{U}}_{\beta\ell}$ is a diagonal matrix. They are written as

$$\bar{I}_{h,k}^{m\alpha E} = \begin{cases} \bar{I}^{m\alpha E}(\theta_k, -(h-1)\Delta d) & 0^\circ < \theta_k < 90^\circ \\ \bar{I}^{m\alpha E}(\theta_k, -h\Delta d) & 90^\circ < \theta_k < 180^\circ \end{cases} \tag{1-2.24}$$

$$\bar{\bar{U}}_{\beta\ell} = [\exp(-\Delta d\beta_1 \sec\theta_\ell), \exp(-\Delta d\beta_2 \sec\theta_\ell), \exp(-\Delta d\beta_3 \sec\theta_\ell), \exp(-\Delta d\beta_4 \sec\theta_\ell)] \tag{1-2.25}$$

$$\bar{\bar{M}}_{k\ell}^{m\alpha E} = (1+\delta_m)\pi \cdot \bar{\bar{M}}^{m\alpha E}(\theta_k, \theta_\ell) \qquad k = 1,2,\cdots N, \ell = 1,2,\cdots N \tag{1-2.26}$$

where $\bar{\bar{M}}^{m\alpha E}(\theta_k, \theta_\ell)$ is the Fourier transform of Eq.(1-2.16).

The boundary condition, Eq.(1-2.7), becomes as

$$\begin{cases} \bar{I}_{0,\ell}^{(0)m\alpha E} = \dfrac{1}{(1+\delta_m)\pi} \bar{I}_0^E \delta(\theta - \theta_0) & \ell = N/2+1,\cdots N \\ \bar{I}_{H+1,\ell}^{(n)m\alpha E} = \bar{\bar{R}}_{12\ell}^E \cdot \bar{I}_{H,N+1-\ell}^{(n)m\alpha E} & \ell = 1,2,\cdots N/2 \end{cases} \tag{1-2.27}$$

where

$$\bar{\bar{R}}_{12\ell}^E = \bar{\bar{R}}_{12}^E(\theta_\ell), \qquad \ell = 1,2,\cdots N/2 \tag{1-2.28}$$

Eqs.(1-2.22~26) present an iterative approach to high-order scattering solution of VRT equation. Iterative steps are as follows:

(1) Using Eqs.(1-2.25, 2.26) and (1-2.28), calculate $\bar{\bar{U}}_{\beta\ell}$, $\bar{\bar{M}}_{k\ell}^{m\alpha E}$ and $\bar{\bar{R}}_{12\ell}^E$,

(2) By using Eqs.(1-2.22, 2.27) and $\bar{I}_0^E$, calculate $\bar{I}_{h,k}^{(0)m\alpha E}$, sequentially, from each layer $h = 1,2,\cdots,H$ downwards, and then from $h = H,\cdots,2,1$ upwards.

(3) From Eqs.(1-2.23a, 2.23b), calculate $\bar{I}_{h,k}^{(1)m\alpha E}$. From the step (2), obtain

$\bar{I}_{1,k}^{(1)m\alpha E}(z=0)$. Then, using Eqs.(1-2.19, 2-8), obtain the first-order $\bar{I}_s^{(1)}(\theta,\varphi,0)$:

$$\bar{I}_s^{(1)}(\theta_k,\varphi,0)=\overline{\overline{E}}\cdot\sum_{m=0}^{\infty}[\bar{I}_{1,k}^{(1)mcE}\cos(m\varphi)+\bar{I}_{1,k}^{(1)msE}\sin(m\varphi)] \tag{1-2.29}$$

(4) Repeat the step (3), the second- and higher-order scattering intensities $\bar{I}_s^{(n)}(\theta_k,\varphi,0)$ can be further calculated.

It can be seen that n-th order solution only takes n times iterations of the first-order solution. Conventionally, high-order solution needs multi-fold integrations over θ and φ [Tsang et al. 1984, 1991]. It makes the formulation very complicated and actually cannot be carried out.

For passive VRT, it is only required to add thermal emission in Eq.(1-2.18) to the RHS of Eq.(1-2.22) as the zeroth-order solution.

1-2.3 Numerical Results and Discussions

A. Homogeneous scattering media

Consider a vertically polarized wave with an unit intensity be incident upon a layer of random scatter medium. Frequency is 5 *GHz*, incident angle $\theta_0=30^\circ$ and $\varphi_0=0^\circ$. The layer with the thickness $d=200cm$ is stratified into 200 thin layers, i.e. $\Delta d=1cm$. The scatterers are identical spheroids with the semi-axes $a=b=0.1cm$, $c=2.5cm$ and dielectric constant $22.5+7.5i$, the fractional volume is 0.0055. Spatial orientation of the Euler angles over $\beta\in(0^\circ,60^\circ)$ and $\gamma\in(0^\circ,360^\circ)$ is assumed uniform. The dielectric constant of underlying space is taken as $15.0+2.0i$. In calculations, the scattering angle θ is discretized into 30 portions, i.e. each sub-angle is 6°. The cutoff number of the Fourier series is $m=2$ (See Fig. 1-2.1).

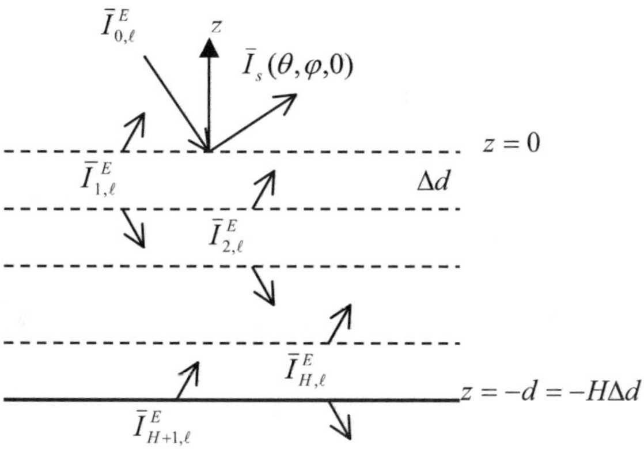

Fig.1-2.1. Geometry of the problem

Figs. 1-2.2(a), 2.2(b) show the first-order bistatic scattering intensities ($\theta,\varphi=180^\circ$), co-polarized I_v and depolarized I_h respectively, and comparisons with the first-order Mueller matrix solution [Jin 1994, Tsang et al. 1985]. It can be seen that two results are exactly identical. Actually, because Eq. (1-2.10) of a single thin layer is identical to the first-order Mueller matrix formulation without reflecting interface, the first-order scattering intensity of the whole media layer is contributed by all thin sub-layers independently (attenuation is considered).

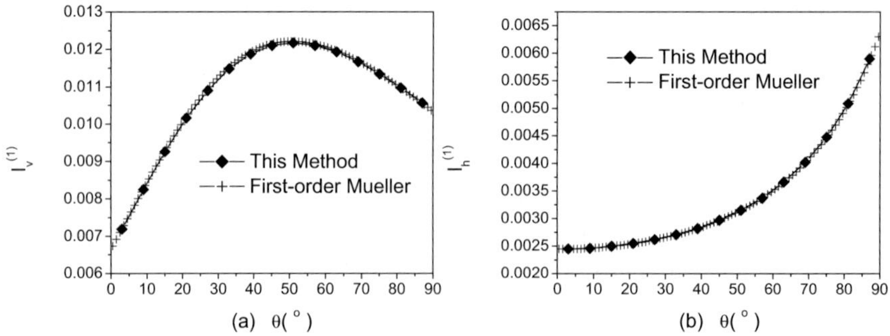

Fig.1-2.2. The first-order scattering intensities: (a) co-polarized; (b) depolarized

It can be illustrated by Fig.1-2.3(a). In other words, the number of the layer stratification and the depth of each thin layer would not affect the final result of whole media layer in the first order.

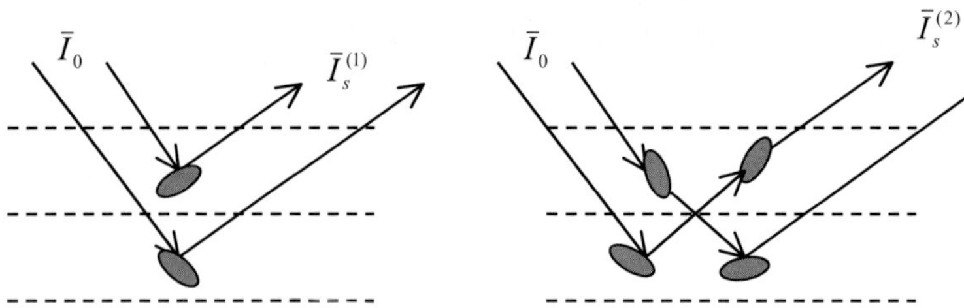

(a) first-order (b) second-order

Fig.1-2.3. The first- and second-order solutions

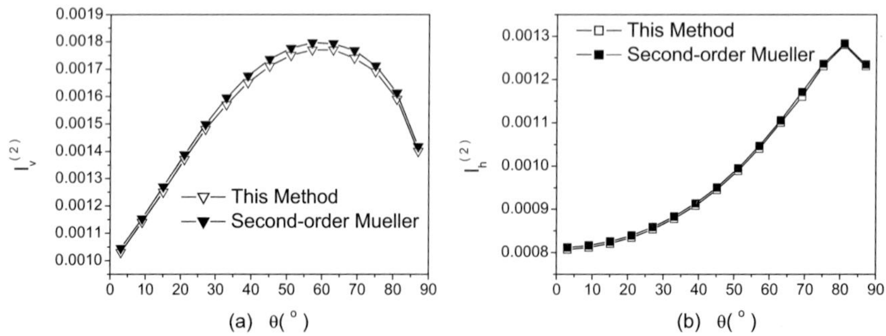

Fig.1-2.4. Second-order solution for a half space random medium: (a) co-polarized, (b) depolarized

To calculate second order solution, we consider a thick layer with total thickness 400 *m* and each thin layer $\Delta d = 0.1cm$, i.e. totally 4000 thin layers. The second order solutions of both co-polarized I_v and depolarized I_h are compared with the second-order Mueller matrix of a half space [Tsang et al. 1984, 1991] in Figs. 1-2.4(a), 2.4(b). It can be seen that two results are well matched with very trivial deviation. This deviation is due to the fact that second-order scattering within every thin layer is actually neglected in the derivation. Fig. 3(b) explained this point. It is known that this error is only at the order $(\Delta\tau)^2$. In our cases, it yields iterative results smaller than second order Mueller result of half space about 1%. As the optical depth of each thin layer is small enough, the result can be very accurate.

Figs. 1-2.5(a), 2.5(b) show bistatic scattering, co-polarized and depolarized of the first, second, and third order solutions for the case of the albedo 0.1. It can be seen that second and third order solutions can be higher than 10% of the first order solution. Thus, it is certain that high order solution is not always negligible.

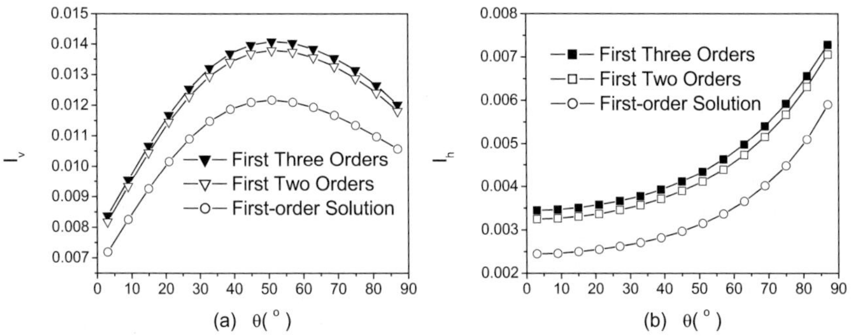

Fig.1-2.5. The first-, second- and third-order scattering intensities: (a) co-polarized; (b) deporlarized

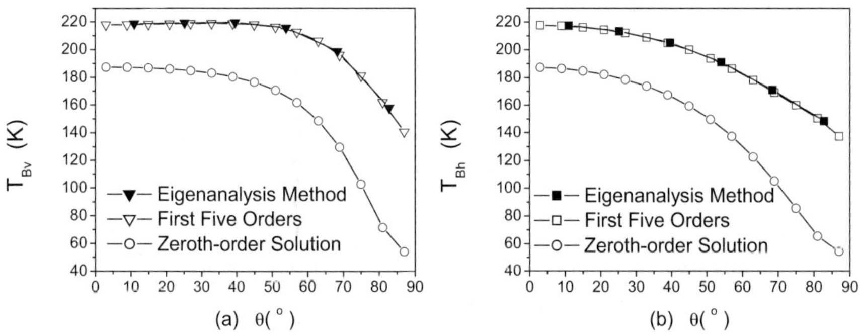

Fig.1-2.6. Brightness temperature of dense spherical particles of first five orders

We now consider thermal emission of VRT for a layer of dense spherical particles [Jin 1994]. The particle radius is $a = 0.06cm$, fractional volume is 0.3, dielectric constant $\varepsilon = 3.2 + 0.0032i$, the layer thickness $d = 20cm$. The dielectric constant of underlying space is $\varepsilon_2 = 3.2 + 0.1i$. The physical temperatures of the medium and underlying space are both 263 K. Frequency is $f = 37GHz$. Figs.1-2.6(a), 2.6(b) show vertically and horizontally brightness temperatures T_{Bv} and T_{Bh} of our iterative solution up to the fourth order, respectively, and comparison with the discrete-ordinate and eigen-analysis method [Jin 1994]. They are well matched.

A small error might be expected as θ approaches extreme 90°, because the approximation $\exp(-\Delta\tau\sec\theta)\approx 1-\Delta\tau\sec\theta$ was used in the derivation of each thin layer.

B. Inhomogeneous scattering media

Now suppose to consider scattering from inhomogeneous random media with varied particle sizes. The axis *c* of spheroids is increasing from $c=1.5cm$ at the top *z*=0 linearly to *c*= 2.5*cm* at the bottom *z*=-*d*. Other axes $a=b=0.1cm$ are kept. The layer thickness *d*=200*cm*. The dielectric constant is $\varepsilon=20.5+9.0i$, the fractional volume is 0.002, other parameters are the same as Fig.1-2.5.

Figs.1-2.7(a) and 1-2.7(b) show bistatic scattering intensities of co-polarized I_v and depolarized I_h ($\theta,\varphi=180^\circ$), respectively. It is interesting to see that as scattering angle θ increases, both I_v and I_h of inhomogeneous medium gradually deviate from the case of homogeneous *c*=2.5 *cm*, and become closer to the case of homogeneous *c*=1.5 *cm*. It is well understandable that as scattering angle θ increases, scattering from the particles at the top layers becomes dominant.

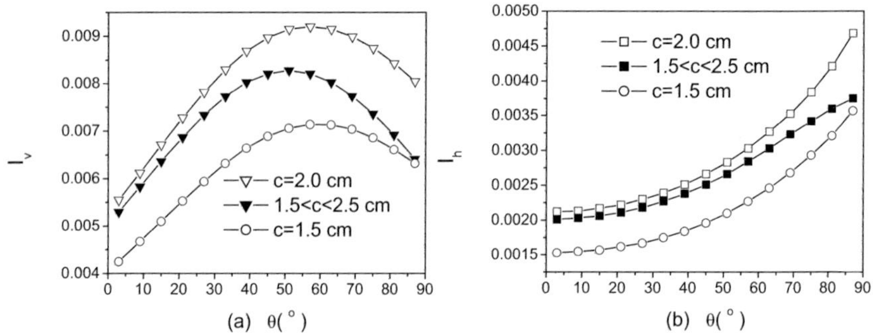

Fig.1-2.7. Bistatic scattering intensities from inhomogeneous and homogeneous media: (a) co-polarized; (b) depolarized

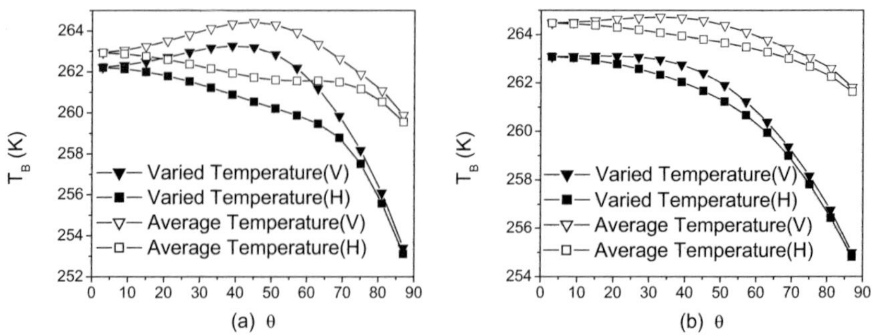

Fig.1-2.8. Brightness temperature of dense spherical particles with temperature profile

Now calculate brightness temperature of a layer of dense spherical particles with temperature profile from 258 *K* at the top *z*=0 linearly increasing to 273 *K* at the bottom *z*=-*d*. The particle radius $a = 0.06cm$, fractional volume is 0.3. The layer thickness $d = 50cm$, the dielectric constant of underlying space is $\varepsilon_2 = 3.2 + 0.1i$ and temperature is 273 *K*. Frequency is $f = 19GHz$

Figs. 1-2.8(a), 1-2.8(b) show the brightness temperatures up to the fourth order solutions. Fig.1-2.8(a) is for the dielectric constant of particles $\varepsilon = 3.2 + 0.032i$, and Fig. 1-2.8(b) is for $\varepsilon = 3.2 + 0.05i$. Comparing with the results of the model with uniform temperature 265.5 *K*, it is interesting to see how the brightness temperature of inhomogeneous media deviates from homogeneous model. As the observation angle increases, the difference between inhomogeneous and homogeneous models is very significant, especially for the particles with high imagery part of dielectric constant. It gives an idea how big change might happen for moistured snowpack with temperature profile.

1-3 An Approach of the Three-Dimensional Vector Radiative Transfer Equation (3D-VRT)

To describe multiple scattering, absorption and emission of radiance intensity, the vector radiative transfer (VRT) equation of the Stokes vector $\bar{I}$ has been studied and applied in broad areas. Conventional VRT is usually for the models of parallel-layered media, i.e. one-dimensional (1D) VRT equation ($d\bar{I}/dz$) [Jin 1994, Tsang and Kong 1990]. There have been some studies of 2D or 3D scalar RT equations, such as the Monte Carlo method and multi-modes approach for radiance of atmospheric discrete clouds [Barker 1992, Evans 1993, Dittmann 1997, Stephens 1986], the discrete-ordinate and finite difference method for the problems of heat transfer [Truelove 1988, Maruyama et al. 2002] and neutron transport etc. However, in all of these approaches it has been necessary to assume that the host medium itself is homogeneous, i.e. extinction, scattering and phase functions are considered to be independent of location within the medium. This fact simplifies the problem somewhat as the inhomogeneity in RT and then requires only a suitable treatment of the lateral sides of the medium and their associated boundary conditions. In advances of polarimetric and radiometric observation in remote sensing and the imaginary technology with an improved high resolution, development of 3D-VRT ($d\bar{I}/dx$, $d\bar{I}/dy$, $d\bar{I}/dz$) model for scattering simulation and detection technology from spatially inhomogeneous scatter media such as vegetation canopy, embedded target etc. becomes of great interest [Sun and Ranson 1995, Gastellu et al. 1996]. However, modeling and numerical solution of 3D-VRT for inhomogeneous scatter media remains to be studied.

To solve 3D-VRT equation for spatially inhomogeneous scatter media and obtain high-order scattering and emission, the finite enclosure of the scatter media is first geometrically divided into many thin boxes, slicing the media in both the vertical z and horizontal (x,y) directions. The zeroth order emission of each thin box and an iterative approach of high-order radiative transfer via the Mueller matrix solution of VRT in all media boxes are applied to deriving high-order scattering and emission of whole inhomogeneous scatter media. High order scattering and brightness temperature of the inhomogeneous scatter media in 3D geometry can be numerically calculated. According to the spatial resolution of the observation to divide the finite enclosure of scatter media, this approach presents numerical method to obtain high-order scattering and emission of the 3-D models for inhomogeneous scatter media such as heterogeneous vegetation canopy and embedded alien target [Jin and Liang 2004].

1-3.1 The 3D – VRT Equation

A geometric model is shown in Fig.1-3.1, where non-spherical particles are randomly and non-uniformly distributed within random limited space, and an alien target lays on the ground surface. It can be the model for tree canopy and alien target embedded. Suppose that the finite enclosure of the media has the length and width as W_1 and W_2, in the x and y directions respectively, and the height d in the z direction. Note that in this model the top and lateral boundaries of the enclosure are not solid, and the underneath of the bottom surface is a homogeneous half-space (land-) medium. The alien target can have different dielectric constant or physical temperature from the underlying surface.

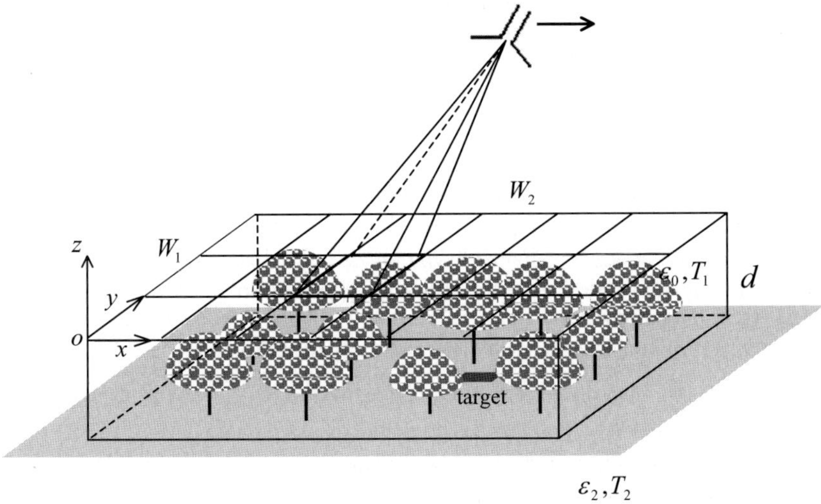

Fig.1-3.1. A geometric model for inhomogeneous scatter media

The 3D-VRT of the random scatter media [Maruyama et al. 2002, Gastellu et al. 1996, Chandrsekhar 1960] is written as

$$\begin{aligned}
&(\cos\theta\frac{\partial}{\partial z}+\sin\theta\cos\varphi\frac{\partial}{\partial x}+\sin\theta\sin\varphi\frac{\partial}{\partial y})\bar{I}(\theta,\varphi,x,y,z)\\
&=-\overline{\overline{\kappa}}_e(\theta,\varphi,x,y,z)\cdot\bar{I}(\theta,\varphi,x,y,z)+\overline{\overline{\kappa}}_a(\theta,\varphi,x,y,z)\cdot C\overline{T}_1(x,y,z)\\
&+\int_0^{2\pi}\mathrm{d}\varphi'\int_0^{\pi}\overline{\overline{P}}(\theta,\varphi,\theta',\varphi',x,y,z)\cdot\bar{I}(\theta',\varphi',x,y,z)\sin\theta'\mathrm{d}\theta' \quad ,
\end{aligned} \tag{1-3.1}$$

where $\overline{\overline{\kappa}}_e$, $\overline{\overline{\kappa}}_a$ are, respectively, the extinction and absorption matrices, $\overline{\overline{P}}$ is the phase matrix, and $\overline{T}_1$ is the physical temperature of the scatter media.

The boundary conditions at the top ($z=0$) and bottom ($z=-d$) surfaces are written as

$$\bar{I}(\pi-\theta,\varphi,x,y,z=0)=0\,, \tag{1-3.2a}$$

$$\bar{I}(\theta,\varphi,x,y,z=-d)=\overline{\overline{R}}_{12}(\theta)\cdot\bar{I}(\pi-\theta,\varphi,x,y,z=-d)+\overline{\overline{T}}_{12}(\theta)\cdot C\overline{T}_2\,, \tag{1-3.2b}$$

where $\overline{\overline{R}}_{12}(\theta), \overline{\overline{T}}_{12}(\theta)$ are the reflectivity and transmittivity matrices of the bottom surface, respectively. Here, the subscript 12 denotes "between" the random media (1) and underlying medium (2). T_2 is the physical temperature of the bottom medium. Eq.(1-3.2a) means that there is no downwards emission incidence at $(\pi-\theta,\varphi)$ from up-space ($z>0$) to the top surface ($z=0$). Eq. (1-3.2b) indicates that the upwards radiance at the bottom ($z=-d$) is contributed by reflection of downwards radiance of the random media and transmitted emission from the underlying medium.

The boundary conditions at the lateral surfaces at $x=0$, $x=W_1$, $y=0$ and $y=W_2$ are respectively written as

$$\begin{cases}\bar{I}(\pi-\theta,\varphi,x=0,y,z)=0\\ \bar{I}(\theta,\varphi,x=0,y,z)=\overline{\overline{T}}_{12}(\theta)\cdot C\overline{T}_2\end{cases} \qquad 0^\circ<\varphi<90^\circ, 270^\circ<\varphi<360^\circ, \tag{1-3.2c}$$

$$\begin{cases} \bar{I}(\pi-\theta,\varphi,x=W_1,y,z)=0 \\ \bar{I}(\theta,\varphi,x=W_1,y,z)=\bar{\bar{T}}_{12}(\theta)\cdot C\bar{T}_2 \end{cases} \qquad 90^\circ<\varphi<270^\circ, \tag{1-3.2d}$$

$$\begin{cases} \bar{I}(\pi-\theta,\varphi,x,y=0,z)=0 \\ \bar{I}(\theta,\varphi,x,y=0,z)=\bar{\bar{T}}_{12}(\theta)\cdot C\bar{T}_2 \end{cases} \qquad 0^\circ<\varphi<180^\circ, \tag{1-3.2e}$$

$$\begin{cases} \bar{I}(\pi-\theta,\varphi,x,y=W_2,z)=0 \\ \bar{I}(\theta,\varphi,x,y=W_2,z)=\bar{\bar{T}}_{12}(\theta)\cdot C\bar{T}_2 \end{cases} \qquad 180^\circ<\varphi<360^\circ. \tag{1-3.2f}$$

Eq.(1-3.2c) indicates that there is no downwards incidence from outside (defined by the azimuth φ region: $0^\circ<\varphi<90^\circ, 270^\circ<\varphi<360^\circ$) to enter the lateral side ($x=0$) of the random media, and the upwards radiance from outside comes from the transmitting emission from the underlying half-space medium. Same physical meanings are described by Eqs. (1-3.2d,e,f) for other lateral sides.

Generally, $\bar{\bar{\kappa}}_e$ is non-diagonal for non-uniformly (the Euler angles β,γ) [Jin 1994] oriented scatterers. To find the matrix $\bar{\bar{E}}$ and its inverse $\bar{\bar{E}}^{-1}$, $\bar{\bar{\kappa}}_e$ can be diagnosed for random scatterers as [Jin 1994, Tsang and Kong 1990] :

$$\bar{\bar{\beta}}(\theta,\varphi,x,y,z)=\bar{\bar{E}}^{-1}(\theta,\varphi,x,y,z)\cdot\bar{\bar{\kappa}}_e(\theta,\varphi,x,y,z)\cdot\bar{\bar{E}}(\theta,\varphi,x,y,z), \tag{1-3.3}$$

where the ii-th elements of the diagonal $\bar{\bar{\beta}}$ is denoted as β_i, $i=1,2,3,4$. It can be known that β_i is the eigen-values of $\bar{\bar{\kappa}}_e$, and $\bar{\bar{E}}$ is composed by the eigen-vectors of $\bar{\bar{\kappa}}_e$. All formulations of $\bar{\bar{\kappa}}_e$, $\bar{\bar{\beta}}$ and $\bar{\bar{E}}$ can be found in Jin (1994), Tsang et al. (1985).

Left-multiplying $\bar{\bar{E}}^{-1}$ on the both sides of Eq. (1-3.1), it yields

$$\begin{aligned} &(\cos\theta\frac{\partial}{\partial z}+\sin\theta\cos\varphi\frac{\partial}{\partial x}+\sin\theta\sin\varphi\frac{\partial}{\partial y})\bar{I}^E(\theta,\varphi,x,y,z) \\ &=-\bar{\bar{\beta}}(\theta,\varphi,x,y,z)\cdot\bar{I}^E(\theta,\varphi,x,y,z)+\bar{\bar{\kappa}}_a^E(\theta,\varphi,x,y,z)\cdot C\bar{T}_0^E(x,y,z) \\ &\quad+\int_0^{2\pi}d\varphi'\int_0^{\pi}\bar{\bar{P}}^E(\theta,\varphi,\theta',\varphi',x,y,z)\cdot\bar{I}^E(\theta',\varphi',x,y,z)\sin\theta'd\theta', \end{aligned} \tag{1-3.4}$$

where

$$\bar{I}^E(\theta,\varphi,x,y,z)=\bar{\bar{E}}^{-1}(\theta,\varphi,x,y,z)\cdot\bar{I}(\theta,\varphi,x,y,z), \tag{1-3.5a}$$

$$\bar{T}_0^E(x,y,z)=\bar{\bar{E}}^{-1}(\theta,\varphi,x,y,z)\cdot\bar{T}_0^E(x,y,z), \tag{1-3.5b}$$

$$\bar{\bar{P}}^E(\theta,\varphi;\theta',\varphi',x,y,z)=\bar{\bar{E}}^{-1}(\theta,\varphi,x,y,z)\cdot\bar{\bar{P}}(\theta,\varphi;\theta',\varphi',x,y,z)\cdot\bar{\bar{E}}(\theta,\varphi,x,y,z), \tag{1-3.5c}$$

$$\bar{\bar{\kappa}}_a^E(\theta,\varphi,x,y,z)=\bar{\bar{E}}^{-1}(\theta,\varphi,x,y,z)\cdot\bar{\bar{\kappa}}_a(\theta,\varphi,x,y,z)\cdot\bar{\bar{E}}(\theta,\varphi,x,y,z). \tag{1-3.5d}$$

For convenience, all notations of E would not be especially indicated in next derivations.

Let's slice the media enclosure into many thin-slabs with the thickness Δd along the z direction, and denote the slabs by the subscripts $\ell=1,2,\cdots,L$ from the top to the bottom. According to the spatial resolution Δh, dividing the (x,y) plane of the media enclosure to form many thin rectangular boxes with the length-width Δh and thickness Δd as illustrated in Fig.1-3.2. Because all boxes are very thin, we can assume that the medium within each box is homogeneous. But different boxes can be different, e.g. with different particles, different particle's fractional volumes, or different physical temperatures, etc. It is noted that as the $\bar{\bar{\kappa}}_e$ and $\bar{\bar{P}}$ of random particles have been calculated, the approach of VRT of random discrete particles is the same as one of continuous random media no matter what size the particles are comparing with Δd.

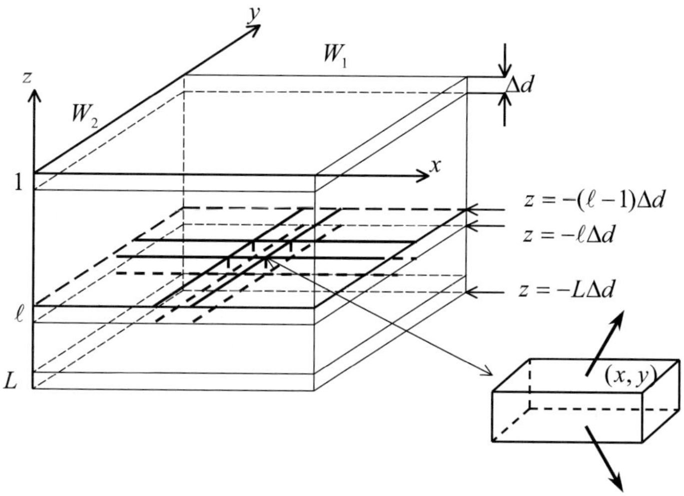

Fig.1-3.2. Dividing the scatter media into multi-boxes

As illustrated in Fig.1-3.2, the radiance intensity of the ℓ-th slab is defined as

$$\bar{I}_\ell(\theta,\varphi,x,y) = \begin{cases} \bar{I}(\theta,\varphi,x,y,z=-(\ell-1)\Delta d) & 0^\circ < \theta < 90^\circ \\ \bar{I}(\theta,\varphi,x,y,z=-\ell\Delta d) & 90^\circ < \theta < 180^\circ \end{cases} \quad (1\text{-}3.6)$$

where $\bar{I}_\ell(\theta,\varphi,x,y)$ is the radiance intensity in the direction (θ,φ) from the box of the ℓ-th slab whose center is defined at (x,y).

1-3.2 Scattering and Radiative Transfer of 3D-Thin Boxes

Because Δd is very small ($\Delta d << \Delta h$), change of the radiance intensity through the lateral sides is always much smaller than one through the top and bottom surfaces. Thus, based on VRT Eq. (1-3.1), the changes of up-going ($0^\circ < \theta < 90^\circ$) and down-going ($90^\circ < \theta < 180^\circ$) radiance intensities through the box of the ℓ-th slab whose center is defined at (x,y) are derived as follows

$$\bar{I}_\ell^{(n)}(\theta,\varphi,x+\Delta x,y+\Delta y) = \bar{I}_{\ell+1}^{(n)}(\theta,\varphi,x,y)\exp[-\bar{\beta}_\ell(\theta,\varphi,x,y)\Delta d\sec\theta] + \bar{I}_{\ell s}^{(n)}(\theta,\varphi,x+\Delta x,y+\Delta y)\ , \quad (1\text{-}3.7a)$$

$$\bar{I}_\ell^{(n)}(\theta,\varphi,x+\Delta x,y+\Delta y) = \bar{I}_{\ell-1}^{(n)}(\theta,\varphi,x,y)\exp[\bar{\beta}_\ell(\theta,\varphi,x,y)\Delta d\sec\theta] + \bar{I}_{\ell s}^{(n)}(\theta,\varphi,x+\Delta x,y+\Delta y)\ , \quad (1\text{-}3.7b)$$

where $\bar{I}_{\ell s}^{(n)}$ is the radiance intensity (zero-th order emission and multiple scattering) happening in the box self, where the superscript n indicates the iteration number. $\bar{\beta}_\ell = [\beta_{1\ell},\beta_{2\ell},\beta_{3\ell},\beta_{4\ell}]$ is the vector of diagonal $\bar{\bar{\beta}}$ of the ℓ-th slab.

Radiance transfering from the $(\ell-1)$-th and $(\ell+1)$-th slabs can present the following relations :

$$\Delta x = \tan\theta\cos\varphi\cdot\Delta d\ , \qquad \Delta y = \tan\theta\sin\varphi\cdot\Delta d\ . \quad (1\text{-}3.8)$$

The zero-th order emission of the box self is

$$\bar{I}_{\ell s}^{(n=0)}(\theta,\varphi,x,y)=\frac{1-\exp(-\bar{\beta}_\ell\Delta d\,|\sec\theta\,|)}{\bar{\beta}_\ell}\bar{\bar{\kappa}}_{a\ell}(\theta,\varphi,x,y)C\bar{T}_{0\ell}(x,y)\,. \tag{1-3.9a}$$

Multiple scattering from all directions under up-going and down-going incidences upon the box is

$$\begin{aligned}&\bar{I}_{\ell s}^{(n>0)}(\theta,\varphi,x,y)\\&=\int_0^{2\pi}\mathrm{d}\varphi'\int_{\pi/2}^{\pi}\bar{\bar{M}}_\ell(\theta,\varphi,\theta',\varphi',x',y')\cdot\bar{I}_{\ell-1}^{(n-1)}(\theta',\varphi',x',y')\sin\theta'\mathrm{d}\theta'\\&+\int_0^{2\pi}\mathrm{d}\varphi''\int_0^{\pi/2}\bar{\bar{M}}_\ell(\theta,\varphi,\theta'',\varphi'',x'',y'')\cdot\bar{I}_{\ell+1}^{(n-1)}(\theta'',\varphi'',x'',y'')\sin\theta''\mathrm{d}\theta''\end{aligned} \tag{1-3.9b}$$

and the relations are also given as

$$x'=x+\frac{\Delta x}{2}+\frac{\Delta d\tan\theta'\cos\varphi'}{2},\quad y'=y+\frac{\Delta y}{2}+\frac{\Delta d\tan\theta'\sin\varphi'}{2}, \tag{1-3.10a}$$

$$x''=x+\frac{\Delta x}{2}-\frac{\Delta d\tan\theta''\cos\varphi''}{2},\quad y''=y+\frac{\Delta y}{2}-\frac{\Delta d\tan\theta''\sin\varphi''}{2}. \tag{1-3.10b}$$

The Mueller matrix of the ℓ-th slab, $\bar{\bar{M}}_\ell$, is approximated by the first order Mueller matrix as :

$$M_{ij\ell}(\theta,\varphi,\theta',\varphi',x,y)\approx M_{ij\ell}^{(1)}(\theta,\varphi,\theta',\varphi' x,y)$$

$$=\frac{P_{ij\ell}(\theta,\varphi,\pi-\theta',\varphi' x,y)\sec\theta}{\beta_{j\ell}\sec\theta'+\beta_{i\ell}\sec\theta}\begin{cases}1-\exp[-\Delta d(\beta_{j\ell}\sec\theta'+\beta_{i\ell}\sec\theta)]\\ \qquad 0^\circ<\theta<90^\circ,0^\circ<\theta'<90^\circ\\ \exp(-\beta_{j\ell}\Delta d\sec\theta')-\exp(\beta_{i\ell}\Delta d\sec\theta)\\ \qquad 90^\circ<\theta<180^\circ,0^\circ<\theta'<90^\circ\\ \exp(\beta_{j\ell}\Delta d\sec\theta')-\exp(-\beta_{i\ell}\Delta d\sec\theta)\\ \qquad 0^\circ<\theta<90^\circ,90^\circ<\theta'<180^\circ\\ 1-\exp[\Delta d(\beta_{j\ell}\sec\theta'+\beta_{i\ell}\sec\theta)]\\ \qquad 90^\circ<\theta<180^\circ,90^\circ<\theta'<180^\circ\end{cases}, \tag{1-3.11}$$

where the subscripts $i,j=1, 2, 3, 4$.

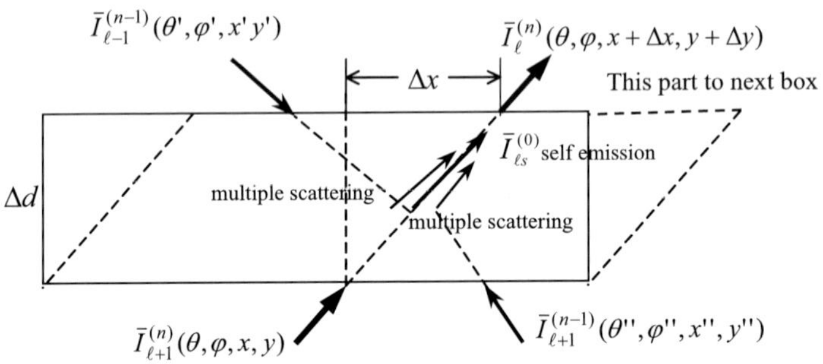

Fig.1-3.3. Illustration of radiative transfer through a box in the zox section

From Fig. 1-3.3 and Eq. (1-3.7a), it can be seen that the radiance $\bar{I}_{\ell+1}^{(n)}(\theta,\varphi,x,y)$ of the box at the center (x,y) becomes $\bar{I}_{\ell}^{(n)}(\theta,\varphi,x+\Delta x,y+\Delta y)$ after propagating and attenuating through this box medium. $\bar{I}_{\ell}^{(n)}(\theta,\varphi,x+\Delta x,y+\Delta y)$ will be distributed and enter into the boxes centers of the (ℓ-1)-th slab based on the ray-projected areas.

The thin-lines with arrow in Fig. 1-3.3 indicate the zero-th order emission $\bar{I}_{\ell s}^{(0)}$ and multiple scattering from the up and bottom interfaces of the box.

Suppose $\Delta h = 100cm$, $\Delta d = 1cm$, the calculation of Eq. (1-3.8) yields $\Delta x = 16.5cm$ and $\Delta y = 9.5cm$ for the direction $\theta = 87^\circ$ and $\varphi = 30^\circ$ (noted that the maximum $\theta_M = \cot^{-1}(2\Delta d/\Delta h) \approx 89^\circ$). The result shows that 0.756 of $\bar{I}_\ell^{(n)}(\theta,\varphi,x+\Delta x,y+\Delta y)$ remained within the box of the center (x,y), and only 0.079, 0.016, 0.149 of it enter into the neighbor boxes of the centers at $(x,y+\Delta h)$, $(x+\Delta h,y+\Delta h)$, $(x+\Delta h,y)$, respectively. Thus, the radiance intensity through the lateral sides of each box is much smaller than one through the top and bottom interfaces due to very small Δd, and $\bar{I}_{\ell s}^{(n)}$ of Eq. (1-3.7) can be calculated by Eqs. (1-3.9a, 1-3.9b).

Now the steps to calculate brightness temperature emitted from the top interface $\bar{T}_B(\theta,\varphi,x,y,z=0)$ is summarized as follows :

(1) By using Eqs. (1-3.9a, 1-3.11), the zero-th order emission $\bar{I}_{\ell s}^{(n=0)}(\theta,\varphi,x,y)$ and the Mueller matrix $\overline{\overline{M}}_\ell(\theta,\varphi,\theta',\varphi',x,y)$ of each box are calculated;

(2) By using the boundary condition (2), calculate the radiative transfer Eq. (1-3.7) from the top surface, the slabs $\ell = 1,2,\cdots,L$ sequentially, and finally to the bottom surface. Adding the emission and reflection of the bottom surface, calculate the radiative transfer Eq.(1-3.7) from the bottom surface, the slabs $\ell = L, L-1,\cdots,1$ sequentially, and finally to the top surface. Thus, $\bar{I}_{\ell=1}^{(n=0)}(\theta,\varphi,x,y)$ is obtained ;

(3) By using Eq. (1-3.9b), calculate the (n+1)-th (n=0,...) order iteration to obtain $\bar{I}_{\ell s}^{(n+1)}(\theta,\varphi,x,y)$ of each box ;

(4) Repeating the steps of (2) and (3) to the N-th iteration, the calculation is finished when $\bar{I}_{\ell=1}^{(N)}(\theta,\varphi,x,y)$ is small enough.

Let $\bar{I}_{\ell=1}(\theta,\varphi,x,y)$ is the sum of all iterations $\bar{I}_{\ell=1}^{(n)}(\theta,\varphi,x,y)$ ($n = 0,1,\cdots,N$), the brightness temperature observed in the up-space is

$$\bar{T}_B(\theta,\varphi,x,y,z=0) = \frac{1}{C}\bar{I}_{\ell=1}(\theta,\varphi,x,y), \qquad 0^\circ < \theta < 90^\circ. \qquad (1\text{-}3.12)$$

1-3.3 Numerical Results

A. Homogeneous media

First, compare the results of 3D-VRT (as scatterers are uniformly distributed within the rectangular enclosure) with conventional 1D-VRT for a homogeneous scatter medium. Suppose that the radiometer's frequency is $3GHz$, the scatter particles are prolate spheroids with the semi-radii $a = b = 0.1\,cm$ and $c = 2.5\,cm$, the dielectric constant of the particle is $\varepsilon_s = 22 + 5i$, the fractional volume is 0.0075, the dielectric constant of the bottom medium is $\varepsilon_2 = 8 + 1i$. The physical temperatures $T_1 = T_2 = 284\,K$.

Both the length and width of the rectangular enclosure are $15m$, and the depth d=1m, respectively. Divide the rectangular enclosure to form thin multi boxes with $\Delta h = 1\,m$, $\Delta d = 0.01\,m$ (i.e. 100 thin slabs), and take calculations at discrete angles $\Delta\theta = 9^\circ$, $\Delta\varphi = 18^\circ$.

From Eqs.(1-3.7, 1-3.9), it can be seen that rigorous calculation requires to storage all radiance at different angles (θ,φ) and different centers (x,y) of the boxes. It needs the storage memory about 144M. When the media enclosure becomes large, the storage memory would be tremendously increased. To reduce such requirement, we propose to use a parabola line to approximately match the radiance of each slab along all direction (θ,φ), and only storage some coefficients of the matching line.

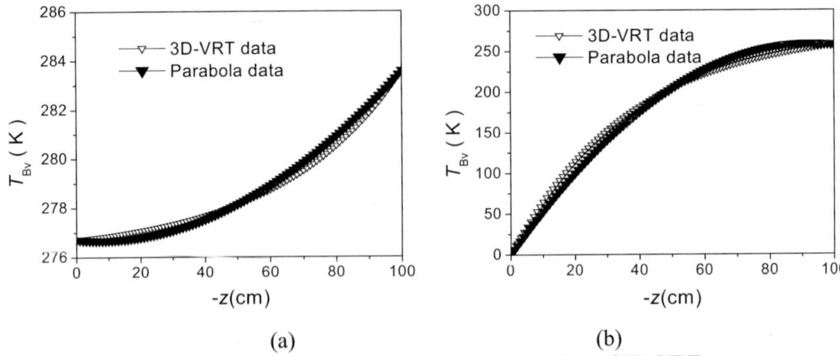

Fig. 1-3.4. Parabola data to match the calculation of 3D-VRT

As shown in Figs.1-3.4(a,b), three matching points of the parabola line are chosen at $z=0,-d/2,-d$. It only needs to storage 3 coefficients from 100 data, and significantly reduces the storage memory to 4*M*. For example, Figs.1-3.4(a,b) show the data of 3D-VRT from the slabs, respectively, along the line $(\theta=58^\circ,\varphi=9^\circ)$ and $(\pi-\theta=58^\circ,\varphi=9^\circ)$, and good matching by a parabola line.

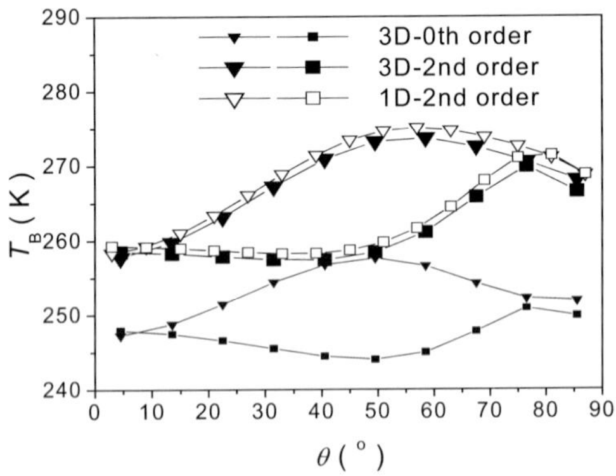

Fig.1-3.5. Brightness temperature from a homogeneous scatter medium

Fig.1-3.5 presents the zero-th order (small black points) and N=2 (big black points) iterative brightness temperatures observed towards the center ($x=7.5\,m,\ y=7.5\,m,\ z=0$). Calculation shows that N=2 iteration is good enough in this 3 *GHz* case. Difference of the radiance of the zero-th order and N=2 results is significant as about 10K. Because the brightness temperature observed at this center won't be significantly affected by the radiance afar from the lateral sides, as we believed, the result of 3D-VRT is well matched to the result of conventional 1D-VRT (1D means that the enclosure becomes infinite) [Liang and Jin 2003].

It numerically validates our 3D-VRT codes.

B. High-orders scattering and emission from inhomogeneous scatter media

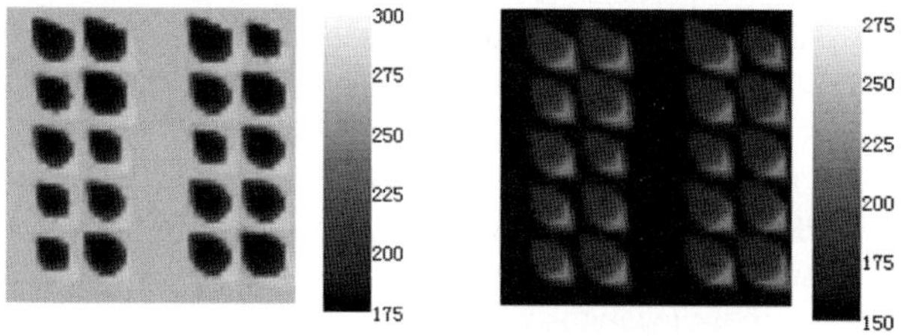

(a) N=0

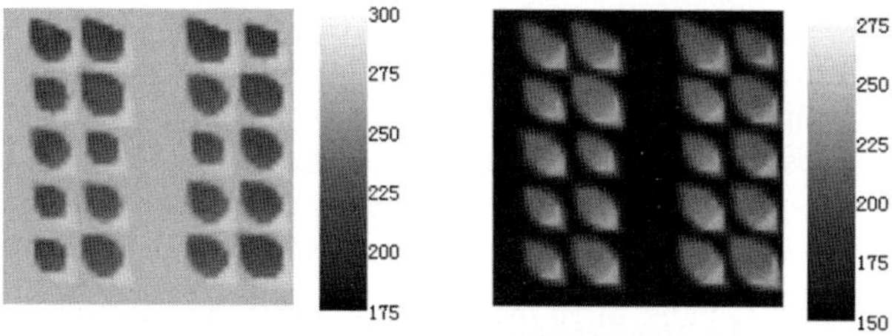

(b) N=1 iteration order

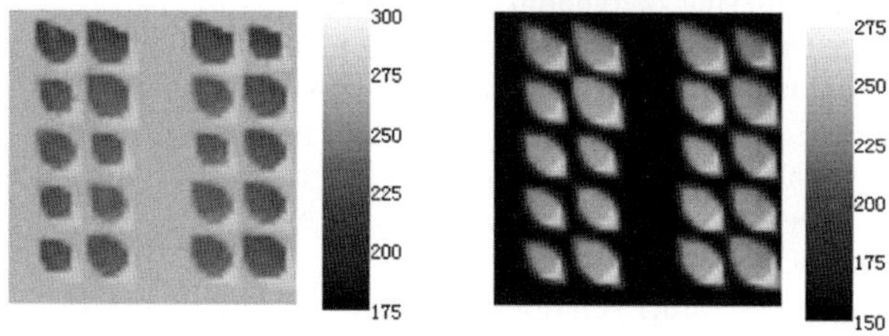

(c) N=2 iteration order

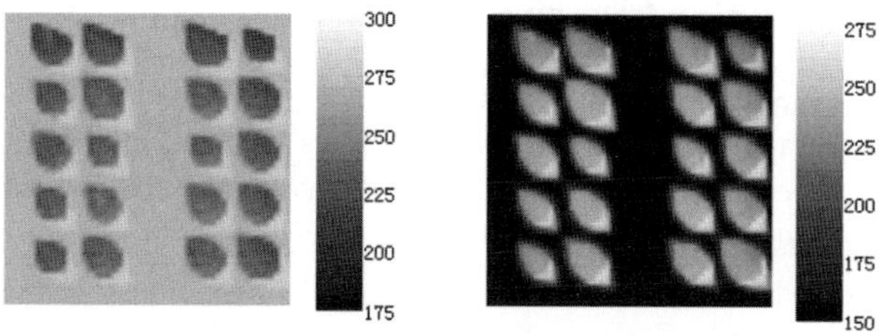

(d) N=3 iteration order

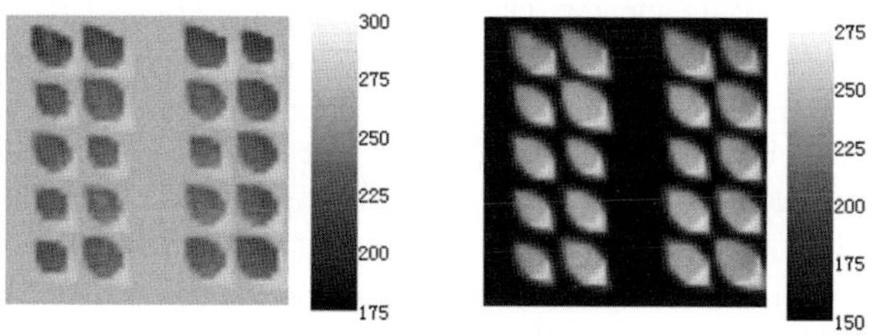

(e) N=4 iteration order

Fig. 1-3.6 High-orders from N=0 to N=4 scattering and emission at 5GHz from inhomogeneous canopy

Now suppose the rectangular enclosure of Fig.1-3.1 has both the length and width $55\,m$, and the depth d=4m. Spheroid particles are randomly distributed within geometrical cones. The radius and height of the cones are randomly takes the values of $3 \sim 4\,m$ from Monte Carlo realization, and the cone radius is equal to height only for simplicity. Distance between the cones is assumed be 10m. The spatial orientation distribution of random particles is uniformly over $\beta \in (30^\circ, 60^\circ)$ and $\gamma \in (0^\circ, 360^\circ)$. Both the physical temperatures of random scatterers and ground surface are 284 K. Other parameters are the same as Fig.1-3.5.

It is noted that by using the parabola line to reduce the storage (about 58M), the matching points should be chosen at those locations where there are scatter particles.

Suppose that the radiometer at 5GHz with the spatial resolution Δh=1m at the angle $\theta = 58^\circ, \varphi = 315^\circ$ observes each box one by one. Figs. 1-3.6(a-e) show the T_{Bv}, T_{Bh} solution from the zero-th order to N=4 iteration, sequentially.

It can be seen that higher-order iterative brightness temperature from those locations of random scatter particles significantly enhances the zero-th order emission. Brightness temperature from some parts within the cone locations are higher than the radiance from bottom surface medium because the emission is contributed by random particles, scattering reflected from the bottom surface and the emission from the bottom medium. However, there are also some parts within the cone locations where brightness temperatures become lower because stronger backscattering of random particles darkens the emission from the bottom medium. On the other hand, because the horizontally polarized brightness temperature from flat bottom surface is low, random particles always enhance brightness temperature within those cone locations.

It is interesting to note that the maximum increase of the 1-st order iteration can reach 50K, and then each order iterative solutions sequentially increase as 15K, 5K, and 1K. Higher iteration is not necessary.

Calculation of the zero-th order solution by using the Pentium 1.8 GHz PC takes about 10 min, but higher order solutions would take much more as 10 hr.

C. *Scattering and emission from an embedded alien target*

Let's consider that there is an alien target laying over the ground and embedded under inhomogeneous tree canopy.

Suppose the radiometer's frequency 3GHz, the sizes of oblate spheroids for tree leaves model are $a = b = 1.5\ cm$, $c = 0.05\ cm$, fractional volume 0.001, dielectric constant $\varepsilon_s = 15 + 5i$, spatial orientations are uniform over $\beta \in (30^\circ, 60^\circ)$, $\gamma \in (0^\circ, 360^\circ)$. The crowns of tree canopy are modeled as semi-spheroids, whose radii of the bottoms randomly take the value within 4~6m in Monte Carlo realization, and tree heights are assumed as 3/4 of respective bottom radius. All tree bottoms are simply assumed on the same level over the ground at height 2m. Physical temperatures of tree canopy and ground surface are 293K. The observation angle is $\theta = 49.5^\circ, \varphi = 315^\circ$. In computation, both the length and width of each small box are 50 cm, and the height 1.5cm. The resolution of the radiometer is 1m, i.e. is coarser than the small box size. Scatterings from small boxes within the resolution are averaged. The canopy enclosure covers the area 55m×55m, dielectric constant of ground surface is $\varepsilon_2 = 15 + 2i$.

An alien target with the rectangular sizes 3m×3m lays over the ground, and its dielectric constant $\varepsilon_{t\arg} = 80 + 10i$ is much different from the ground surface, e.g. water or conductor.

Figs1-3.7(a,b,c)showthe T_{Bv}, T_{Bh} solution from the zero-th to N=2 iteration, sequentially.

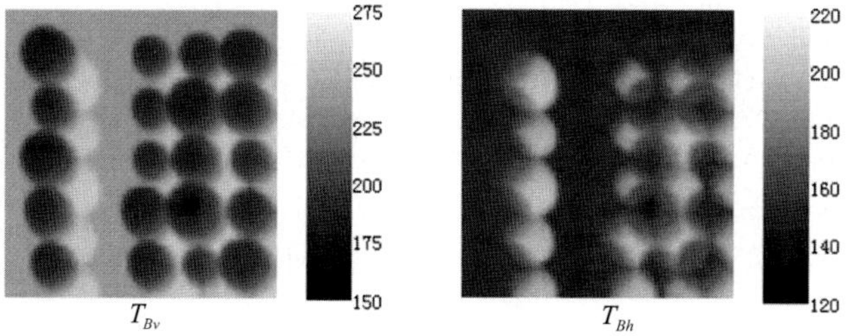

(a) Zero-th order

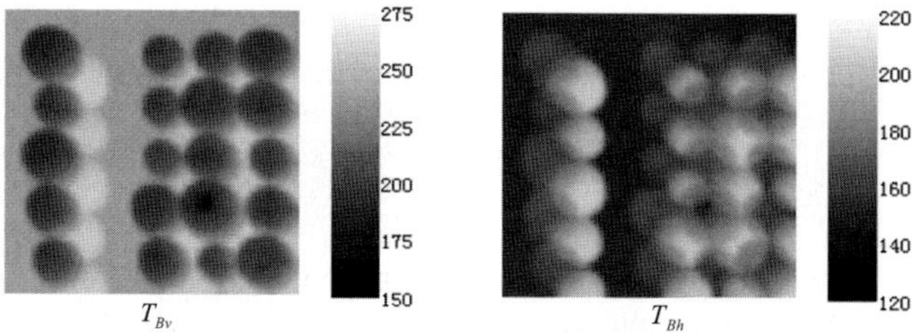

(b) N=1 order iteration

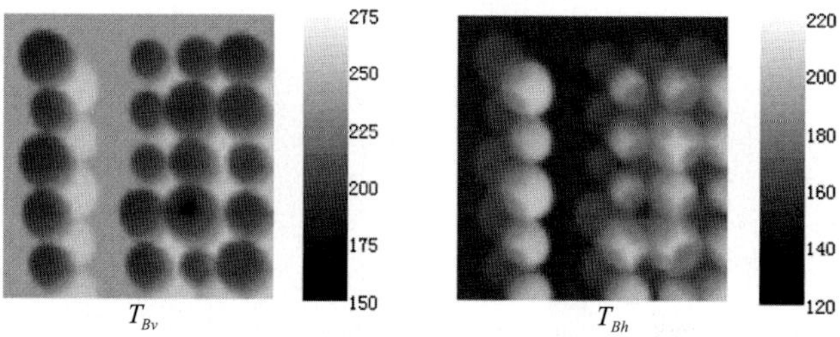

(c) N=2 order iteration

Fig. 1-3.7. Brightness temperature from tree canopy and an alien target at 3 GHz, N=0~2 orders

It can be seen that bright spots come from stronger reflected emission from interaction of tree canopy and ground surface. Much lower emission from the target area, where the dielectric constant is much larger than surrounding areas, can be well identified at this spatial resolution 1m, and be darken by high-order scattering.

D. Different resolutions for target detection

To see the resolution in observations, we consider another example when the alien target has much higher physical temperature 30K than the surrounding ground surface. The radiometer's frequency is 2GHz, the T_{Bv}, T_{Bh} solution only need N=1 iteration order. The ground surface has $\varepsilon_2 = 6 + 1i$. Other parameters are the same as Fig.1-3.7.

Figs. 1-3.8(a,b,c) give the T_{Bv}, T_{Bh} solutions at N=1 order for the radiometer's resolution Δh=1m, 2m, and 4m, respectively. It can be seen that variation of brightness temperature in whole area becomes a little smooth because coarser resolution takes the average of brightness temperature over a larger resolution area and whole image becomes blurred. Actually, at much coarser resolution Δh=4m, the target cannot be identified.

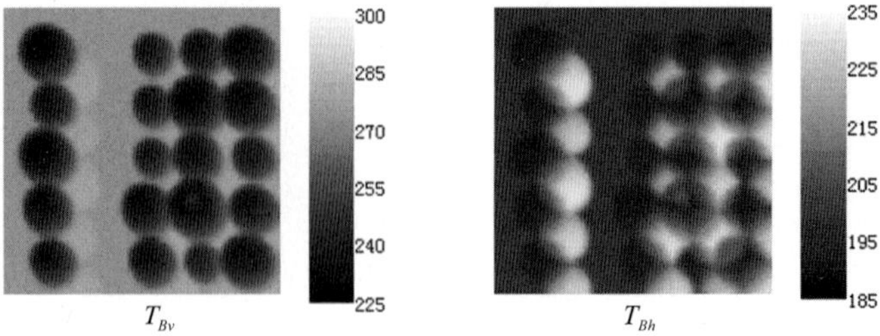

(a) resolution= 1*m*, *N*=1

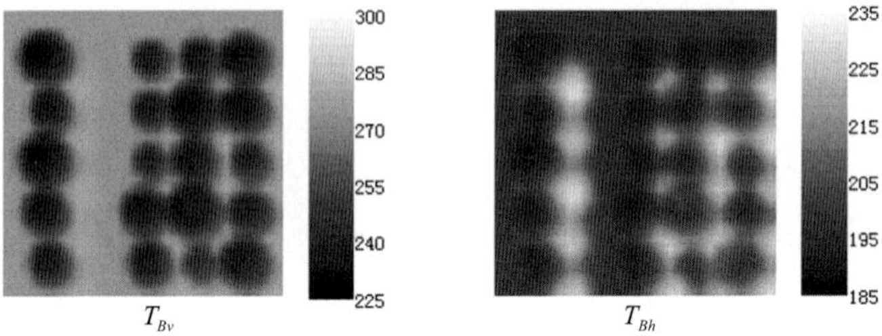

(b) resolution =2*m*, *N*=1

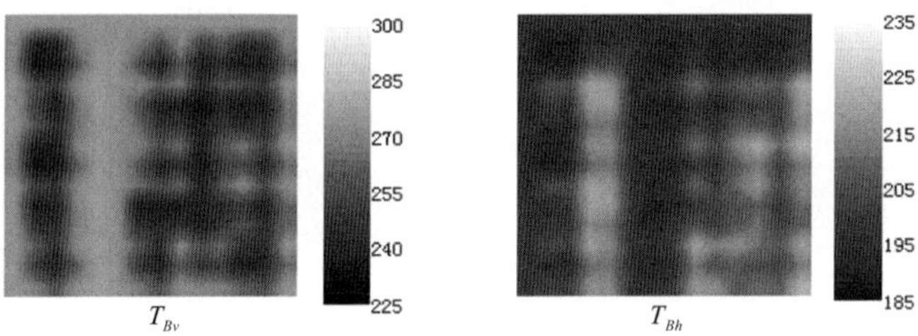

(c) resolution =4*m*, *N*=1.

Fig.1-3.8. Brightness temperature for different resolution at 2 *GHz*, *N*=1 order iteration

To study the visibility, two images at $\Delta h = 1m$ with and without the target are subtracted as shown in Figs.1-3.9.

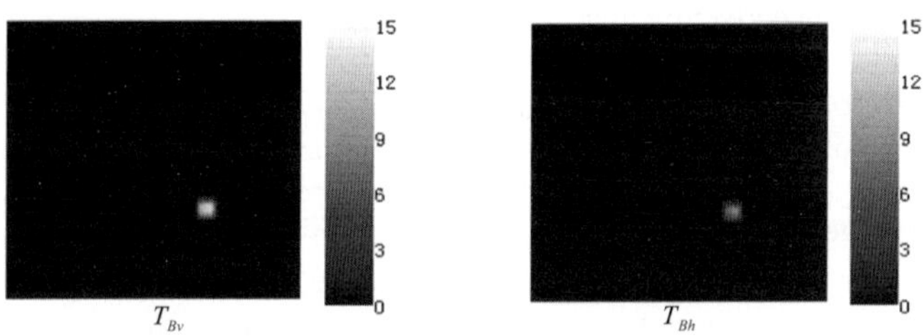

Fig.1-3.9. subtracting two images at $\Delta h = 1m$ with and without the target

Define the visibility as the maximum of the brightness temperatures difference ΔT_B. Fig. 1-3.9 shows that identification of ΔT_B becomes worse when the resolution is coarser

It is noted that at coarser resolution, the pixels number becomes less. For example, there are 16 pixels of a picture within 1m resolution, but only 1 pixel within 4m resolution. Another way to define the visibility is to sum all ΔT_B within the resolution. Fig. 1-3.9(b) shows $\sum \Delta T_B$ vs. resolution Δh.

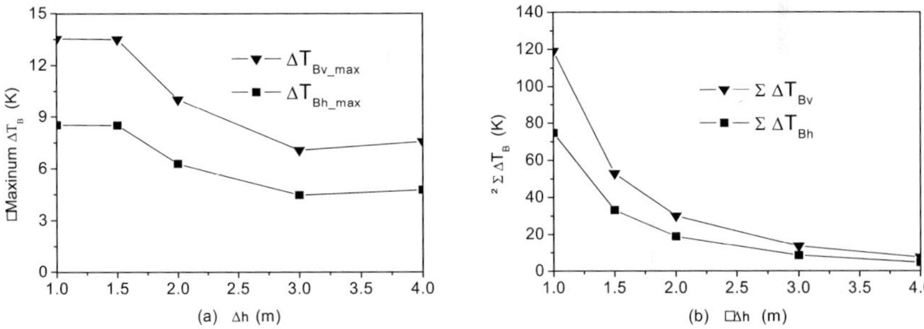

Fig.1-3.9. Visibility vs resolution

1-4 References

Andrew, B. and R.D. Mark (2000), "Recent changes in the microwave scattering properties of the Antarctic ice sheet", *IEEE Transactions on Geoscience and Remote Sensing*, 37(4): 1810-1820.

Barichello, L.B. *et al.* (2000), "Particular solutions for the discrete-ordinates method", *Journal of Quantitative Spectroscopy and Radiative Transfer*, 64: 219

Barker, H.W. (1992), "Solar radiative transfer through clouds possessing isotropic variable extinction coefficient", *Quandary Journal of Royal Meteorology Society*, 118: 1145-1162.

Chandrasekhar, S. (1960), *Radiative Transfer*, New York: Dover.

Dittmann, O.J. (1997), "The influence of cloud geometry on continuum polarization", *Journal of Quantitative Spectroscopy and Radiative Transfer*, 57: 249-263.

Evans, K.F. (1993), "Two-dimensional radiative transfer in cloudy atmospheres-the spherical harmonic spatial grid method", *Journal of Atmospheric Science,* 50: 3111-3124.

Gastellu-Etchegorry, J.P. *et al.* (1996), "Modeling Radiative Transfer in Heterogeneous 3-D Vegetation Canopies", *Remote Sensing of Environment*, 58: 131-156.

Ishimaru, A. *et al.* (1982), "Multiple scattering calculation of rain effects", *Radio Science*, 17: 1425-1432.

Jin, Y.Q. (1994), *Electromagnetic Scattering Modelling for Quantitative Remote Sensing*, Singapore: World Scientific.

Jin, Y.Q. (2000), *Information of Electromagnetic Scattering and Radiative Transfer in Natural Media*, Beijing: Science Press.

Jin, Y.Q. and Z. Liang (2003), "Iterative solution of multiple scattering and emission from inhomogeneous scatter media", *Journal of Applied Physics*, 93(4): 2251-2256.

Jin, Y.Q. and Z. Liang (2004), "Iterative inversion from the multi-order Mueller matrix solution of vector radiative transfer equation for a layer of random spheroids", *Journal of Quantitative Spectroscopy and Radiate Transfer*, 83(4): 303-311.

Jin, Y.Q. and Z. Liang (2004), "An approach of the three-dimensional vector radiative transfer equation for inhomogeneous scatter media", *IEEE Transactions on Geoscience and Remote Sensing*, 42(2): 355-360.

Jin, Y,Q. and Z. Liang (2005), "Scattering and emission from inhomogeneous vegetation canopy and alien target by using three-dimensional vector radiative transfer (3D-VRT) equation", *Journal of Quantitative Spectroscopy and Radiative Transfer*, 92(3): 261-274.

Liang, Z. and Y.Q. Jin (2003), "Iterative approach of high-order scattering solution for vector radiative transfer of inhomogeneous random media", *Journal of Quantitative Spectroscopy and Radiative Transfer*, 77: 1-12.

Maruyama, S. *et al.* (2002) "Improvement of computational time in radiative heat transfer of three-dimensional participating media using the radiation element method", *Journal of Quantitative Spectroscopy and Radiative Transfer*, 73: 239-248.

She, Z. (2002), "Three-dimensional space-borne synthetic aperture radar (SAR) imaging with multiple pass processing", *International Journal of Remote Sensing*, 23: 4357-4382.

Siewert, C.E. (2000), "A concise and accurate solution to Chandrasekhar's basic problem in radiative transfer", *Journal of Quantitative Spectroscopy and Radiative Transfer*, 64: 109-118.

Stephens, G.L. (1986), "Radiative transfer in spatially heterogeneous two dimensional anisotropically scattering media", *Journal of Quantitative Spectroscopy and Radiative Transfer*, 36: 51-67.

Sun, G. and K.J. Ranson (1995), "A three-dimensional radar backscatter model of forest canopies", *IEEE Transactions on Geoscience and Remote Sensing*, 33: 372-382.

Truelove, J.S. (1988), "3-D Radiation in absorbing-emitting-scattering media using the discrete ordinates approximation", *Journal of Quantitative Spectroscopy and Radiative Transfer*, 39: 27-31.

Tsang, L., J. A. Kong and R. Shin (1984), "Radiative transfer theory for active remote sensing of a layer of nonspherical particles", *Radio Science*, 19: 629-639.

Tsang, L., J.A. Kong, and R. Shin (1985), *Theory of Microwave Remote Sensing*, New York: John Wiley.

Tsang, L. and J.A. Kong (1990), *Polarimetric Remote Sensing*, New York: Elsevier, 1990

Tsang, L. and K. H. Ding (1991), "Polarimetric signatures of a layer of random nonspherical discrete scatterers overlying a homogeneous half-space based on first- and second-order vector radiative transfer theory", *IEEE Transactions on Geoscience and Remote Sensing*, 29: 242-250.

Tsang, L. and J.A. Kong (2001), *Scattering of Electromagnetic Waves—Numerical Simulations*, Mass: John Wiley.

Chapter 2

Fully Polarimetric Scattering of Non-Spherical Scatter Media 34

2-1 Polarimetric Scattering Indexes and Information Entropy of the SAR Imagery for Surface Monitoring

Synthetic aperture radar (SAR) imagery technology is one of most important advances in space-borne microwave remote sensing during recent decades. Completely polarimetric scattering from complex terrain surfaces can be measured [Zebker and Van Zyl 1991]. Fully understanding and retrieving information from polarimetric scattering signatures of natural media have become a key issue for the SAR remote sensing and its broad applications.

During recent years there has been extensive research on polarimetric scattering for SAR imagery [Evans et al. 1988]. Numerical simulation of polarimetrc scattering from natural media via the Mueller matrix solution has been studied [Tsang et al. 1985, Jin 1994a]. The coherency matrix and eigen-analysis of information entropy have been studied for polarimetry of natural media in SAR imagery [Cloude 1986, Jin and Cloude 1994].

To indicate the difference between vertically (*vv*) and horizontally (*hh*) co-polarized backscatterings, the polarization index was usually defined to classify the polarized scattering signature from different natural media.

However, the entropy and eigen-analysis need full Mueller matrix measurement and have not been directly related with the backscattering signatures. In other words, how to present the entropy and eigenanalysis merely from backscattering measurement has not been demonstrated.

In this session, the relationship between the entropy and co-polarized, cross-polarized backscattering indexes is established. Thus, it can now combine the Mueller matrix solution, the entropy of the coherency matrix, and measurements of the polarization indexes together and yield a quantitative description of the natural media. As examples, this theory is applied to an AirSAR image and field measurements for surface monitoring and classification [Jin and Chen 2002].

2-1.1 The Mueller and Coherency Matrices with the Eigen-Analysis

As a polarized wave $\overline{E}_{inc}(\chi,\psi)$ is incident upon the natural media, the scattering field is written as

$$\begin{bmatrix} E_{vs} \\ E_{hs} \end{bmatrix} = \frac{e^{ikr}}{r} \begin{bmatrix} F_{vv} & F_{\mathrm{vh}} \\ F_{hv} & F_{\mathrm{hh}} \end{bmatrix} \cdot \begin{bmatrix} E_{vi} \\ E_{hi} \end{bmatrix} \equiv \frac{e^{ikr}}{r} \overline{\overline{F}} \cdot \overline{E}_{inc}(\chi,\psi) \tag{2-1.1}$$

where 2×2-*D* (Dimensional) complex scattering amplitude function $\overline{\overline{F}}$ can be measured by the polarimetry. The incident polarization is indicated by the elliptic and orientation angles (χ,ψ). Using the Mueller matrix solution of VRT equation [Tsang et al., 1985, Jin 1994a] and Eq. (2-1.1), the scattered Stokes vector (four Stokes parameters) can be obtained or measured as

$$\overline{I}_s(\theta,\phi) = \overline{\overline{M}}(\theta,\phi;\pi-\theta_0,\phi_0) \cdot \overline{I}_i(\chi,\psi) \tag{2-1.2}$$

Making use of the Mueller matrix solution of vector radiative transfer theory, the theoretical model and numerical simulation of polarimetric scattering from the layering scatter media have been studied [Jin et al. 1999]. The Mueller matrix $\overline{\overline{M}}$ is constructed by the non-diagonal extinction matrix $\overline{\overline{\kappa}}_e$ and the functions $< F_{pq}F_{st}^* >$, $p,q,s,t=v,h$. The co-polarized and cross-polarized backscattering coefficients, σ_c and σ_x, polarization degree m_s for scattered Stokes echo with partial polarization and other functions can be numerically calculated.

In either the measurement or theoretical simulation, a key issue is to understand the scattering mechanism through the Mueller matrix solution. The Mueller matrix is a 4×4-*D* real matrix with complex eigenvalues and eigenvectors. To be physically realizable, this matrix must satisfy the Stokes criterion together with several other restrictive conditions [Van de Mee and Hovenier 1992]. Unfortunately, however, these restrictions do not have any direct physical interpretation in terms of the eigenstructure of the Mueller matrix. The coherency matrix $\overline{\overline{C}}$ is applied to the study of polarimetric scattering of SAR image [Cloude 1986, Jin and Cloude 1994]. Define the scattering vector as

$$\begin{aligned}\bar{t} &= \tfrac{1}{2}[F_{vv}+F_{hh}, F_{vv}-F_{hh}, F_{vh}+F_{hv}, i(F_{vh}-F_{hv})]^T \\ &\equiv \tfrac{1}{2}[A,B,C,iD]^T\end{aligned} \tag{2-1.3}$$

where *A, B, C, D* are sequentially defined in Eq.(2-1.3). The coheremcy matrix is defined as follows [Cloude 1986, Jin and Cloude, 1994]

$$\overline{\overline{C}} = <\bar{t}\,\bar{t}^+> = \frac{1}{4}\begin{bmatrix} <AA^*> & <AB^*> & <AC^*> & <AD^*> \\ <BA^*> & <BB^*> & <BC^*> & <BD^*> \\ <CA^*> & <CB^*> & <CC^*> & <CD^*> \\ <DA^*> & <DB^*> & <DC^*> & <DD^*> \end{bmatrix} \tag{2-1.4}$$

The eigenvalue and eigenvector of the coherency matrix is defined as

$$\overline{\overline{C}} = <\bar{t}\,\bar{t}^+> = \sum_{i=1}^{4}\lambda_i <\bar{t}_i\,\bar{t}_i^+> \tag{2-1.5}$$

where λ_i and $\bar{t}_i$ are the *i*-th eigen-value and eigen-vector, respectively. All eigenvalues are real and non-negative, and $\lambda_1 \ge \lambda_2 \ge \lambda_3 \ge \lambda_4$.

The coherency matrix is also related with the Mueller matrix as

$$M_{i+1,s+1} = \sum_t\sum_u \frac{1}{2}\mathrm{Trace}(\overline{\overline{\sigma}}_i\,\overline{\overline{\sigma}}_t\,\overline{\overline{\sigma}}_s\overline{\overline{\sigma}}_u)[\overline{\overline{C}}]_{tu} \tag{2-1.6}$$

where $\overline{\overline{\sigma}}_i, i = 0,1,2,3$ are the Pauli matrices [Cloude 1986].

Usually, the case of $F_{vh} = F_{hv}$ is simply assumed. The entropy H is defined by the eigenvalues of the coherency matrix as

$$H = -\sum_{i=1}^{3} P_i \log_3 P_i \tag{2-1.7}$$

where the functions P_i (i=1,2,3) are expressed by the eigenvalues as

$$P_i = \lambda_i / (\lambda_1 + \lambda_2 + \lambda_3) \tag{2-1.8}$$

The entropy is an important feature since it relates the randomness of scatter media with other physical parameters, such as canopy configuration, biomass, etc. However, to construct the eigenvalues λ_i, functions P_i, and the entropy H need full Mueller matrix measurement as shown in Eq. (2-1.4), and have not been directly related with the measured echo power, i.e. polarimetric σ_c or σ_x. A question is to be answered that when the backscattering σ_c and σ_x are measured, what should the H be.

The coherency matrix is 4×4-*D* positive semidefinite Hermitian and as such has real non-negative eigenvalues and complex orthogonal eigenvectors. Amplitude and difference of four eigen-values are functionally related with polarimetric scattering process of the terrain canopy. Eigen-analysis of the coherency matrix yields better physical insight into the polarimetric scattering mechanisms than does the Mueller matrix and further, it can be employed to physically identify a Mueller matrix by virtue of the semi-definite nature of the corresponding coherency matrix.

2-1.2 Relationship between the Eigenvalues, Entropy and Polarization Indexes

For the first-order solution in small albedo, cross polarization is small and correlations such as the terms $<CA^*>, <CB^*>, <DA^*>, <DB^*>$ etc. in Eq.(2-1.4) are always very small, and might be neglected in our following derivations. Actually, for weak assumption of azimuthal symmetric orientation, these correlations have been proved to be zero [Nighiem et al. 1992]. Applying this approximation to Eq. (2-1.4), the eigenvalues of $\overline{\overline{C}}$ have been derived as [Jin and Cloude 1994]

$$\lambda_{1,2} = \frac{1}{8}[<|A|^2> + <|B|^2> \pm \sqrt{(<|A|^2> - <|B|^2>)^2 + 4 <A^*B><AB^*>}] \tag{2-1.9}$$

$$\lambda_3 = \frac{1}{4} <|C|^2> \tag{2-1.10}$$

It is interesting to see that if $<AB^*><A^*B>=<AA^*><BB^*>$ at an ordered case, it yields $\lambda_2 = 0$.

Substituting A,B,C and D of Eq. (2-1.3) into Eqs. (2-1.9, 2-1.10), we newly derived the eigen-values in the power terms of $<|F_{vv}|^2>, <|F_{hh}|^2>, <|F_{hv}|^2>$

$$\lambda_1 = \frac{1}{2}(<|F_{vv}|^2> + <|F_{hh}|^2>) - \lambda_2 \tag{2-1.11}$$

$$\lambda_2 = \frac{1}{2}\frac{<|F_{vv}|^2><|F_{hh}|^2>}{(<|F_{vv}|^2> + <|F_{hh}|^2>)}(1-\Delta) \tag{2-1.12}$$

$$\lambda_3 = <|F_{hv}|^2> \tag{2-1.13}$$

where the configuration parameter

$$\Delta = \frac{<F_{vv}F_{hh}^*><F_{vv}^*F_{hh}>}{<|F_{vv}|^2><|F_{hh}|^2>} \tag{2-1.14}$$

varies from 0 of totally random media to 1 of ordered media, i.e. no-random case.

It can be seen that the first eigen-value λ_1 now indicates the total *vv* and *hh* co-polarized power subtracting their coherence defined by the parameter Δ of Eq.(2-1.14), which varies from 1 at ordered case to 0 in totally random medium. The second eigen-value λ_2 now takes into account for the *vv* and *hh* coherent power. As Δ approaches to one, $\lambda_2 \to 0$ for ordered media, and as Δ approaches to zero, λ_2 would increase to indicate the loss of the *vv* and *hh* power coherency for totally random media. The third eigen-value λ_3 is now due to depolarization caused by the media randomness.

From Eq.(2-1.2), it is obtained that the polarized backscattering coefficient is

$$\sigma_{pq} = 4\pi \cos\theta <|F_{pq}|^2> \tag{2-1.15}$$

where pq = *vv, hh, hv*. Thus, Eqs. (2-1.11-14) and (2-1.15) link the eigen-analysis with the measurements of $\sigma_{vv}, \sigma_{hh}, \sigma_{vh}$.

Substituting Eq. (2-1.15) into Eqs. (2-1.11-13) and Eq.(2-1.8), the functions $P_i, i = 1,2,3$, of Eqs. (2-1.7, 2-1.8) are then derived as

$$P_1 = 1 - P_2 - P_3 \tag{2-1.16}$$

$$P_2 = X(1-\Delta)(1-\delta) \tag{2-1.17}$$

$$P_3 = \delta(1-\delta) \quad << 1 \tag{2-1.18}$$

where

$$X \equiv \frac{\sigma_{hh}/\sigma_{vv}}{(1+\sigma_{hh}/\sigma_{vv})^2} \tag{2-1.19}$$

$$\delta = \frac{2\sigma_{hv}}{\sigma_{vv} + \sigma_{hh}} \tag{2-1.20}$$

Note that all functions $P_i, i = 1,2,3$, are now calculated by polarized backscattering coefficient σ_{pq}, $p,q = v,h$, not by Eq. (2-1.8) from full Mueller matrix solution. Then, the entropy H of Eq. (2-1.7) is calculated using P_i of Eqs.(2-1.16~20). It can be seen that the function P_1 is related with the total *vv* and *hh* co-polarized power $\sigma_{hh}+\sigma_{vv}$; P_2 is due to the difference between σ_{hh} and σ_{vv}; and P_3 is due to depolarization σ_{hv}. They are modulated by the media configuration and randomness via the parameters Δ and δ. As the medium becomes more random or disordered, δ increases and Δ approaches zero. Vice versa, as the medium becomes more ordered, δ decreases and Δ increases. Thus, the relationship between the entropy H and backscattering measurements σ_{pq} (pq = *vv*, *hh*, *hv*) is established.

Define the co-polarized and cross polarized indexes, respectively, as

$$CPI \equiv 10\log_{10}(\sigma_{hh}/\sigma_{vv}) \ (dB) \tag{2-1.21}$$

$$XPI \equiv \delta \tag{2-1.22}$$

The question in Section 2 is answered that the entropy H is directly related with backscattering indexes *CPI* and *XPI* via Eqs. (2-1.16~22).

2-1.3 Demonstration with AirSAR Imagery

Fig. 2-1.1 presents the theoretical relationship between the entropy H and index *CPI* with variations of the parameters Δ and δ. The entropy H is calculated by Eqs.(2-1.16~20), i.e. via $\sigma_{hh}, \sigma_{vv}, \sigma_{hv}$ and variations of Δ and δ. It can be seen that as the medium becomes more random (δ increases) and *CPI* becomes smaller, H will be increased. Vice versa, as the medium becomes more ordered and *CPI* becomes larger, H will be decreased. The bold line in Fig. 2-1.1 is for the case of $\Delta = 0$ and $\delta = 0$. More randomness will shift *H–CPI* upwards. Or *vice versa*, the ordered or less random medium will move *H-CPI* downwards. The distance of the line *H-CPI* from this solid line indicates how much random the medium might be.

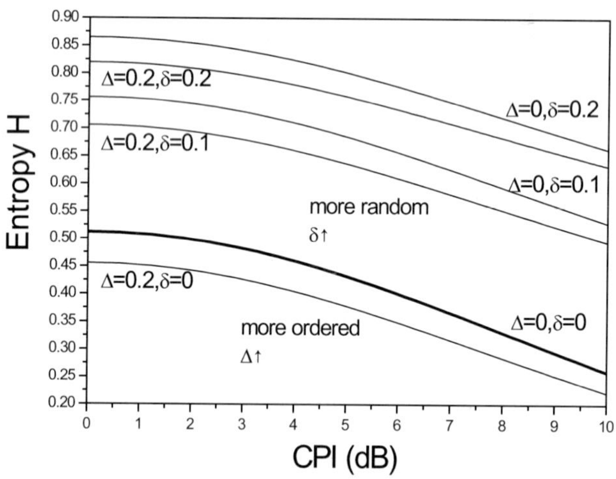

Fig,2-1.1. Entropy H vs. Index *CPI* with variations of Δ and δ

Fig. 2-1.2 presents an AirSAR image of total power of σ_{vv}, σ_{hh} and σ_{vh} at the L band over a Jack Pine area of Canada (near the Prince Albert National Park, about $103.33°W, 53.9°N$). The image data from some typical areas such as the lake, an island surface, sparse trees, and thick forest (see the frames in the figure) are chosen, and their *H*, *CPI* and *XPI* are compared.

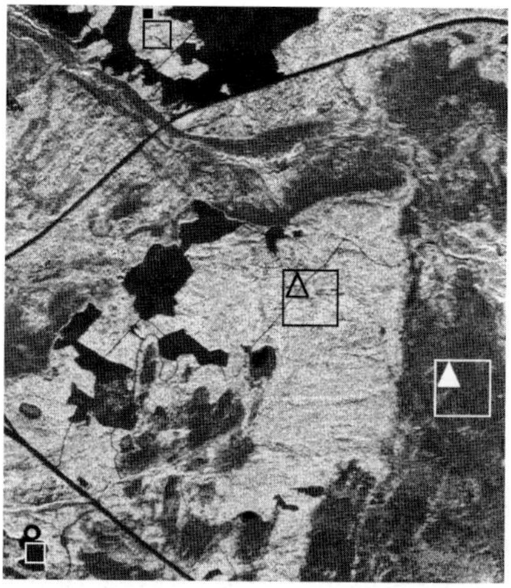

Fig.2-1.2. An AirSAR image of total power $\sigma_{vv}, \sigma_{hh}, \sigma_{vh}$ at L band

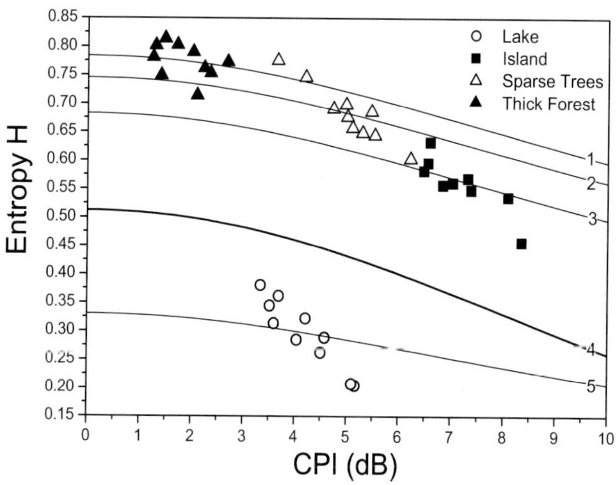

Fig.2-1.3. Entropy *H* and index *CPI* for surface classification from AirSAR data
Line 1: $\Delta = 0.257, \delta = 0.176$, Line 2: $\Delta = 0.318, \delta = 0.156$, Line 3: $\Delta = 0.365, \delta = 0.118$
Line 4: $\Delta = 0., \delta = 0.$, Line 5: $\Delta = 0.731, \delta = 0.0265$

To demonstrate the *H-CPI* relationship from the AirSAR imagery, Fig. 2-1.3 shows discrete points of the *H-CPI* from four regions of Fig. 2-1.2. The *H* is first rigorously calculated by polarimetric data of the Mueller matrix, i.e. the eigenvalues of coherency matrix

$\overline{\overline{C}}$ of Eq. (2-1.4). Using the AirSAR data $\sigma_{vv}, \sigma_{hh}, \sigma_{vh}$ and the averaged δ in the respective region, the lines are calculated by Eqs. (2-1.16~20) with appropriate Δ to match the center of discrete data. All δ and Δ are listed in the figure title. The bold line is for the case of $\Delta = 0$ and $\delta = 0$. The lines to match discrete data indicate how far the data in the region are from the case of $\Delta = 0, \delta = 0$, and how random the media are. It can be seen that the *H-CPI* can be well applied to surface classification.

Fig. 2-1.4 gives a 3-D graph of *H* versus two indexes (*CPI* and *XPI*) of Eqs. (2-1.21,22). It can be seen that the medium randomness yields high *H*, decreases *CPI*, and increases *XPI*. More information for surface classification can be inverted from this 3-*D* graph. The discrete points in Fig. 2-1.4 are the averaged data chosen from four regions of Fig. 2-1.2, respectively.

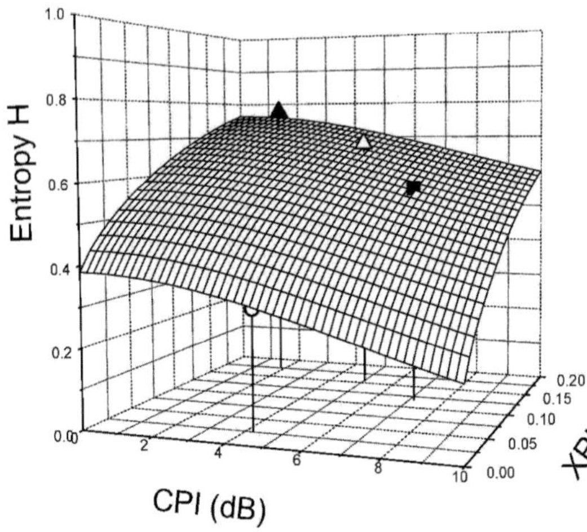

Fig. 2-1.4. Entropy *H* vs. two indexes (*CPI* and *XPI*). The curved surface is for $\Delta = 0.3$

It is certainly known that polarimetric echoes from terrain canopy are contributed by volumetric scattering of random non-spherical particles and rough surface scattering. Any configuration change of volumetric or surface scatter conditions can make variation of the *H-CPI, XPI.* To see an example during the period of crop growth, Figs.2-1.5 and 2-1.6 present the *H-CPI* and *H-*(*CPI, XPI*) from the corn field measurements $\sigma_{vv}, \sigma_{hh}, \sigma_{vh}$ and δ, even no fully polarimetric data is available at this case. The measurements were made by the scatterometer at 4.6*GHz* and at observation angle 40 ° during corn growth from May 18 to July 11, 1988 [Matzler and Wegmuller 1993]. The layer thickness of corn field is *d* (*cm*). At the beginning of corn growth, e.g. *d*<*30* cm, scattering is governed by the underlying rough surface. The *CPI* index might be also small due to scattering of very roughly cultivated surface. As the canopy grows, the rough surface might be gradually smoothed or shadowed by scatterers. More scatterers of the canopy enhance volumetric scattering and yield higher *H*. Specific orientation of non-spherical scatterers might also lightly increase *CPI*. As the canopy becomes much thicker and more random, e.g. *d*>100 cm, totally random medium would increase *H* and decrease *CPI*. The nonmonotonic trend of the *H-CPI, H-CPI, XPI* during corn growth, depending upon variation of volumetric and surface configurations, can be seen in

Figs. 2-1.5 and 2-1.6. The lines with different δ and Δ indicate chracteristics of the corn growth.

The parameter δ in Figs. 2-1.5 and 2-1.6 is given by the measurements of σ_{vv}, σ_{hh} and σ_{vh} via Eqs. (2-1.16~20). As $\delta_{\max} = 0.25$ is reached, the medium is assumed be totally random and $\Delta = 0$ is defined. Thus, it empirically yields $\Delta = 1 - \delta / 0.25$ for calculations in Figs. 2-1.5 and 2-1.6.

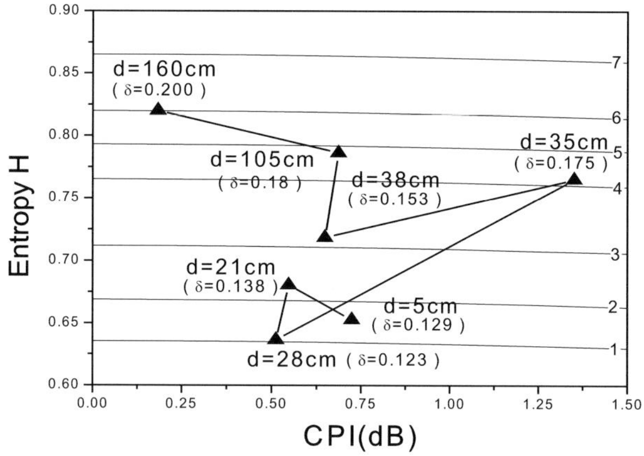

Fig. 2-1.5. Entropy *H-CPI* from the corn field measurements
Line 1: $\Delta = 0.5, \delta = 0.12$, Line 2: $\Delta = 0.45, \delta = 0.13$, Line 3: $\Delta = 0.4, \delta = 0.15$
Line 4: $\Delta = 0.4, \delta = 0.15$, Line 5: $\Delta = 0.3, \delta = 0.2$, Line 6: $\Delta = 0.2, \delta = 0.2$, Line 7: $\Delta = 0., \delta = 0.2$

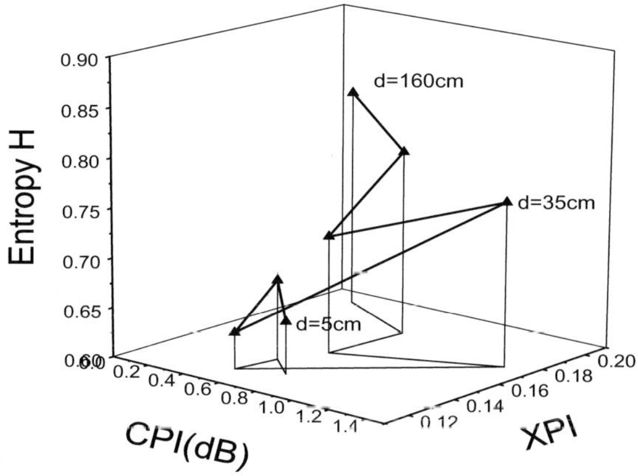

Fig.2-1.6. Entropy *H*- two indexes (*CPI, XPI*) from the corn field measurements

Numerical simulation of polarimetric scattering from a layer of non-spherical particles over a rough surface [Jin et al. 1999] can be also applied to verification of the functional dependence of the *H* upon *CPI* and *XPI.*

2-2 Temporal Mueller Matrix Solution for Polarimetric Scattering from Inhomogeneous Random Media

With the advance of radar polarimetry and imagery of SAR technology [Zebker et al. 1991, Boerner et al. 1997], the study of completely polarimetric scattering of the terrain surface has become of great importance. The Mueller matrix solution of vector radiative transfer has been studied for polarimetric scattering from a layer of random non-spherical particles [Sanh et al. 2000,2001, Jin 1994]. However, as the incidence is a plane pulse, it leads the problem of the pulse echoes for polarimetric scattering from random scatterers media and the temporal Mueller matrix solution.

There are a few works on the pulse echo from random media. For example, using two-frequency mutual coherence function, Hong and Ishimaru (1977), Liu and Yeh (1978), Ito (1980) studied the scalar pulse wave propagation in random media. Oguchi and Ito (1990) obtained the numerical solutions of the two frequency vector radiative transfer equation in the case where a circularly polarized plane pulse was normally incident upon a half space of spherical raindrops. Whitman et al. (1996) investigated the pulse propagation in a strongly forward scattering random medium. All aforementioned studies are limited to discussion of homogeneous random media. Using the ray tracing method and range calculation, Sun (2000) presented a model of forest canopy with vertical structure and calculated scattering for simulation of lidar returns. However, completely polarimetric scattering echo from inhomogeneous random media has remained to be further studied.

This session presents a theoretical model of inhomogeneous layering random media and derives the temporal Mueller matrix solution of vector radiative transfer for polarimetric scattering of random non-spherical scatterers. Co-polarized and cross-polarized bistatic and back-scattering are numerically calculated. Variations of the shape and intensity of the polarimetric echoes well depict the inhomogeneous fraction profile, and the functional dependence upon layering thickness, incident angle and other parameters are discussed [Jin and Chang 2002, Chang and Jin 2002].

2-2.1 Radiative Transfer in Inhomogeneous Random Scatterers Media

The N-layers model is shown in Fig.2-2.1. The fractional volume of non-spherical scatterers in the n-th layer is assumed as f_{sn} ($n = 1,2, \ldots, N$). The layer-(N+1) is assumed be the air region over an underlying surface with the permittivity ε_b. Let a polarized plane pulse be incident at the angle of $(\pi-\theta_0, \varphi_0)$. The time-dependent vector radiative transfer (VRT) equations for the up-going wave $\bar{I}_n(\theta,\varphi,z,t)$ and the down-going wave $\bar{I}_n(\pi-\theta,\varphi,z,t)$ in the layer-n are written as ($0 \le \theta < \pi/2$)

$$\cos\theta \frac{\partial}{\partial z}\bar{I}_n(\theta,\varphi,z,t) = -\bar{\bar{\kappa}}_{en}(\theta,\varphi)\cdot\bar{I}_n(\theta,\varphi,z,t) + \bar{S}_n(\theta,\varphi,z,t) - \frac{1}{c_0}\frac{\partial}{\partial t}\bar{I}_n(\theta,\varphi,z,t) \tag{2-2.1a}$$

$$-\cos\theta \frac{\partial}{\partial z}\bar{I}_n(\pi-\theta,\varphi,z,t) = -\bar{\bar{\kappa}}_{en}(\pi-\theta,\varphi)\cdot\bar{I}_n(\pi-\theta,\varphi,z,t) + \bar{W}_n(\theta,\varphi,z,t) - \frac{1}{c_0}\frac{\partial}{\partial t}\bar{I}_n(\pi-\theta,\varphi,z,t) \tag{2-2.1b}$$

where the subscript n denotes the n-th layer, $z \in [d_{n-1}, d_n]$, $\bar{\bar{\kappa}}_{en}$ is the extinction matrix in the n-th layer, c_0 is the light velocity, and the multiple scattering function

$$\overline{S}_n(\theta,\varphi,z,t)=\int_0^{\pi/2}d\theta'\int_0^{2\pi}d\varphi'\sin\theta'[\overline{\overline{P}}_n(\theta,\varphi;\theta',\varphi')\cdot\overline{I}_n(\theta',\varphi',z,t) + \overline{\overline{P}}_n(\theta,\varphi;\pi-\theta',\varphi')\cdot\overline{I}_n(\pi-\theta',\varphi',z,t)] \quad (2\text{-}2.2)$$

and the function $\overline{W}_n(\theta,\varphi,z,t)$ is the same as $\overline{S}_n(\theta,\varphi,z,t)$ with all θ replaced by $\pi-\theta$. $\overline{\overline{P}}_n$ is the phase matrix, (θ,φ) and (θ',φ') are the incident and scattered direction angles in the n-th layer, respectively.

The boundary conditions are written as

$$\overline{I}_1(\pi-\theta,\varphi,z=0,t)=\overline{I}_0(t)\delta(\cos\theta-\cos\theta_0)\delta(\varphi-\varphi_0) \quad (2\text{-}2.3\text{a})$$

$$\overline{I}_{n+1}(\pi-\theta,\varphi,z=-d_n,t)=\overline{I}_n(\pi-\theta,\varphi,z=-d_n,t) \qquad (n=1,\ldots,N\text{-}1) \quad (2\text{-}2.3\text{b})$$

$$\overline{I}_N(\theta,\varphi,z=-d_N,t)=\overline{\overline{R}}(\theta)\cdot\overline{I}_N(\pi-\theta,\varphi,z=-d_N,t-\frac{2\Delta d_{N+1}\sec\theta}{c_0}) \quad (2\text{-}2.3\text{c})$$

$$\overline{I}_n(\theta,\varphi,z=-d_n,t)=\overline{I}_{n+1}(\theta,\varphi,z=-d_n,t) \qquad (n=N\text{-}1,\ldots,1) \quad (2\text{-}2.3\text{d})$$

where $\overline{I}_0(t)$ is the incident Stokes vector, and $\overline{\overline{R}}(\theta)$ is the reflectivity matrix of the underlying surface at $z=-d_{N+1}$. Δd_{N+1} $(=d_{N+1}-d_N)$ is the thickness of the layer-(N+1).

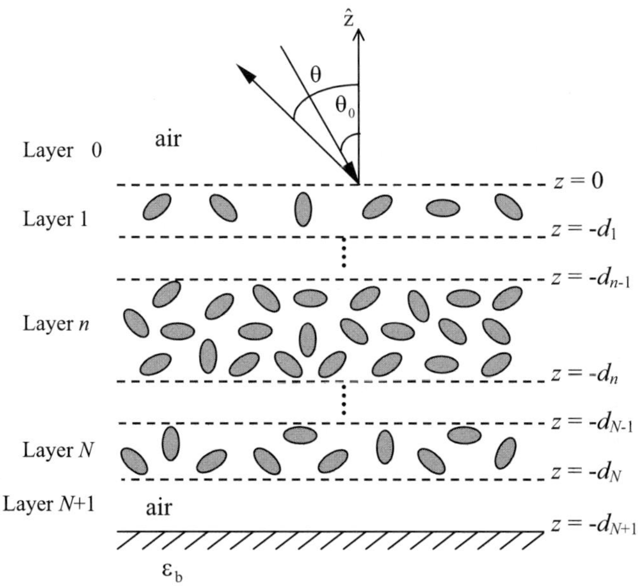

Fig.2-2.1. Model of inhomogeneous layering random media

Introducing the exponential Fourier series representations for the time dependence of Eqs. (2-2.1)~(2-2.3)

$$\overline{I}(\theta,\varphi,z,t)=\frac{1}{2\pi}\int_{-\infty}^{\infty}\overline{I}(\theta,\varphi,z,\omega)e^{j\omega t}d\omega \quad (2\text{-}2.4)$$

it yields the VRT equations of (2-2.1a,1b) as

$$\cos\theta\frac{\partial}{\partial z}\overline{I}_n(\theta,\varphi,z,\omega)=-[\overline{\overline{\kappa}}_{en}(\theta,\varphi)+\frac{j\omega}{c_0}\overline{\overline{U}}]\cdot\overline{I}_n(\theta,\varphi,z,\omega)+\overline{S}_n(\theta,\varphi,z,\omega) \quad (2\text{-}2.5\text{a})$$

$$-\cos\theta\frac{\partial}{\partial z}\overline{I}_n(\pi-\theta,\varphi,z,\omega)=-[\overline{\overline{\kappa}}_{en}(\pi-\theta,\varphi)+\frac{j\omega}{c_0}\overline{\overline{U}}]\cdot\overline{I}_n(\pi-\theta,\varphi,z,\omega)+\overline{W}_n(\theta,\varphi,z,\omega) \quad (2\text{-}2.5\text{b})$$

where $\overline{\overline{U}}$ is the 4×4-D (dimensional) unit matrix, and Eq.(2-2.2) becomes

$$\overline{S}_n(\theta,\varphi,z,\omega)=\int_0^{\pi/2}d\theta'\int_0^{2\pi}d\varphi'\sin\theta'[\overline{\overline{P}}_n(\theta,\varphi;\theta',\varphi')\cdot\overline{I}_n(\theta',\varphi',z,\omega)+\overline{\overline{P}}_n(\theta,\varphi;\pi-\theta',\varphi')\cdot\overline{I}_n(\pi-\theta',\varphi',z,\omega)] \quad (2\text{-}2.6)$$

and $\overline{W}_n(\theta,\varphi,z,\omega)$ is obtained in the same way. It can be seen that the term of $j\omega\overline{\overline{U}}/c$ takes into account for time delay of wave propagation through the random media in the Fourier domain.

The boundary conditions Eqs. (2-2.3a~d) are rewritten as

$$\overline{I}_1(\pi-\theta,\varphi,z=0,\omega)=\overline{I}_0(\omega)\delta(\cos\theta-\cos\theta_0)\delta(\varphi-\varphi_0) \quad (2\text{-}2.7a)$$

$$\overline{I}_{n+1}(\pi-\theta,\varphi,z=-d_n,\omega)=\overline{I}_n(\pi-\theta,\varphi,z=-d_n,\omega) \qquad (n=1,\ldots,N\text{-}1) \quad (2\text{-}2.7b)$$

$$\overline{I}_N(\theta,\varphi,z=-d_N,\omega)=\overline{\overline{R}}(\theta)\cdot\overline{I}_N(\pi-\theta,\varphi,z=-d_N,\omega)\cdot\exp(-j\omega\frac{2\Delta d_{N+1}\sec\theta}{c_0}) \quad (2\text{-}2.7c)$$

$$\overline{I}_n(\theta,\varphi,z=-d_n,\omega)=\overline{I}_{n+1}(\theta,\varphi,z=-d_n,\omega) \qquad (n=N\text{-}1,\ldots,1) \quad (2\text{-}2.7d)$$

Given the eigenvalues of $\overline{\overline{\kappa}}_{en}(\pi-\theta,\varphi)$ be $\xi_{ni}(\pi-\theta,\varphi)$, $i=1,2,3,4$, and the corresponding eigenvector matrix be $\overline{\overline{E}}(\pi-\theta,\varphi)$ (Tsang et al. 2000, p.37-40, and Ito 1980), it is easy to understand that the eigenvalues $\zeta_{ni}(\pi-\theta,\varphi)$ of $[\overline{\overline{\kappa}}_{en}(\pi-\theta,\varphi)+j\omega\overline{\overline{U}}/c_0]$ are equal to ($\xi_{ni}(\pi-\theta,\varphi)+j\omega/c_0$) and the eigenvector matrix is also $\overline{\overline{E}}(\pi-\theta,\varphi)$.

The formal general solution of Eq. (2-2.5b) can be written as (Tsang et al. 2000, p.124)

$$\overline{I}_n(\pi-\theta,\varphi,z,\omega)=\overline{\overline{E}}(\pi-\theta,\varphi)\overline{\overline{D}}[\zeta_n(\pi-\theta,\varphi)z\sec\theta]\cdot\overline{C}_n+\sec\theta\overline{\overline{E}}(\pi-\theta,\varphi)\cdot\int_z^{-d_{n-1}}\overline{\overline{D}}[\zeta_n(\pi-\theta,\varphi)(z-z')\sec\theta]\overline{\overline{E}}^{-1}(\pi-\theta,\varphi)\cdot\overline{W}_n(\theta,\varphi,z',\omega)dz' \quad (2\text{-}2.8)$$

where $\overline{\overline{D}}$ is a 4×4-D diagonal matrix with the ii-th element of $\exp[\zeta_{ni}(\pi-\theta)z\sec\theta]$, $\overline{C}_n$ are the unknown column vectors to be determined by the boundary conditions (2-2.7a,b).

Let the eigenvalues of the matrix $[\overline{\overline{\kappa}}_{en}(\theta,\varphi)+j\omega\overline{\overline{U}}/c_0]$ be $\zeta_{ni}(\theta,\varphi)=\xi_{ni}(\theta,\varphi)+j\omega/c_0$, $i=1,2,3,4$, (where $\xi_{ni}(\theta,\varphi)$ are the eigenvalues of the matrix $\overline{\overline{\kappa}}_{en}(\theta,\varphi)$), and the eigenvector matrix $\overline{\overline{E}}(\theta,\varphi)$, correspondingly, the formal general solution of Eq.(2-2.5a) is obtained as

$$\overline{I}_n(\theta,\varphi,z,\omega)=\overline{\overline{E}}(\theta,\varphi)\overline{\overline{D}}[-\zeta_n(\theta,\varphi)z\sec\theta]\cdot\overline{C}'_n+\sec\theta\overline{\overline{E}}(\theta,\varphi)\cdot\int_{-d_n}^{z}\overline{\overline{D}}[-\zeta_n(\theta,\varphi)(z-z')\sec\theta]\overline{\overline{E}}^{-1}(\theta,\varphi)\cdot\overline{S}_n(\theta,\varphi,z',\omega)dz' \quad (2\text{-}2.9)$$

where $\overline{C}'_n$ are the unknown column vectors to be determined by the boundary conditions Eqs. (2-2.7c,2-2.7d).

As $\overline{C}_1$ is first obtained by the boundary condition (2-2.7a), others $\overline{C}_2$, …, and $\overline{C}_N$ are then determined in turn by the boundary conditions (2-2.7b). Next, $\overline{C}'_N$ is first determined by Eq. (2-2.7c) after $\overline{C}_N$ is obtained, the others $\overline{C}'_{N-1}$, …, and $\overline{C}'_1$ are then in turn determined by Eq. (2-2.7d).

As $\overline{C}_n$ and $\overline{C}'_n$, $n=1,\ldots,N$, are now determined, the formal solutions of $\overline{I}_n(\theta,\varphi,z,\omega)$ and $\overline{I}_n(\pi-\theta,\varphi,z,\omega)$ of Eqs. (2-2.8, 9) are obtained.

Taking the integral-independent terms of the formal solutions of Eqs.(2-2.8, 2-2.9) as the zeroth-order solutions $\overline{I}_n^{(0)}(\theta,\varphi,z,\omega)$ and $\overline{I}_n^{(0)}(\pi-\theta,\varphi,z,\omega)$, and substituting them into the multiple scattering terms $\overline{W}_1(\theta,\varphi,z,\omega)$,…, $\overline{W}_N(\theta,\varphi,z,\omega)$ and $\overline{S}_1(\theta,\varphi,z,\omega)$,…, $\overline{S}_N(\theta,\varphi,z,\omega)$ of $\overline{I}_1(\theta,\varphi,z,\omega)$ (where $z\in[-d_1,0]$), the first-order solution $\overline{I}_1^{(1)}(\theta,\varphi,z,\omega)$ at z=0 can be obtained

$$\bar{I}_1^{(1)}(\theta,\varphi,z=0,\omega)=\sum_{n=1}^{N}\{\sec\theta\overline{\overline{E}}(\theta,\varphi)\overline{\overline{D}}[-\sum_{m=1}^{n}\xi_m(\theta,\varphi)\Delta d_m\sec\theta]$$
$$\cdot\int_{-d_n}^{-d_{n-1}}dz'D[\xi_n(\theta,\varphi)(z'+d_n)\sec\theta]\overline{\overline{E}}^{-1}(\theta,\varphi)\overline{\overline{P}}_n(\theta,\varphi;\pi-\theta_0,\varphi_0)\overline{\overline{E}}(\pi-\theta_0,\varphi_0)$$
$$\cdot\overline{\overline{D}}[-\sum_{m=1}^{n}\xi_m(\pi-\theta_0,\varphi_0)\Delta d_m\sec\theta_0+\xi_n(\pi-\theta_0,\varphi_0)(z'+d_n)\sec\theta_0]$$
$$\cdot\overline{\overline{E}}^{-1}(\pi-\theta_0,\varphi_0)\cdot\bar{I}_0(\omega)\exp[j\omega\frac{z'(\sec\theta+\sec\theta_0)}{c_0}]\}$$
$$+\sum_{n=1}^{N}\{\sec\theta\overline{\overline{E}}(\theta,\varphi)\overline{\overline{D}}[-\sum_{m=1}^{n}\xi_m(\theta,\varphi)\Delta d_m\sec\theta]$$
$$\cdot\int_{-d_n}^{-d_{n-1}}dz'\overline{\overline{D}}[\xi_n(\theta,\varphi)(z'+d_n)\sec\theta]\overline{\overline{E}}^{-1}(\theta,\varphi)\overline{\overline{P}}_n(\theta,\varphi;\theta_0,\varphi_0)\overline{\overline{E}}(\theta_0,\varphi_0)$$
$$\cdot\overline{\overline{D}}[-\sum_{m=n}^{N}\xi_m(\theta_0,\varphi_0)\Delta d_m\sec\theta_0-\xi_n(\theta_0,\varphi_0)(z'+d_{n-1})\sec\theta_0]$$
$$\cdot\overline{\overline{E}}^{-1}(\theta_0,\varphi_0)\overline{\overline{R}}(\theta_0)\overline{\overline{E}}(\pi-\theta_0,\varphi_0)\overline{\overline{D}}[-\sum_{m=1}^{N}\xi_m(\pi-\theta_0,\varphi_0)\Delta d_m\sec\theta_0]$$
$$\cdot\overline{\overline{E}}^{-1}(\pi-\theta_0,\varphi_0)\cdot\bar{I}_0(\omega)\exp[-j\omega\frac{2d_{N+1}\sec\theta_0+z'(\sec\theta_0-\sec\theta)}{c_0}]\}$$
$$+\sum_{n=1}^{N}\{\sec\theta\overline{\overline{E}}(\theta,\varphi)\overline{\overline{D}}[-\sum_{m=1}^{N}\xi_m(\theta,\varphi)\Delta d_m\sec\theta]\overline{\overline{E}}^{-1}(\theta,\varphi)\overline{\overline{R}}(\theta)\overline{\overline{E}}(\pi-\theta,\varphi)$$
$$\cdot\overline{\overline{D}}[-\sum_{m=n}^{N}\xi_m(\pi-\theta,\varphi)\Delta d_m\sec\theta]\int_{-d_n}^{-d_{n-1}}dz'\overline{\overline{D}}[-\xi_n(\pi-\theta,\varphi)(z'+d_{n-1})\sec\theta]$$
$$\cdot\overline{\overline{E}}^{-1}(\pi-\theta,\varphi)\overline{\overline{P}}_n(\pi-\theta,\varphi;\pi-\theta_0,\varphi_0)\overline{\overline{E}}(\pi-\theta_0,\varphi_0)$$
$$\cdot\overline{\overline{D}}[-\sum_{m=1}^{n}\xi_m(\pi-\theta_0,\varphi_0)\Delta d_m\sec\theta_0+\xi_n(\pi-\theta_0,\varphi_0)(z'+d_n)\sec\theta_0]$$
$$\cdot\overline{\overline{E}}^{-1}(\pi-\theta_0,\varphi_0)\cdot\bar{I}_0(\omega)\exp[-j\omega\frac{2d_{N+1}\sec\theta+z'(\sec\theta-\sec\theta_0)}{c_0}]\}$$
$$+\sum_{n=1}^{N}\{\sec\theta\overline{\overline{E}}(\theta,\varphi)\overline{\overline{D}}[-\sum_{m=1}^{N}\xi_m(\theta,\varphi)\Delta d_m\sec\theta]\overline{\overline{E}}^{-1}(\theta,\varphi)\overline{\overline{R}}(\theta)\overline{\overline{E}}(\pi-\theta,\varphi)$$
$$\cdot\overline{\overline{D}}[-\sum_{m=n}^{N}\xi_m(\pi-\theta,\varphi)\Delta d_m\sec\theta]\int_{-d_n}^{-d_{n-1}}dz'\overline{\overline{D}}[-\xi_n(\pi-\theta,\varphi)(z'+d_{n-1})\sec\theta]$$
$$\cdot\overline{\overline{E}}^{-1}(\pi-\theta,\varphi)\overline{\overline{P}}_n(\pi-\theta,\varphi;\theta_0,\varphi_0)\overline{\overline{E}}(\theta_0,\varphi_0)$$
$$\cdot\overline{\overline{D}}[-\sum_{m=n}^{N}\xi_m(\theta_0,\varphi_0)\Delta d_m\sec\theta_0-\xi_n(\theta_0,\varphi_0)(z'+d_{n-1})\sec\theta_0]$$
$$\cdot\overline{\overline{E}}^{-1}(\theta_0,\varphi_0)\overline{\overline{R}}(\theta_0)\overline{\overline{E}}(\pi-\theta_0,\varphi_0)\overline{\overline{D}}[-\sum_{m-1}^{N}\xi_m(\pi-\theta_0,\varphi_0)\Delta d_m\sec\theta_0]$$
$$\cdot\overline{\overline{E}}^{-1}(\pi-\theta_0,\varphi_0)\cdot\bar{I}_0(\omega)\exp[-j\omega\frac{(z'+2d_{N+1})(\sec\theta+\sec\theta_0)}{c_0}]\}\qquad(2\text{-}2.10)$$

Making the inverse Fourier transformation of $\bar{I}_1^{(1)}(\theta,\varphi,z=0,\omega)$ of Eq.(2-2.10), the first-order scattering intensity observed in the layer-0 can be obtained as follows

$$\bar{I}_1^{(1)}(\theta,\varphi,z=0,t)=\sum_{n=1}^{N}\bar{I}_n^{PS}+\sum_{n=1}^{N}\bar{I}_n^{BP}+\sum_{n=1}^{N}\bar{I}_n^{PB}+\sum_{n=1}^{N}\bar{I}_n^{BPB}\qquad(2\text{-}2.11)$$

Four terms of RHS of Eqs. (2-2.11) represent, respectively, four scattering processes (as illustrated in Fig.2-2.2) (It has also been appeared in Eq. (2-2.10) in the ω-domain). $\bar{I}_n^{PS}$ describes the particle scattering (*PS*) in the *n*-th layer, $\bar{I}_n^{BP}$ indicates a single boundary reflection and particle scattering (*BP*), $\bar{I}_n^{PB}$ takes account of the particle scattering and boundary reflection (*PB*), and $\bar{I}_n^{BPB}$ includes the boundary reflection, particle scattering and boundary reflection again (*BPB*).

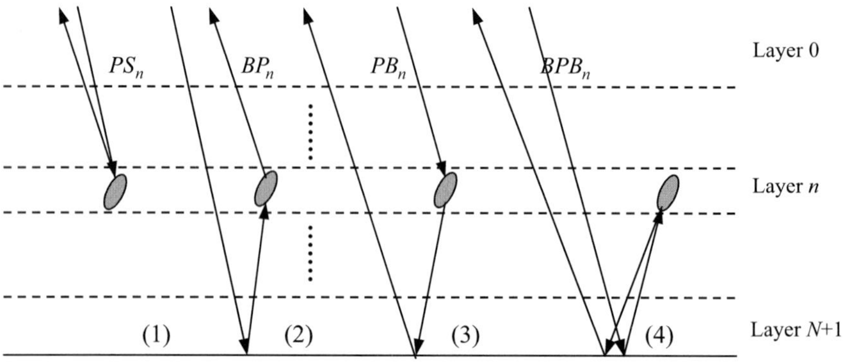

Fig.2-2.2. Four scattering processes in the respective layers

They are explicitly formulated as

$$\bar{I}_n^{PS}(\theta,\varphi,z=0,t)=\sec\theta\overline{\overline{E}}(\theta,\varphi)\overline{\overline{D}}[-\sum_{m=1}^{n}\xi_m(\theta,\varphi)\Delta d_m\sec\theta]\int_{-d_n}^{-d_{n-1}}dz'D[\xi_n(\theta,\varphi)(z'+d_n)\sec\theta]$$
$$\cdot\overline{\overline{E}}^{-1}(\theta,\varphi)\overline{\overline{P}}_n(\theta,\varphi;\pi-\theta_0,\varphi_0)\overline{\overline{E}}(\pi-\theta_0,\varphi_0)$$
$$\cdot\overline{\overline{D}}[-\sum_{m=1}^{n}\xi_m(\pi-\theta_0,\varphi_0)\Delta d_m\sec\theta_0+\xi_n(\pi-\theta_0,\varphi_0)(z'+d_n)\sec\theta_0]$$
$$\cdot\overline{\overline{E}}^{-1}(\pi-\theta_0,\varphi_0)\cdot\bar{I}_0[t+z'(\sec\theta+\sec\theta_0)/c_0] \tag{2-2.12a}$$

$$\bar{I}_n^{BP}(\theta,\varphi,z=0,t)=\sec\theta\overline{\overline{E}}(\theta,\varphi)\overline{\overline{D}}[-\sum_{m=1}^{n}\xi_m(\theta,\varphi)\Delta d_m\sec\theta]\int_{-d_n}^{-d_{n-1}}dz'\overline{\overline{D}}[\xi_n(\theta,\varphi)(z'+d_n)\sec\theta]$$
$$\cdot\overline{\overline{E}}^{-1}(\theta,\varphi)\overline{\overline{P}}_n(\theta,\varphi;\theta_0,\varphi_0)\overline{\overline{E}}(\theta_0,\varphi_0)$$
$$\cdot\overline{\overline{D}}[-\sum_{m=n}^{N}\xi_m(\theta_0,\varphi_0)\Delta d_m\sec\theta_0-\xi_n(\theta_0,\varphi_0)(z'+d_{n-1})\sec\theta_0]$$
$$\cdot\overline{\overline{E}}^{-1}(\theta_0,\varphi_0)\overline{\overline{R}}(\theta_0)\overline{\overline{E}}(\pi-\theta_0,\varphi_0)\overline{\overline{D}}[-\sum_{m=1}^{N}\xi_m(\pi-\theta_0,\varphi_0)\Delta d_m\sec\theta_0]$$
$$\cdot\overline{\overline{E}}^{-1}(\pi-\theta_0,\varphi_0)\cdot\bar{I}_0\{t-[2d_{N+1}\sec\theta_0+z'(\sec\theta_0-\sec\theta)]/c_0\} \tag{2-2.12b}$$

$$\bar{I}_n^{PB}(\theta,\varphi,z=0,t)=\sec\theta\overline{\overline{E}}(\theta,\varphi)\overline{\overline{D}}[-\sum_{m=1}^{N}\xi_m(\theta,\varphi)\Delta d_m\sec\theta]\overline{\overline{E}}^{-1}(\theta,\varphi)\overline{\overline{R}}(\theta)\overline{\overline{E}}(\pi-\theta,\varphi)$$
$$\cdot\overline{\overline{D}}[-\sum_{m=n}^{N}\xi_m(\pi-\theta,\varphi)\Delta d_m\sec\theta]\int_{-d_n}^{-d_{n-1}}dz'\overline{\overline{D}}[-\xi_n(\pi-\theta,\varphi)(z'+d_{n-1})\sec\theta]$$
$$\cdot\overline{\overline{E}}^{-1}(\pi-\theta,\varphi)\overline{\overline{P}}_n(\pi-\theta,\varphi;\pi-\theta_0,\varphi_0)\overline{\overline{E}}(\pi-\theta_0,\varphi_0)$$

$$\cdot \overline{\overline{D}}[-\sum_{m=1}^{n}\xi_m(\pi-\theta_0,\varphi_0)\Delta d_m \sec\theta_0 + \xi_n(\pi-\theta_0,\varphi_0)(z'+d_n)\sec\theta_0]$$

$$\cdot \overline{\overline{E}}^{-1}(\pi-\theta_0,\varphi_0)\cdot \bar{I}_0\{t-[2d_{N+1}\sec\theta + z'(\sec\theta-\sec\theta_0)]/c_0\}$$

(2-2.12c)

$$\bar{I}_n^{BPB}(\theta,\varphi,z=0,t) = \sec\theta \overline{\overline{E}}(\theta,\varphi)\overline{\overline{D}}[-\sum_{m=1}^{N}\xi_m(\theta,\varphi)\Delta d_m \sec\theta]\overline{\overline{E}}^{-1}(\theta,\varphi)\overline{\overline{R}}(\theta)\overline{\overline{E}}(\pi-\theta,\varphi)$$

$$\cdot \overline{\overline{D}}[-\sum_{m=n}^{N}\xi_m(\pi-\theta,\varphi)\Delta d_m \sec\theta]\int_{-d_n}^{-d_{n-1}} dz'\overline{\overline{D}}[-\xi_n(\pi-\theta,\varphi)(z'+d_{n-1})\sec\theta]$$

$$\cdot \overline{\overline{E}}^{-1}(\pi-\theta,\varphi)\overline{\overline{P}}_n(\pi-\theta,\varphi;\theta_0,\varphi_0)\overline{\overline{E}}(\theta_0,\varphi_0)$$

$$\cdot \overline{\overline{D}}[-\sum_{m=n}^{N}\xi_m(\theta_0,\varphi_0)\Delta d_m \sec\theta_0 - \xi_n(\theta_0,\varphi_0)(z'+d_{n-1})\sec\theta_0]$$

$$\cdot \overline{\overline{E}}^{-1}(\theta_0,\varphi_0)\overline{\overline{R}}(\theta_0)\overline{\overline{E}}(\pi-\theta_0,\varphi_0)\overline{\overline{D}}[-\sum_{m=1}^{N}\xi_m(\pi-\theta_0,\varphi_0)\Delta d_m \sec\theta_0]$$

$$\cdot \overline{\overline{E}}^{-1}(\pi-\theta_0,\varphi_0)\cdot \bar{I}_0[t-(z'+2d_{N+1})(\sec\theta+\sec\theta_0)/c_0]$$

(2-2.12d)

where $\Delta d_m = d_m - d_{m-1}$ denotes the thickness of the m-th layer, $m = 1, 2, \ldots, N$, and $d_0 = 0$.

2-2.2 Time-Dependent Mueller Matrix for Inhomogeneous Random Media

Let the incident plane pulse be the Gaussian with the polarization (χ,ψ) as

$$\bar{I}_0(t) = I_0(t)\bar{I}_{i0}(\chi,\psi) \tag{2-2.13}$$

where

$$I_0(t) = \frac{1}{\sqrt{2\pi}\sigma}\exp(-\frac{t^2}{2\sigma^2}) \tag{2-2.14}$$

where $\sigma = T_w/\sqrt{2\ln 2}$, $2T_w$ is the half-power pulse-width. $\bar{I}_{i0}(\chi,\psi)$ is the normalized incident Stokes vector as

$$\bar{I}_{i0} = [(1-\cos 2\chi\cos 2\psi)/2,\ \ (1+\cos 2\chi\cos 2\psi)/2,\ \ -\cos 2\chi\sin 2\psi,\ \ \sin 2\chi]^T \tag{2-2.15}$$

where $\chi \in [-45°,45°]$ and $\psi \in [0°,180°]$ are the elliptic and orientation angles of the incident polarization, respectively. As $\chi = 0°$, and $\psi = 0°$ or 180°, it denotes a horizontally polarized wave. As $\chi = 0°$ and $\psi = 90°$, it is vertically polarized. As $\chi = \pm 45°$, it is circularly polarized. Other cases are the elliptic polarization.

Substituting Eqs. (2-2.13)~(2-2.15) into Eq. (2-2.11) and using the integrand [Oguchi 1990, p.303]

$$\int dx \exp(-ax^2 - 2bx - c) = \frac{1}{2}\sqrt{\frac{\pi}{a}}\exp\left(\frac{b^2-ac}{a}\right)erf\left(\sqrt{a}x + \frac{b}{\sqrt{a}}\right) + \text{constant} \tag{2-2.16}$$

the first Mueller matrix solution is obtained as

$$\bar{I}_s(\theta,\varphi,z=0,t) = \overline{\overline{M}}(\theta,\varphi;\pi-\theta_0,\varphi_0;t)\cdot \bar{I}_{i0}(\chi,\psi) \tag{2-2.17}$$

where the time-dependent $\overline{\overline{M}}(\theta,\varphi;\pi-\theta_0,\varphi_0;t)$ includes the contributions from all N layers of scatterers and takes into account for four scattering processes. Its ℓj-th elements ($\ell, j = 1,\ldots,4$) are derived as

$$M_{\ell j}(\theta,\varphi;\pi-\theta_0,\varphi_0;t) = \sum_{n=1}^{N} PS_n + \sum_{n=1}^{N} BP_n + \sum_{n=1}^{N} PB_n + \sum_{n=1}^{N} BPB_n \tag{2-2.18}$$

where the functions PS_n, BP_n, PB_n and BPB_n are contributed by the scattering processes described by Fig. 2-2.2.

$$PS_n = \sec\theta \sum_{k,i} E_{\ell k}(\theta,\varphi)\cdot\{\overline{\overline{D}}[-\sum_{m=1}^{n}\xi_m(\theta,\varphi)\Delta d_m \sec\theta + \xi_n(\theta,\varphi)d_n \sec\theta]$$
$$\cdot\overline{\overline{E}}^{-1}(\theta,\varphi)\overline{\overline{P}}_n(\theta,\varphi;\pi-\theta_0,\varphi_0)\overline{\overline{E}}(\pi-\theta_0,\varphi_0)$$
$$\cdot\overline{\overline{D}}[-\sum_{m=1}^{n}\xi_m(\pi-\theta_0,\varphi_0)\Delta d_m \sec\theta_0 + \xi_n(\pi-\theta_0,\varphi_0)d_n \sec\theta_0]\}_{ki}$$
$$\times\frac{1}{2R_1}\{\exp[\frac{2\sigma^2 B_{n1}^2}{R_1^2} - C_1]\times[erf(\frac{\sqrt{2}\sigma B_{n1}}{R_1} - \frac{R_1 d_{n-1}}{\sqrt{2}\sigma}) - erf(\frac{\sqrt{2}\sigma B_{n1}}{R_1} - \frac{R_1 d_n}{\sqrt{2}\sigma})]\}_{ki}$$
$$\cdot\{\overline{\overline{E}}^{-1}(\pi-\theta_0,\varphi_0)\}_{ij}$$

(2-2.19a)

$$BP_n = \sec\theta \sum_{k,i} E_{\ell k}(\theta,\varphi)\cdot\{\overline{\overline{D}}[-\sum_{m=1}^{n}\xi_m(\theta,\varphi)\Delta d_m \sec\theta + \xi_n(\theta,\varphi)d_n \sec\theta]$$
$$\cdot\overline{\overline{E}}^{-1}(\theta,\varphi)\overline{\overline{P}}_n(\theta,\varphi;\theta_0,\varphi_0)\overline{\overline{E}}(\theta_0,\varphi_0)$$
$$\cdot\overline{\overline{D}}[-\sum_{m=n}^{N}\xi_m(\theta_0,\varphi_0)\Delta d_m \sec\theta_0 - \xi_n(\theta_0,\varphi_0)d_{n-1} \sec\theta_0]\}_{ki}$$
$$\times\frac{1}{2|R_2|}\{\exp[\frac{2\sigma^2 B_{n2}^2}{R_2^2} - C_2]\times[erf(\frac{\sqrt{2}\sigma B_{n2}}{|R_2|} - \frac{|R_2| d_{n-1}}{\sqrt{2}\sigma}) - erf(\frac{\sqrt{2}\sigma B_{n2}}{|R_2|} - \frac{|R_2| d_n}{\sqrt{2}\sigma})]\}_{ki}$$
$$\cdot\{\overline{\overline{E}}^{-1}(\theta_0,\varphi_0)\overline{\overline{R}}(\theta_0)\overline{\overline{E}}(\pi-\theta_0,\varphi_0)\overline{\overline{D}}[-\sum_{m=1}^{N}\xi_m(\pi-\theta_0,\varphi_0)\Delta d_m \sec\theta_0]\overline{\overline{E}}^{-1}(\pi-\theta_0,\varphi_0)\}_{ij}$$

(2-2.19b)

$$PB_n = \sec\theta \sum_{k,i}\{\overline{\overline{E}}(\theta,\varphi)\overline{\overline{D}}[-\sum_{m=1}^{N}\xi_m(\theta,\varphi)\Delta d_m \sec\theta]\overline{\overline{E}}^{-1}(\theta,\varphi)\overline{\overline{R}}(\theta)\overline{\overline{E}}(\pi-\theta,\varphi)\}_{\ell k}$$
$$\cdot\{\overline{\overline{D}}[-\sum_{m=n}^{N}\xi_m(\pi-\theta,\varphi)\Delta d_m \sec\theta - \xi_n(\pi-\theta,\varphi)d_{n-1}\sec\theta]$$
$$\cdot\overline{\overline{E}}^{-1}(\pi-\theta,\varphi)\overline{\overline{P}}_n(\pi-\theta,\varphi;\pi-\theta_0,\varphi_0)\overline{\overline{E}}(\pi-\theta_0,\varphi_0)$$
$$\cdot\overline{\overline{D}}[-\sum_{m=1}^{n}\xi_m(\pi-\theta_0,\varphi_0)\Delta d_m \sec\theta_0 + \xi_n(\pi-\theta_0,\varphi_0)d_n \sec\theta_0]\}_{ki}$$
$$\times\frac{1}{2|R_2|}\{\exp[\frac{2\sigma^2 B_{n3}^2}{R_2^2} - C_3]\times[erf(\frac{\sqrt{2}\sigma B_{n3}}{|R_2|} - \frac{|R_2| d_{n-1}}{\sqrt{2}\sigma}) - erf(\frac{\sqrt{2}\sigma B_{n3}}{|R_2|} - \frac{|R_2| d_n}{\sqrt{2}\sigma})]\}_{ki}$$
$$\cdot\{\overline{\overline{E}}^{-1}(\pi-\theta_0,\varphi_0)\}_{ij}$$

(2-2.19c)

$$BPB_n = \sec\theta \sum_{k,i}\{\overline{\overline{E}}(\theta,\varphi)\overline{\overline{D}}[-\sum_{m=1}^{N}\xi_m(\theta,\varphi)\Delta d_m \sec\theta]\overline{\overline{E}}^{-1}(\theta,\varphi)\overline{\overline{R}}(\theta)\overline{\overline{E}}(\pi-\theta,\varphi)\}_{\ell k}$$
$$\cdot\{\overline{\overline{D}}[-\sum_{m=n}^{N}\xi_m(\pi-\theta,\varphi)\Delta d_m \sec\theta - \xi_n(\pi-\theta,\varphi)d_{n-1}\sec\theta]$$
$$\cdot\overline{\overline{E}}^{-1}(\pi-\theta,\varphi)\overline{\overline{P}}_n(\pi-\theta,\varphi;\theta_0,\varphi_0)\overline{\overline{E}}(\theta_0,\varphi_0)$$
$$\cdot\overline{\overline{D}}[-\sum_{m=n}^{N}\xi_m(\theta_0,\varphi_0)\Delta d_m \sec\theta_0 - \xi_n(\theta_0,\varphi_0)d_{n-1} \sec\theta_0]\}_{ki}$$

$$\times\frac{1}{2R_1}\{\exp[\frac{2\sigma^2B_{n4}^2}{R_1^2}-C_4]\times[erf(\frac{\sqrt{2}\sigma B_{n4}}{R_1}-\frac{R_1d_{n-1}}{\sqrt{2}\sigma})-erf(\frac{\sqrt{2}\sigma B_{n4}}{R_1}-\frac{R_1d_n}{\sqrt{2}\sigma})]\}_{ki}$$

$$\cdot\{\overline{\overline{E}}^{-1}(\theta_0,\varphi_0)\overline{\overline{R}}(\theta_0)\overline{\overline{E}}(\pi-\theta_0,\varphi_0)\overline{\overline{D}}[-\sum_{m=1}^{N}\xi_m(\pi-\theta_0,\varphi_0)\Delta d_m\sec\theta_0]\overline{\overline{E}}^{-1}(\pi-\theta_0,\varphi_0)\}_{ij}$$

(2-2.19d)

where i,j,ℓ,k take 1,2,3,4. erf (•) is the error function. The eigen-matrix $\overline{\overline{E}}$ and its inverse $\overline{\overline{E}}^{-1}$, and the diagonal matrix $\overline{\overline{D}}$ are related with the forward scattering amplitude functions of the scatterers. The phase matrix $\overline{\overline{P}}_n$ are constructed by the functions of $N_{0n}<f_{pq}f_{st}^*>$ where $p,q,s,t = v,h$, and N_{0n} is the number of the particles per unit volume in the n-th layer, f_{pq}, f_{st}^* indicate the scattering amplitude functions of the particles and its conjugation, the angular brackets <•> denote the averaging of Euler angles orientation. The explicit expressions of $\overline{\overline{E}}$, $\overline{\overline{D}}$ and $\overline{\overline{P}}$ of small particles can be found in Tsang et al. (2000, p.16,39,40). And the functions in Eqs. (2-2.19a~d) are denoted as

$$R_1=(\sec\theta+\sec\theta_0)/c_0 \quad (2\text{-}2.20a)$$

$$R_2=(\sec\theta-\sec\theta_0)/c_0 \quad (2\text{-}2.20b)$$

$$B_{n1}=[R_1t/\sigma^2-\xi_{nk}(\theta,\varphi)\sec\theta-\xi_{ni}(\pi-\theta_0,\varphi_0)\sec\theta_0]/2 \quad (2\text{-}2.21a)$$

$$B_{n2}=[R_2(t-2d_{N+1}\sec\theta_0/c_0)/\sigma^2-\xi_{nk}(\theta,\varphi)\sec\theta+\xi_{ni}(\theta_0,\varphi_0)\sec\theta_0]/2 \quad (2\text{-}2.21b)$$

$$B_{n3}=-[R_2(t-2d_{N+1}\sec\theta/c_0)/\sigma^2-\xi_{nk}(\pi-\theta,\varphi)\sec\theta+\xi_{ni}(\pi-\theta_0,\varphi_0)\sec\theta_0]/2 \quad (2\text{-}2.21c)$$

$$B_{n4}=-[R_1(t-2d_{N+1}R_1)/\sigma^2-\xi_{nk}(\pi-\theta,\varphi)\sec\theta-\xi_{ni}(\theta_0,\varphi_0)\sec\theta_0]/2 \quad (2\text{-}2.21d)$$

$$C_1=t^2/(2\sigma^2) \quad (2\text{-}2.21e)$$

$$C_2=(t-2d_{N+1}\sec\theta_0/c_0)^2/(2\sigma^2) \quad (2\text{-}2.21f)$$

$$C_3=(t-2d_{N+1}\sec\theta/c_0)^2/(2\sigma^2) \quad (2\text{-}2.21g)$$

$$C_4=(t-2d_{N+1}R_1)^2/(2\sigma^2) \quad (2\text{-}2.21h)$$

As the scattered angle $\theta\to\theta_0$ (e.g. in the specular or backward direction), it yields $R_2\to0$, the related terms of Eqs. (2-2.19b,2-2.19c) have to be rewritten as (see Appendix 2-2A)

$$\lim_{|R_2\to0|}\frac{1}{2|R_2|}\exp[\frac{2\sigma^2B_{n2}^2}{R_2^2}-C_2]\times[erf(\frac{\sqrt{2}\sigma B_{n2}}{|R_2|}-\frac{|R_2|d_{n-1}}{\sqrt{2}\sigma})-erf(\frac{\sqrt{2}\sigma B_{n2}}{|R_2|}-\frac{|R_2|d_n}{\sqrt{2}\sigma})]$$

$$=\frac{1}{\sqrt{2\pi}\sigma}\exp[-\frac{(t-2d_{N+1}\sec\theta_0/c_0)^2}{2\sigma^2}]\times\{\exp\{-d_{n-1}[\xi_{nk}(\theta,\varphi)\sec\theta-\xi_{ni}(\theta_0,\varphi_0)\sec\theta_0]\}$$

$$-\exp\{-d_n[\xi_{nk}(\theta,\varphi)\sec\theta-\xi_{ni}(\theta_0,\varphi_0)\sec\theta_0]\}\}/[\xi_{nk}(\theta,\varphi)\sec\theta-\xi_{ni}(\theta_0,\varphi_0)\sec\theta_0]$$

(2-2.22a)

$$\lim_{|R_2\to0|}\frac{1}{2|R_2|}\exp[\frac{2\sigma^2B_{n3}^2}{R_2^2}-C_3]\times[erf(\frac{\sqrt{2}\sigma B_{n3}}{|R_2|}-\frac{|R_2|d_{n-1}}{\sqrt{2}\sigma})-erf(\frac{\sqrt{2}\sigma B_{n3}}{|R_2|}-\frac{|R_2|d_n}{\sqrt{2}\sigma})]$$

$$=\frac{1}{\sqrt{2\pi}\sigma}\exp[-\frac{(t-2d_{N+1}\sec\theta/c_0)^2}{2\sigma^2}]\times\{\exp\{-d_{n-1}[\xi_{ni}(\pi-\theta_0,\varphi_0)\sec\theta_0-\xi_{nk}(\pi-\theta,\varphi)\sec\theta]\}$$

$$-\exp\{-d_n[\xi_{ni}(\pi-\theta_0,\varphi_0)\sec\theta_0-\xi_{nk}(\pi-\theta,\varphi)\sec\theta]\}\}$$

$$/[\xi_{ni}(\pi-\theta_0,\varphi_0)\sec\theta_0-\xi_{nk}(\pi-\theta,\varphi)\sec\theta]$$

(2-2.22b)

As $t \to 0$, $c_0 \to \infty$ (i.e. a plane pulse becomes a continuous plane wave), and $N = 1$, $d_{N+1} = d_N$ (i.e. inhomogeneous random media degenerates to homogeneous case), the temporal Muller matrix of Eq.(2-2.18) can return to conventional Mueller matrix in Tsang et al. (2000, p.126) (see the Appendix 2-2A).

2-2.3 Polarimetric Bistatic and Back-Scattering Pulses Responses

Using the temporal Mueller matrix solution, polarimetric bistatic and back-scattering from multilayering random media of non-spherical scatterers under a Gaussian plane pulse incidence can be numerically simulated. The averaging result of many realizations in real field measurements statistically presents the scattered Stokes parameters. In our simulations of the Mueller solution, only scattering from the terrain canopy is taken into account, the propagation and fading effect through the turbid atmosphere are not yet discussed.

The polarized bistatic scattering coefficient is defined as

$$\gamma_{pq}(\theta_s,\varphi_s;\pi-\theta_0,\varphi_0) = 4\pi\frac{\cos\theta_s I_{sp}(\theta_s,\varphi_s)}{\cos\theta_0 I_{i0q}}, \qquad p,q = v,h \tag{2-2.23}$$

where v, h denote vertical and horizontal polarization, respectively, I_{i0q} is the q-polarized incident intensity, I_{sp} is the p-polarized scattered intensity.

The polarized back-scattering coefficient is defined as

$$\begin{aligned}\sigma_{pq}(\theta_0,\pi+\varphi_0;\pi-\theta_0,\varphi_0) &= \cos\theta_0\gamma_{pq}(\theta_0,\pi+\varphi_0;\pi-\theta_0,\varphi_0)\\ &= 4\pi\cos\theta_0\frac{I_{sp}(\theta_0,\pi+\varphi_0)}{I_{i0q}}\end{aligned} \tag{2-2.24}$$

The co-polarized back-scattering coefficient is

$$\sigma_c(\theta_0) = 4\pi\cos\theta_0 P_n \tag{2-2.25}$$

where P_n is the co-polarized backscattering power

$$\begin{aligned}P_n = \frac{1}{2}[&I_{vs}(1-\cos 2\chi\cos 2\psi) + I_{hs}(1+\cos 2\chi\cos 2\psi)\\ &+U_s\cos 2\chi\sin 2\psi + V_s\sin 2\chi]\end{aligned} \tag{2-2.26}$$

The cross-polarized back-scattering coefficient is

$$\sigma_x(\theta_0) = 4\pi\cos\theta_0(I_{vs} + I_{hs} - P_n) \tag{2-2.27}$$

In following calculations, the pulse echoes from four-layered (i.e. N+1 = 4, and Layer-4 is an air region) media of random disk-like particles are numerically simulated. The carrier frequency is 5.3 *GHz*, and the Gaussian pulse with a half-power width 2.0*ns*. For convenient programming, all particles in the media are simply assumed be identical in shape and size (the radius 1.0*cm*, the thickness 0.025*cm*) with the dielectric constant $(14.1 + j4.5)\varepsilon_0$ and uniformly spatial orientaion of the Euler angles $\beta \in (0°,180°)$, $\gamma \in (0°,360°)$. Actually, the scatter particles in the layering media can have different shape and size, and can be spatially oriented with non-uniform orientation distribution or around particular direction. In such case, it is believed that the polarimetric scattering from the media under a pulse incidence can be of greatly helpful to identify the particular locations of those scatter particles.

The dielectric constant of the underlying medium ε_b is assumed as $(6 + i0.5)\varepsilon_0$. The incident pulse is horizontally polarized ($\chi = 0°, \psi = 0°$) and $\theta_0 = 0°$, $\varphi_0 = 0°$. The particles are sparsely distributed and their sizes are small in the low frequency limit. The generalized Rayleigh-Gans approximation is applied to derive the scattering amplitude functions of random small particles [Jin 1994].

In following figures, the solid curves denote the σ_c of inhomogeneous three-layered media and the dash curves are the σ_c from an overall homogenous layer of random disks (i.e. N = 1). The dotted lines indicate the fractional volume profiles of random disks. To show the shape of the pulse echoes and compare with the fraction profile in the same figure, the time delay $t(s)$ of the echoes is transformed by the height as

$$Height(m) = d_4(m) - t(s) \times c_0(m/s)/2 \tag{2-2.28}$$

The fractional volume of an overall homogeneous layer is averaged as

$$f_s = \sum_{n=1}^{N} (d_n - d_{n-1}) f_{sn} / d_N \tag{2-2.29}$$

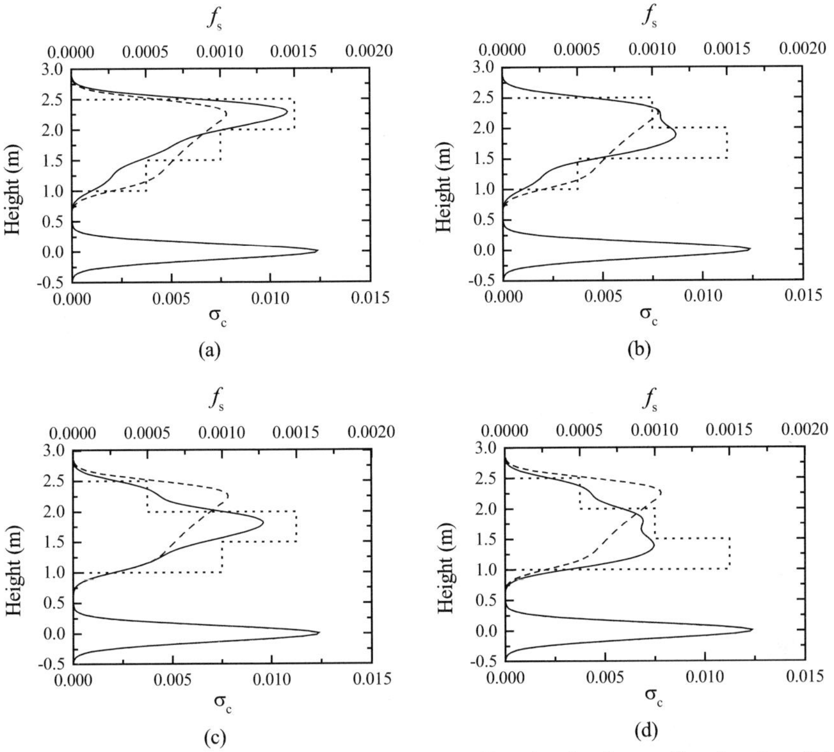

Fig. 2-2.3 σ_c vs. height with indication of layering fraction profile of random disks (N = 3, $d_1 = 0.5m$, $d_2 = 1.0m$, $d_3 = 1.5m$ and $d_4 = 2.5m$, i.e. total thickness = 2.5m)

Figs. 2-2.3(a~d) simulate temporal σ_c vs. height with comparison of four different layering profiles of f_{s1}, f_{s2}, f_{s3}: (a) 0.0015, 0.001, 0.0005; (b) 0.001, 0.0015, 0.0005; (c) 0.0005, 0.0015, 0.001; (d) 0.0005, 0.001, 0.0015, respectively. The layer depths are $d_1 = 0.5m$, $d_2 = 1.0m$, $d_3 = 1.5m$ and $d_4 = 2.5m$, respectively, i.e. total layer thickness is 2.5m.

It can be seen that the patterns of pulse echoes with height well depict the vertical fraction profiles of random disks. A separate peak is coming from reflection of the underlying surface. The return from the underlying surface can be strong due to the sparse scatterers and the thin layer depth. If the fractional volumes become larger, the peak due to the bottom reflection would be shadowed. The results for cross-polarized case are similar but the values are far smaller than those of co-polarized case. Pulse distortion is due to multi-path interaction described by PS_n, BP_n, PB_n and BPB_n in Eqs. (2-2.18, 2-2.19), which certainly depends upon the media parameters, not only fraction profile, but also the scatter particles shape, size, orientation, dielectric properties, layer thickness, underlying surface, etc.

Figs. 2-2.4(a-d) show σ_c vs. height for total layer thickness 5.0*m*, where there are four different fraction profiles (corresponding to Figs. 2-2.3a~d) and $d_1 = 1.0m$, $d_2 = 2.0m$, $d_3 = 3.0m$ and $d_4 = 5.0m$, respectively. Comparing with Fig. 2-2.3, the pulse echoes shift to the top layer due to more scatterers at the top crown and the more scatterers can darken the bottom returns.

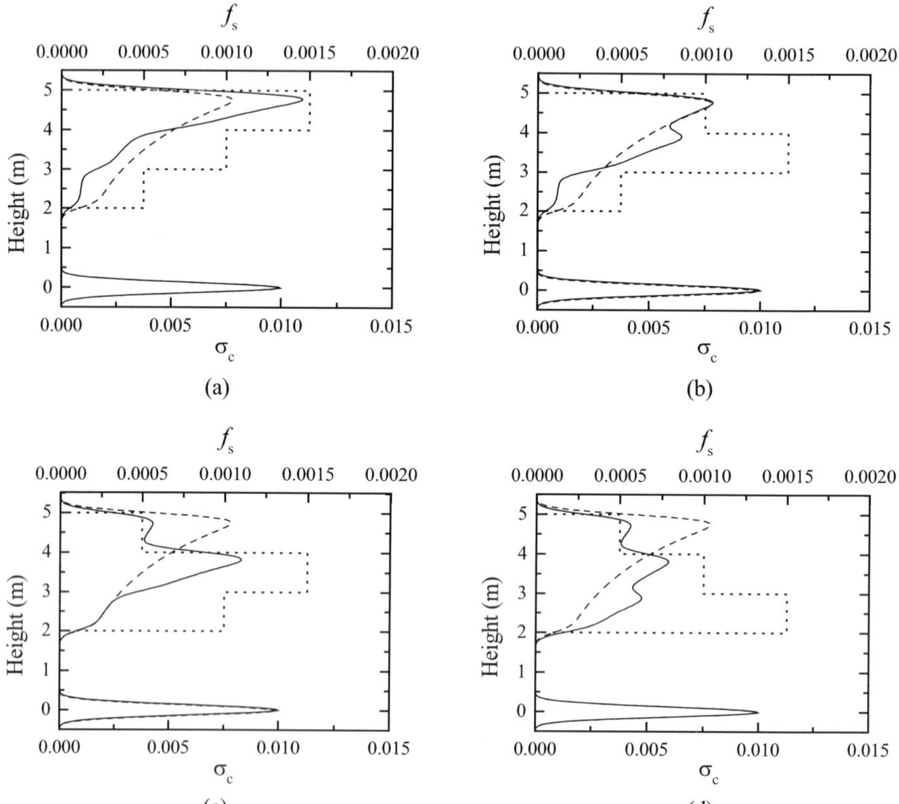

F

Fig.2-2.4. σ_c vs. height with indication of layering fraction profile of random disks (N= 3, $d_1 = 1.0m$, $d_2 = 2.0m$, $d_3 = 3.0m$ and $d_4 = 5.0m$, i.e. total thickness = 5.0*m*)

Fig.2-2.5 presents the σ_c for a total thickness 10*m* where $d_1 = 2.0m$, $d_2 = 4.0m$, $d_3 = 6.0m$ and $d_4 = 10.0m$, respectively. The fractional volumes of Figs. 2-2.5(a~d) also correspond to Figs. 2-2.3(a~d), respectively. It can be seen that the pulse echoes are mainly contributed by the scatterers of the top crown and the bottom return is significantly shadowed.

Fig. 2-2.6 shows the σ_c for different incident angles $\theta_0 = 0°, 40°, 60°$, respectively. All parameters are the same as Fig. 2-2.4(d) (total layer thickness = 5.0*m*). It can be seen that as θ_0 increases, the layering media becomes thicker for wave penetration and the bottom return becomes weaker. The time delay difference Δt between the bottom returns for different incident angles θ_{10}, θ_{20} can be derived as a function of the total layer thickness d

$$\Delta t = t_2 - t_1 = \frac{2d(\sec\theta_{20} - \sec\theta_{10})}{c_0} \tag{2-2.30}$$

In Fig. 2-2.6, there are $t_1 = 43.6ns$ for $\theta_{10} = 40^0$ and $t_2 = 66.6ns$ for $\theta_{20} = 60^0$. It is interesting to see that Eq. (2-2.30) can invert $d = 4.964m \approx 5.0m$.

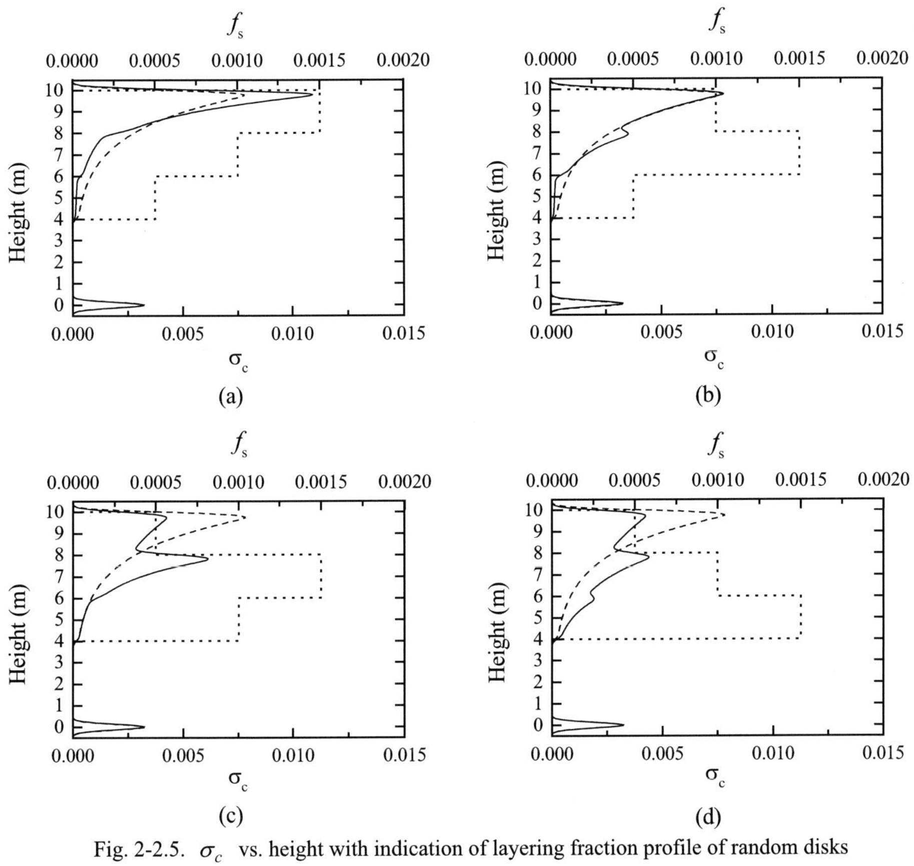

Fig. 2-2.5. σ_c vs. height with indication of layering fraction profile of random disks ($N = 3$, $d_1 = 2.0m$, $d_2 = 4.0m$, $d_3 = 6.0m$ and $d_4 - 10.0m$, i.e. total thickness = 10.0*m*)

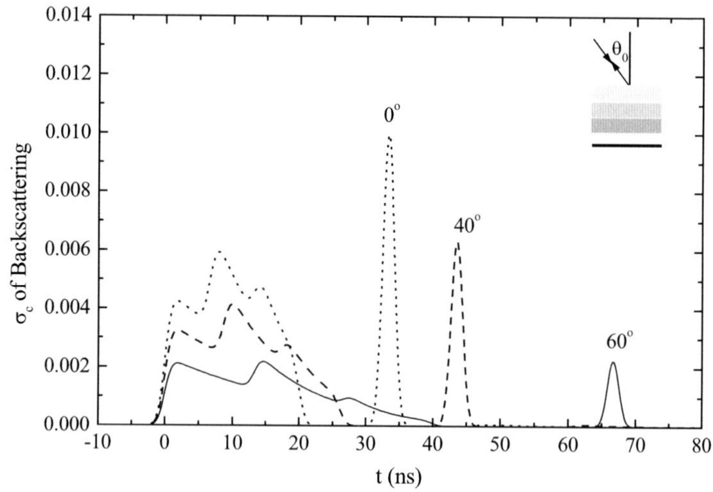

Fig. 2-2.6. Temporal variation of backscattering σ_c for different incidence θ_0 $(0^\circ, 40^\circ, 60^\circ)$

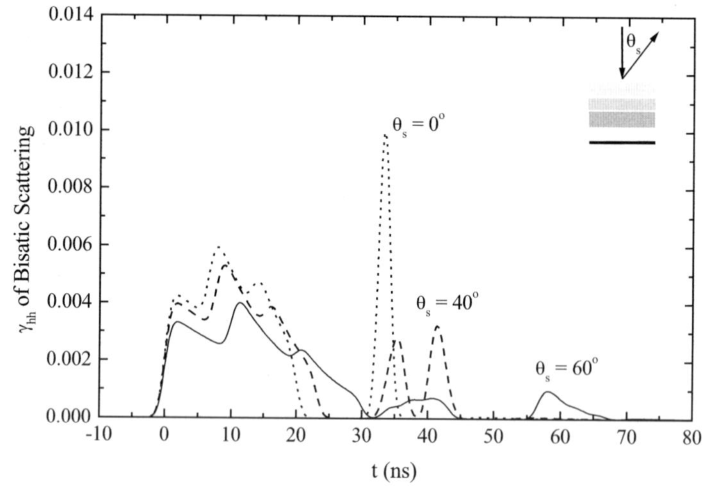

Fig. 2-2.7. Temporal variation of bistatic scattering ($\theta_s = 0^\circ, 40^\circ, 60^\circ$) coefficients γ_{hh} at $\theta_0 = 0^\circ$

Fig. 2-2.7 presents the pulse echoes of bistatic scattering at $\theta_s = 0^\circ, 40^\circ, 60^\circ$ with incident $\theta_0 = 0^\circ$. All parameters are the same as Fig. 2-2.4(d). As θ_s becomes larger, the pulse shape is broadened since the media seem thicker for wave penetration. It is noted that the bistatic scattering shows two peak tails due to the interactions of random disks and underlying surface. The first tail is due to single reflection from the surface to particles and scattering. The second tail results from scattering of the particles to the surface and reflecting. In backscattering, these two scattering processes take the same path and have the same time-delay. So only one peak tail appears as $\theta_s = \theta_0 (= 0^\circ)$ in Fig. 2-2.7.

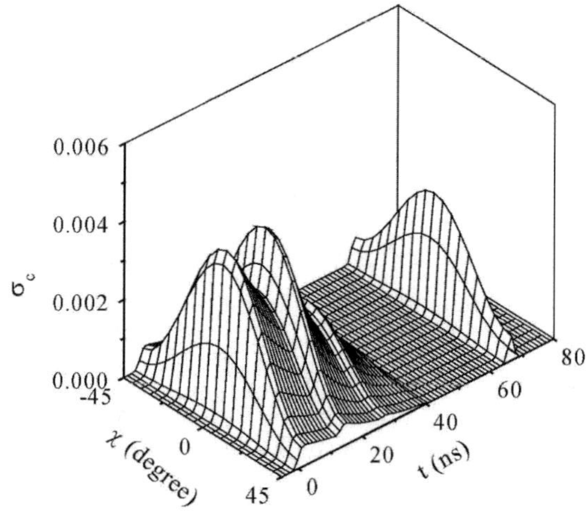

(a)

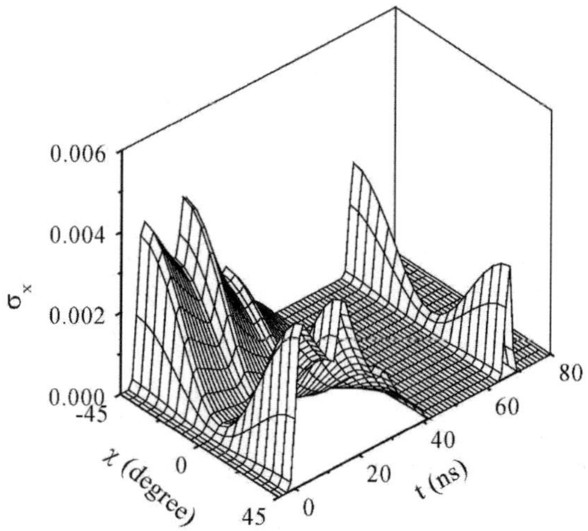

(b)

Fig.2-2.8. σ_c, σ_x vs. (χ , t) at the orientation angle $\psi = 0^\circ$

Figs. 2-2.8(a,b) show the co-polarized and cross-polarized backscattering coefficients σ_c, σ_x vs. (χ, t), respectively, with the orientation angle of the incident polarization $\psi = 0^\circ$. All parameters are the same as Fig. 2-2.5(d). It demonstrates temporal variations of the σ_c and σ_x for different incidence polarization (e.g. horizontal $\chi = 0^\circ$, circular $\chi = \pm 45^\circ$ and elliptic χ=others).

Appendix 2-2A

According to the Taylor's formula (Oguchi 1990, p.14)

$$f(x+h)=f(x)+\sum_{m=1}^{\infty}\frac{1}{m!}f^{(m)}(x)h^m \tag{2-2A.1}$$

The error function $erf(x+h)$ is expanded as

$$erf(x+h)=erf(x)+\sum_{m=1}^{\infty}\frac{1}{m!}erf^{(m)}(x)h^m \tag{2-2A.2}$$

where $erf^{(m)}(\cdot)$ denotes the m-th-order derivative of the error function and can be written as (Oguchi 1990, p.298)

$$erf^{(m)}(x)=(-1)^{m-1}\frac{2}{\sqrt{\pi}}H_{m-1}(x)\exp(-x^2) \tag{2-2A.3}$$

where the Hermite polynomials $H_{m-1}(x)$ can be written in terms of (Oguchi 1990, p.775)

$$H_{m-1}(x)=\sum_{\ell=0}^{[m/2]}(-1)^{\ell}\frac{(m-1)!}{\ell!(m-1-2\ell)!}(2x)^{m-1-2\ell} \tag{2-2A.4}$$

From (2A-2,3,4), it can be obtained that

$$erf(x+h)=erf(x)+\frac{2}{\sqrt{\pi}}\exp(-x^2)\sum_{m=1}^{\infty}\frac{(-1)^{m-1}}{m!}h^m\sum_{\ell=0}^{[m/2]}(-1)^{\ell}\frac{(m-1)!}{\ell!(m-1-2\ell)!}(2x)^{m-1-2\ell} \tag{2-2A.5}$$

and then

$$\begin{aligned}&\exp(x^2)\cdot[erf(x+h_1)-erf(x+h_2)]\\&=\frac{2}{\sqrt{\pi}}\sum_{m=1}^{\infty}\frac{(-1)^{m-1}}{m!}(h_1^m-h_2^m)\sum_{\ell=0}^{[m/2]}(-1)^{\ell}\frac{(m-1)!}{\ell!(m-1-2\ell)!}(2x)^{m-1-2\ell}\end{aligned} \tag{2-2A.6}$$

Taking $x=\sqrt{2}\sigma B_{n2}/|R_2|$, $h_1=-|R_2|d_{n-1}/(\sqrt{2}\sigma)$, $h_2=-|R_2|d_n/(\sqrt{2}\sigma)$ and substituting them into (2A-6) , it is obtained that

$$\begin{aligned}&\frac{1}{2|R_2|}\exp[\frac{2\sigma^2B_{n2}^2}{R_2^2}-C_2]\times[erf(\frac{\sqrt{2}\sigma B_{n2}}{|R_2|}-\frac{|R_2|d_{n-1}}{\sqrt{2}\sigma})-erf(\frac{\sqrt{2}\sigma B_{n2}}{|R_2|}-\frac{|R_2|d_n}{\sqrt{2}\sigma})]\\&=\frac{1}{\sqrt{2\pi}\sigma}\exp(-C_2)\sum_{m=1}^{\infty}\frac{(2B_{n2})^{m-1}}{m!}[(d_n)^m-(d_{n-1})^m]\sum_{\ell=0}^{[m/2]}(-1)^{\ell}\frac{(m-1)!}{\ell!(m-1-2\ell)!}\left(\frac{2\sqrt{2}\sigma B_{n2}}{|R_2|}\right)^{-2\ell}\end{aligned} \tag{2-2A.7}$$

Because

$$\lim_{|R_2|\to 0,\ell\ge 1}\left(\frac{2\sqrt{2}\sigma B_{n2}}{|R_2|}\right)^{-2\ell}=0 \tag{2-2A.8}$$

and keeping the $\ell=0$ term of Eq.(2A-7), it yields

$$\begin{aligned}&\lim_{|R_2\to 0|}\frac{1}{2|R_2|}\exp[\frac{2\sigma^2B_{n2}^2}{R_2^2}-C_2]\times[erf(\frac{\sqrt{2}\sigma B_{n2}}{|R_2|}-\frac{|R_2|d_{n-1}}{\sqrt{2}\sigma})-erf(\frac{\sqrt{2}\sigma B_{n2}}{|R_2|}-\frac{|R_2|d_n}{\sqrt{2}\sigma})]\\&=\lim_{|R_2|\to 0}\frac{1}{\sqrt{2\pi}\sigma}\exp(-C_2)\sum_{m=1}^{\infty}\frac{(2B_{n2})^{m-1}}{m!}[(d_n)^m-(d_{n-1})^m]\\&=\frac{1}{\sqrt{2\pi}\sigma}\exp[-\frac{(t-2d_{N+1}\sec\theta_0/c_0)^2}{2\sigma^2}]\\&\times\sum_{m=1}^{\infty}\frac{[-\xi_{nk}(\theta,\varphi)\sec\theta+\xi_{ni}(\theta_0,\varphi_0)\sec\theta_0]^{m-1}}{m!}\cdot[(d_n)^m-(d_{n-1})^m]\end{aligned}$$

(2-2A.9)

Using $e^x = \sum_{m=0}^{\infty} \frac{x^m}{m!}$, $\frac{1-e^x}{x} = -\sum_{m=1}^{\infty} \frac{x^{m-1}}{m!}$, we obtain

$$\lim_{|R_2 \to 0|} \frac{1}{2|R_2|} \exp[\frac{2\sigma^2 B_{n2}^2}{R_2^2} - C_2] \times [erf(\frac{\sqrt{2}\sigma B_{n2}}{|R_2|} - \frac{|R_2| d_{n-1}}{\sqrt{2}\sigma}) - erf(\frac{\sqrt{2}\sigma B_{n2}}{|R_2|} - \frac{|R_2| d_n}{\sqrt{2}\sigma})]$$

$$= \frac{1}{\sqrt{2\pi}\sigma} \exp[-\frac{(t - 2d_{N+1} \sec\theta_0 / c_0)^2}{2\sigma^2}] \times \{\exp\{-d_{n-1}[\xi_{nk}(\theta,\varphi)\sec\theta - \xi_{ni}(\theta_0,\varphi_0)\sec\theta_0]\}$$
$$- \exp\{-d_n[\xi_{nk}(\theta,\varphi)\sec\theta - \xi_{ni}(\theta_0,\varphi_0)\sec\theta_0]\}\} / [\xi_{nk}(\theta,\varphi)\sec\theta - \xi_{ni}(\theta_0,\varphi_0)\sec\theta_0]$$

(2-2A.10)

Similarly, it can be obtained that

$$\lim_{|R_2 \to 0|} \frac{1}{2|R_2|} \exp[\frac{2\sigma^2 B_{n3}^2}{R_2^2} - C_3] \times [erf(\frac{\sqrt{2}\sigma B_{n3}}{|R_2|} - \frac{|R_2| d_{n-1}}{\sqrt{2}\sigma}) - erf(\frac{\sqrt{2}\sigma B_{n3}}{|R_2|} - \frac{|R_2| d_n}{\sqrt{2}\sigma})]$$

$$= \frac{1}{\sqrt{2\pi}\sigma} \exp[-\frac{(t - 2d_{N+1} \sec\theta / c_0)^2}{2\sigma^2}] \times \{\exp\{-d_{n-1}[\xi_{ni}(\pi-\theta_0,\varphi_0)\sec\theta_0 - \xi_{nk}(\pi-\theta,\varphi)\sec\theta]\}$$
$$- \exp\{-d_n[\xi_{ni}(\pi-\theta_0,\varphi_0)\sec\theta_0 - \xi_{nk}(\pi-\theta,\varphi)\sec\theta]\}\}$$
$$/[\xi_{ni}(\pi-\theta_0,\varphi_0)\sec\theta_0 - \xi_{nk}(\pi-\theta,\varphi)\sec\theta] \quad (2\text{-}2A.11)$$

$$\lim_{R_1 \to 0} \frac{1}{2R_1} \exp[\frac{2\sigma^2 B_{n1}^2}{R_1^2} - C_1] \times [erf(\frac{\sqrt{2}\sigma B_{n1}}{R_1} - \frac{R_1 d_{n-1}}{\sqrt{2}\sigma}) - erf(\frac{\sqrt{2}\sigma B_{n1}}{R_1} - \frac{R_1 d_n}{\sqrt{2}\sigma})]$$

$$= \frac{1}{\sqrt{2\pi}\sigma} \exp(-\frac{t^2}{2\sigma^2}) \times \{\exp\{-d_{n-1}[\xi_{nk}(\theta,\varphi)\sec\theta + \xi_{ni}(\pi-\theta_0,\varphi_0)\sec\theta_0]\}$$
$$- \exp\{-d_n[\xi_{nk}(\theta,\varphi)\sec\theta + \xi_{ni}(\pi-\theta_0,\varphi_0)\sec\theta_0]\}\}$$
$$/[\xi_{nk}(\theta,\varphi)\sec\theta + \xi_{ni}(\pi-\theta_0,\varphi_0)\sec\theta_0] \quad (2\text{-}2A.12)$$

$$\lim_{R_1 \to 0} \frac{1}{2R_1} \exp[\frac{2\sigma^2 B_{n4}^2}{R_1^2} - C_4] \times [erf(\frac{\sqrt{2}\sigma B_{n4}}{R_1} - \frac{R_1 d_{n-1}}{\sqrt{2}\sigma}) - erf(\frac{\sqrt{2}\sigma B_{n4}}{R_1} - \frac{R_1 d_n}{\sqrt{2}\sigma})]$$

$$= \frac{1}{\sqrt{2\pi}\sigma} \exp(-\frac{t^2}{2\sigma^2}) \times \{\exp\{d_n[\xi_{nk}(\pi-\theta,\varphi)\sec\theta + \xi_{ni}(\theta_0,\varphi_0)\sec\theta_0]\}$$
$$- \exp\{d_{n-1}[\xi_{nk}(\pi-\theta,\varphi)\sec\theta + \xi_{ni}(\theta_0,\varphi_0)\sec\theta_0]\}\}$$
$$/[\xi_{nk}(\pi-\theta,\varphi)\sec\theta + \xi_{ni}(\theta_0,\varphi_0)\sec\theta_0] \quad (2\text{-}2A.13)$$

Substituting above Eqs. (2-2A.10~2-2A.13) into Eqs. (2-19a~d), it can be seen that if let $t \to 0$, $c_0 \to \infty$ (i.e. $R_1, R_2 \to 0$), i.e. a plane pulse becomes a continuous plane wave, and $N = 1$, $d_{N+1} = d_N$, i.e. inhomogeneous random media degenerates to homogeneous case, the temporal Muller matrix of Eq. (2-2.18) can return to conventional Mueller matrix in Tsang et al. (1990, p.126).

2-3 Pulse Echoes from a Inhomogeneous Layered Canopy and Random Targets Beneath the Canopy

The previous session presents the temporal Mueller matrix solution of radiative transfer when a polarized pulse is incident upon multilayering media of random non-spherical scatterers. Co-polarized and cross-polarized echoes depict the inhomogeneous fraction profile of random scatterers. It is applicable to estimation of the canopy height, which has been extensively studied by SAR tomography and interferometry [Cloude and Papathanassiou 1998, Papathanassiou and Cloude 2001, Reigber and Moreira 2000, and others].

The effective height of inhomogeneous canopy is determined by the fraction profile of sparsely random scatterers. Meanwhile, polarized scattering from the model of random metal alien targets beneath the layered canopies is simulated for possible identification of anomalous scatterers [Jin et al. 2003]. A model is shown in Fig.2-3.1 as below.

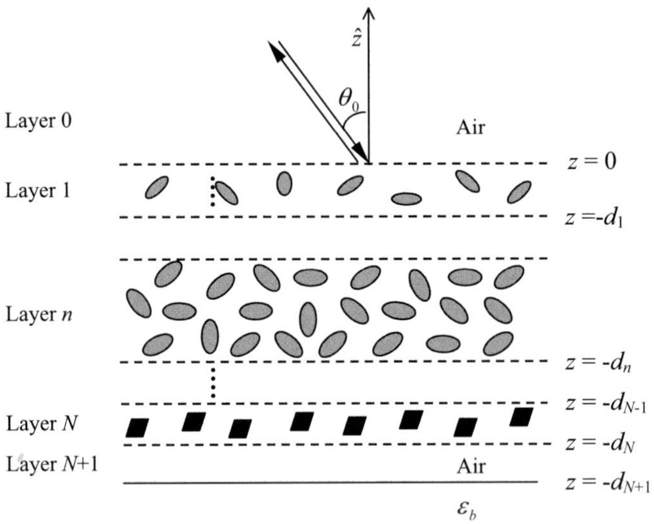

Fig. 2-3.1. Model of inhomogeneous layered canopy and random metal targets beneath canopies

2-3.1 Effective Height of Vegetation Canopy

One of the most important purposes for land remote sensing is to invert the height of the vegetation canopy such as forestry [Cloude and Papathanassiou 1998, Papathanassiou and Cloude 2001, Reigber and Moreira 2000, and others]. Using a first-order scattering model applied to fractal-generated tree structures and the equivalence relation between an INSAR and a Δk-radar, Sarabandi and Lin (2000) recently calculated the statistics of the scattering phase-center height. But the scattering phase center is a strong function of the inhomogeneous and sparsely distributed random scatterers. In fact, the height inverted by this method is an effective height, which is always lower than the real height. In other words, the height of tree canopy is defined not only by trunk or branch geometric lengths, but also by fractional profile. Up to now, there seems no general approach to define and determine the effective height of the inhomogeneous vegetation canopy and its relationship to the real geometric height and fractional profile.

The temporal variation of the polarized echoes can well depict spatially fractional profile of random scatterers. Its temporal pattern corresponds to the spatial fraction and might yield the effective height of inhomogeneous canopy. In order to compare the temporal variation of the echoes with the spatially fraction profile of random scatterers, the time delay of the impulse response t is converted to the height from the underlying surface h as follows

$$h = d_{N+1} - t \times c_0 \cos(\theta_0)/2 \tag{2-3.1}$$

Taking the backscattering coefficient $\sigma(h_i)$ at the height h_i, the effective height is defined as

$$H_{eff} = \sum_{i=1}^{m} \sigma(h_i) h_i \Big/ \sum_{i=1}^{m} \sigma(h_i) \tag{2-3.2}$$

where m is the number of the height samples, which are chosen at equal intervals.

In following examples, the carrier frequency is 5.3 *GHz*, and the Gaussian pulse with a half-power width 2.0*ns*. Incident angle is ($\theta_0 = 40^\circ, \varphi_0 = 0^\circ$). All scatter particles of disk-like spheroids are assumed identical in shape and size (the radius 1.0*cm*, the thickness 0.025*cm*) with the dielectric constant $(14.1 + i4.5)\varepsilon_0$ and uniformly spatial orientaion of the Euler angles are over $\beta \in (0^\circ, 180^\circ)$, $\gamma \in (0^\circ, 360^\circ)$. The dielectric constant of the underlying land surface is $\varepsilon_b = (11.5 + i2.5)\varepsilon_0$. The generalized Rayleigh-Gans approximation is applied to derive the scattering amplitude functions of small disks [Jin et al. 1999].

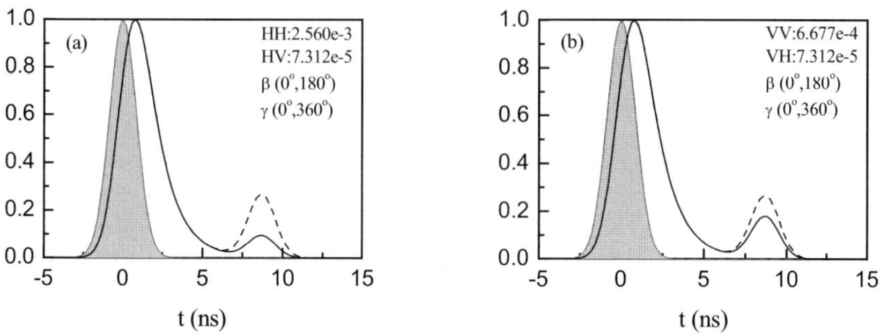

In specular direction: (a) *hh* and *hv*-polarized (b) *vv* and *hv*-polarized

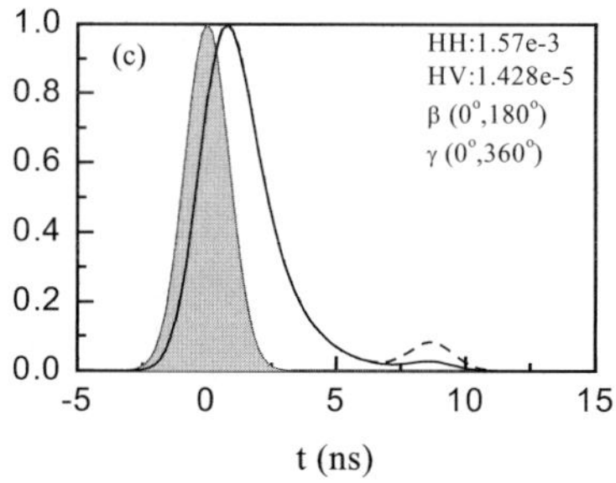

(c) *hh* and *hv*-polarized backscattering

Fig. 2-3.2. Polarimetric scattering from a layer of uniformly oriented scatterers

Firstly, to see polarimetric *hh, vv, hv*-polarized (pol) scattering, Figs. 2-3.2(a~c) show, respectively, *hh, hv, vv*-pol scattering in the specular direction ($\theta_s = \theta_0, \varphi_s = \varphi_0$) and *hh, hv*-pol in backscattering ($\theta_s = \theta_0, \varphi_s = \pi + \varphi_0$) from a layer of uniformly oriented random particles. The fractional volume of the particles is 0.005 and the layer depth is 100*cm*. In these figures, the normalized co-pol (*hh* or *vv*, solid line) and cross-pol (*hv*, dashed line) echoes are compared in shapes with the incident pulse (shadowing area). The actual peak amplitudes of the co-pol and cross-pol echoes are listed in the figures.

It can be seen that both co-pol and cross-pol echoes take time delay and pulse broadening. The peak tails in both co-pol and cross-pol echoes are most likely due to the reflections from the underlying surface. In other words, if the scatterers can more or less shadow the reflection from the underlying surface, the peak tail would not be significant. Furthermore, if the scatterers are not uniformly oriented over the azimuth γ range, the scatterers are actually laid in specific direction. It can especially enhance the distortion of cross-pol echo.

To understand the scattering mechanism of the pulse wave, the contributions of four scattering processes for co-pol and cross-pol echoes are, separately, presented in Figs.2-3.3(a, b), respectively. All parameters are the same as Fig. 2-3.2. It can be seen that the single reflections from the underlying surface (indicated by (2), (3)) make the peak tails in both co-pol and cross-pol echoes. The contributions of the double reflections of the underlying surface are very trivial in these examples.

It can be seen in Fig.2-3.2(c) that distortion of backscattering seems much smaller than that in specular direction, especially for cross-pol. It also can be further studied that cross-pol scattering of non-uniform γ orientation has especially longer time delay and more pulse broadening.

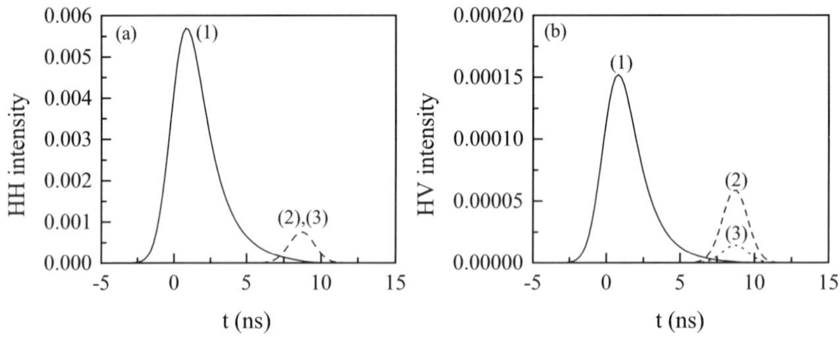

Fig. 2-3.3. Contributions of four scattering processes

Now, we turn to inhomogeneous canopy.

Figs. 2-3.4(a,b,c) show numerical results of co-pol backscattering coefficients σ_{hh} and σ_{vv}, and cross-pol backscattering σ_{hv} from 4 layers of the random disk-like particles (i.e. N+1 = 4, and Layer-4 is an air region). From Eqs. (2-3.19, 2-3.20) with m=15 interval heights, the effective heights for 3 different fraction profiles are calculated. Total canopy thickness is 5m with 3 different fraction profiles of f_{s1}, f_{s2}, f_{s3}: (1) 0.004, 0.002, 0.001; (2) 0.002, 0.004, 0.001; (3) 0.001, 0.002, 0.004, respectively. The layers thickness are $\Delta d_1 = 1.0m$, $\Delta d_2 = 1.0m$, $\Delta d_3 = 1.0m$ and $\Delta d_4 = 2.0m$, respectively, i.e. total layer thickness is d_{N+1}=5m.

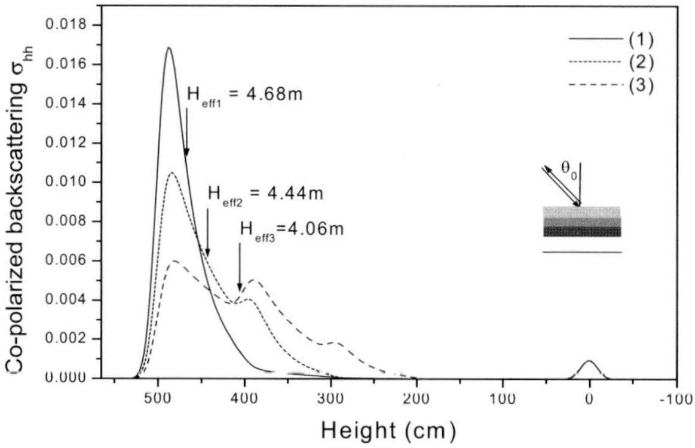

(a) inversion of the effective height from σ_{hh}.

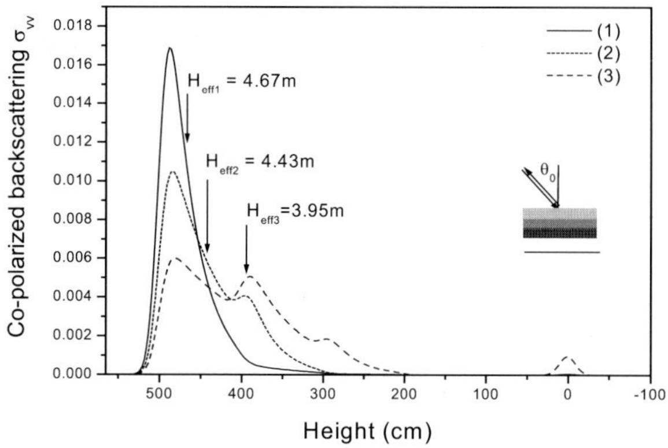

(b) inversion of the effective height from σ_{vv}.

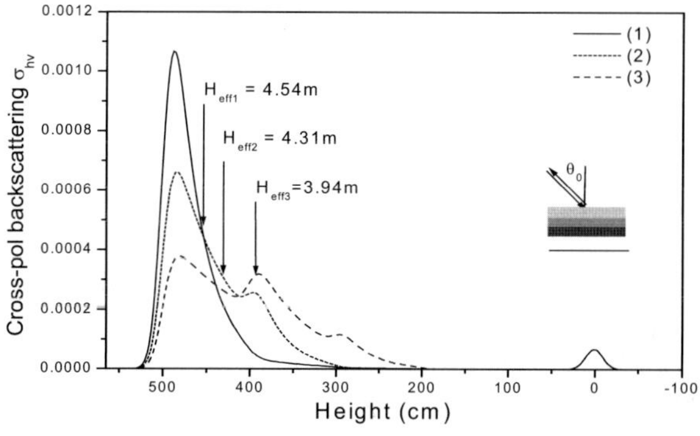

(c) inversion of the effective height from σ_{hv}.

Fig. 2-3.4. The effective heights of the vegetation canopy with different fraction profiles

It can be seen that temporal variation of the echoes intensity well depicts different inhomogeneous fraction profile. Meanwhile, we can see that the effective heights are related to the respective fraction profile and are all lower than the real heights (5*m*). Different canopy can have different effective height because the electromagnetic wave propagating through the random media of sparsely distributed scatterers undergoes different scattering, absorption and extinction coefficients. So the depth of the penetration and the temporal pattern of the echoes are both changed.

Co-pol and cross-pol backscattering present very similar effective heights from the figures (a,b,c). Generally, linearly co-pol backscattering σ_{hh} or and σ_{vv} is always much larger than cross-pol backscattering, e.g. σ_{hv}. Which one, σ_{hh} or σ_{vv}, is more significant depends on the scatterers' shape, e.g. disks or needles.

In our model of Fig. 2-3.1, the underlying surface is simply as a smooth ground, and there is no direct ground-surface scattering in VRT solutions, Eqs. (2-3.11, 2-3.12). However, even the rough surface might, more or less, diffuse or distort final scattering echoes pattern, contribution from direct ground scattering can be identified and would not cause essential difference from the above approach to invert the effective canopy height.

2-3.2 Anomalous Scattering from Random Metal Targets beneath Canopies

We use rectangular metal plates as the anomalous scatter targets beneath the canopies. As shown in Fig. 2-3.1, the *N*-th layer is the anomalous scatter targets layer. The normal direction of the metal plate is $\hat{z}_b$, the length and width sizes are *a*, *b*, respectively. Under the physical optics approximation, the forward and backward scattering amplitude functions of a metal plate are [Ulaby et al. 1990]:

$$\text{Forward: } f_{vv}(\hat{k}_s=\hat{k}_i,\hat{k}_i)=f_{hh}(\hat{k}_s=\hat{k}_i,\hat{k}_i)=-iab(\hat{z}_b\cdot\hat{k}_i)/\lambda_0 \tag{2-3.22}$$

$$\text{Backward: } f_{vv}(-\hat{k}_i,\hat{k}_i)=f_{hh}(-\hat{k}_i,\hat{k}_i)=-iab(\hat{z}_b\cdot\hat{k}_i)\frac{\sin(k_0a\hat{k}_i\cdot\hat{x}_b)\sin(k_0b\hat{k}_i\cdot\hat{y}_b)}{\lambda_0(k_0a\hat{k}_i\cdot\hat{x}_b)(k_0b\hat{k}_i\cdot\hat{y}_b)} \tag{2-3.23}$$

where $\hat{k}_i$ is the incident direction vector. The relationship between $\hat{x}_b,\hat{y}_b,\hat{z}_b$ of the scatter plate and the principle coordinates ($\hat{x},\hat{y},\hat{z}$) can be found in Jin (1994), Tsang et al. (1985).

In the following calculation, the size of the anomalous scatterers (i.e. the metal plates as example) are identical, $a\times b$ =2$cm\times$2cm with the thickness 0.25*cm*. The layer depths are: $\Delta d_1=1.0m$, $\Delta d_2=1.0m$, anomalous scatter layer $\Delta d_3=0.1m$, and air region $\Delta d_4=1m$, i.e. the total media thickness is 4.1*m*. The fractional volumes f_{s1}, f_{s2}, f_{s3} are 0.0010, 0.0015, 0.0010, respectively. Other parameters are the same as Fig. 2-3.2.

At the incident angle $\theta_0=40^\circ$, Fig. 2-3.5 shows the comparison of the temporal variation of backscattering σ_{hh} for two cases: one is for random metal plates, which are laid horizontally (i.e. all $\hat{z}_b=\hat{z}$) in the 3rd layer, and another for random dielectric disks with the same fractional volume in the same layer.

It can be seen that backscattering from the metal plates is significantly enhanced after 15*ns*, and much greater than the case of dielectric disks. The enhanced echo during 15~20 *ns* is dominantly contributed by the term $\overline{I}_n^{PS}$ in Eq. (2-3.8) (i.e. the term of PS_n in Eq. (2-3.12)). Making use of this anomalous pattern, the anomalous targets under vegetation canopy might be identified. In other words, the echoes pattern might provide a way to analyze scatterers types. Meanwhile, the scattering from anomalous targets layer also shadows the terms of $\bar{I}_n^{BP}, \bar{I}_n^{PB}$ in Eq. (2-3.8) or BP_n, PB_n in Eq.(2-3.12) around 33*ns*. The $\bar{I}_n^{BPB}$ around 50 ~ 60 *ns* are very trivial. Analysis of the echoes pattern and time delay can be applied for identification of targets to avoid false alarm due to multiple scattering and surface reflections.

If the underlying surface is randomly rough, direct ground scattering might diffuse the peak tail enhanced by the metal targets beneath the canopy. But, as long as this enhancement is high enough, anomaly alarm due to metal targets can be identified even at existence of rough surface scattering.

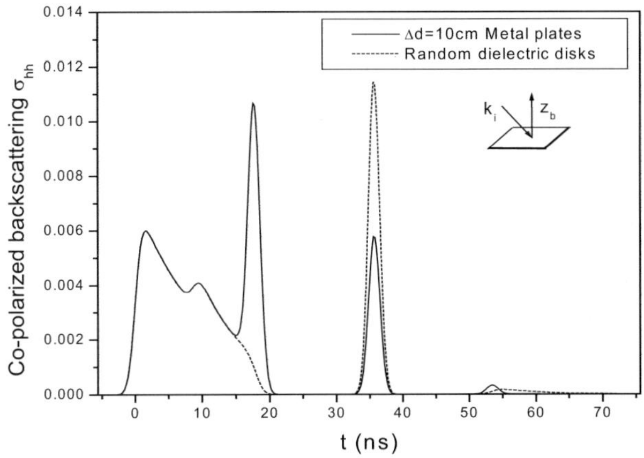

Fig. 2-3.5. Comparison of temporal σ_{hh} for anomalous metal plates and dielectric disks

Fig. 2-3.6 gives another example when the 3-rd layer of metal targets becomes much thicker, $\Delta d_3 = 50\,cm$. The incident angle is also $\theta_0 = 40^\circ$.

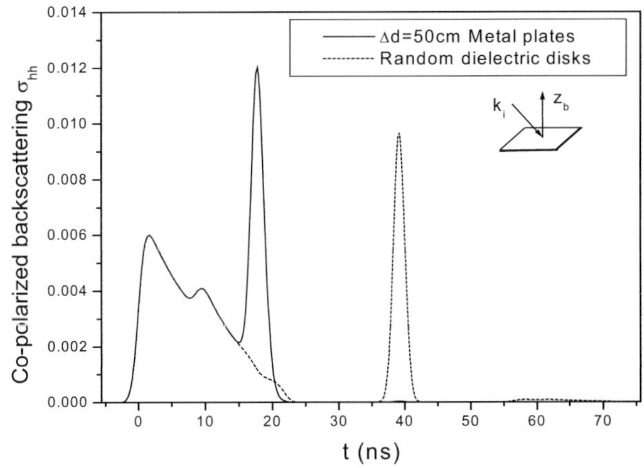

Fig. 2-3.6. Comparison of temporal σ_{hh} for anomalous metal plates and dielectric disks

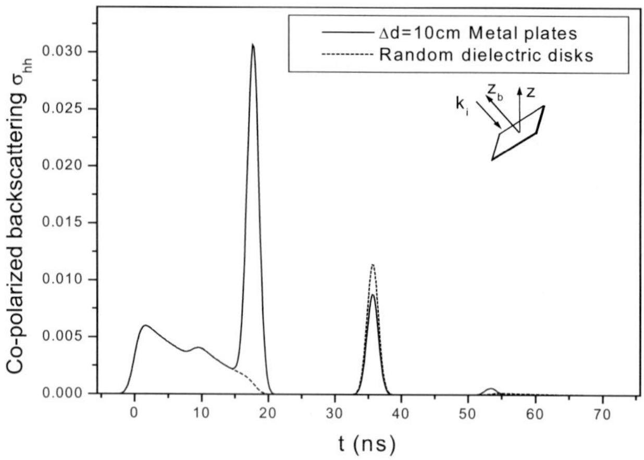

Fig. 2-3.7. Comparison of temporal σ_{hh} for anomalous metal plates and dielectric disks

It can be seen that the echoes from the metal targets layer become extremely high because the number of the metal targets is increased. Another significant effect is that the responses from the underlying surface i.e. $\bar{I}_n^{BP}$, $\bar{I}_n^{PB}$ and $\bar{I}_n^{BPB}$ in Eq. (2-3.8) or BP_n, PB_n, BPB_n in Eq. (2-3.12) are almost totally shadowed. When the anomalous scatter layer becomes thick enough, it is believed that the penetration of the incident waves is restricted and the responses from BP_n, PB_n, and BPB_n, especially, are negligible. Anomalous $\bar{I}_n^{PS}$ can be individually identified from those particle-surface iterations, $\bar{I}_n^{BP}$, $\bar{I}_n^{PB}$ and $\bar{I}_n^{BPB}$.

Fig. 2-3.7 shows an example of uniformly tilted metal plates in the 3rd layer. The spatial orientation of all metal plates happens to be orthogonal to the incident wave direction.

It is interesting to see that the echoes become anomalously great because the electro-magnetic wave is directly reflected when all plates are uniformly tilted in this way. In other words, the term $\bar{I}_n^{PS}$ in Eq. (2-3.8) is dominantly increased when all metal plates are oriented in this specular direction.

2-4 Retrievals of Underlying Surface Roughness and Moisture from Polarimetric Pulse Echoes

One of the most important purposes for land remote sensing is to retrieve the land moisture. Variation of land moisture (via water content and soil texture) changes soil permittivity, and affects scattering observation over terrain surfaces. Because land surfaces in nature are always rough and are often covered by vegetation canopy, the roughness and canopy coverage, as well as soil moisture, simultaneously affect scattering observation. It is difficult to identify which one (moisture, roughness or both, or vegetation canopy) cause scattering variation. How to invert the soil moisture and surface roughness, with or without vegetation canopy, from remote sensing has been onc of the most interesting problems to be resolved.

In the previous sessions, the pulse echoes from inhomogeneous random media are obtained by using the temporal Mueller matrix solution. It has been seen that the peak tails in pulse echoes are due to reflections from the underlying surface. In this session, the underlying boundary surface in the model of Fig. 2-2.1 is taken as a randomly rough surface. Co-polarized tail peaks are then employed for simultaneous retrievals of the underlying surface roughness and moisture when vegetation canopy is presented [Jin et al. 2004].

2-4.1 Polarimetric Bistatic Scattering Pulses Responses

We consider the example of the pulse echoes from four-layered media (i.e. $N+1=4$, the layers 1-3 are random disk-like particles, and the layer-4 is an air region). The carrier frequency is 5.3 *GHz*, and the Gaussian pulse with a half-power width 2.0*ns*. As a simple case (but not necessary), all particles are identical in shape and size (the radius 1.0*cm*, the thickness 0.025*cm*) with the dielectric constant $(14.1+i4.5)\varepsilon_0$ and uniformly spatial orientaion of the Euler angles $\beta\in(0^\circ,180^\circ)$, $\gamma\in(0^\circ,360^\circ)$. Each layer has different fractional volume. The dielectric constant of the underlying medium, ε_b, is determined by the volumetric moisture m_v [Wang and Choudhury 1981]. The generalized Rayleigh-Gans approximation is applied to derive the scattering amplitude functions of small disks.

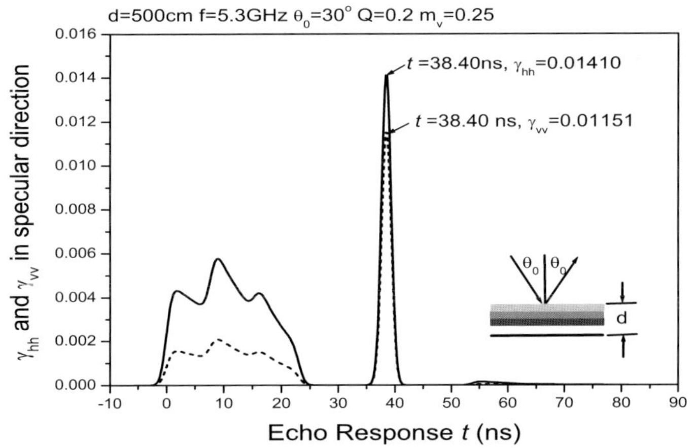

Fig.2-4.1. Temporal bistatic scattering γ_{hh} and γ_{vv} at specular direction ($\theta_0=30^\circ$)

Fig. 2-4.1 and Fig. 2-4.2 demonstrate temporal bistatic scattering in the specular direction, γ_{hh} of solid lines and γ_{vv} of dotted lines, for two incident angles $(\theta_0 = 30^\circ)$ and $(\theta_0 = 60^\circ)$, respectively. The stratified 4-layer canopy has the fractions, respectively, of f_{s1}, f_{s2}, f_{s3}: (a) 0.0005, 0.001, 0.0015. The layer depths are $d_1 = 1m$, $d_2 = 2m$, $d_3 = 3$m and $d_4 = 5m$, respectively, i.e. the total canopy thickness is 5*m*.

It can be seen that the echoes well depict the stratified fraction distribution. The time delay difference between two echoes for different incident angles θ_{10}, θ_{20} can also be used to invert the total layer thickness.

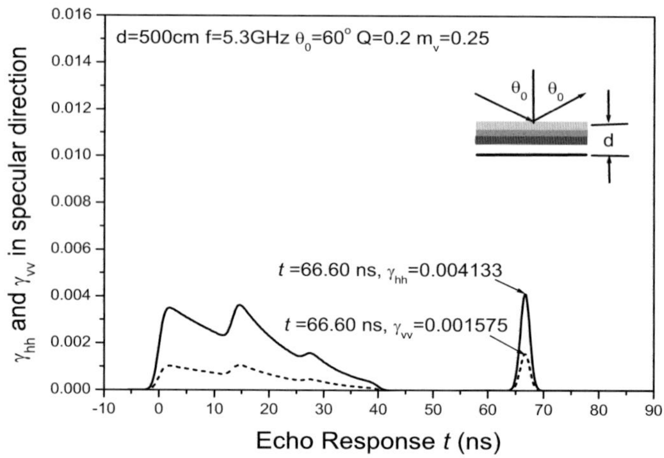

Fig. 2-4.2. Temporal bistatic scattering γ_{hh} and γ_{vv} at specular direction ($\theta_0 = 60^\circ$)

2-4.2 Retrievals of Underlying Surface Roughness and Moisture

From Figs. (2-4.1, 2-4.2), it can be seen that the pulse echoes can identify the different, volumetric or surface, scattering mechanism. Numerical results also demonstrate that these tails are mainly contributed by the terms of BP_n and PB_n in Eq. (2-2.18).

In the specular direction and for totally random, sparse distribution of spheroids, making the approximation of $\overline{\overline{D}} \approx \overline{\overline{I}}$ in Eqs. (2-2.19b,c) and taking the phase functions

$$P_{11}(\theta_0, \phi_0; \pi - \theta_0, \phi_0) = P_{22}(\theta_0, \phi_0; \pi - \theta_0, \phi_0) \tag{2-4.1}$$

it yields

$$R(\theta_0) \equiv \frac{\gamma_{vv}(\theta_0, \phi_0; \pi - \theta_0, \phi_0)\,|_{at\ tail\ peak}}{\gamma_{hh}(\theta_0, \phi_0; \pi - \theta_0, \phi_0)\,|_{at\ tail\ peak}} \approx \frac{r_v(\theta_0)}{r_h(\theta_0)} \tag{2-4.2}$$

where r_v, r_h are the *v*- and *h*–polarized reflectivity of the underlying rough surface. To take into account for surface roughness, r_v and r_h are empirically written by Wang and Choudhury (1981) as

$$r_v(\theta_0) = [(1-Q)r_{v0}(\theta_0) + Qr_{h0}(\theta_0)]\exp(-4k^2\delta^2\cos^2\theta_0) \tag{2-4.3a}$$

$$r_h(\theta_0) = [Qr_{v0}(\theta_0) + (1-Q)r_{h0}(\theta_0)]\exp(-4k^2\delta^2\cos^2\theta_0) \tag{2-4.3b}$$

where r_{v0}, r_{h0} are the *v*- and *h*–polarized reflectivity of the flat surface, Q is the characteristic parameter to describe or evaluate surface roughness (from Q=0 for smooth surface to Q=0.5 for extremely rough surface, which describes the mixing of energies in orthogonal polarization states). An empirical expression for Q and the surface variance δ may be given by $Q = 0.35[1-\exp(-0.6\delta^2 f)]$, where the frequency f in GHz. We take Q as the representation or measure of surface roughness.

Substituting Eqs. (2-4.3a,b) into (2-4.2), it leads that Q is determined by

$$Q = \frac{R_1 - R_{10}}{(1-R_{10})(1+R_1)} \tag{2-4.4}$$

where $R_1 = R(\theta_{01})$ is measured from the tail peaks defined by Eq. (2-4.20) at a specular angle θ_{01}, its twin of flat surface $R_{10} = r_{v0}(\theta_{01})/r_{h0}(\theta_{01})$ is the function of m_v and can be calculated by $\varepsilon_b(m_v)$.

If two observations at different specular angles θ_{01}, θ_{02} are measured, two ratios R_1, R_2 are obtained. It yields that

$$\frac{R_1 - R_{10}}{(1-R_{10})(1+R_1)} = \frac{R_2 - R_{20}}{(1-R_{20})(1+R_2)} \quad , \tag{2-4.5}$$

because both RHS and LHS of Eq. (2-4.5) are equal to Q due to derivation of (2-4.4).

Thus, the next task is how to develop an iterative approach to find the best surface moisture m_v, which yields the surface dielectric constant $\varepsilon_b(m_v)$, then $r_{v0}(\varepsilon_b)$ and $r_{h0}(\varepsilon_b)$ to meet Eq. (2-4.5).

Now, taking 25 sets of the parameters m_v^T (0.10, 0.15, 0.20, 0.25, 0.30) and Q^T (0.10, 0.15, 0.25, 0.30), which are seen as the ground truth data with the superscript T, we make numerical simulation of the Mueller matrix solution and obtain $R_1(\theta_{01}=30°)$ and $R_2(\theta_{02}=60°)$ of Eq. (2-4.2). Meanwhile, by making use of iterative m_v, ε_b and $R_0(r_{v0}, r_{h0})$ can be calculated.

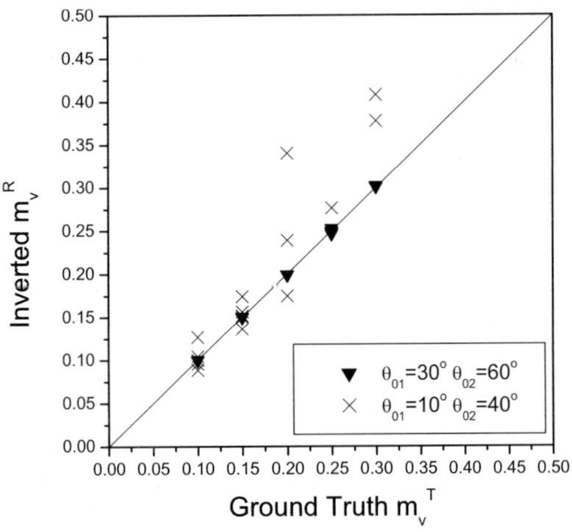

Fig. 2-4.3. Retrieved m_v^R and comparison with ground truth m_v^T

Numerical iterations for solving Eq. (2-4.5) is developed for retrievals of m_v and Q.

The retrieved m_v^R and Q^R and comparisons with previous ground truth m_v^T and Q^T are presented in Figs. 2-4.3, 2-4.4 as the solid points. It can be seen that the moisture m_v and roughness Q are well inverted for all 25 cases. Some other cases such as different fraction distribution and layer thickness are also tested to well retrieve m_v and roughness Q.

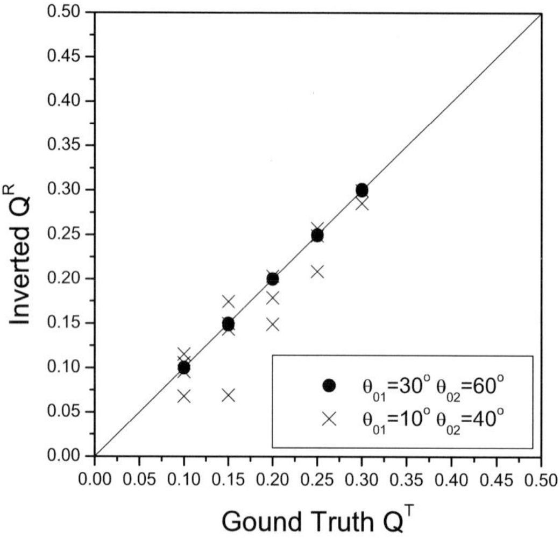

Fig. 2-4.4. Retrieved Q^R and comparison with ground truth Q^T

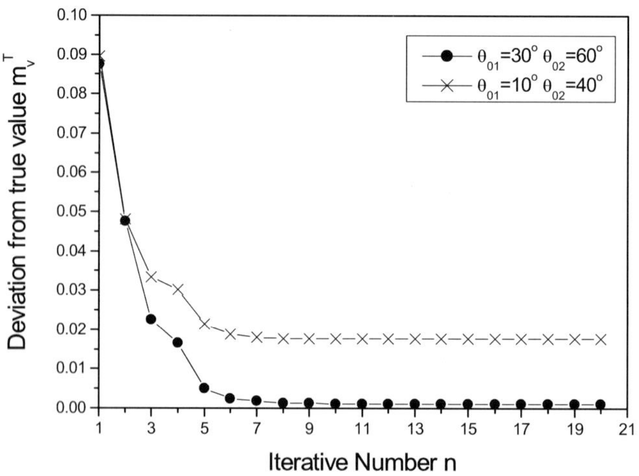

Fig. 2-4.5. Deviation from the true value m_v^T in iterations

However, if the incident angle is small such as less than 20°, the retrievals are not good.

The cross points in Figs. 2-4.4, 2-4.5 are for the case of $\theta_{01}=10^\circ$, $\theta_{02}=40^\circ$. A reason for it is that the differences between the *v*- and *h*-polarized reflectivity's $r_v(\theta_0)$ and $r_h(\theta_0)$ of the rough surface, and between $r_{v0}(\theta_0)$ and $r_{h0}(\theta_0)$ of the flat surface are not well identified at small θ_0. It yields not accurate $R(\theta_0)$ and $R_0(\theta_0)$.

Fig. 2-4.5 shows the averaged deviation $(1/25)\sum_{j=1}^{25}|m_v^R(j)-m_v^T(j)|$ during iterations, where the subscript *j* indicate the *j*-th set of m_v^T and Q^T. It can be seen that convergence for the case of $R_1(30^\circ)$ and $R_2(60^\circ)$ is very good. But, it would not be well converged for the case of $R_1(10^\circ)$ and $R_2(40^\circ)$ as explained in Figs. 2-4.4 and 2-4.5.

The retrieval of surface roughness and moisture in this paper is essentially based on the ability to separate the tail peaks from co-polarized echoes in specular direction. Real ground problem has been remained for the data availability.

2-5 Statistics of Four Stokes Parameters in Multi-look Polarimetric SAR Imagery

In recent decades, the imaging radar technology has been able to measure completely polarimetric scattering, and four complex scattering amplitude functions, $f_{ab}, a,b=v,h$ indicating vertical and horizontal polarizations respectively, are available [Zebker and Van Zyl 1991]. Full understanding of the statistics of polarimetric SAR data has become of great importance for image speckle reduction, data verification, target identification and surface classification [Sheen and Johnston 1992, Lee et al. 1999a]. As the area illuminated by the radar contains many uniformly distributed random scatterers, the multilook covariance matrix can be assumed to have complex Wishart distribution [Lee *et al.* 1994]. The statistics of the scattered intensity, amplitude ratio and phase term from multilook polarimetric SAR data have been studied by Lee *et al.* (1999b). Frery *et al.* (1997) presented a clutter model for the amplitude statistics from extremely heterogeneous background, e.g. city.

The Stokes vector, or four Stokes parameters, has been studied for partially polarized waves propagating through random media [Chansakhar, 1965, Jin 1994]. Barakat (1987) first described the temporal statistics of the Stokes parameters of partially polarized light, which is applicable to single-look SAR data. Touzi and Lopes (1996) provided a numerically tractable extension to Barakat's work to include the effective phase difference by deriving the probability density functions (*PDF*'s) of the multilook Stokes vector as a function of coherent electric fields and complex correlation coefficient.

The aim of this session is to analytically formulate the statistics of four Stokes parameters for multi-look SAR imagery, and to validate SAR multi-channel measurements over different terrain surfaces. The *PDF* of the first Stokes parameter, *I*, is found to be that given by Touzi and Lopes (1996). However, the *PDF* of the second Stokes parameter, *Q*, is much simpler than the series summation of Touzi and Lopes (1996). Under some reasonable approximations, validated by real SAR data, the *PDF*'s of the third and forth Stokes parameters, *U* and *V*, can be formulated analytically to replace the integral forms of Touzi and Lopes (1996). These *PDF*'s have explicit functional dependence upon the look number and mean values of the respective Stokes parameters [Jin and Zhang 2002]. To validate these statistics, the AirSAR data at multi-P, L, C bands over some typical areas such as water, sparse trees and thick forest were chosen for comparison. In addition, an effective number of looks might be assigned to provide an excellent match with real data.

2-5.1 Multilook Covariance Matrix and the Complex Wishart Distribution

As the complex scattering amplitude matrix $f_{ab}, a,b=v,h$, is measured in polarimetric SAR observation, the 4×4 dimensional real Mueller matrix can be constructed. For the data obtained from the Jet Propulsion Laboratory (JPL), the Mueller matrix is symmetrical and is compressed into a 10 bytes blocks [Dubois and Norikane 1987]. The Mueller representation of the SAR data is usually averaged by multiple pixels in both azimuth and range directions to form the multi-look SAR data and to reduce the image speckle. The statistical distribution for the multi-look Mueller formulation needs to be refined.

The *n*-look covariance matrix is usually defined as (Lee *et al.* 1994)

$$\overline{\overline{Z}} = \frac{1}{n}\sum_{k=1}^{n} < \overline{u}(k)\overline{u}(k)^T > \tag{2-5.1}$$

where the vector $\overline{u}(k)$ is the k-th one-look sample; its element f_1, f_2 and f_3 denote f_{hh}, f_{vv} or f_{hv}; the superscript T denotes complex conjugate transpose; and the angular brackets < > means the ensemble average. The dimension of vector $u(k)$ can be taken as 1, 2 or 3. To simplify the following derivation, let $\overline{\overline{A}} = n\overline{\overline{Z}}$. Then the matrix $\overline{\overline{A}}$ has a complex Wishart distribution [Goodman 1963, Lee *et al.* 1994]

$$P_A(\overline{\overline{A}}) = \frac{|\overline{\overline{A}}|^{n-q}\exp[-Tr(\overline{\overline{C}}^{-1}\overline{\overline{A}})]}{K(n,q)|\overline{\overline{C}}|^n} \tag{2-5.2}$$

where $K(n,q) = \pi^{q(q-1)/2}\Gamma(n)\Gamma(n-1)\cdots\Gamma(n-q+1)$, q is the dimention of the vector $u(k)$, the complex covariance matrix $\overline{\overline{C}} = E[\overline{u}\,\overline{u}^T]$, $E[v]$ is the expected value of the random variable v, $|\overline{\overline{C}}|$ is the determinant of $\overline{\overline{C}}$. Applying the complex Wishart distribution, the *PDF*'s of four multilook Stokes parameters can be derived.

2-5.2 The PDF's of Four Stokes Parameters

The stokes vector is usually defined as [Jin 1994]

$$\overline{I} = \begin{bmatrix} I \\ Q \\ U \\ V \end{bmatrix} = \begin{bmatrix} <|E_h|^2> + <E_v|^2> \\ <|E_h|^2> - <E_v|^2> \\ 2<\mathrm{Re}(E_h E_v^*)> \\ 2<\mathrm{Im}(E_h E_v^*)> \end{bmatrix} \tag{2-5.3}$$

where E_v, E_h are the v, h-polarized electric fields, respectively.

The scattered Stokes vector, $\overline{I}_s$, can be obtained by the Mueller matrix $\overline{\overline{M}}$

$$\overline{I}_s = \overline{\overline{M}} \cdot \overline{I}_i \tag{2-5.4}$$

where the subscripts s and i denote scattered and incident, respectively. The elements of the Mueller matrix can be expressed by the elements of the scattering amplitude functions [Jin 1994]. The Mueller matrix of multilook SAR data is the average of single-look Mueller matrices

$$\overline{\overline{M}}^{(n)} = \frac{1}{n}\sum_{k=1}^{n}\overline{\overline{M}}(k) \tag{2-5.5}$$

where n is the look number and k indicates the k-th single-look. The normalized incident Stokes vector is written as

$$\overline{I}_i = \begin{bmatrix} 1 \\ \cos 2\chi \cos 2\psi \\ -\cos 2\chi \sin 2\psi \\ \sin 2\chi \end{bmatrix} \tag{2-5.6}$$

where χ, ψ are the elliptical and orientation angles of incident wave, respectively. For $\chi = 0°$, the incident radiation is linearly polarized (horizontaly polarized (h-pol) when $\psi = 0°$, and verticaly polarized (v-pol) when $\psi = 90°$). For $\chi = 45°$, it is circularly polarized.

As the incident wave is a-linearly polarized (a=h or v), the scattered Stokes parameters of n-looks SAR imagery are written as follows

$$I = \frac{1}{2n}\sum_{k=1}^{n}[< f_{aa}(k) f_{aa}^*(k) > + < f_{ab}(k) f_{ab}^*(k) >] \tag{2-5.7a}$$

$$Q = \frac{\pm 1}{2n} \sum_{k=1}^{n} [< f_{aa}(k) f_{aa}^*(k) > - < f_{ab}(k) f_{ab}^*(k) >] \tag{2-5.7b}$$

$$U = \frac{1}{n} \sum_{k=1}^{n} < \text{Real}[f_{aa}(k) f_{ab}^*(k)] > \tag{2-5.7c}$$

$$V = -\frac{1}{n} \sum_{k=1}^{n} < \text{Imag}[f_{aa}(k) f_{ab}^*(k)] > \tag{2-5.7d}$$

where the signs $\pm$ in Eq. (2-5.7b), respectively, indicate + for *h*-pol and – for *v*-pol. Since every Stokes parameter Eqs. (2-7a~d) is expressed by only two scattering amplitude functions, let $q = 2$ and

$$\overline{\overline{A}} = \begin{bmatrix} A_{11} & A_{12} \\ A_{21} & A_{22} \end{bmatrix}, \ \overline{\overline{C}} = \begin{bmatrix} C_{11} & \sqrt{C_{11}C_{22}}\,|\rho_c|\,e^{i\theta} \\ \sqrt{C_{11}C_{22}}\,|\rho_c|\,e^{-i\theta} & C_{22} \end{bmatrix} \tag{2-5.8}$$

where

$$A_{11} = \sum_{k=1}^{n} < f_{aa}(k) f_{aa}^*(k) >, \ A_{22} = \sum_{k=1}^{n} < f_{ab}(k) f_{ab}^*(k) > \tag{2-5.9a}$$

$$A_{21}^* = A_{12} = \sum_{k=1}^{n} < f_{aa}(k) f_{ab}^*(k) > \tag{2-5.9b}$$

with $C_{11} = E[A_{11}]$, $C_{22} = E[A_{22}]$ and $\rho_c = E[A_{12}]/\sqrt{C_{11}C_{22}}$ is the complex correlation coefficient for $< f_{aa}(k) f_{ab}^*(k) >, a \neq b$.

Comparing Eq. (2-5.8) with Eqs. (2-5.7a~d), we have

$$C_{11} = E[I] \pm E[Q] \tag{2-5.10a}$$

$$C_{22} = E[I] \mp E[Q] \tag{2-5.10b}$$

$$|\rho_c| = \sqrt{E^2[U] + E^2[V]} / \sqrt{E^2[I] - E^2[Q]} \tag{2-5.10c}$$

The joint *PDF* of R_1, R_2 has been derived by Lee *et al.* (1994) as

$$P(R_1, R_2) = \frac{n^{n+1} (R_1 R_2)^{\frac{n-1}{2}} \exp[-\frac{n(R_1 C_{22} + R_2 C_{11})}{C_{11} C_{22} (1 - |\rho_c|^2)}]}{(C_{11} C_{22})^{\frac{n+1}{2}} \Gamma(n) (1 - |\rho_c|^2) |\rho_c|^{n-1}} I_{n-1}\left(2n \sqrt{\frac{R_1 R_2}{C_{11} C_{22}}} \frac{|\rho_c|}{1 - |\rho_c|^2}\right) \tag{2-5.11}$$

where $R_1 = A_{11}/n$, $R_2 = A_{22}/n$. $I_\upsilon(x)$ is the modified first kind and υ-th order Bessel function.

From Eqs. (2-5.7a,b), $I = (R_1 + R_2)/2$ and $Q = \pm (R_1 - R_2)/2$, the joint *PDF* of I and Q is written as follows

$$P(I, Q) = \frac{2n^{n+1} (I^2 - Q^2)^{\frac{n-1}{2}}}{(C_{11} C_{22})^{\frac{n+1}{2}} \Gamma(n) (1 - |\rho_c|^2) |\rho_c|^{n-1}} \tag{2-5.12}$$

$$\times \exp[-\frac{n(C_{11} + C_{22}) I \mp n(C_{11} - C_{22}) Q}{C_{11} C_{22} (1 - |\rho_c|^2)}] I_{n-1}\left(2n \sqrt{\frac{I^2 - Q^2}{C_{11} C_{22}}} \frac{|\rho_c|}{1 - |\rho_c|^2}\right)$$

The first Stokes parameter *I* is total scattered power. Using the integration identiy (2-5A.1) in the Appendis 2-5A and integrating Eq. (2-5.12) with respect to *Q*, we obtain the *PDF* of *I* as

$$P_I(I)=\int_{-I}^{I}P(I,Q)dQ\ =\int_{-1}^{1}I\,P(I,xI)dx$$

$$=\frac{\sqrt{\pi}(2n)^{n+\frac{1}{2}}I^{n-\frac{1}{2}}\exp[-\frac{n(C_{11}+C_{22})I}{C_{11}C_{22}(1-|\rho_c|^2)}]I_{n-\frac{1}{2}}[\frac{n\sqrt{(C_{11}-C_{22})^2+4C_{11}C_{22}|\rho_c|^2}}{C_{11}C_{22}(1-|\rho_c|^2)}I]}{\Gamma(n)\sqrt{C_{11}C_{22}(1-|\rho_c|^2)}[(C_{11}-C_{22})^2+4C_{11}C_{22}|\rho_c|^2]^{\frac{n}{2}-\frac{1}{4}}}$$

(2-5.13)

Substituting Eqs.(2-5.10a~c) into Eq. (2-5.13), the *PDF* of *I* is finally obtained as follows

$$P_I(I)=\frac{2\sqrt{\pi}n^{n+\frac{1}{2}}I^{n-\frac{1}{2}}\exp[-\frac{2nI}{E[I](1-p^2)}]I_{n-\frac{1}{2}}[\frac{2npI}{E[I](1-p^2)}]}{\Gamma(n)(E[I])^{n+\frac{1}{2}}(1-p^2)^{\frac{1}{2}}p^{n-\frac{1}{2}}} \tag{2-5.14}$$

where

$$p=\sqrt{E^2[Q]+E^2[U]+E^2[V]}/E[I] \tag{2-5.15}$$

and is defined as the polarization degree to describe partial polarization of scattered intensity propagating through random media.

Normalizing *I* with respect to *E*[*I*], we obtain the *PDF* of normalized first Stokes parameter, $S_0=I/E[I]$, for the linear incidence as

$$P_{S_0}(S_0)=\frac{2\sqrt{\pi}n^{n+\frac{1}{2}}S_0^{n-\frac{1}{2}}\exp[-\frac{2nS_0}{1-p^2}]\mathrm{I}_{n-\frac{1}{2}}[\frac{2np}{1-p^2}S_0]}{\Gamma(n)(1-p^2)^{\frac{1}{2}}p^{n-\frac{1}{2}}} \tag{2-5.16}$$

This *PDF* has been derived in Touzi and Lopes (1996, Eq.(47)).

Applying the integration identity (2-5A.2) in the Appendix 2-5A and integrating Eq. (2-5.12) with respect to *I,* we obtain

$$P_Q(Q)=\int_{|Q|}^{\infty}P(I,Q)dI\ =\int_{1}^{\infty}|Q|P(x|Q|,Q)dx$$

$$=\frac{(2n)^{n+\frac{1}{2}}|Q|^{n-\frac{1}{2}}\exp[\pm\frac{n(C_{11}-C_{22})Q}{C_{11}C_{22}(1-|\rho_c|^2)}]K_{n-\frac{1}{2}}[\frac{n\sqrt{(C_{11}+C_{22})^2-4C_{11}C_{22}|\rho_c|^2}}{C_{11}C_{22}(1-|\rho_c|^2)}|Q|]}{\sqrt{\pi}\Gamma(n)\sqrt{C_{11}C_{22}(1-|\rho_c|^2)}[(C_{11}+C_{22})^2-4C_{11}C_{22}|\rho_c|^2]^{\frac{n}{2}-\frac{1}{4}}}$$

(2-5.17)

where $K_\upsilon(x)$ is the modified second kind and υ -th order Bessel function. Substituting Eqs. (2-5.10a~c) into Eq. (2-5.17), the *PDF*'s of the second Stokes parameter *Q* and its nomalized parameter $S_1=Q/E[I]$ can be obtained. The *PDF* of S_1 for the linearly polarized incidence is written as

$$P_{S_1}(S_1)=\frac{2n^{n+\frac{1}{2}}|S_1|^{n-\frac{1}{2}}\exp[\frac{2nqS_1}{1-p^2}]\mathrm{K}_{n-\frac{1}{2}}[\frac{2n\sqrt{1-u^2-v^2}|S_1|}{1-p^2}]}{\sqrt{\pi}\Gamma(n)(1-p^2)^{\frac{1}{2}}(1-u^2-v^2)^{\frac{n}{2}-\frac{1}{4}}} \tag{2-5.18}$$

where $q=E[Q]/E[I]$, $u=E[U]/E[I]$ and $v=E[V]/E[I]$. It is noted that Eq. (2-5.18) for normalized second Stokes parameter is much simpler than the series summation of

Touzi and Lopes (1996, Eq.(55)).

Let complex A_{12} in Eq. (2-5.9) be expressed by $A_{12R}+iA_{12I}$, then Eq.(2-5.2) can become

$$P_A(A_{11},A_{22},A_{12R},A_{12I})=\frac{2[A_{11}A_{22}-(A_{12R}^2+A_{12I}^2)]^{n-2}}{\pi\Gamma(n)\Gamma(n-1)(1-|\rho_c|^2)^n C_{11}^n C_{22}^n}$$
$$\cdot\exp[-\frac{A_{11}C_{22}+A_{22}C_{11}-2|\rho_c|\sqrt{C_{11}C_{22}}(A_{12R}\cos\theta+A_{12I}\sin\theta)}{C_{11}C_{22}(1-|\rho_c|^2)}] \tag{2-5.19}$$

where θ is the phase of the complex correlation coefficient ρ_c.

Using the integration identity (2-5A.3) in the Appendix 2-5A and integrating Eq.(2-5.19) with respect to A_{11} for $n>1$, we obtain

$$P(A_{22},A_{12R},A_{12I})=\int_{\frac{A_{12R}^2+A_{12I}^2}{A_{22}}}^{\infty}P_A(A_{11},A_{22},A_{12R},A_{12I})dA_{11}$$
$$=\frac{2A_{22}^{n-2}\exp[-\frac{C_{11}A_{22}+C_{22}(A_{12R}^2+A_{12I}^2)/A_{22}-2|\rho_c|\sqrt{C_{11}C_{22}}(A_{12R}\cos\theta+A_{12I}\sin\theta)}{C_{11}C_{22}(1-|\rho_c|^2)}]}{\pi\Gamma(n)(1-|\rho_c|^2)C_{11}\ C_{22}^n} \tag{2-5.20}$$

Again, using the integration identity given by Eq. (2-5A.4) in the Appendix 2-5A and integrating Eq. (2-5.20) over A_{22}, we obtain

$$P(A_{12R},A_{12I})=\frac{4\exp[2|\rho_c|(A_{12R}\cos\theta+A_{12I}\sin\theta)/\sqrt{C_{11}C_{22}}(1-|\rho_c|^2)]}{\pi\Gamma(n)(1-|\rho_c|^2)C_{11}^{\frac{(n+1)}{2}}C_{22}^{\frac{(n+1)}{2}}} \tag{2-5.21}$$
$$\cdot(A_{12R}^2+A_{12I}^2)^{\frac{n-1}{2}}K_{n-1}[2\sqrt{(A_{12R}^2+A_{12I}^2)/C_{11}C_{22}(1-|\rho_c|^2)^2}]$$

Since ρ_c takes into account for correlation of $<f_{aa}(k)f_{ab}^*(k)>, a\neq b$, its magnitude at a-linearly polarized incidence is always small when comapred with C_{11}, C_{22}. It can yield the approximations

$$A_{12R},A_{12I}<<\sqrt{C_{11}C_{22}},\ \frac{2|\rho_c|A_{12I}\sin\theta}{\sqrt{C_{11}C_{22}}(1-|\rho_c|^2)}\approx 0 \text{ and } \frac{2|\rho_c|A_{12R}\cos\theta}{\sqrt{C_{11}C_{22}}(1-|\rho_c|^2)}\approx 0\ . \tag{2-5.22}$$

These approximations can be validated by SAR image data, as an example, chosen from Fig. 2-5.3, the forest area C. It is expected that these approxiamtions in Eq.(2-5.22) to be valid for natural terrain at low frequency such as P,L,C bands. Man-mde structures may provide pathological cases where they fail. From Fig. 2-5.1, it is demonstrated that

$$\exp[2|\rho_c|A_{12I}\sin\theta/\sqrt{C_{11}C_{22}}(1-|\rho_c|^2)]\approx 1 \tag{2-5.23a}$$
$$\exp[2|\rho_c|A_{12R}\cos\theta/\sqrt{C_{11}C_{22}}(1-|\rho_c|^2)]\approx 1 \tag{2-5.23b}$$

with very small deviation.

Substituting Eq. (2-5.23a) and the identity (2-5A.5) of the Appendix 2-5A into Eq. (2-5.21), we obtain the integration with respect to A_{12I} as

$$P_{A_{12R}}(A_{12R})=\frac{2|A_{12R}|^{n-\frac{1}{2}}\exp[\frac{2|\rho_c|\cos\theta A_{12R}}{\sqrt{C_{11}C_{22}}(1-|\rho_c|^2)}]K_{n-\frac{1}{2}}(\frac{2|A_{12R}|}{\sqrt{C_{11}C_{22}}(1-|\rho_c|^2)})}{\sqrt{\pi}\Gamma(n)(1-|\rho_c|^2)^{\frac{1}{2}}C_{11}^{\frac{2n+1}{4}}C_{22}^{\frac{2n+1}{4}}} \tag{2-5.24}$$

Since $A_{12R} = \sum_{1}^{n} < \text{Real}(f_{aa}(k) f_{ab}^{*}(k)) >$, the *PDF* of the third Stokes parameter, $U = A_{12R}/n$ is written as follows

$$P_U(U) = \frac{2n^{n+\frac{1}{2}} |U|^{n-\frac{1}{2}} \exp[\frac{2n|\rho_c|\cos\theta}{\sqrt{C_{11}C_{22}}(1-|\rho_c|^2)} U] K_{n-\frac{1}{2}}(\frac{2n|U|}{\sqrt{C_{11}C_{22}}(1-|\rho_c|^2)})}{\sqrt{\pi}\Gamma(n)(1-|\rho_c|^2)^{\frac{1}{2}} C_{11}^{\frac{2n+1}{4}} C_{22}^{\frac{2n+1}{4}}} \quad \text{(2-5.25)}$$

Substituting Eqs. (2-5.10a~c) into Eq.(2-5.25), the *PDF*'s of the third Stokes parameter U and its nomalized parameter $S_2 = U/E[I]$ can be obtained. The *PDF* of S_2 for the linearly polarized incidence is

$$P_{S_2}(S_2) = \frac{2n^{n+\frac{1}{2}} |S_2|^{n-\frac{1}{2}} \exp(\frac{2n\cos\theta\sqrt{u^2+v^2}}{1-p^2} S_2) K_{n-\frac{1}{2}}(\frac{2n\sqrt{1-q^2}|S_2|}{1-p^2})}{\sqrt{\pi}\Gamma(n)(1-p^2)^{\frac{1}{2}}(1-q^2)^{\frac{n}{2}-\frac{1}{4}}} \quad \text{(2-5.26)}$$

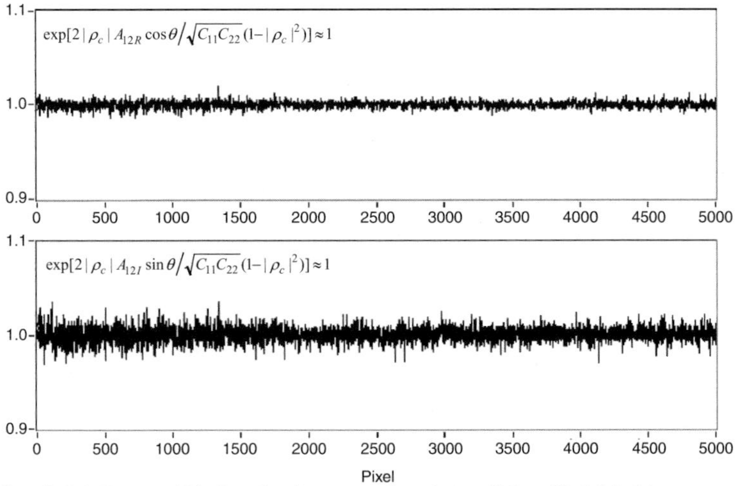

Fig. 2-5.1. Data validation for the approximations of Eqs. (2-5.24a,b)

Similarly, employing Eq. (2-5.23b) and integrating Eq. (2-5.21) with respect to A_{12R}, the *PDF*'s of the forth Stokes parameter V and its normalized parameter $S_3 = V/E[I]$ can be obtained. The *PDF* of S_3 for the linearly polarized incidence is written as

$$P_{S_3}(S_3) = \frac{2n^{n+\frac{1}{2}} |S_3|^{n-\frac{1}{2}} \exp(\frac{2n\sqrt{u^2+v^2}\sin\theta}{1-p^2} S_3) K_{n-\frac{1}{2}}(\frac{2n\sqrt{1-q^2}|S_3|}{1-p^2})}{\sqrt{\pi}\Gamma(n)(1-p^2)^{\frac{1}{2}}(1-q^2)^{\frac{n}{2}-\frac{1}{4}}} \quad \text{(2-5.27)}$$

The *PDF*'s of normalized third and fourth Stokes parameters, Eqs. (2-5.26, 2-5.27), are different from the integral forms of Touzi and Lopes (1996, their Eqs. (64,65), respectively), which cannot be analytically solved. Our approximations in Eqs. (2-5.23a,b) makes Eqs. (2-5.26, 2-5.27) analytically formulated.

Fig. 2-5.2 gives the *PDF*'s of S_0, S_1, S_2, S_3 for different look number n. It can be seen that increasing the look number makes the statistic distributions sharper and less skewed.

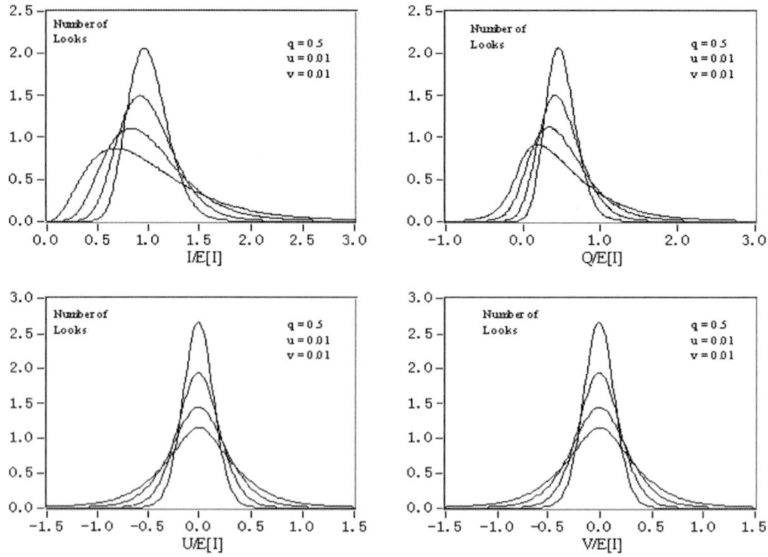

Fig. 2-5.2. The *PDF*'s of S_0, S_1, S_2, S_3 for different look number

The variance of the distribution is usually defined as

$$\delta^2 = \int (x - \bar{x})^2 PDF(x) dx \tag{2-5.28}$$

where $x = I, Q, U, V$. The skewness, as another important characteristic property of the distribution, is defined as

$$\gamma = \int (x - \bar{x})^3 PDF(x) dx \tag{2-5.29}$$

The skewness characterizes the asymmetry degree around the mean value. A positive value of skewness corresponds to the distribution with asymmetric tail extending to the right of the mean. A negative value of skewness corresponds to the distribution with asymmetric tail extending to the left.

2-5.3 Comparision with AirSAR Image Data

Three typical sites from a 16-looks AirSAR image are chosen to demonstrate the statistics of four Stokes parameters. The image is the AirSAR data acquired on August 12, 1993 over the Jack Pine area of Canada near the Prince Albert National Park (about $W103.33°, N53.9°$). The near range incident angle is $29.18°$. The range and azimuthal pixel spacing is 6.66*m* and 8.22*m*, respectively. A part of the image (1024×1024 pixels) is shown in Fig. 2-5.3, where the red, green and blue colors in the colorful image picture represent the total power at the P, L and C, respectively. The data histogram at three sites, marked by the squares A, B and C, sequentially, are chosen to compare with the theoretical *PDF* statistics. Our theoretical *PDF*'s are calculated by using the mean values of respective Stokes parameters measurements. Site A is a area of water. At site B, the backscattering at P band is higher than that at C band. Site B is most likely a sparse growth of trees. Backscattering from a tree is usually coming from direct scattering from the tree, interaction of tree trunk with the ground surface, and the rough land-surface. The interaction of tree trunk with the ground surface at lower P band is usually more significant than that at higher C band [Freeman and Derden 1996], and it yields higher scattering at P band. As the tree canopy becomes thick however, volumetric scattering from tree crowns are dominant and yields stronger scattering at C band than at P band, as happened at site C.

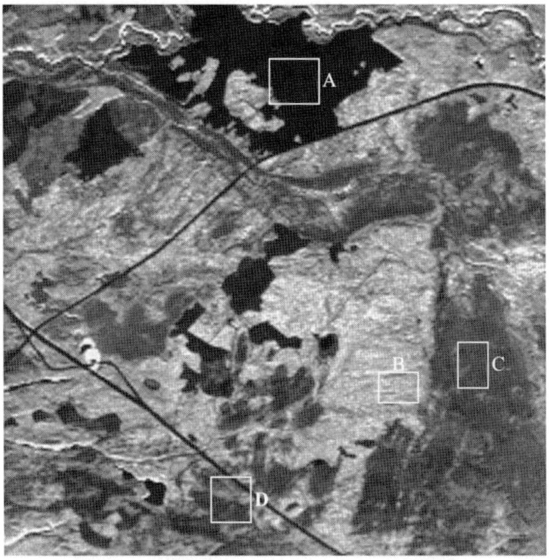

Fig. 2-5.3. The AirSAR image

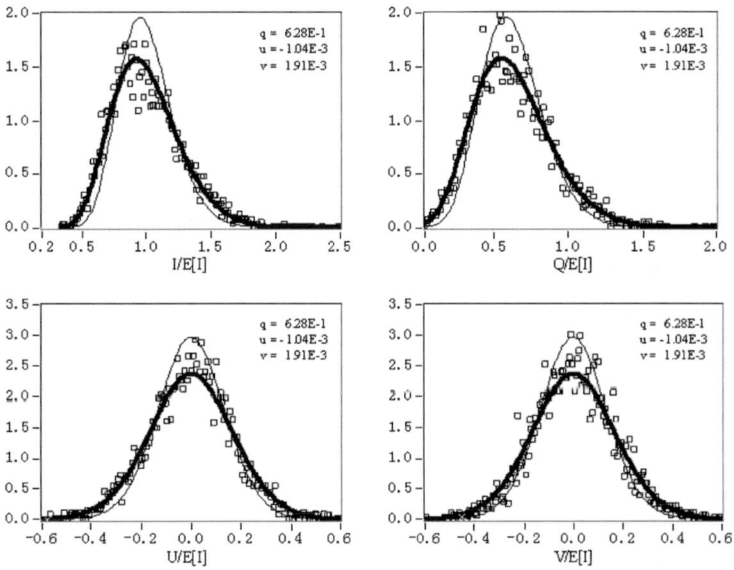

Fig. 2-5.4. The *PDF*'s of S_0, S_1, S_2, S_3 at C band on C Site

Figs. 2-5.4(a-d) show the data histogram of four nomalized Stokes parameters, S_0, S_1, S_2, S_3, at C band and on site C. The discrete points are the SAR data and the thin solid lines are the theoretical *PDF*'s with n=16. Both the lines and the discrete points are well matched. Note however that an effective looks number of n=10, instead of n=16, might give better match, which is indicated by the bold lines in Figs. 2-5.4(a-d). It is most likely due to the assumption of the Wishart distribution for uniformly random scatter media. Although the data are called 16 look data, they are actually the result of a 16 sample (pixel) integration in range and azimuth. Adjacent pixels are correlated if the data is Nyquist sampled. The *PDF*s are based on *N* independent samples of effective looks (corresponding to resolution cells, not pixels). Thus for this case the sampling theorem dictates that $N<16$. It is not yet known for certain , however, how to choose the effective number.

Figs. 2-5.5(a~d) show the statistics of four Stokes parameters, *I,Q,U,V*, at the C band at the water site A, the sparse tree site B, and the thick forest site C. The theoretical statistics are well matched with the SAR data for all different types of surface, especially by using the bold lines of the effective look number. It sounds reasonable to assign different effective look number for different surface types. Also, it might be a potential application to surface classification.

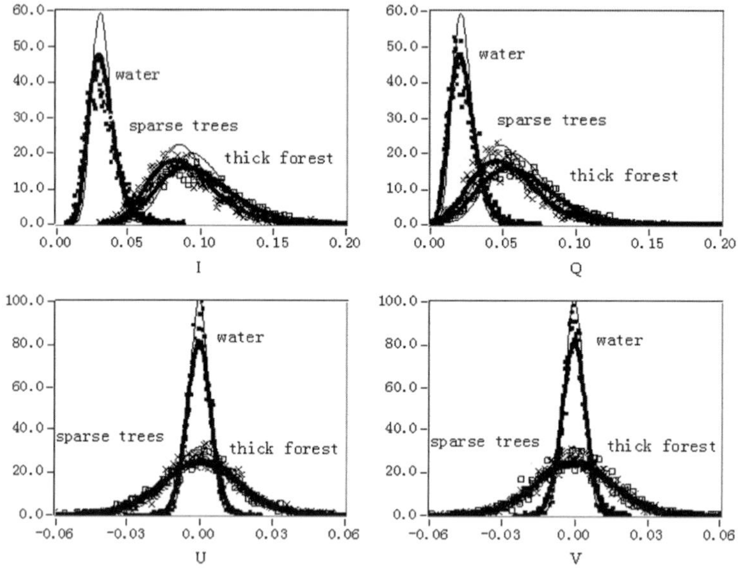

Fig. 2-5.5. The *PDF*'s of *I, Q, U, V* at C band on multi-Sites A, B and C

Figs. 2-5.6(a-d) and Figs. 2-5.7(a~d) give the *PDF*'s of *I,Q,U,V*, at multi-P,L,C bands at Sites B and C, respectively. It is interesting to see the difference between these figures. At Site B of sparse trees, stronger scattering at lower P band than higher C band gives statistic patterns as the sequences of C, L, P from the left to right in Figs. 2-5.6(a, b) and from the top to bottom in Figs.2-5.6(c, d). As a comparison, at Site C of thick forest with stronger scattering from tree crowns at C band, the statistic patterns turns the sequences as P, L, C from the left to right in Figs.2-5.7(a, b) and from the top to bottom in Figs. 2-5.7(c, d).

The variance, δ^2, of Eq. (2-5.28) and skewness, γ, of Eq. (2-5.29), on Sites A, B and C at multi- P, L and C bands are listed in Tables 2-5.1 and 2-5.2. It can be seen that the area of thick forest has larger variance and is more skewed at C band, and the area of sparse trees has larger variance and more skewed at P band. It is expected that the variance and skewness become larger as the amount of scattering increases.

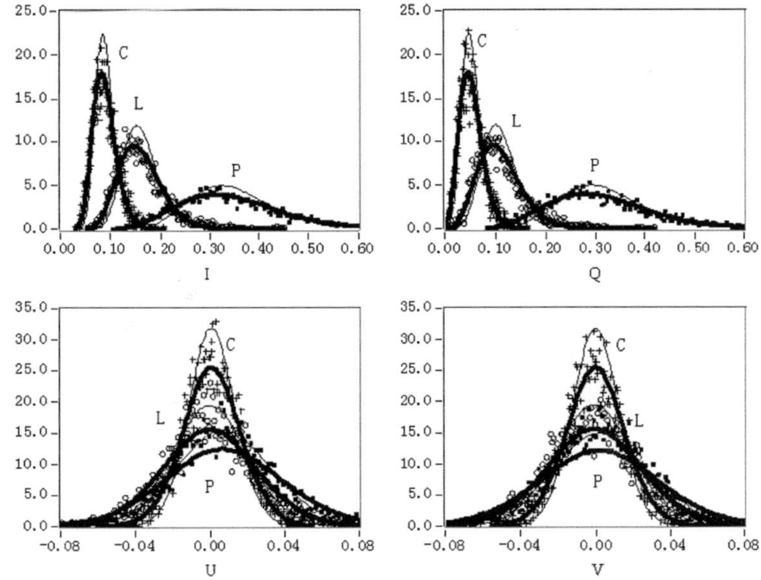

Fig. 2-5.6. The *PDF*'s of *I, Q, U, V* on Site B at multi-P, L, C bands

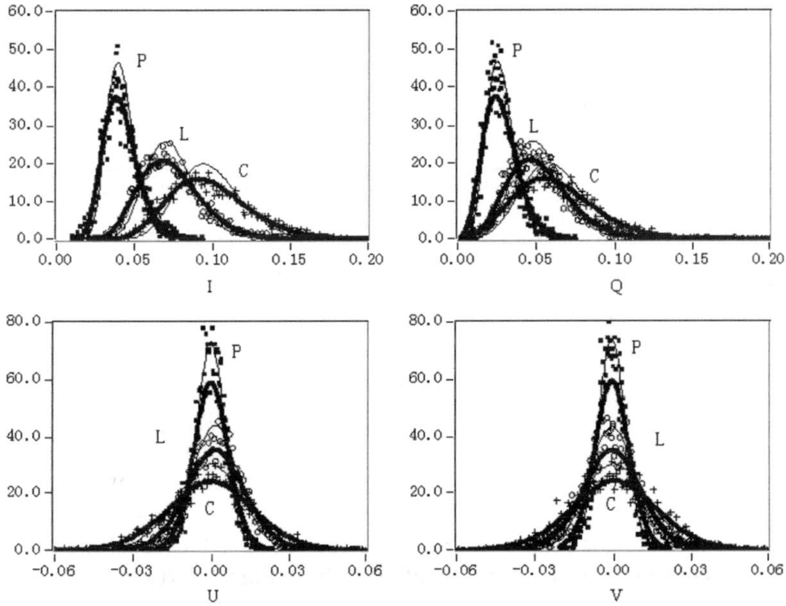

Fig. 2-5.7. The *PDF*'s of *I, Q, U, V* on Site C at multi-P,L,C bands

Table 2-5.1.

	Water	Sparse Trees	Thick Forest
C Band	0.119E-3	0.618E-3	0.753E-3
L Band	0.026E-3	2.432E-3	0.417E-3
P Band	0.008E-3	1.336E-2	0.126E-3

(a) Variance δ^2 of I

	Water	Sparse Trees	Thick Forest
C Band	0.091E-3	0.549E-3	0.675E-3
L Band	0.016E-3	2.431E-3	0.376E-3
P Band	0.007E-3	1.270E-2	0.100E-3

(b) Variance δ^2 of Q

	Water	Sparse Trees	Thick Forest
C Band	0.024E-3	0.242E-3	0.284E-3
L Band	0.003E-3	0.592E-3	0.117E-3
P Band	0.001E-3	0.760E-3	0.036E-3

(c) Variance δ^2 of U

	Water	Sparse Trees	Thick Forest
C Band	0.025E-3	0.250E-3	0.265E-3
L Band	0.003E-3	0.631E-3	0.115E-3
P Band	0.001E-3	0.687E-3	0.033E-3

(d) Variance δ^2 of V

Table 2-5.2.

	Water	Sparse Trees	Thick Forest
C Band	0.117E-5	0.951E-5	1.490E-5
L Band	0.018 E-5	1.378E-4	0.743E-5
P Band	0.005 E-5	1.524E-3	0.087E-5

(a) Skewness γ of I

	Water	Sparse Trees	Thick Forest
C Band	0.075E-5	0.819E-5	1.155E-5
L Band	0.008E-5	1.483E-4	0.611E-5
P Band	0.004E-5	1.366E-3	0.080E-5

(b) Skewness γ of Q

	Water	Sparse Trees	Thick Forest
C Band	0.001E-5	0.023E-5	0.021E-5
L Band	0.000E-5	0.088E-5	0.005E-5
P Band	0.000E-5	0.448E-5	0.000E-5

(c) Skewness γ of U

	Water	Sparse Trees	Thick Forest
C Band	0.001E-5	0.028E-5	0.021E-5
L Band	0.000E-5	0.220E-5	0.004E-5
P Band	0.000E-5	-0.006E-5	0.001E-5

(d) Skewness γ of V

Appendix 2-5A

$$\int_{-1}^{1}(1-x^2)^{\frac{\upsilon}{2}}\exp(-\alpha x)I_\upsilon(\beta\sqrt{1-x^2})dx = \sqrt{2\pi}\,\beta^\upsilon(\alpha^2+\beta^2)^{-\frac{\upsilon}{2}-\frac{1}{4}}I_{\upsilon+\frac{1}{2}}(\sqrt{\alpha^2+\beta^2}) \quad (2\text{-}5A.1)$$

From Gradshteyn and Ryzhik (1980, page 721: Eq.6.645-2)

$$\int_{1}^{\infty}(x^2-1)^{\frac{\upsilon}{2}}\exp(-\alpha x)J_\upsilon(\beta\sqrt{x^2-1})dx = \sqrt{\frac{2}{\pi}}\beta^\upsilon(\alpha^2+\beta^2)^{-\frac{\upsilon}{2}-\frac{1}{4}}K_{\upsilon+\frac{1}{2}}(\sqrt{\alpha^2+\beta^2}) \quad (2\text{-}5A.2)$$

From Gradshteyn and Ryzhik (1980, page 318: Eq.3.382-2)

$$\int_{u}^{\infty}(x-u)^\upsilon e^{-\mu x}dx = \mu^{-\upsilon-1}e^{-u\mu}\Gamma(\upsilon+1) \quad (2\text{-}5A.3)$$

From Gradshteyn and Ryzhik (1980, Pagc 340: Eq.3.471-9)

$$\int_0^\infty x^{\nu-1} e^{-\frac{\beta}{x}-\gamma x} dx = 2\left(\frac{\beta}{\gamma}\right)^{\frac{\nu}{2}} K_\nu(2\sqrt{\beta\gamma}) \qquad (2\text{-}5A.4)$$

From Gradshteyn and Ryzhik (1980, P.705: Eq.6.596-3)

$$\int_0^\infty K_\nu(\alpha\sqrt{x^2+z^2}) \frac{x^{2\mu+1}}{\sqrt{(x^2+z^2)^\nu}} dx = \frac{2^\mu \Gamma(\mu+1)}{\alpha^{\mu+1} z^{\nu-\mu-1}} K_{\nu-\mu-1}(\alpha z) \qquad (2\text{-}5A.5)$$

2-6 Image Simulation of High Resolution Radar for Fully Polarimetric Scattering from Heterogeneous Canopy Surface

Polarimetric imaginary of high resolution SAR is one of most important advance in microwave remote sensing [Zebker and Van Zyl 1991]. It has been extensively applied to surface classification, inversion of geophysical and hydrological parameters, and target detection, etc. Using electromagnetic theory, surface canopy is modeled as randomly oriented non-spherical scatterers, e.g. leaves as disks or needles, branches as needles, and trunks as vertical cylinders, etc. [Jin et al. 1999]. Employing the 4×4 dimensional Mueller matrix solution, fully polarized backscattering of the scattered Stokes vector with four Stokes parameters can be obtained. However, this modeling is usually suitable for uniformly stratified canopy covers. The problem becomes more complex if the three-dimensional (3D) structure and non-uniform distribution of tree canopy, such as tree number, location, gaps, clustering, etc., are taken into account. Image texture demonstrates spatial variation of radar echo to canopy heterogeneity. Sensitivity of backscattering on canopy heterogeneity is a important information for image analysis and applications such as geology study and target detection in the Earth surface observation.

Simulation of radar image yields a direct approach for analysis of surface heterogeneity. Sun and Ranson (1995) divided a scatter into cells according to a certain resolution and, using the ray tracing method and range calculation, to assign backscattering from the cells within the same range intervals to the image pixels. This mapping yields the image simulation with spatial variation. However, image simulation of heterogeneous scatterers has not been studied yet, and moreover, fully polarimetric scattering has not been taken into account.

This session, using fully polarimetric Mueller matrix solution [Jin 1994, Jin et al. 1999] and ray tracing method, simulates radar images for fully polarimetric scattering from heterogeneous canopy surface. Non-uniform 3D structures and non-uniform distribution of multi-components of surface canopy are taken into account. Simulated image texture demonstrates spatial variation of radar echoes due to canopy 3D-heterogeneity, and is also compared with an AirSAR image at P,L,C bands [Jin and Zhang 1999].

2-6.1 Image Simulation of High Resolution Radar

As a incident wave $\overline{E}_i$ with the polarization defined by the elliptic angle χ and orientation angle ψ is incident upon a layer model of the vegetation canopy, the Mueller matrix solution has been obtained [Jin 1994]. The Mueller matrix is contributed by five scattering mechanisms such as:

$$\overline{\overline{M}} = \sum_{n=1}^{5} \overline{\overline{M}}_n \tag{2-6.1}$$

where $\overline{\overline{M}}_1$ is direct volumetric scattering from tree crown of leaves and branches, $\overline{\overline{M}}_2$ is direct surface scattering from underlying land surface (in previous sessions, this part was not taken into account), $\overline{\overline{M}}_3$ is direct volumetric scattering from trunks; $\overline{\overline{M}}_4$ is coming from interaction of crown- underlying land surface; $\overline{\overline{M}}_5$ is from interaction of vertical trunks-underlying surface. For any polarized incidence, the co-polarized (σ_c) and cross-polarized (σ_d) backscattering can be obtained. It is important to simulate the radar image of non-uniformly 3D structure.

Fig. 2-6.1 shows a scene of trees canopy with 3D structure. Non-uniform distribution such as tree location, gaps and clustering etc., should be sensitive to affect radar image texture. The whole space of canopy cover is divided into cells, and the cell to be as gap, crown or trunk, depends on its spatial position. As electromagnetic wave propagates through a cell, attenuation has been taken into account by calculation of the extinction through crown cells or along the path between the trees.

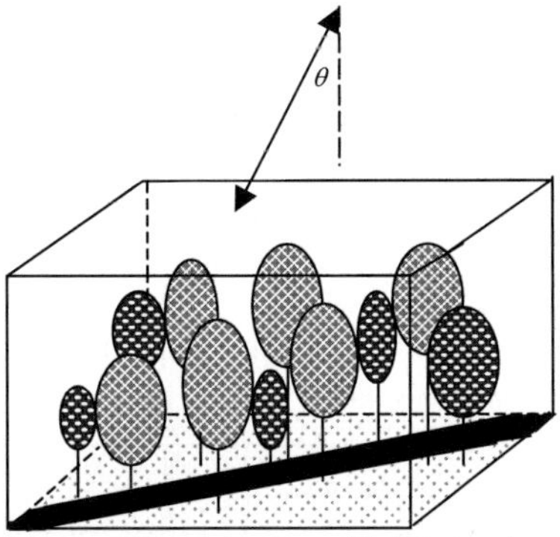

Fig. 2-6.1. A scene of trees canopy with 3D structure

We consider a *x-y* field of 100*m*×100*m*, where 1000 trees are randomly located determined by the Monte-Carlo method. Since trees cannot overlap each other, the new tree is omitted if its location with the distance of previously produced tree is less than 1.5*m*; and new tree is assumed as small one if the distance is larger than 1.5*m* and less than 2.5*m*; and new tree is assumed as big one if the distance is larger than 2.5*m*. A spatial distribution of tree location is thus randomly produced. The radius of big trees is assumed as 50*cm* and the height is randomly chosen between 14.4*m* and 21.6*m*.

The radius of small trees is as 25*cm* and the height is between 7.2*m* and 10.8*m*. The dielectric constant of trunks is $(15+5i)\varepsilon_0$ following Matzler (1994). The crown shape is simply assumed as spheroid (it can be assumed as any shape, actually). The radius and height of crowns follow the empirical formulations of Sun (1998) to have the half width of spheroid as r=0.15607×H, and a half length as b=0.22517×H, where H is tree height. Trees are assumed as conifer, and leaves have the radius 1*mm* and length 5*cm*, the dielectric constant $(22.5+i7.5)\varepsilon_0$, following Matzler (1994). The dielectric constant of underlying media is taken as $(15+i2)\varepsilon_0$, the surface roughness is assumed with variance 1*cm* and correlation length 10*cm*. The Euler angles of spatial orientation of leaves are assumed as γ uniformly over the range $(0^\circ, 360^\circ)$ and β over $(0^\circ, 80^\circ)$.

Fig. 2-6.2 shows a simulation method for a single tree. Spatial resolution is assumed as 0.5*m*. Thus, the crown and trunk are divided into the cells of $(0.5\ m)^3$. As radar wave hits on the crown, direct volumetric scattering of the crown cells are calculated by using the Mueller matrix

$\overline{\overline{M}}_1$, and, following ray tracing within the same range intervals, are incoherently added to map into the pixels of range A. As radar waves emerges from the crown and hits the underlying surface to make the shadowing range B, it will yield attenuation of surface scattering calculated by $\overline{\overline{M}}_2$. As the waves hit the trunk, direct volumetric scattering from the trunk is calculated by $\overline{\overline{M}}_3$, and is mapped into the line *ao*, (which is generally very small in that specular direction). Interaction of crown-underlying surface is calculated by $\overline{\overline{M}}_4$, and is mapped into the area O. Interaction of trunk-underlying surface is calculated by $\overline{\overline{M}}_5$ and has been proved to have the same range of the tree base o [Sun and Ranson 1995]. All back-scatterings associated with respective $\overline{\overline{M}}_n (n=1,\cdots,5)$ are assigned to the image pixels according to the range of the radar signals. It yields a image simulation. This approach is also applicable to simulate image of artificial target embedded in tree canopy.

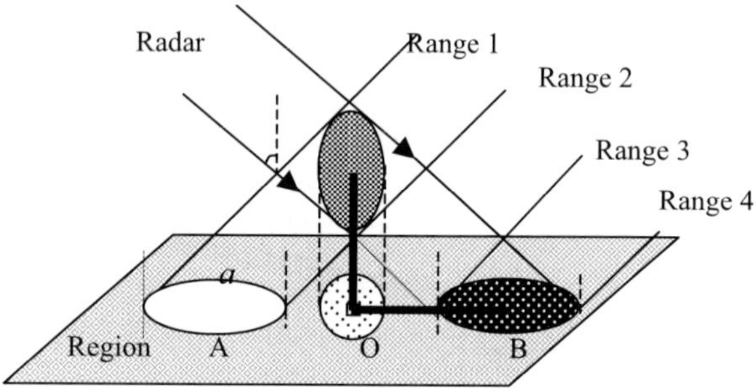

Fig. 2-6.2. Image simulation for a single tree

2-6.2 Image Simulation of Non-Uniform Surface Canopy

Fig. 2-6.3 presents a simulated radar image of total echo power of *hh, vv, hv* polarization at C band with incidence 30° for a tree distribution produced in the last session.

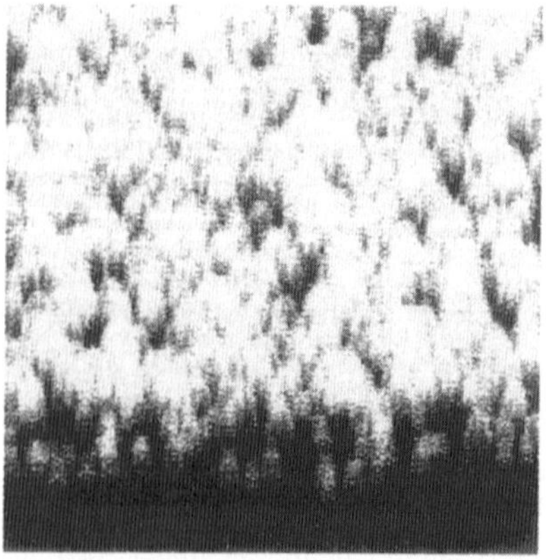

Fig. 2-6.3. A simulated radar image of total *vv,hh,hv* echo power

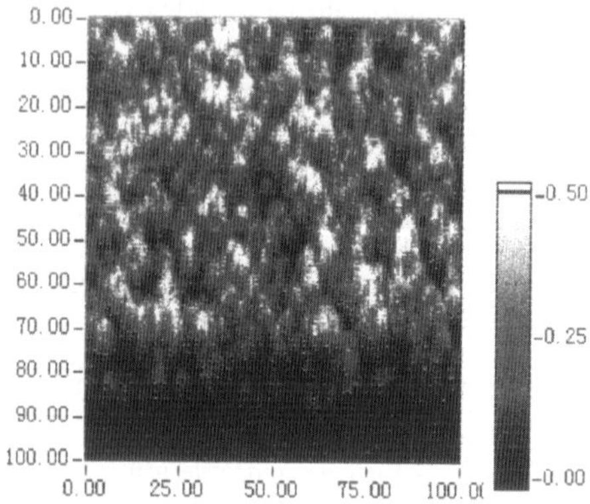

Fig. 2-6.4. Cross-polarizedσ_d of circularity polarized incidence

In the color image Fig. 2-6.3, red color is used for σ_{hh}, green is for σ_{vv} and blue is for σ_{hv}. Bright color indicates the echo strength. Yellow color is produced as three colors are mixed. From Fig. 2-6.3, image texture demonstrates spatial heterogeneity.

It is known from calculation of Eq. (2-6.2) that cross-polarized backscattering σ_d for circularity polarized incidence ($\chi=\pm 45^\circ$) is stronger than co-polarized σ_c. Fig. 2-6.4 shows a simulated image of σ_d at C band with incidence 30°. Random trees can produce strong cross-polarized σ_d for circularity polarized incidence. It is also suggested to use the circularity polarized incidence to detect the target which might produce significant cross-polarized echo.

Generally speaking, *hh*-copolarized scattering σ_{hh} of crown-surface and trunk-surface are much stronger than σ_{vv}, almost higher than 20*dB* and 8*dB*, respectively. Thus, change of the underlying surface, e.g. its dielectric constant, will especially affect σ_{hh}. Fig. 2-6.5(a) is a simulated image of *hh,vv,hv* total echo power as the underlying media is land with $(15+i2)\varepsilon_0$.

As a comparison, Fig. 2-6.5(b) shows the image as the underlying media is replaced by water with $(72+i22)\varepsilon_0$. It can be seen that much contributions enhanced (about 18*dB*) due to interaction of crown-surface (as region O in Fig. 2-6.2). In the color image, this contribution enhances red color of σ_{hh}, and whole picture tends to more yellow. This image might illustrate the case of flooding under canopy covers.

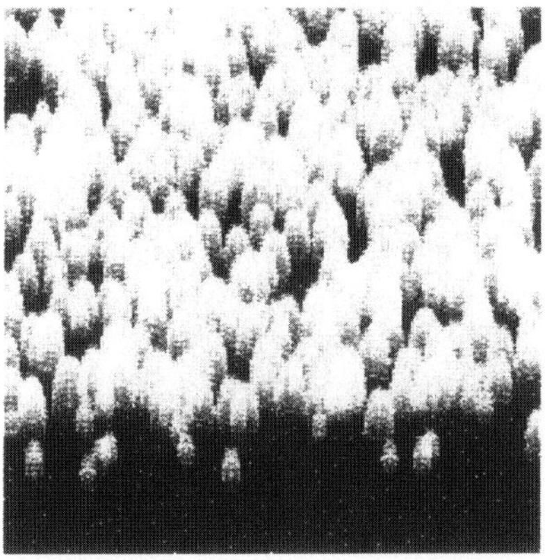

Fig. 2-6.5(a). Image with underlying land

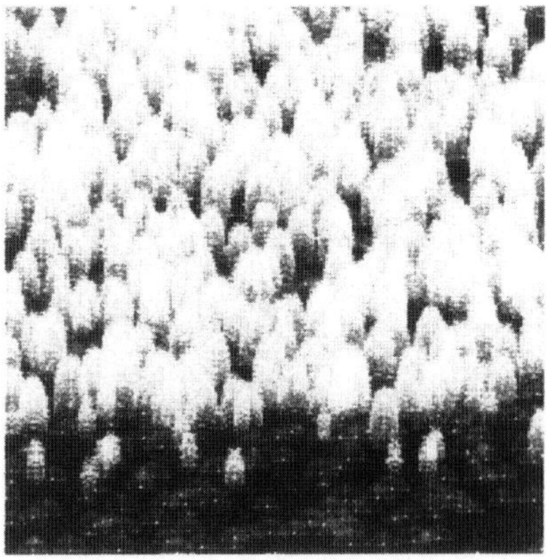

Fig. 2-6.5(b). Image with underlying water

Now, there is a road of 8.5*m* width located diagonally in the field. Figs. 2-6.6(a,b) give the simulated images of *hh, vv, hv* total echo power at C band with incidence of 30° and 60°, respectively. Shadowing due to tree crowns over the road detection can be seen. As incident angle increases, the ranges from all cells of the crown become close. It yields smaller shadows and the road becomes detectable.

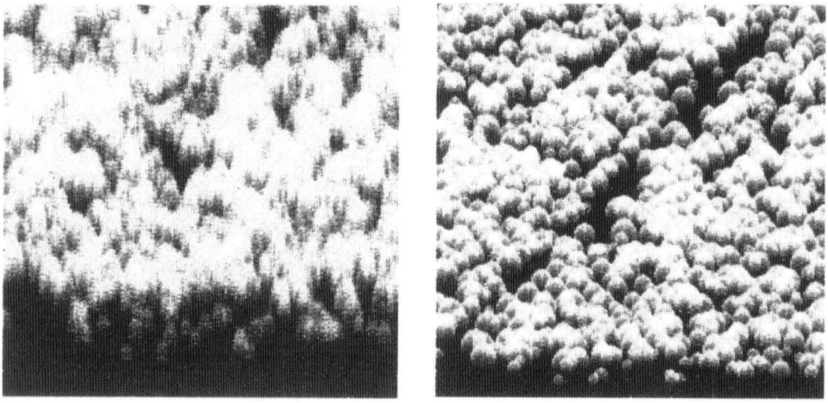

(a) incident angle 30° (b) incident angle 60°

Fig. 2-6.6. Image of tree and road

Fig. 2-6.7(a) gives a part of the AirSAR image of the *hh, vv, hv* total power at P, L, C bands measured over a field of the Price Albert National Park of Canada ($104.7°W$, $53.9°N$) in January 27, 1994. Spatial resolution is $8m$.

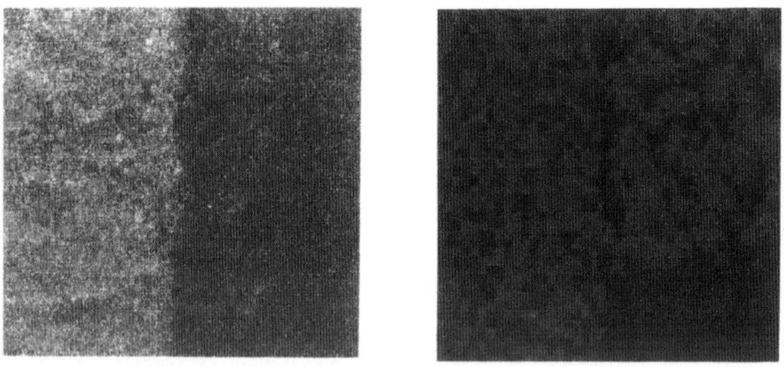

(a) An AirSAR image. (b) An simulated image.

Fig. 2-6.7. The AirSAR image and its simulated image

In the color picture, red color is the echo at P band, green color is at L band, and blue color is at C band. Generally, as frequency increases, scattering becomes stronger. That is the reason why scattering of the right half (thick tree area) in Fig. 2-6.7(a) is dominated by scattering at high C band (blue color). But, on the left half of Fig. 2-6.7(a), scattering at low P band is stronger than L and C bands, and color picture shows red dominance. Since there are less trees and homogeneous and dissipative covers (such as wet snow), volumetric scattering is smaller at all P, L, C bands, and main scattering is coming from the underlying bottom media. Let the dielectric constant of uniform cover medium assumed as $\varepsilon' + i\varepsilon''$ and layer depth as d_o. The attenuation of incident and re-bounced (scattered) waves through the homogeneous layer is written as: $A = \exp(-4k''d_o \sec\theta)$, where the imaginary part of the wave number of the homogeneous medium $k''(\approx 0.5(2\pi v/30)\varepsilon''/\sqrt{\varepsilon'})$ is proportional to the frequency ν (*GHz*). Thus, the electromagnetic waves at the C band is much seriously attenuated comparing with low P band, and scattering is mainly dominated by the waves at the P band. That is a reason why the left half of Fig. 2-6.7(a) becomes red.

Fig. 2-6.7(b) shows a simulation for this case where the right half is scattering of thick trees dominated at C band, and the left half is scattering from the bottom rough surface with attenuation through the homogeneous cover.

2-7 References

Barakat, R. (1987), "Statistics of the Stokes parameters", *Journal of Optical Society of American A*, 4(7): 256-1263.

Boerner, W.M., H. Mott and E. Luneburg (1997), "Polarimetry in remote sensing: basic and applied concepts", *Proceedings International Geoscience and Remote Sensing Symposium,* (IGARSS'1997, Singapore: Piscataway), 3: 1401-1403.

Chandrasekhar, S. (1960), *Radiative Transfer*, New York: Dover Pub.

Chang, M. and Y.Q. Jin (2002), "The temporal Mueller matrix solution for polarimetric scattering from a layer of random non-spherical particles with a pulse incidence", *Journal of Quantitative Spectroscopy and Radiative Transfer*, 74: 339-353.

Chang, M. and Y.Q. Jin Y.Q. (2003), "Temporal Mueller matrix solution for polarimetric scattering from inhomogeneous random media of non-spherical scatterers under a pulse incidence", *IEEE Transactions on Antenna and Propagation*, 51(4): 820-832.

Cloude, S.R. (1986), "Group theory and polarization algebra", *Optik*, 75(1): 26-36.

Cloude, S.R. and K.P. Papathanassiou (1998), "Polarimetric SAR interferometry", *IEEE Transactions on Geoscience and Remote. Sensing*, 36(5): 1551-1565.

Dubois, P.C. and L. Norikane (1987), "Data volume reduction for image radar polarimetry", *Proceedings International Geoscience and Remote Sensing Symposium* (IGARSS'87, Ann Arbor MI), 691-696.

Evans, D.L., T.G. Farr, J.J. Van Zyl and H.J. Zebker (1988), "Image radar polarimetry: analysis and application", *IEEE Transactions on Geoscience and Remote Sensing*, 26(5): 714-789.

Freeman, A. and S.L. Durden (1996), "A three-component scattering model for polarimetric SAR data", *IEEE Transactions on Geoscience and Remote Sensing*, 36(3): 963-973.

Frery, A.C. and H.J. Müller (1997), "A model for extremely heterogeneous clutter", *IEEE Transactions on Geoscience and Remote Sensing*, 35(3): 648-659.

Fung, A.K. (1994), *Microwave Scattering and Emission Models and Their Applications*, Mass: Artech.

Goodman, N.R. (1963), "Statistic analysis based on a certain multivariate complex Gaussian distribution (An introduction)", *Annual Mathematical Statistics*, 34(152): 152-180.

Gradshteyn, I.S. and I.M. Ryzhik (1980), *Table of integrals, Series, and Products* , New York: Academic Press.

Hong, S.T. *et al.* (1977), "Plane wave propagation through random media", *IEEE Transactions on Antennas and Propagation,* 25(6): 822-828.

Ishimaru, A. (1978), *Wave Propagation and Scattering in Random Media*, New York: Academic Press.

Ito, S. (1980), "On the theory of pulse wave propagation in media of discrete random Scatterers", *Radio Science*, 15(5): 893-901.

Jin, Y.Q. (1992), "A Mueller matrix approach to complete polarimetric scattering from a layer of non-uniformly oriented, non-spherical Scatterers", *Journal of Quantitative Spectroscopy and Radiative Transfer*, 48(3): 295-306.

Jin, Y.Q. (1994), *Electromagnetic Scattering Modelling for Quantitative Remote Sensing* , Singapore: World Scientific.

Jin, Y.Q. (1994), "Polarimetric scattering from a layer of random clustters of small spheroids", *IEEE Transaction on Antenna and Propagation*, 42(8): 1138-1144.

Jin, Y.Q. and Cloude S. R. (1994), "Numerical eigenanalysis of the coherency matrix for a layer of random nonspherical scatterers", *IEEE Transactions on Geoscience and Remote Sensing*, 32(6): 1179-1185.

Jin, Y.Q., N.X. Zhang and M. Chang (1999), "Simulation of polarimetric scattering from tree canopy and cigenanalysis of microwave SAR imagery", *Chinese Journal of Electronics*, 8(3): 306-312.

Jin, Y.Q., N.X. Zhang and M. Chang (1999), "The Mueller and coherency matrices solution for polarimetric scattering simulation for tree canopy in SAR imaging at C band", *Journal of Quantitative Spectroscopy and Radiative Transfer*, 63(2): 599-612.

Jin, Y.Q. (2000), *Information of Electromagnetic Scattering and Radiative Transfer in Natural Media*, Beijing: Science Press.

Jin, Y.Q. and W. Zhang (2000), "Image simulation of high-resolution radar for fully polarimetric scattering from heterogeneous canopy surface", *Chinese Journal of Electronics*, 9(3): 332-336.

Jin, Y.Q. and N.X. Zhang (2002), "Statistics of four Stokes parameters in multi-look polarimetric SAR imagery", *Canadian Journal of Remote Sensing*, 28(4): 610-619.

Jin, Y.Q. and F. Chen (2002), "Polarimetric scattering indexes and information entropy of the SAR imagery for surface monitoring", *IEEE Transactions on Geoscience and Remote Sensing*, 40(11): 2502-2506.

Jin, Y.Q. and F. Chen (2003), "Scattering simulation for inhomogeneous layered canopy and random Targets beneath canopies by using the Mueller matrix solution of the pulse radiative transfer", *Radio Science*, 38(6): 1107-1116.

Jin, Y.Q., F. Chen and M. Chang (2004), "Retrievals of underlying surface roughness and moisture for stratified vegetation canopy using polarized pulse echoes in the specular direction", *IEEE Transactions on Geoscience and Remote Sensing*, 42(2): 426-433.

Lee, J.S., K.W. Hoppel, S.A. Mango and A.R. Miller (1994), "Intensity and phase statistics of multilook polarimetric and interferometric SAR imagery", *IEEE Transactions on Geoscience and Remote Sensing*, 32(5): 1017-1027.

Lee, J.S., M.R. Grunes and G. De Grandi (1999a), "Polarimetric SAR speckle filtering and its implication for classification", *IEEE Transactions on Geoscience and Remote Sensing*, 37(5): 2363-2373.

Lee, J.S. *et al.* (1999b), "Unsupervised classification using polarimetric decomposition and the complex Wishart classifier", *IEEE Transactions on Geoscience and Remote Sensing*, 37(5): 2249-2257.

Liu, C.H. and K.C. Yeh (1978), "Pulse propagation in random media", *IEEE Transactions on Antennas and Propagation*, 26(4): 561-566.

Matzler, C. and U. Wegmuller (1993), *Active and Passive Microwave Signature Catalogue (2-12 GHz)*, Institute of Applied Physics, University of Berne.

Matzler, C. (1994), "Microwave(1-100GHz) dielectric model of leaves", *IEEE Transactions on Geoscience and Remote Sensing*, 32(5): 947-949.

Nighiem, S.V., S.H. Yueh, R. Kwok and F.K. Li (1992), "Symmetry properties on polarimetric remote sensing", *Radio Science*, 27(5): 693-701.

Oguchi, T. and S. Ito (1990), "Multiple scattering effects on the transmission and reflection of millimeter pulse waves in rain", *Radio Science.*, 25(3): 205-216.

Papathanassiou, K.P., and S.R. Cloude (2001), "Single baseline polarimetric SAR interferometry", *IEEE Transactions on Geoscience and Remote Sensing*, 39(11): 2352-2363.

Reigber, A. and A. Moreira (2000), "First demonstration of airborne SAR tomography using multi- baseline L band data", *IEEE Transactions on Geoscience and Remote Sensing*, 38(5): 2142-2152.

Sarabandi, K. and Y.C. Lin (2000), "Simulation of interferometric SAR response for characterizing the scattering phase center statistics of forest canopics", *IEEE Transactions on Geoscience and Remote Sensing*, 38(1): 115-125.

Sheen, D.R. and L.P. Jonhston (1992), "Statistical and spatial properties of forest clutter measured with polarimetric synthetic aperture radar (SAR)", *IEEE Transactions on Geoscience and Remote Sensing*, 30(3): 578-588.

Sun, G. and K.J. Ranson (1995), "A three-dimensional radar backscatter model of forest canopies", *IEEE Transactions on Geoscience and Remote Sensing*, 33(2): 372-382.

Sun, G. and K.J. Ranson (1998), "Radar modelling of forest spatial patterns", *International Journal of Remote Sensing*, 19(9): 1769-1791.

Sun, G., and K.J. Ranson (2000), "Modeling lidar return from forest canopies, *IEEE Transactions on Geoscience and Remote Sensing*, 38(6): 2617-2626.

Touzi, R. and A. Lopes (1996), "Statistics of the Stokes parameters and of the complex coherence parameters in one-look and multilook speckle fields", *IEEE Transactions on Geoscience and Remote Sensing*, 34(2): 519-531.

Treuhaft, R.N., B.E. Law and G.P. Asner (2000), "Vertical structure of vegetated land surfaces from interferometric and polarimetric radar", *Radio Science.*, 35(1): 141-177.

Tsang, L., J.A. Kong and R. Shin (1985), *Theory of Microwave Remote Sensing*, New York: John Wiley.

Tsang, L., J.A. Kong and K.H. Ding (2000), *Scattering of Electromagnetic Waves: Theories and Applications*, New York: John Wiley.

Tsang, L., J.A. Kong, K.H. Ding and C.O. Ao (2001), *Scattering of Electromagnetic Waves: Numerical Simulations*, New York: John Wiley.

Tsang, L. and J.A. Kong (2001), *Scattering of Electromagnetic Waves: Advanced Topics*, New York: John Wiley.

Ulaby, F.T. and M.C. Dobson (1989), *Handbook of Radar Scattering Statistics for Terrain* , Norwood Mass.: Artech House.

Ulaby, F.T., and C. Elachi (1990), *Radar Polarimetry for Geoscience Applications*, Mass: Artech House, 53-67.

Van der Mee, C.V.M. and J.W. Hovenier (1992), "Structure of matrices transforming Stokes parameters", *Journal of Mathematical Physics*, 33(10): 3574-3584.

Wang, J.R. and B.J. Choudhury (1981), "Remote sensing of soil moisture over bare fields at 1.4 GHz frequency", *Journal of Geophysical. Research,* 86(4): 5277-5282.

Whitman, G.M., F. Schwering, A. Triolo *et al.* (1996), "A transport theory of pulse propagation in a strongly forward scattering random medium", *IEEE Transactions on Antennas and Propagation*, 44(1): 118-127.

Zebker, H.A. and J.J. Van Zyl (1991), "Image radar polarimetry: a review", *Proceedings of IEEE*, 79(11): 1583-1605.

Chapter 3

SAR Imagery and Applications 91

3-1 Terrain Topographic Inversion from Single-Pass Polarimetric SAR Image Data

Fully polarimetric SAR imagery has been extensively studied for terrain surface classification. The 2×2-D (Dimensional) complex scattering amplitude functions F_{pq} (*p,q*=*v,h*), and 4×4-D real Mueller matrix M_{ij} (*i,j*=1,…,4) can be measured [Jin 1994]. Co-polarised or cross-polarised back-scattering signature is the function of the incidence wave with the ellipticity angle χ and orientation angle ψ. Recently, polarimetric SAR image data has been utilized to generate digital surface elevation and to invert terrain topography. When the terrain surface is flat, polarimetric scattering signature has the maximum largely at the orientation angle $\psi = 0$. However, it has been shown that as the surface is tilted, the orienattion angle ψ at the maximum of co-polarized (co-pol) or cross-polarized (cross-pol) signature can shift from $\psi = 0$. This shift can be applied to convert the surface slopes. Making use of the assumption of real co-pol and zero cross-pol scattering amplitude functions [Lee et al. 1998], the ψ shift is expressed by the real scattering amplitude functions. Since both the range and azimuth angles are coupled, two- or multi-pass SAR image data are required for solving two unknowns of the surface slopes. This approach has been well demonstrated for inversion of digital elevation mapping (DEM) and terrain topography by using airborne SAR data [Schuler et al. 1998, 2000, Lee et al. 2000].

However, scattering signature is an ensemble average of echo power from random scatter media. Measurable Stokes parameters as the polarized scattering intensity should be directly related to the ψ shift. In this ssession, using the Mueller matrix solution, the ψ shift is newly derived as a function of three Stokes parameters, I_{vs}, I_{hs}, U_s, which are measurable by the polarimetric SAR imagery. This formula is implied to be more general than that in Lee et al. (1998) and Schuler (2000).

Using the Euler angles transformation between the principal and local coordinates, the orientation angle ψ is related with both the range and azimuth angles, β and γ, of the tilted surface pixel and radar viewing geometry. These results are consistent with Lee et al. (1998) and Schuler (2000), but are more general.

It is proposed that the linear texture of tilted surface alignment is used to specify the azimuth angle γ. The adaptive threshold method and image morphological thinning algorithm are applied to determine the azimuth angle γ from image linear textures. Thus, the range angle β is then solved, and both β and γ are utilized to obtain the azimuth slope and range slope. Then, the full multi-grid algorithm is employed to solve the Poisson equation of DEM and produce the terrain topography from a single pass Polarimetric SAR image [Jin and Luo 2003].

3-1.1 The Shift of Orientation Angle as a Function of Stokes Parameters

Consider a polarized electromagnetic wave incident upon the terrain surface at $(\pi - \theta_i, \varphi_i)$. Incidence polarization is defined by the ellipticity angle χ and orientation angle ψ [Jin 1994]. The 2×2-D complex scattering amplitude functions are obtained from polarimetric measurement as

$$\overline{E}_s(\theta_s, \varphi_s) = \frac{e^{ikr}}{4\pi r} \overline{\overline{F}}(\theta_s, \varphi_s; \pi - \theta_i, \varphi_i) \cdot \overline{E}_i(\pi - \theta_i, \varphi_i; \chi, \psi) \tag{3-1.1}$$

where the elements of scattering amplitude matrix, F_{pq} (*p,q*=*v,h*), take account for volumetric and surface scattering from layered scatter media. In SAR imagery, $\overline{\overline{F}}$ in Eq. (3-1.1) is measured and is used to construct the real Mueller matrix as follows

$$\bar{I}_s(\theta_s,\varphi_s)=\overline{\overline{M}}(\theta_s,\varphi_s;\pi-\theta_i,\varphi_i)\cdot\bar{I}_i(\pi-\theta_i,\varphi_i;\chi,\psi) \qquad (3\text{-}1.2)$$

where the normalized incident Stokes vector is written as

$$\begin{aligned}\bar{I}_i(\chi,\psi)&=[I_{vi},I_{hi},U_i,V_i]^T\\&=\left[0.5(1-\cos 2\chi\cos 2\psi),0.5(1+\cos 2\chi\cos 2\psi),-\cos 2\chi\sin 2\psi,\sin 2\chi\right]^T\end{aligned} \qquad (3\text{-}1.3)$$

The elements of the Mueller matrix, M_{ij} $(i,j=1,\ldots,4)$, are expressed by the ensemble averages of real or imagery parts of $<F_{pq}F_{st}^*>$ $(p,q,s,t=v,h)$. It is noted that co-pol terms $<|F_{pp}|^2>$ are always much larger than cross-pol terms, such as $<F_{pq}F_{st}^*>$ $(s\neq t)$.

The co-pol backscattering coefficient $\sigma_c(\theta_i,\pi-\varphi_i;\pi-\theta_i,\varphi_i)$ can be obtained as

$$\sigma_c=4\pi\cos\theta_i P_n \qquad (3\text{-}1.4)$$

where

$$P_n=0.5\left[I_{vs}(1-\cos 2\chi\cos 2\psi)+I_{hs}(1+\cos 2\chi\cos 2\psi)+U_s\cos 2\chi\sin 2\psi+V_s\sin 2\chi\right] \qquad (3\text{-}1.5)$$

When the terrain surface is flat, co-pol backscattering σ_c versus the incidence polarization (χ,ψ) has the maximum at $\psi=0$ [Lee et al. 1998]. However, it has been shown that as the surface is tilted, the orientation angle ψ for the σ_c maximum is shifted from $\psi=0$ [Lee at al. 1998, 2000, Schuler et al. 1998, 2000].

Let $\partial P_n/\partial\psi=0$ at the maximum σ_c and $\chi=0$ of symmetric case, from Eq.(3-1.5), it yields

$$0=-(I_{hs}-I_{vs})\sin 2\psi+U_s\cos 2\psi+0.5(I_{hs}+I_{vs})'+0.5U_s'\sin 2\psi+0.5(I_{hs}-I_{vs})'\cos 2\psi \qquad (3\text{-}1.6)$$

where the superscript of prime denote $\partial/\partial\psi$. Calculations of Eq. (3-1.2) and comparison with the terms on RHS (right hand side) of Eq. (3-1.6) can show that

$$0.5(I_{hs}+I_{vs})'\sim(M_{13}+M_{23})\cos 2\psi\sim(\mathrm{Re}<F_{vv}F_{vh}^*>+\mathrm{Re}<F_{hv}F_{hh}^*>)\cos 2\psi \qquad (3\text{-}1.7a)$$

$$0.5U_s'\sim M_{33}\cos 2\psi\sim\mathrm{Re}<F_{vv}F_{hh}^*+F_{vh}F_{hv}^*>\cos 2\psi \qquad (3\text{-}1.7b)$$

$$(I_{hs}-I_{vs})\sim 0.5(M_{11}+M_{22})\cos 2\psi\sim 0.5(<|F_{vv}|^2>+<|F_{hh}|^2>)\cos 2\psi \qquad (3\text{-}1.7c)$$

$$0.5(I_{hs}-I_{vs})'\sim 0.5(M_{13}-M_{23})\cos 2\psi\sim 0.5(\mathrm{Re}<F_{vv}F_{vh}^*>-\mathrm{Re}<F_{hv}F_{hh}^*>)\cos 2\psi \qquad (3\text{-}1.8a)$$

$$U_s\sim 0.5(M_{31}+\mathrm{M}_{32})\sim\mathrm{Re}<F_{vv}F_{hv}^*>+\mathrm{Re}<F_{vh}F_{hh}^*> \qquad (3\text{-}1.8b)$$

It can be seen that (3-1.7a) and (3-1.7b) are much less than (3-1.7c), and (3-1.8a) is much less than (3-1.8b), so the last three terms on RHS of Eq. (3-1.6) are now neglected. Thus, it yields the ψ shift at the σ_c maximum expressed by the Stokes parameters as follows

$$\tan 2\psi=\frac{U_s}{I_{hs}-I_{vs}} \qquad (3\text{-}1.9)$$

It can be seen that the third Stokes parameter $U_s\neq 0$ does cause the ψ shift.

By the way, if both U_s and $I_{hs}-I_{vs}$ approach zero, e.g. scattering from uniformly oriented scatterers or isotropic scatter media such as thick vegetation canopy, ψ cannot be well defined by Eq.(3-1.9).

Furthermore, fom Eq.(3-1.2), it yields

$$U_s\sim 0.5(M_{31}+M_{32})=\mathrm{Re}(<F_{vv}F_{hv}^*>+<F_{vh}F_{hh}^*>) \qquad (3\text{-}1.10a)$$

$$I_{hs}-I_{vs}\sim 0.5(M_{22}-M_{11})=0.5(<|F_{hh}|^2>-<|F_{vv}|^2>) \qquad (3\text{-}1.10b)$$

It is interesting to see that if let

$$\mathrm{Re}(<F_{vv}F_{hv}^*>+<F_{vh}F_{hh}^*>)=2F_{hv}F_0 \text{ and } 0.5(<|F_{hh}|^2>-<|F_{vv}|^2>)\sim F_0(F_{hh}-F_{vv}),$$

where $F_0=F_{vv}=F_{hh}$ is assumed, Eq. (3-1.9) can be reduced to the result of Schuler et al. (2000),

$$\tan 2\psi=\frac{2F_{vh}}{F_{hh}-F_{vv}} \tag{3-1.11}$$

Using the cross-pol backscattering coefficient σ_x, the same result can be obtained.

3-1.2 The Range and Azimuth Angles from the Euler Angles Transformation

The polarization vectors $\hat{h}_i,\hat{v}_i$ of the incident wave at $(\pi-\theta_i,\phi_i)$ in the principle coordinate ($\hat{x},\hat{y},\hat{z}$) are defined as [Jin 1994]

$$\hat{h}_i=\frac{\hat{z}\times\hat{k}_i}{|\hat{z}\times\hat{k}_i|}=-\sin\phi_i\hat{x}+\cos\phi_i\hat{y} \text{ and } \hat{v}_i=\hat{h}_i\times\hat{k}_i \tag{3-1.12}$$

where the incident wave vector

$$\hat{k}_i=\sin\theta_i\cos\phi_i\hat{x}+\sin\theta_i\sin\phi_i\hat{y}-\cos\theta_i\hat{z} \tag{3-1.13}$$

As the pixel surface has local slope as shown in Fig. 3-1.1, the polarization vectors should be re-defined following the local normal vector $\hat{z}_b$ as follows

$$\hat{h}_b=\frac{\hat{z}_b\times\hat{k}_{ib}}{|\hat{z}_b\times\hat{k}_{ib}|} \text{ and } \hat{v}_b=\hat{h}_b\times\hat{k}_{ib} \tag{3-1.14}$$

By using the transformation of the Euler angles (β,γ) between two coordinates ($\hat{x},\hat{y},\hat{z}$) and $(\hat{x}_b,\hat{y}_b,\hat{z}_b)$ [Jin 1994], it has

$$\hat{x}=\cos\gamma\cos\beta\hat{x}_b+\sin\gamma\hat{y}_b-\cos\gamma\sin\beta\hat{z}_b \tag{3-1.15a}$$

$$\hat{y}=-\sin\gamma\cos\beta\hat{x}_b+\cos\gamma\hat{y}_b+\sin\gamma\sin\beta\hat{z}_b \tag{3-1.15b}$$

$$\hat{z}=\sin\beta\hat{x}_b+\cos\beta\hat{z}_b \tag{3-1.15c}$$

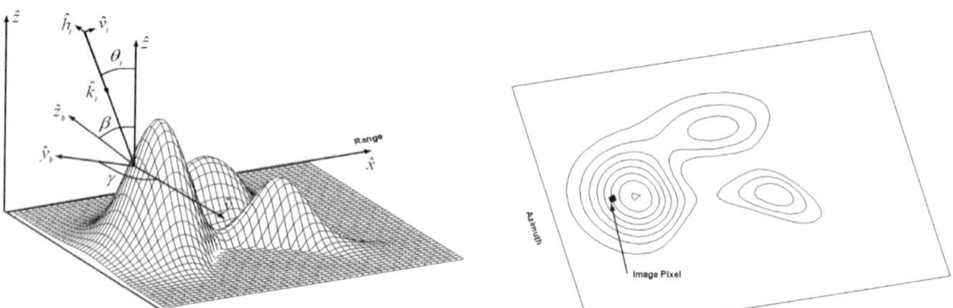

Fig.3-1.1. Incidence upon a pixel of tilted surface and the contour to be inverted

Substituting above relations into Eqs. (3-1.12, 3-1.13), it yields

$$\hat{h}_i=-\cos\beta\sin(\gamma+\phi_i)\hat{x}_b+\cos(\gamma+\phi_i)\hat{y}_b+\sin\beta\sin(\gamma+\phi)\hat{z}_b \tag{3-1.16}$$

$$\begin{aligned}\hat{k}_{ib}=&(\cos\beta\sin\theta_i\cos(\gamma+\phi_i)-\sin\beta\cos\theta_i)\hat{x}_b+\sin\theta_i\sin(\gamma+\phi_i)\hat{y}_b\\&-[\sin\beta\sin\theta_i\cos(\gamma+\phi_i)+\cos\beta\cos\theta_i]\hat{z}_b\end{aligned} \tag{3-1.17}$$

Using Eq.(3-1.17) to Eq.(3-1.14), the polarization vector $\hat{h}_b$ for local surface pixel is written as

$$\hat{h}_b = \frac{a\hat{x}_b + b\hat{y}_b}{\sqrt{a^2 + b^2}} \tag{3-1.18}$$

where

$$a = -\sin\theta_i \sin(\gamma + \phi_i) \tag{3-1.19a}$$

$$b = \sin\theta_i \cos\beta \cos(\gamma + \phi_i) - \sin\beta \cos\theta_i \tag{3-1.19b}$$

Thus, Eqs. (3-1.16, 3-1.18) yield the orientation angle as

$$\cos\psi = \hat{h}_b \cdot \hat{h}_i = \frac{\cos\beta \sin\theta_i - \cos(\gamma + \phi_i)\sin\beta \cos\theta_i}{\sqrt{a^2 + b^2}} \tag{3-1.20a}$$

$$\tan\psi = \frac{\tan\beta \sin(\gamma + \phi_i)}{\sin\theta_i - \cos\theta_i \tan\beta \cos(\gamma + \phi_i)} \tag{3-1.20b}$$

Taking $\phi_i = 0$, i.e. $\hat{x}$ is the range direction and $\hat{y}$ is the azimuth direction, Eq. (3-1.20b) is reduced to Schuler et al. (2000). Note that the azimuth angle in Lee et al. (1998) was defined by the angle between $\hat{\rho}$ and $\hat{y}$. Thus, the range and azimuth slopes of the pixel surface can be defined as Lee et al. (1998)

$$S_R = \tan\beta \cos\gamma$$
$$S_A = \tan\beta \sin\gamma \tag{3-1.21}$$

Since a single ψ shift cannot simultaneously determine two unknowns of β and γ in Eq. (3-1.20b), two- or multi-pass SAR image data are usually needed. It has been discussed in Lee et al. (1998).

3-1.3 The Azimuth Angle of Every Pixel in a SAR Image

If only the single-pass SAR image data are available, one of two unknown angles, β or γ, should be first determined. The azimuth alignment of tilted surface pixels can be visualized as a good indicator of the azimuth direction. We apply the adaptive threshold method and image morphological thinning algorithm [Castleman 1996] to specify the azimuth angle in all SAR image pixels. The algorithm contains the following steps:

(a) Make speckle filtering over the entire image.

(b) Apply the adaptive threshold method [Chan et al. 1998] to produce a binary image (see Fig. 3-1.2(a)). The global threshold value is not adopted because of the heterogeneity of the image pixels.

(c) Apply the image morphological processing for the binary image, remove those isolated pixels and fill small holes. Referring to the part of binary's "1" as the foreground and the part of binary's "0" as the background, the edges from the foreground are extracted (see Fig. 3-1.2(b)).

(d) Each pixel on the edge is set as the center of a square 21×21 window (see Fig. 3-1.2(c)), and a curve segment through the centered pixel is then obtained. Then, applying the polynomial algorithm for fitting curve segment in the least-squares sense, the tangential slope of the centered pixel is obtained (see Fig. 3-1.2(d)). It yields the azimuth angle of the centered pixel. Make a mark on that pixel so that it won't be calculated in the next turn.

(e) Removing the edge in Step (d) from the foreground, a new foreground is formed. Repeat Step (d) until the azimuth angle of every pixel in the initial foreground is determined.

(f) Make the complementary binary image, i.e. the initial background now becomes the foreground (see the gray line in Fig. 3-1.2(e)). Then, the Steps (d) and (e) are repeated to this complementary image until the azimuth angle of every pixel in the initial background is determined.

This approach provides a supplementary information to firstly determine the angle γ

over whole image area if there is no other information available.

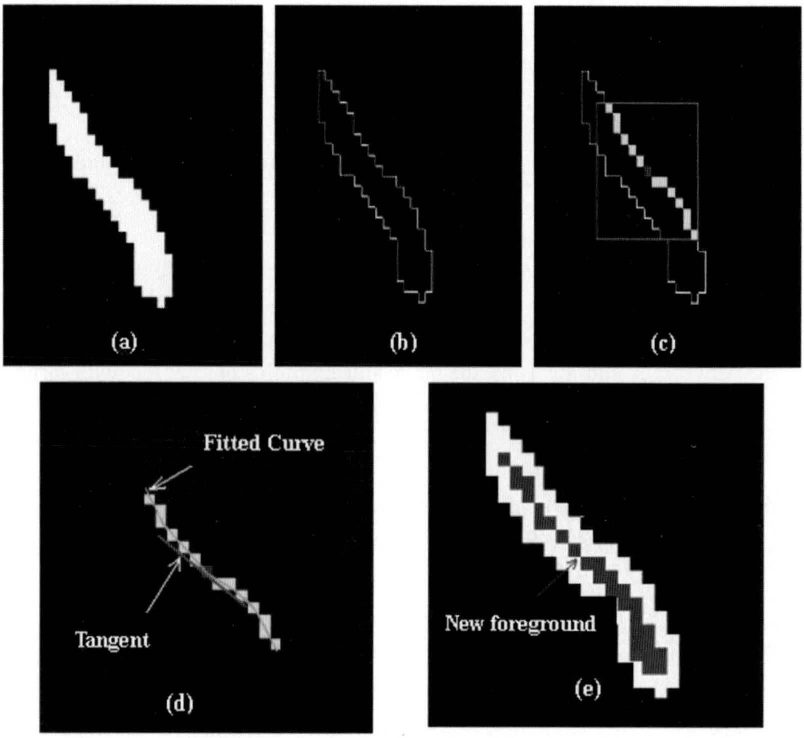

Fig. 3-1.2. Morphological thinning algorithm to determine the azimuth angle

3-1.4 An Example of Algorithm Implementation

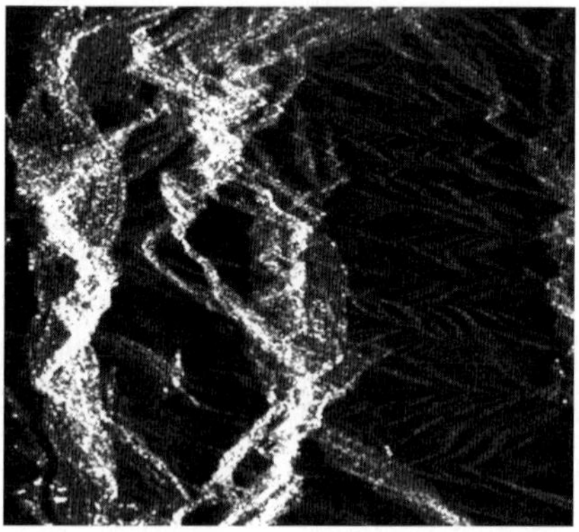

Fig.3-1.3. SAR image of total power *hh, vv, hv* at L band

As an example for DEM inversion, the L-band polarimetric SAR data on October 4, 1994 over Taiwan, China measured by SIR-C of the Cal-Tech JPL is investigated as shown in Figure

3-1.3. The center location is about $(120.964^\circ E, 23.487^\circ N)$ under $\theta_i = 32.85^\circ$, and the pixel size is 12.5×12.5 m^2. Fig. 3-1.3 is a 512px × 512px part from the original SAR image. The center location is $(120.8626^\circ E,\ 23.6659^\circ N)$ situated in the Ali Mountain of Taiwan.

To focus on the prominent topographic structure for DEM, some trivial undulations are omitted. Note that the azimuth angle γ is the angle between the slope direction $\hat{y}_b$ and the azimuth direction $\hat{y}$, as shown in Fig. 3-1.1.

As the azimuth angle γ of each pixel is obtained by the adaptive threshold method and thinning algorithm as described in the above steps, the β angle of each pixel can be determined by Eq. (3-1.20b). Taking $\phi_i = 0$, it yields

$$\tan\beta = \frac{\tan\psi \sin\theta_i}{\sin\gamma + \tan\psi \cos\theta_i \cos\gamma} \tag{3-1.22}$$

where the orientation angle ψ is calculated using Eq. (3-1.9), while the incident angle θ_i is determined by the SAR view geometry.

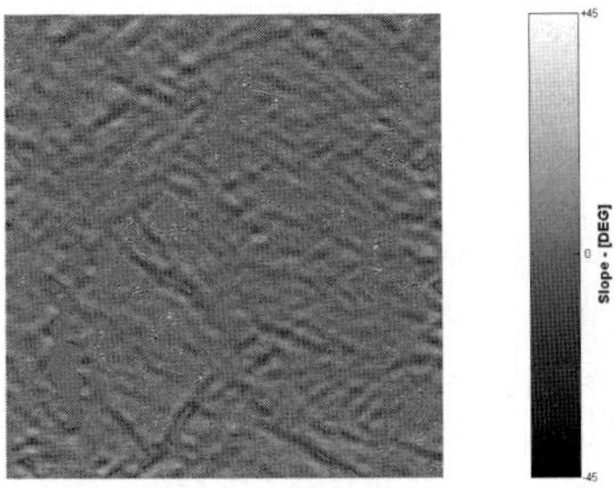

Fig. 3-1.4(a). Azimuth slope

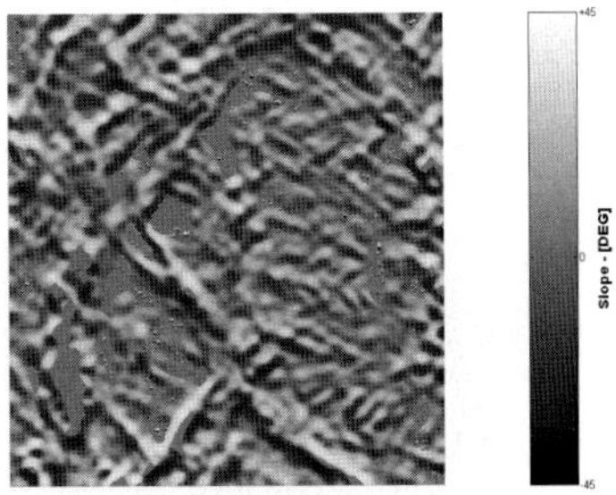

Fig. 3-1.4(b). Range slope

Substituting β and γ into Eq. (3-1.21), the azimuth slope S_A and range slope S_R are obtained. Following the approach of Ghiglia and Romero (1994), Pritt and Shipman (1994), Demmel (1997), we can utilize the two slopes to invert the terrain topography and DEM. The slopes S_A and S_R of all pixels in the SAR image are calculated as shown in Figs. 3-1.4(a) and 3-1.4(b).

The DEM can be generated by solving the Poisson equation for a $M \times N$ rectangular grid area [Schuler et al. 2000]. The approach is similar to the phase unwrapping of INSAR [Pritt 1996, Takajo and Takahashi 1998, Ghiglia and Romero 1994]. The Poisson equation can be written as

$$\nabla^2 \phi(x, y) = \rho(x, y) \tag{3-1.23}$$

where ∇^2 is the Laplace operator $\partial^2/\partial x^2 + \partial^2/\partial y^2$. The source function $\rho(x, y)$ consists of the surface curvature calculated by the slopes $S(x, y)$. Taking the $\hat{x}$-direction as the range direction and $\hat{y}$-direction as the azimuth direction, the discrete values of the source function can be derived (the superscript A and R represent the azimuth and range directions, respectively) as follows

$$\rho_{ij}^R = S_{ij}^R - S_{i-1,j}^R,\quad \rho_{ij}^A = S_{ij}^A - S_{i,j-1}^A \quad (0 \le i \le M, 0 \le j \le N) \tag{3-1.24}$$

Assuming that the errors in the input slopes are Gaussian random number, the integrated elevation values also have Gaussian errors. Therefore, a χ^2 statistics is a good measure of matching the solution to the exact value. This is also the standard assumption of the phase unwrapping algorithm of INSAR. χ^2 is given by

$$\chi^2 = \sum_{i,j} (\phi_{ij} - \phi_{i-1,j} - S_{ij}^R)^2 + \sum_{i,j} (\phi_{ij} - \phi_{i,j-1} - S_{ij}^A)^2 \qquad (3\text{-}1.25)$$

The differences between the partial derivatives of the solution ϕ_{ij} and partial derivatives defined by Eq. (3-1.24) should be minimized. Differentiating χ^2 with respect to ϕ_{ij} and setting the results equal to zeros, it yields

$$(\phi_{i+1,j} - 2\phi_{ij} + \phi_{i-1,j}) + (\phi_{i,j+1} - 2\phi_{ij} + \phi_{i,j-1}) = \rho_{ij} \qquad (3\text{-}1.26)$$

Eq. (3-1.26) presents the Poisson equation to be solved.

The discrete source function ρ_{ij} is defined by

$$\rho_{ij} = (S_{i+1,j}^R - S_{ij}^R) + (S_{i,j+1}^A - S_{ij}^A) \qquad (3\text{-}1.27)$$

In INSAR approach [Lee et al. 1998, Schuler et al. 1998], the ADI (Alternating Direction Iteration) algorithm was applied to generate the DEM. In this session, the FMG (full multi-grid) algorithm [Pritt and Shipman 1994, Demmel 1997] is employed to solve the Poisson equation. The benefits of FMG are due to its rapid-convergence, robustness and low computation load. For a $M \times N$ rectangular grid, the FMG method has computations on the order of $(MN)\ln(MN)$. The detailed steps of FMG are referred to Pritt and Shipman (1994).

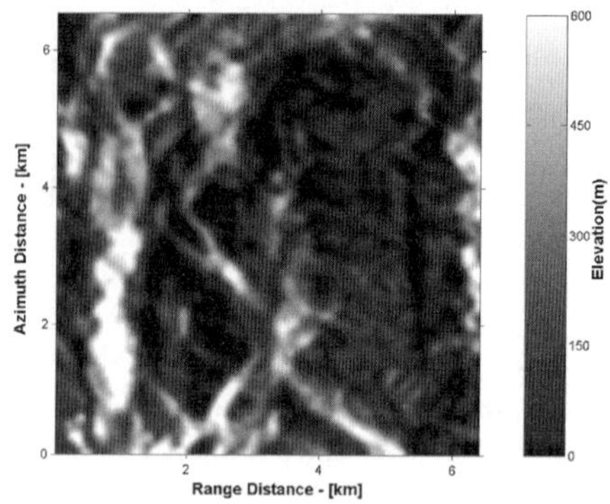

Fig. 3-1.5. The DEM image

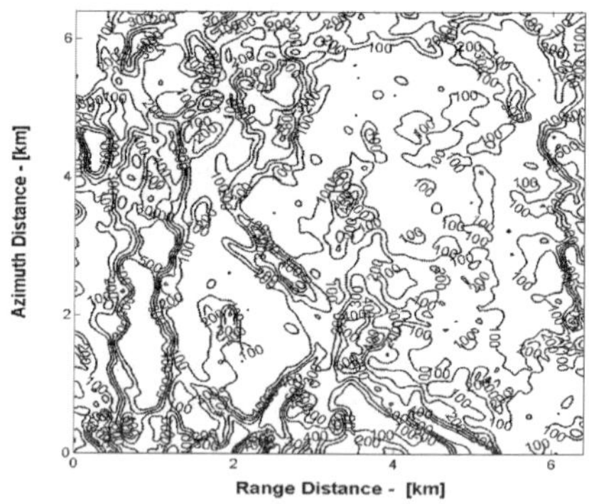

Fig. 3-1.6. The contour map

Fig. 3-1.7. A picture of inverted DEM

Fig. 3-1.5 is the inverted DEM. Fig. 3-1.6 is the contour map, and Fig. 3-1.7 presents the DEM picture.

It won't be surprised that there is a similarity between the inverted topography (Fig. 3-1.5) and the SAR image (Fig. 3-1.3). Also, the inverted contour map has been validated in comparison to the map of China Integrative Atlas (1990). But, because more accurate DEM as ground truth is not yet available, the inverted topography has been only roughly verified.

3-2 Change Detection of Multi-temporal ERS-2 SAR Images by Using Two-Thresholds EM and MRF Algorithms

The development of the urban areas is a great topic for human civilization in this century. Changes in the urban area, e.g. buildings, roads, greenbelts, botanic zoology, water body, population migration, and local climate etc. need to be accurately assessed. Multi-temporal observations of space-borne remote sensing imagery provide fast and practical technical means for surveying and assessing such vast changes. One direct application is to detect and classify the information on changes in the terrain surfaces [Rignot and van Zyl 1993, Singh 1989, Bruzzone and Prieto 2000, Merril and Jiajun 1998, Kasetkasem and Varshney 2002, Zhang et al. 2002]. The advancement of the SAR imagery technology, such as SIR-C SAR, ERS-1,2, RADARSAT during the latest twenty years, has provided the high-resolution and all-weather image data. It is very helpful to retrieve accurate information on terrain surfaces change.

It would be laborious to make an intuitive assessment for a huge amount of multi-temporal image data over a vast area. Such assessment based on qualitative gray-level analysis is not accurate and might lose some important information. How to detect and automatically analyze information on change in the terrain surfaces is a key issue in remote sensing. There are two usual methods: the supervised and unsupervised approaches. The former needs a mass of multi-temporal ground truth data for training samples in a classification procedure. The latter performs change detection by making a direct comparison of the multi-temporal remote sensing images without relying on any additional information. Our discussion in this session adopts the unsupervised approach.

Rignot and van Zyl (1993) studied change detection from the difference of two SAR images. Singh (1989) performed a statistical analysis of the change vector of infrared multi-channel data for surface change detection. Based on Bayes theory and Markov models, Bruzzone and Prieto (2000) proposed an automatic algorithm known as expectation maximum (EM) and Markov Random Field (MRF) analysis. Two classifications are made: changed and unchanged regimes. Their approach, however, has not been validated by using real SAR observations. Moreover, so-called changed might have different meaning. For example, the backscattering was reduced when an area of roughly distributed buildings became a flat grassland, or vice versa: backscattering was increased when a flat surface turned to a group of buildings or many trees. A ±dB change in backscattering might contain different content.

In this session, two-thresholds EM and MRF algorithm is developed to detect the change direction of backscattering enhanced, reduced and unchanged regimes from the SAR difference image [Jin et al. 2004]. As an example, two ERS-2 SAR images in 1996 and 2002 over Shanghai city are analyzed. The results illustrate well the direction of change in the Shanghai area during those years. It is proposed that multi-thresholds EM and MRF analysis may become traceable when multi-polarization, multi-channels, multi-sensors multi-temporal image data become available.

3-2.1 The EM Algorithm

Consider two co-registered images X_1 and X_2 with size $I \times J$ at two different times (t_1, t_2). Their difference image is $X = (X_2 - X_1)$ (It also can take $X = X_2 / X_1$). Let X be a random variable to represent the value of $I \times J$ pixels in $X_D = \{X(i,j), 1 \le i \le I, 1 \le j \le J\}$.

Change detection and analysis for a difference image is usually to distinguish two opposite classes: the unchanged class ω_n and the changed class ω_c. The probability density function *p(X)* of the pixel value in the difference image X_D was modeled as a mixture distribution of two density components associated with ω_n and ω_c, i.e.

$$p(X) = p(X \mid \omega_n)P(\omega_n) + p(X \mid \omega_c)P(\omega_c) \tag{3-2.1}$$

where $P(\omega_n)$, $P(\omega_c)$ of classes ω_n and ω_c are the *a priori* probabilities. The conditional probability density function is $p(X \mid \omega_k)$ $(k = n, c)$. The unsupervised estimations of $p(X \mid \omega_n)$, $p(X \mid \omega_c)$, $P(\omega_n)$ and $P(\omega_c)$ can be performed by the expectation maximum (EM) algorithm.

The EM algorithm is a general approach to maximum likelihood (ML) estimation for incomplete data problems [Fukunaga 1990, Dempster et al. 1977, Moon 1996, Render and Walker 1984, Shahshahani and Landgrebe 1994]. It consists of an expectation step and a maximization step. The expectation step is computed with respect to the unknown underlying variable using the estimates of the parameters conditioned by observations. The maximization step provides new estimates of the parameters. Assume that both $p(X \mid \omega_n)$ and $p(X \mid \omega_c)$ are modeled by Gaussian distributions. The density function associated with the class ω_n is described by the mean μ_n and the variance σ_n^2, and the density function associated with the class ω_c is described by μ_c and σ_c^2. It can be proved that the equations for estimating the statistical parameters and the *a priori* probability for the class ω_n are the following [Bruzzone and Prieto 2000, Render and Walker 1984]

$$P^{t+1}(\omega_n) = \frac{\sum_{X(i,j)\in X_D} \frac{P^t(\omega_n)p^t(X(i,j) \mid \omega_n)}{p^t(X(i,j))}}{IJ} \tag{3-2.2}$$

$$\mu_n^{t+1} = \frac{\sum_{X(i,j)\in X_D} \frac{P^t(\omega_n)p^t(X(i,j) \mid \omega_n)}{p^t(X(i,j))} X(i,j)}{\sum_{X(i,j)\in X_D} \frac{P^t(\omega_n)p^t(X(i,j) \mid \omega_n)}{p^t(X(i,j))}} \tag{3-2.3}$$

$$(\sigma_n^2)^{t+1} = \frac{\sum_{X(i,j)\in X_D} \frac{P^t(\omega_n)p^t(X(i,j) \mid \omega_n)}{p^t(X(i,j)} [X(i,j) - \mu_n^t]^2}{\sum_{X(i,j)\in X_D} \frac{P^t(\omega_n)p^t(X(i,j) \mid \omega_n)}{p^t(X(i,j))}} \tag{3-2.4}$$

where the superscripts t and $t+1$ denote the current and next iterations.

Analogously, these equations can also be used to estimate the a priori probability, the mean and the variation of the conditional probability density function for the class ω_c.

The estimates are computed starting from the initial values by iterating the above equations until convergence. It can be proved that, the estimated parameters provide an increase in the log-likelihood function $L(\theta) = \ln p(X_D \mid \theta)$ after each iteration, where $\theta = \{P(\omega_n), P(\omega_c), \mu_n, \mu_c, \sigma_n^2, \sigma_c^2\}$. At convergence, a local maximum of the log-likelihood function is reached [Dempster et al. 1977, Moon 1996].

The initial value of the estimated parameters can be obtained by the analysis of the histogram of the difference image. A pixel subset S_n likely to belong to ω_n and a pixel subset S_c likely to belong to ω_c can be obtained by preliminarily selecting two-threshold T_n and T_c on the histogram. T_n and T_c are expressed as $T_n = M_D(1-\gamma)$ and $T_c = M_D(1+\gamma)$, where $M_D = [\max\{X_D\} + \min\{X_D\}]/2$ as the middle value of $h(X)$. $\gamma \in (0,1)$ is a parameter that defines the range around M_D in which the pixels cannot be easily identified.

Two subsets $S_n = \{X(i,j) \mid X(i,j) < T_n\}$ and $S_c = \{X(i,j) \mid X(i,j) > T_c\}$ are used to compute the initial estimates associated with two classes ω_n and ω_c, respectively.

Under the assumption of interpixels independence and according to the Bayes rule for minimum error, each pixel $X(i,j)$ in the difference image X_D should be assigned to the class ω_k that maximizes the posterior conditional probability, i.e.,

$$\omega_k = \arg \max_{\omega_i \in \{\omega_n, \omega_c\}} \{P(\omega_i \mid X(i,j))\} = \arg \max_{\omega_i \in \{\omega_n, \omega_c\}} \{P(\omega_i) p(X(i,j) \mid \omega_i)\} \tag{3-2.5}$$

This is equivalent to solving the ML boundary T_o for the two classes ω_n and ω_c on the difference image. The optional threshold T_o requires

$$\frac{P(\omega_c)}{P(\omega_n)} = \frac{p(X \mid \omega_n)}{p(X \mid \omega_c)} \tag{3-2.6}$$

Under the assumption of Gaussian distribution, Eq.(3-2.6) yields

$$(\sigma_n^2 - \sigma_c^2)T_o^2 + 2(\mu_n\sigma_c^2 - \mu_c\sigma_n^2)T_o + \mu_c^2\sigma_n^2 - \mu_n^2\sigma_c^2 - 2\sigma_n^2\sigma_c^2 \ln\left[\frac{\sigma_c P(\omega_n)}{\sigma_n P(\omega_c)}\right] = 0 \tag{3-2.7}$$

to obtain optimal T_o.

3-2.2 Two-Thresholds EM

The pixels value of SAR image represents the echo power (backscattering σ^0) of the radar observation. The difference image (i.e. $X = \sigma_D^0 = \sigma_2^0 - \sigma_1^0$) can show where the backscattering is enhanced ($\sigma_D^0 > 0$), unchanged ($\sigma_D^0 \approx 0$) or reduced ($\sigma_D^0 < 0$). This classification is applicable to information retrieval of the change in terrain surfaces. The pixels of the difference image are classified into three classes: σ_D^0-enhanced ω_{c1}, σ_D^0-unchanged ω_n and σ_D^0-reduced ω_{c2}, i.e. three subsets of pixels with $\sigma_D^0 > 0$, approximately 0 and $\sigma_D^0 < 0$.

Thus, the probability density function $p(X)$ of X_D is modeled as a mixture density distribution consisting of three components:

$$p(X) = p(X \mid \omega_{c1})P(\omega_{c1}) + p(X \mid \omega_n)P(\omega_n) + p(X \mid \omega_{c2})P(\omega_{c2}) \tag{3-2.8}$$

T_{o1} is first obtained by application of the EM algorithm to the enhanced class ω_{c1} and no-enhanced class ω_{n1} (ω_n and ω_{c2}). Then, we obtain T_{o2} by applying the EM to the reduced class ω_{c2} and no-reduced class ω_{n2} (ω_n and ω_{c1}), where $\omega_{n1} = \omega_n \cup \omega_{c2}$, $\omega_{n2} = \omega_n \cup \omega_{c1}$. Finally, the two results are superposed to get the final classification.

In Fig. 3-2.1(a), subset S_{n1} likely to belong to class ω_{n1} and subset S_{c1} likely to belong to class ω_{c1}, are obtained by applying two-thresholds T_{n1} and T_{c1} in the enhanced regime ($X \geq 0$) of the histogram $h(X)$. T_{n1} and T_{c1} are expressed as $T_{n1} = M_{D1}(1-\gamma)$ and $T_{c1} = M_{D1}(1+\gamma)$, where M_{D1} is the mean value of the enhanced regime in $h(X)$, i.e. $M_{D1} = [\max\{X_D\}/2]$.

Analogously, subset S_{n2} likely to belong to the class ω_{n2} and subset S_{c2} likely to belong to the class ω_{c2} are obtained in the reduced regime ($X < 0$) in the $h(X)$, where the mean value of the reduced regime in $h(X)$ is $M_{D2} = [\min\{X_D\}/2]$. In the present, $\gamma = 0.5$ is chosen as a middle value of (0,1).

Fig. 3-2.1(b) shows the grayscale histogram of the difference image (shown in Fig. 3-2.5) from the ERS-2 SAR image data around the Yan Zhong greenbelt of Shanghai downtown center. Three classes of the probability distributions are outlined in this figure. The histogram is normalized in accordance with the probability density distributions.

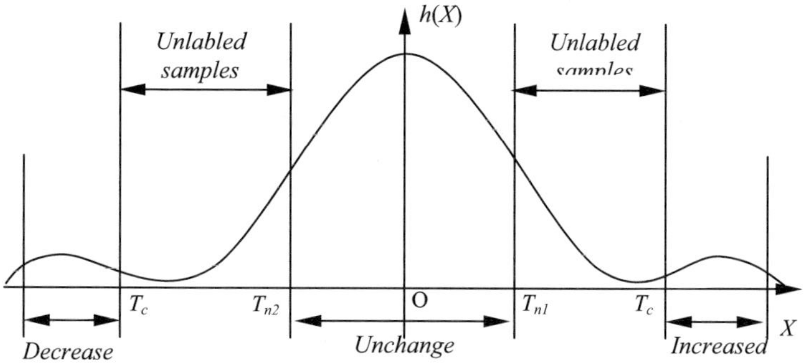

(a) the histogram of the difference image and thresholds

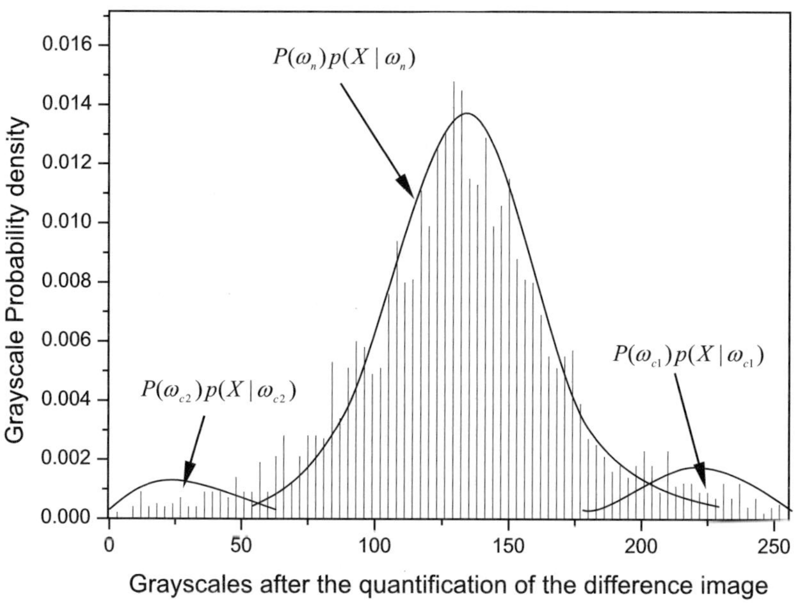

(b) the normalized histogram for 3 classes from real difference image data

Fig. 3-2.1. The histogram of the difference image

The initial statistical parameters of the subsets of $S_{n1} = \{X(i,j) \mid 0 < X(i,j) < T_{n1}\}$, $S_{c1} = \{X(i,j) \mid X(i,j) > T_{c1}\}$ are first derived, and then derive the parameters of $S_{n2} = \{X(i,j) \mid T_{n2} < X(i,j) < 0\}$, $S_{c2} = \{X(i,j) \mid X(i,j) < T_{c2}\}$. The initial statistical parameters are related to the classes ω_{n1} , ω_{c1} and the classes ω_{n2} , ω_{c2} , respectively.

Then, Eqs. (3-2.2~4) are sequentially used to perform the iterations on the four classes, i.e. the *a priori* probability and other statistical parameters [$P(\omega_{n1})$, μ_{n1} , σ_{n1}^2], [$P(\omega_{c1})$, μ_{c1}, σ_{c1}^2], [$P(\omega_{n2})$, μ_{n2}, σ_{n2}^2] and [$P(\omega_{c2})$, μ_{c2}, σ_{c2}^2] . The thresholds T_{o1} and T_{o2} are solved one by one from Eq. (3-2.7).

3-2.3 The Spatial-Textual Classification Based on MRF

As Bruzzone and Prieto (2000) discussed for the MRF approach, the pixels of the image are spatially correlated, i.e. a pixel belonging to the class ω_k is likely to be surrounded by pixels belonging to the same class. To take account of spatial texture may yield more reliable and accurate change detection.

Let the set $C = \{C_\ell, 1 \le l \le L\}$ represent the possible sets of the labels in the difference image, where $C_\ell = \{C_\ell(i,j), 1 \le i \le I, 1 \le j \le J\}$, $C_\ell(i,j) \in \{\omega_{c1}, \omega_{c2}, \omega_n\}$, $L = 3^{IJ}$, I,J is the height and width of difference image respectively.

According to Bayes rule for minimum error, the classification result should maximize the posterior conditional probability,

$$C_k = \arg\max_{C_\ell \in C}\{P(C_\ell \mid X_D)\} = \arg\max_{C_\ell \in C}\{P(C_\ell) p(X_D \mid C_\ell)\} \tag{3-2.9}$$

where X_D is the difference image, $P(C_\ell)$ is the prior model for the class labels, and $p(X_D \mid C_\ell)$ is the joint density function of the pixel values in the difference image given the set of labels C_ℓ. The solution of Eq. (3-2.9) requires the MRF approach to estimate both $P(C_\ell)$ and $p(X_D \mid C_\ell)$.

In order to formulate the problem by using the MRF, it is necessary to introduce the concept of a spatial neighborhood system defining, N . Considering a pixel (i,j) , its neighborhood can be expressed:

$$N(i,j) = \{(i,j) + (v,\varsigma), (v,\varsigma) \in N\} \tag{3-2.10}$$

in our case, $N = \{(\pm1,0),(0,\pm1),(1,\pm1),(-1,\pm1)\}$, i.e. the neighborhood of one pixel is composed of the eight pixels surrounding it.

Let $C_\ell(i,j)$ represent the class label of (i,j), $C_{\ell N}$ represent the class labels of the neighborhood pixels of (i,j), and $C_{\ell|(i,j)}$ represents the class labels of the pixels in the difference image except (i,j). If C_ℓ is taken as a MRF, C_ℓ has two properties as follows:

$$P(C_\ell) > 0, \quad \forall C_\ell(i,j) \in \{\omega_{c1}, \omega_{c2}, \omega_n\} \tag{3-2.11}$$

$$P(C_\ell(i,j) \mid C_{\ell|(i,j)}) = P(C_\ell(i,j) \mid C_{\ell N}) \tag{3-2.12}$$

The second-order neighborhood system and cliques are shown in Fig. 3-2.2.

Cliques refer to the connected sub-graph of neighboring pixels, including the center pixel in the neighborhood system. In our work, only single-pixel and two-pixel cliques are considered. As shown in Fig. 3-2.2(b,c), the single-pixel clique has one form, and the two-pixel clique has four forms. The mathematical expressions are, respectively,

$$L_1 = (i,j), \quad L_2 = \{\{(i,j),(g,h)\} \mid (g,h) \in N_i(i,j)\}$$

According to the Hammersley-Clifford theorem, MRF is equivalent to Gibbs distribution. Therefore, the conditional probability of the class label of (i,j) given the pixel labels elsewhere is expressed as [Bruzzone and Prieto 2000]

$$P(C_\ell(i,j) \mid \{C_\ell(g,h),(g,h) \ne (i,j)\}) = P(C_\ell(i,j) \mid \{C_\ell(g,h),(g,h) \in N(i,j)\})$$

$$= \frac{1}{Z}\exp[-U(C_\ell(i,j) \mid \{C_\ell(g,h),(g,h) \in N(i,j)\})] \tag{3-2.13}$$

where $U(\cdot)$ is the Gibbs energy function, and Z is the normalizing factor,

$$Z=\sum_{(g,h)\in\{(i,j)\cup N(i,j)\}}\exp[-U(C_\ell(i,j)\,|\,\{C_\ell(g,h)\})] \tag{3-2.14}$$

$$\begin{aligned} U(C_\ell\,|\,X_D) &= \sum_{1\le i\le I}\sum_{1\le j\le J} U(C_\ell(i,j)\,|\,C_{\ell|(i,j)},X_D) \\ &= \sum_{1\le i\le I}\sum_{1\le j\le J} U(C_\ell(i,j)\,|\,C_{\ell N},X_D) \end{aligned} \tag{3-2.15}$$

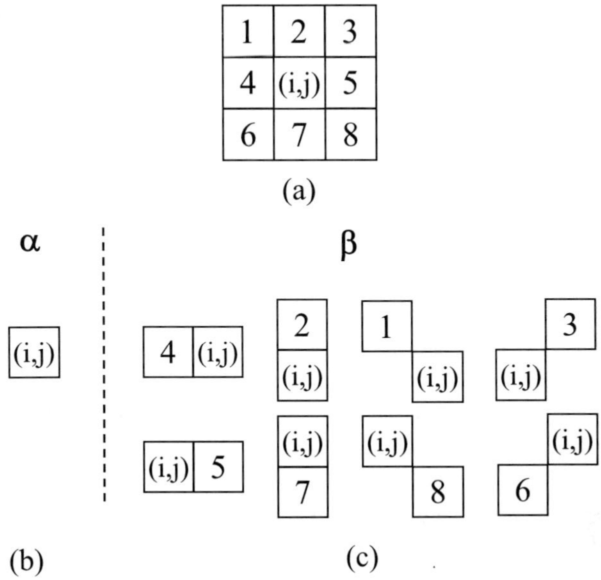

Fig.3-2.2. The Second-order Neighborhood System and Cliques
(a) second-order neighborhood system, (b) single-pixel clique L_1 and its weight α,
(c) two-pixel clique L_2 and its weight β

In the Guassian case

$$\begin{aligned} U(C_\ell(i,j)\,|\,C_{\ell N},X_D) &= \frac{1}{2}\ln|2\pi\sigma^2_{C_\ell(i,j)}| + \frac{(X(i,j)-\mu_{C_\ell(i,j)})^2}{2\sigma^2_{C_\ell(i,j)}} \\ &\quad + v_1(C_\ell(i,j)) + \sum_{(g,h)\in N(i,j)} v_2(C_\ell(i,j),C_\ell(g,h)) \end{aligned} \tag{3-2.16}$$

where

$$v_1(C_\ell(i,j))=\alpha,\quad (i,j)\in L_1 \tag{3-2.17}$$

Here α is the weight value of single-pixel clique, and it is assumed $\alpha = 0$.

$$v_2(C_\ell(i,j),C_\ell(g,h))=\begin{cases} -\beta, & C_\ell(i,j)=C_\ell(g,h), \{(i,j),(g,h)\}\in L_2 \\ \beta, & C_\ell(i,j)\ne C_\ell(g,h), \{(i,j),(g,h)\}\in L_2 \end{cases} \tag{3-2.18}$$

where β is the weight value of two-pixel clique, $0.7<\beta<1.5$, in our calculation $\beta=1.0$.

According to Eq. (3-2.9), the generation of the final change-detection map involves the labeling of all the pixels in the difference image so that the posterior probability is maximized. In terms of Markov models, this is equivalent to the minimization of the following energy function

$$U(C_\ell \mid X_D) = \sum_{1\le i\le I} \sum_{1\le j\le J} [U_I(X(i,j)/C_\ell(i,j))$$

$$+U_C(C_\ell(i,j)/\{C_\ell(g,h),(g,h)\in N(i,j)\}]) \tag{3-2.19}$$

where $U_C(\cdot)$ describes the inter-pixel class dependence, which is given by the last two terms in Eq.(3-2.16). $U_I(\cdot)$ represents the energy function under the assumption of conditional independences, as defined by the first two terms in Eq.(3-2.16).

$$U_I(X(i,j)/C_\ell(i,j)) = \frac{1}{2}\ln\left|2\pi\sigma^2_{C_\ell(i,j)}\right| + \frac{1}{2}(X(i,j)-\mu_{C_\ell(i,j)})^2[\sigma^2_{C_\ell(i,j)}]^{-1} \tag{3-2.20}$$

where $\sigma^2_{C_\ell(i,j)} \in \{\sigma^2_{c1}, \sigma^2_{c2}, \sigma^2_n\}$, $\mu_{C_\ell(i,j)} \in \{\mu_{c1}, \mu_{c2}, \mu_n\}$ have been obtained by the EM algorithm.

Generally, the solution of Eq. (3-2.19) requires an iterative algorithm, e.g. the simulated annealing algorithm. We employ the ICM (iterated conditional modes) algorithm [Besag 1974, 1986], which has been proved computationally economical and converges fast to a local minimum of the energy function. The mathematical expression of the ICM algorithm can be written as

$$C_\ell = ICM(\alpha, \beta, \mu_{C_\ell(i,j)}, \sigma^2_{C_\ell(i,j)}, X_D, C_{\ell 0}) \tag{3-2.21}$$

where $C_{\ell 0}$ is the initial value of C_ℓ.

C_ℓ is obtained by the following steps:

(a) Obtain $C_{\ell 0}$ by the EM algorithm, and initialize C_ℓ with $C_{\ell 0}$.

(b) Calculate new $C_\ell(i,j)$ that minimizes the energy function in Eq.(3-2.19).

(c) Repeat step (b) until convergence is reached.

The convergence condition of ICM algorithm is either iterative cycles exceed 30 (actual converge is much quickly reached) or that fewer than $\sqrt{IJ}/20$ pixels change. The details of ICM algorithm can be found in Besag (1986).

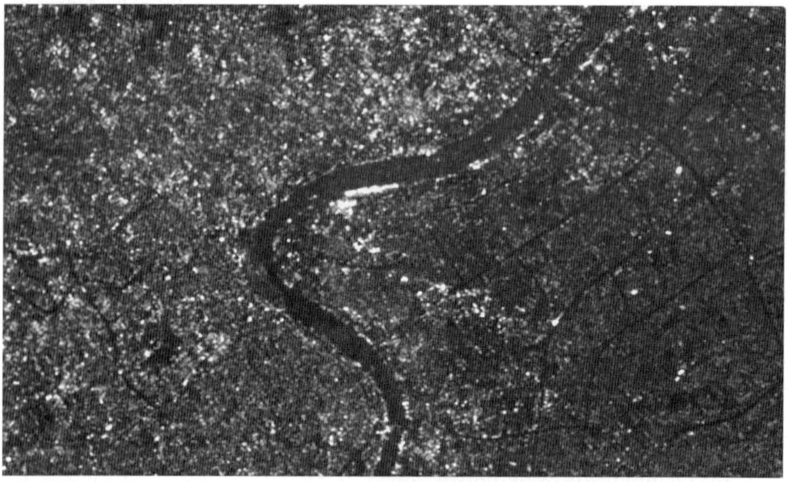

Fig.3-2.3. ERS-2 SAR image on June 4, 1996, Shanghai

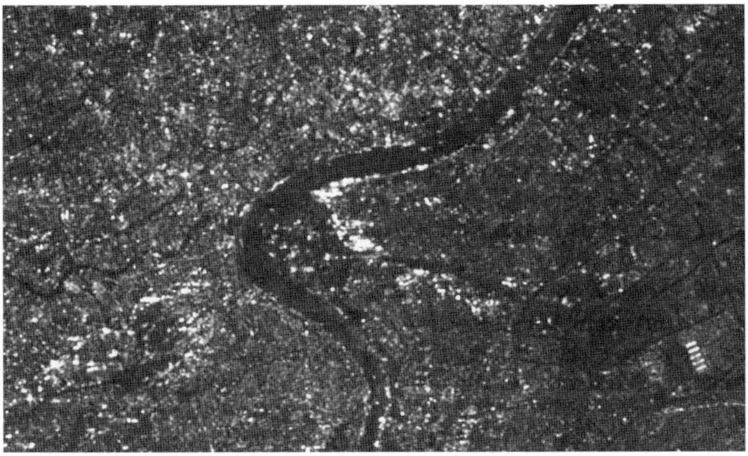

Fig.3-2.4. ERS-2 SAR image on April 9, 2002, Shanghai

3-2.4 An Example for ERS-2 SAR Data

Figs. 3-2.3 and 3-2.4 present two image data of ERS-2 SAR in 4 June 1996 and 9 April 2002 over Shanghai China (~31.3°*N*, 121.4°*E*), respectively. The image resolution is $12.5m \times 12.5m$. The images have been co-registered, and the noise is simply reduced by applying mean filtering (3×3 window size). The images size given in Figs. 3-2.3 and 3-2.4 is 580px×980px. Fig. 3-2.5 shows the difference image.

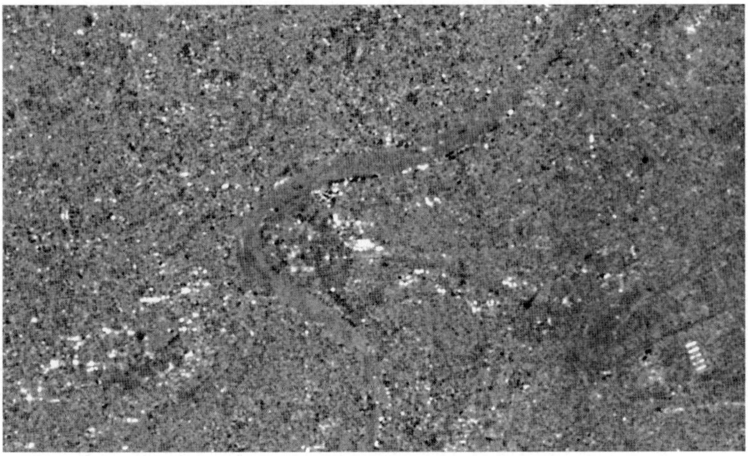

Fig. 3-2.5. The difference image

The white pixels of Fig. 3-2.5 indicate enhanced scattering and the black ones reduced backscattering.

The EM algorithm is applied to the difference image to obtain iteratively the *a prior* probability and other statistical parameters. The iterative process of the first class is shown in Figs. 3-2.6(a)~(c), and the iteration of the second class is very similar.

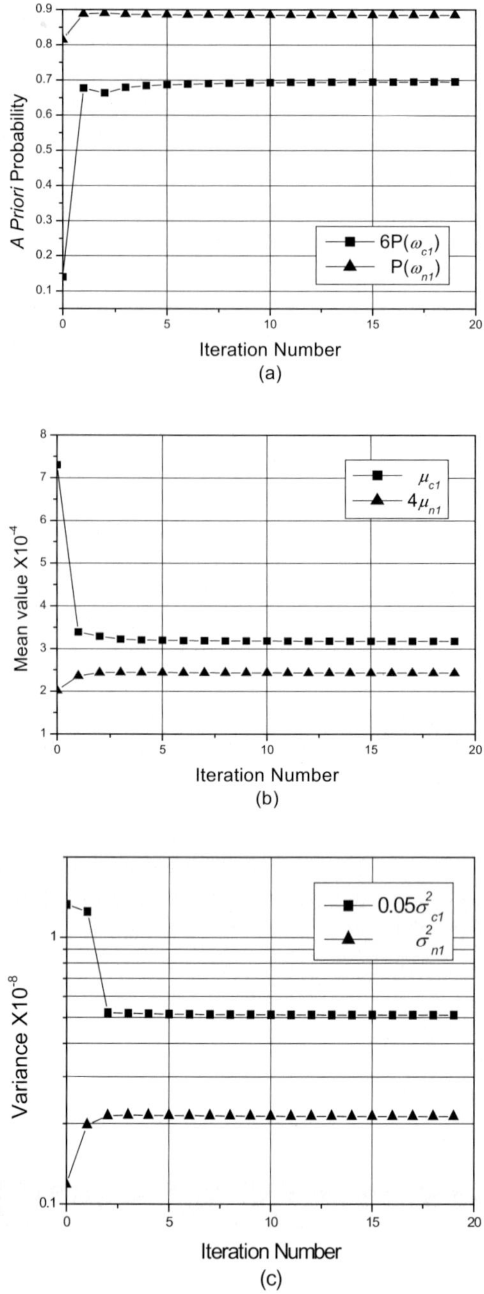

Fig.3-2.6. The parameters after each iteration (γ =0.5)

Two-thresholds T_{o1}， T_{o2} for surface change detection are then computed. Thc present example yields $T_{o1} = 0.946 \times 10^{-4}$， $T_{o2} = -1.013 \times 10^{-4}$ (the maximum $\sigma_D^0|_{\max} = 5.202 \times 10^{-4}$ and the minimum， the minimum $\sigma_D^0|_{\min} = -3.381 \times 10^{-4}$) .

Fig. 3-2.7 shows automatic detection of change direction by utilizing the EM algorithm. The white pixels indicate the σ^0-enhanced places, the black pixels show the σ^0-reduced places, and the gray area is σ^0-unchanged.

Fig. 3-2.8 shows the spatially cotextual change based on the MRF-ICM algorithm. Comparing with Fig. 3-2.7, it can be seen that the pixels of the same class are more merged and the isolated pixels become much fewer.

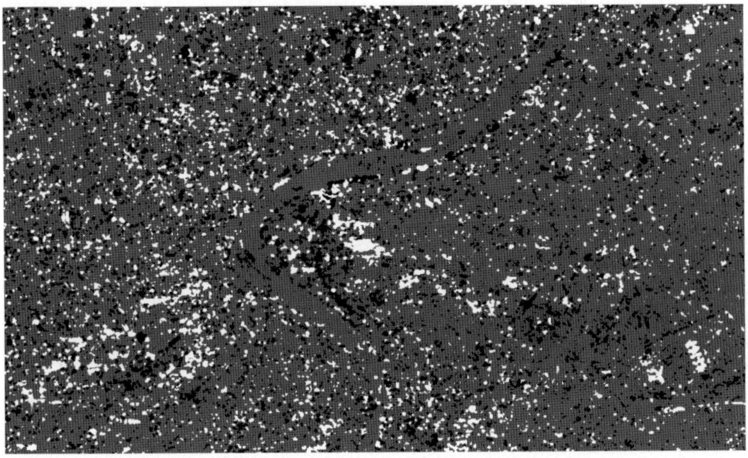

Fig. 3-2.7. Change detection using two-thresholds EM

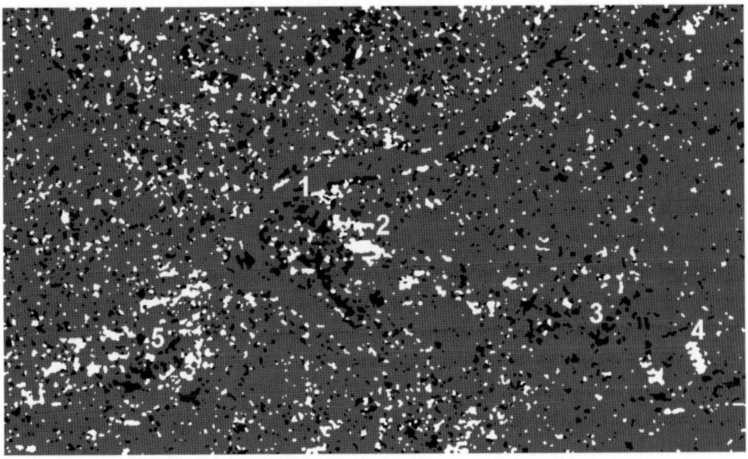

Fig. 3-2.8. Change detection using ICM algorithm based on MRF

In Fig. 3-2.8, the black areas marked by the odd number (1,3,5) are the σ^0-reduced areas, and the white areas marked by the even number (2,4) are the σ^0-enhanced areas. For example, area 1 is the Lu Jia Zui greenbelt, area 3 is the Pu Dong Park, and the area 5 is the Yan Zhong greenbelt, where the flat green surface caused less scattering. But, area 2 is the commercial center at the Lu Jia Zui, where there are many different buildings causing more electromagnetic wave double-reflection and enhance scattering. Area 4 is the location of the new buildings of International Exhibition Center of Shanghai. Areas 1-5 are newly rebuilt after 1996 and are shown well in the change detection data of 2002-1996.

It is interesting to overlap Fig. 3-2.8 over the Shanghai tourist map and identify detailed changes along the city streets during these years as indicated from Fig. 3-2.9.

Fig.3-2.9. Change detection overlapped on the Shanghai tourist map

3.3 Deorientation Theory of Polarimetric Scattering Targets and Application to Terrain Surface Classification

Radar polarimetry has become one of most advanced technologies for terrain surface observation and remote sensing, especially due to the pioneer experiments of SIR-C in 1994 and the coming launch of Radarsat-2 SAR, and other airborne flights such as AIRSAR etc. Classification of complex terrain surfaces using polarimetric SAR imagery is one of most important SAR applications.

Earlier polarimetric classification methods were based on the supervised method of coherent scattering matrix or coherence scattering data of multi-polarization channels [Kong et al. 1990, Zebker et al. 1991]. Van Zyl (1989) developed an unsupervised method of classification based on comparison of the polarization properties of each pixel in the image with the scattering of some target classes, such as odd-, even-number of reflections and diffuse scattering. Cloude and Pottier (1997) presented an unsupervised method based on the entropy H and target decomposition parameters, which extracts the target decomposition parameter α and entropy H from an eigen-analysis of coherency matrix and constructs an unsupervised classification spectrum on the $\alpha - H$ plane. In 1998, Freeman and Durden (1998) suggested a three-component decomposition method of polarimetric data, which introduced a combination of surface, volume and even-bounce scatterings. Lee *et al.* (1999,2001,2004) developed some classification methods combining both the target decomposition and complex Wishart classifier. These methods achieve preliminary classification sets using the $\alpha - H$, $\alpha - H - A$ method or three-component decomposition method, and thereafter iteratively classify these preliminary sets to make final classifications.

Scattering from terrain targets functionally depends on the scatter orientation, shape, dielectric property, scattering mechanism etc. Scatter targets of complex terrain surfaces are often randomly oriented and cause randomly fluctuating echoes. It is difficult to classify randomly oriented and randomly distributed scatter targets. Different scatters with different orientations can make similar scattering, and vice versa, the same scatters with random orientation can make different scattering leading to a confused classification.

In this session, a deorientation concept is introduced to reduce the influence of randomly fluctuating orientation and show the prominence of targets' generic characteristics for terrain surface classification [Xu and Jin 2005]. Based on this deoriented classification, a detailed targets orientation, orderly or randomly, and further classification of the scatter targets can be obtained. The deorientation concept is similar to the slope compensation concept [Schuler et al. 1998,1999 Lee et al. 2000,2002,2003], while the deorientation is applied to all kinds of targets including both targets and background.

Transformation of the target scattering vector to rotate the target along the sight line is presented. Deorientation is introduced to transform the target orientation into such a fixed state with minimization of cross-polarization (min-x-pol). And meanwhile, the deorientation angle ψ is extracted to indicate the angle deviation of the target orientation from this min-x-pol state. The min-x-pol method is somewhat equivalent to Cameron's maximal diagnolization of symmetric target decomposition [Cameron et al. 1990,1996]. His aim is to achieve a maximum degree of diagonalization of scattering matrix, which is coincident to the minimization of x-pol.

A set of new parameters u, v, w from "lexicographical straightforward" vectorization of the target scattering vector is defined to indicate the ratio and phase difference of the two co-polarized (co-pol) backscattering terms, and the significance of the cross polarized (x-pol) term. The relationship between u, v and conventional α is analyzed on the plane of $u - v$.

Based on the eigen-analysis of coherency matrix of the target scattering vectors, this methodology takes account for randomly distributed (non-deterministic) scatter targets including the background. The parameters u,v and ψ are extracted from deorientation of the principal scattering vector of random scatter targets. All these parameters as well as the entropy H are applied to classification of random terrain surface.

Numerical simulation of polarimetric scattering from a single scatterer or a random cloud of small non-spherical particles is analyzed to illustrate the relationships between the physical properties e.g. the Euler angles, shape, dielectric constant etc. and the parameters u,v,ψ. As a vector radiative transfer (VRT) model, polarimetric scattering from a layer of random non-spherical particles above a rough surface is studied to show the effectiveness of the parameters u,v,ψ,H and the capability of u,v,H for classification of complex terrain surfaces.

Based on the numerical simulation, a new unsupervised terrain surfaces classification using deorientation analysis is proposed. As an example, classification of China's Guangdong Huiyang district from a SIR-C polarimetric image is made by using our deorientation method and is compared with conventional $\alpha-H$ method. Another example is the orientation-analysis applied to an AIRSAR image over Canadian Boreal forests, which makes further discrimination over some easily-confused surface types and provides a detailed description of the terrain targets orientation.

3-3.1 Deorientation Theory

A. Target decomposition and parameterization

As an electromagnetic plane wave $\bar{E}_i$ is incident upon a scatter target, the scattering wave $\bar{E}_s$ is written as

$$\bar{E}_s = [S]\cdot\bar{E}_i . \tag{3-3.1}$$

where $[S]$ is the scattering matrix. The target scattering vector $\bar{k}$ is defined by vectorizing $[S]$ [Cloude and Pottier 1996], one of which is "Pauli" vectorization and the other is "Lexicographical straightforward" vectorization as follows

$$\bar{k}_P = \frac{1}{\sqrt{2}}[S_{hh}+S_{vv} \quad S_{hh}-S_{vv} \quad 2S_x]^T , \tag{3-3.2a}$$

$$\bar{k}_L = \left[S_{hh} \quad \sqrt{2}S_x \quad S_{vv}\right]^T , \tag{3-3.2b}$$

where the superscript T denotes the transpose, and the subscripts P, L indicate "Pauli" and "Lexicographic straightforward", respectively. The subscripts h, v denote horizontal and vertical polarization, respectively, and x denotes cross polarization. Here the reciprocity condition, i.e. $S_x \equiv S_{hv} = S_{vh}$, is as usually assumed. The antenna coordinate is adopted which is different from the wave coordinate [Cloude 1986].

The coherency matrix $[T]$ and covariance matrix $[C]$ are defined by the target scattering vector as [Cloude et al. 2001]:

$$[T] = \langle k_P \cdot k_P^+ \rangle, \quad [C] = \langle k_L \cdot k_L^+ \rangle . \tag{3-3.3}$$

which are linked with the Mueller matrix and ensured the integrity of using $[T]$ or $[C]$ to describe the scattering information of scatter targets [Jin and Cloude 1994]. Here $\langle\cdot\rangle$ denotes the ensemble average, and the superscript $+$ denotes the conjugate transpose of complex vector or matrix.

Cloude *et al.* (1996,2001) defined the parameterization of $\bar{k}_P$ in terms of the parameters α, β etc. as

$$\bar{k}_P = \|\bar{k}_P\| \begin{bmatrix} \cos\alpha e^{j\phi_1} & \sin\alpha\cos\beta e^{j\phi_2} & \sin\alpha\sin\beta e^{j\phi_3} \end{bmatrix}^T, \tag{3-3.4a}$$

where the operator $\|\bullet\|$ means the vector norm. It can be seen that the angles α, β represent the relative scale of each complex element of $\bar{k}_P$ in the Pauli space, while the angles ϕ_1, ϕ_2 and ϕ_3 are the phase of each element. The vector $\bar{k}_P$ is shown in Fig. 3-3.1(a).

A similar parameterization of $\bar{k}_L$ is defined as

$$\bar{k}_L = \|\bar{k}_L\| \begin{bmatrix} \sin c\cos a \cdot e^{j\phi_0} & \cos c \cdot e^{j\phi_x} & \sin c\sin a \cdot e^{j(\phi_0+2b)} \end{bmatrix}^T. \tag{3-3.4b}$$

Eqs. (3-3.2b) and (3-3.b) yield

$$\begin{cases} a = \tan^{-1}\left(|S_{vv}|/|S_{hh}|\right) \\ b = \dfrac{1}{2}\arg\left(S_{vv}/S_{hh}\right) \\ c = \cos^{-1}\left(\sqrt{2}|S_x|/\|\bar{k}_L\|\right) \end{cases} \tag{3-3.5}$$

where the operator $\arg(\bullet)$ means the phase angle of complex number. It can be seen in Eq.(3-3.5) that the parameter a indicates the amplitude ratio of co-polarized (co-pol) scatterings S_{vv} and S_{hh}; b indicates the phase difference between S_{vv} and S_{hh}; c indicates the cross-polarized (x-pol) scatterings. The vector $\bar{k}_L$ is also shown in Fig. 3-3.1(b) as a comparison of Fig. 3-3.1(a). Although the orthogonality of $\bar{k}_L$ might not be as good as $\bar{k}_P$, the straightforward vectorization is more intuitive and easier to process. Thus the parameterization of $\bar{k}_L$ will be used in later deorientation analysis, which make it easy to extract information from co-pol scatterings.

From Eqs. (3-3.2~5), two sets of the parameters a, b, c of $\bar{k}_L$ and α, β of $\bar{k}_P$ can be related as

$$\cos\alpha = \sqrt{\frac{1+\sin 2a\cos 2b}{2+2/\tan^2 c}}; \quad \cos\beta = \sqrt{\frac{1-\sin 2a\cos 2b}{1-\sin 2a\cos 2b + 2/\tan^2 c}}. \tag{3-3.6}$$

When the x-pol scattering is small and negligible, it yields $c \approx 90^\circ$, $\beta \approx 0^\circ$ and

$$\cos 2\alpha = \sin 2a\cos 2b \tag{3-3.7}$$

At this time, the parameter α can be determined by the parameters a and b.

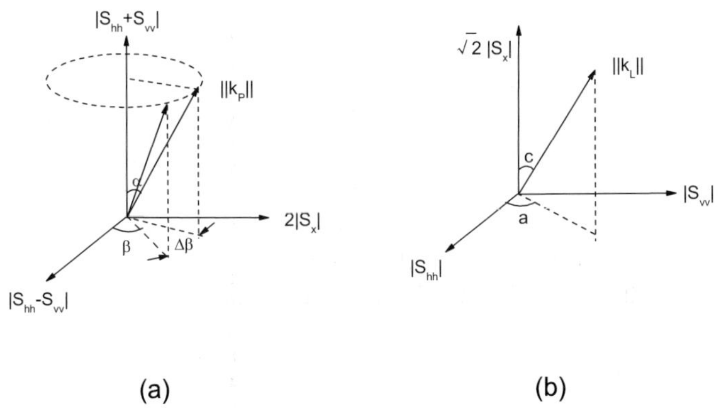

Fig. 3-3.1. Parameterization of the target scattering vectors

B. Extraction of target orientation information

Parameter β in the Pauli vectorization was not used in Cloude's $\alpha - H$ classification. Actually, under certain conditions, β denotes the target orientation on the plane perpendicular to the sight line [Cloude and Pottier 1996] linking the observer and target. The rotation transformation of scattering vector has been studied in many literatures [Lee et al. 2000, Cameron and Leung 1990, Huynen 1970]. Some derivations are presented here as a reference.

Rotating through the angle ψ of the polarization base $(\hat{h},\hat{v})$ along with the sight line, a new polarization base $(\hat{h}',\hat{v}')$ is obtained. Correspondingly, the electrical field vector $\bar{E} = [E_h \;\; E_v]^T$ becomes $\bar{E}'$ with the base $(\hat{h}',\hat{v}')$ as follows

$$\bar{E}' = [E'_h \;\; E'_v]^T = [R]\cdot\bar{E}, \qquad [R] = \begin{bmatrix} \cos\psi & -\sin\psi \\ \sin\psi & \cos\psi \end{bmatrix}. \tag{3-3.8}$$

In the backscattering direction, two bases of the incident polarization and backscattered polarization are the same (in the antenna coordinate), namely $(\hat{h},\hat{v})$. After the rotation of Eq. (3-3.8), Eq.(3-3.1) yields

$$\bar{E}_s' = [R]\cdot[S]\cdot[R]^T\cdot\bar{E}_i' = [S']\cdot\bar{E}_i', \qquad [S'] = [R]\cdot[S]\cdot[R]^T, \tag{3-3.9}$$

where $[S']$ is the new scattering matrix after the ψ rotation. Substituting $[R]$ of (3-3.8) into Eq. (3-3.9), all elements of $[S']$ are expressed as:

$$\begin{aligned} S'_{hh} &= \cos^2\psi S_{hh} + \cos\psi\sin\psi S_{hv} + \cos\psi\sin\psi S_{vh} + \sin^2\psi S_{vv} \\ S'_{hv} &= -\cos\psi\sin\psi S_{hh} + \cos^2\psi S_{hv} - \sin^2\psi S_{vh} + \cos\psi\sin\psi S_{vv} \\ S'_{vh} &= -\cos\psi\sin\psi S_{hh} - \sin^2\psi S_{hv} + \cos^2\psi S_{vh} + \cos\psi\sin\psi S_{vv} \\ S'_{vv} &= \sin^2\psi S_{hh} - \cos\psi\sin\psi S_{hv} - \cos\psi\sin\psi S_{vh} + \cos^2\psi S_{vv} \end{aligned} \tag{3-3.10}$$

From Eqs. (3-3.2),(3-3.10), the relationship between $\bar{k}_P$ and $\bar{k}'_P$ is given as

$$\bar{k}'_P = \frac{1}{\sqrt{2}}\begin{bmatrix} S'_{hh} + S'_{vv} \\ S'_{hh} - S'_{vv} \\ 2S'_{hv} \end{bmatrix} = \begin{bmatrix} 1 & 0 & 0 \\ 0 & \cos 2\psi & \sin 2\psi \\ 0 & -\sin 2\psi & \cos 2\psi \end{bmatrix}\cdot\frac{1}{\sqrt{2}}\begin{bmatrix} S_{hh} + S_{vv} \\ S_{hh} - S_{vv} \\ 2S_{hv} \end{bmatrix} = [U]\cdot\bar{k}_P, \tag{3-3.11}$$

where

$$[U] = \begin{bmatrix} 1 & 0 & 0 \\ 0 & \cos 2\psi & \sin 2\psi \\ 0 & -\sin 2\psi & \cos 2\psi \end{bmatrix}. \tag{3-3.12}$$

This formula for the rotation matrix is the same as that found in references [Lee et al. 2000, Cameron and Leung 1990, Huynen 1970].

It can be understood that the ψ rotation of the polarization base along the sight line is equivalent to the $-\psi$ angle inverse rotation of target，and vice versa. Thus, Eq.(3-3.11) also indicates the target rotation.

From Eqs. (3-3.4) and (3-3.11), those parameters of $\bar{k}_P$ and $\bar{k}'_P$ make the following relationship:

$$\begin{bmatrix} \cos\alpha' e^{j\phi_1'} \\ \sin\alpha'\cos\beta' e^{j\phi_2'} \\ \sin\alpha'\sin\beta' e^{j\phi_3'} \end{bmatrix}\cdot\left\|\bar{k}'_P\right\| = \begin{bmatrix} \cos\alpha\cdot e^{j\phi_1} \\ \sin\alpha\left(\cos 2\psi\cos\beta + \sin 2\psi\sin\beta\cdot e^{j(\phi_3-\phi_2)}\right)\cdot e^{j\phi_2} \\ \sin\alpha\left(-\sin 2\psi\cos\beta + \cos 2\psi\sin\beta\cdot e^{j(\phi_3-\phi_2)}\right)\cdot e^{j\phi_2} \end{bmatrix}\cdot\left\|\bar{k}_P\right\|, \tag{3-3.13}$$

Since the rotation transformation is a unitary transform, the norm of $\bar{k}_p$ is roll-invariant and further α is also roll-invariant [Cloude and Pottier 1996], i.e. $\alpha' = \alpha$. Under the condition of $\phi_3 - \phi_2 \approx 0, \pi$, Eq.(3-3.13) becomes

$$\begin{bmatrix} \cos\alpha' e^{j\phi_1'} \\ \sin\alpha' \cos\beta' e^{j\phi_2'} \\ \sin\alpha' \sin\beta' e^{j\phi_3'} \end{bmatrix} = \begin{bmatrix} \cos\alpha \cdot e^{j\phi_1} \\ \sin\alpha \cos(\beta \mp 2\psi) \cdot e^{j\phi_2} \\ \sin\alpha \sin(\pm\beta - 2\psi) \cdot e^{j\phi_2} \end{bmatrix} . \tag{3-3.14}$$

Therefore, we obtain $\alpha' = \alpha$, $\beta' = \beta - 2\psi$ or $-\beta - 2\psi$, which means that α remains invariant and β changes by 2ψ after the ψ angle rotation of the target along the sight line. Actually the condition of $\phi_3 - \phi_2 \approx 0, \pi$ is the condition of a symmetric target.

C. Deorientation and the minimization of cross polarization

As aforementioned, the parameter α is roll-invariant and fits the requirement of orientation independent. However, α is extracted from the sum of co-pol and contains mixture of various information of the target. We aim to extract different kinds of information more independently, which are expected to give an improvement of the characteristic description and targets classification. Therefore, a general deorientation operation should be applied to scattering data before further extraction of information.

A similar operation i.e. slope compensation has been developed by Schuler et al. (1998,1999), Lee et al. (2000,2002,2003), Hajnsek et al. (2001). This method is suggested to compensate the influence of ground slope variation and also to retrieve the topographic information by using estimated slope data. The importance and effectiveness of slope compensation have been well illustrated and demonstrated by Lee et al. (2000).

The purpose of deorientation introduced here is to reduce or eliminate the fluctuating influence of randomly distributed target orientation on polarimetric scattering, including slope variation of bare earth ground, and to show the prominence of the intrinsic scatter characteristics. Deorientation operation makes two identical targets with different orientations yield the same scattering and thus makes them as the same class, and makes different classes for those different scatter targets which might produce similar scattering before deorientation. We place more emphasis on the generic properties of the scatter targets, e.g. type, shape, etc. as for terrain surface classification.

Ideally, a complete deorientation is preferred. The extraction of target orientation on the plane perpendicular to the sight line is possible to realize. However, the other orthogonal orientation i.e. deviation away from the sight line is extremely difficult or nearly how impossible to extract completely. Therefore, only partial deorientation upon target is implemented in this session.

We make a certain angle ψ_m rotation of the target along the sight line to minimize the cross-polarized scattering (min-x-pol) of new target scattering vector and focus on co-pol scattering terms. Note that the concept of min-x-pol is different from Huynen's optimal polarization [Huynen 1970], because the transformation adopted here preserves the physical meaning, namely spatial rotation of the target along the sight line rather than a pure mathematical transform.

Here, one of the reasons why min-x-pol is adopted, is that by using min-x-pol the information can be concentrated onto the co-pol, and the corresponding parameters, e.g. u, v, become more effective in describing target characteristics and make better classification.

From the third term of $\bar{k}_P'$, x-pol scattering $k_{P,3}'$ in Eq. (3-3.13), min-x-pol yields

$$\min_{\psi} \frac{|k_{P,3}'|}{\|\bar{k}_P'\|} = \min_{\psi} \left| \sin\alpha \left(-\sin 2\psi \cos\beta + \cos 2\psi \sin\beta \cdot e^{j(\phi_3 - \phi_2)} \right) \cdot e^{j\phi_2} \right| . \tag{3-3.15}$$

Let $\frac{\partial |k'_{P,3}|^2}{\partial \psi} = 0$, $\frac{\partial^2 |k'_{P,3}|^2}{\partial \psi^2} > 0$, it yields

$$\begin{cases} \tan 4\psi = \tan 2\beta \cos(\phi_2 - \phi_3), & \cos 2\beta \neq 0 \\ \cos 2\beta \cos 4\psi > 0 \end{cases}. \tag{3-3.16}$$

Eq. (3-3.16) is in agreement with the results of maximal symmetry target decomposition of Cameron et al. (1990) and the slope estimation by circular polarization of Lee et al. (2000) (see Appendix 3-3A). In fact, Cameron's maximal diagnolization of the symmetric target decomposition is presented to obtain the symmetric target component, which is characterized by the scattering matrix diagonalization. This is actually the same as min-x-pol, which also leads to a maximal diagonalization of scattering matrix.

Assuming that $\psi_m \in [0, \pi/2)$, Eq. (3-3.16) is solved and

$$\psi_m = \left[\operatorname{sgn}\{\cos(\phi_2 - \phi_3)\} \frac{2\beta - [2\beta]_\pi + \left[\tan^{-1}\{\tan 2\beta |\cos(\phi_2 - \phi_3)|\}\right]_\pi}{4} \right]_{\frac{\pi}{2}} \tag{3-3.17}$$

where the operator $[x]_y$ means the remainder of x divided by y. The $\operatorname{sgn}\{\bullet\}$ means sign.

Under the condition of $\phi_3 - \phi_2 = 0, \pi$, Eq. (3-3.17) becomes

$$\psi_m = \left[\operatorname{sgn}\{\cos(\phi_2 - \phi_3)\} \cdot \frac{\beta}{2} \right]_{\frac{\pi}{2}}. \tag{3-3.18}$$

This ψ_m satisfies Eq. (3-3.14), and the min-x-pol equals 0 at this time. This agrees with the condition for a symmetric target [Cameron and Leung 1990]. In other words, $\phi_3 - \phi_2$ indicates the degree of non-symmetry or the target helixity [Cloude et al. 2001]. Thus, non-helix surface scattering or volumetric scattering always satisfy the condition $\phi_3 - \phi_2 \approx 0, \pi$.

After ψ_m rotation of the target along the sight line, $\bar{k}_P$ becomes the deoriented $\bar{k}_P^d$ as

$$\bar{k}_P^d = [U_m] \cdot \bar{k}_P, \tag{3-3.19}$$

where the superscript d denotes deoriented. The deorientation matrix $[U_m]$ is the transform matrix in Eq. (3-3.12) with ψ replaced by ψ_m. Similarly, we can obtain the deoriented scattering matrix $[S_d]$ and the other form of scattering vector $\bar{k}_L^d$ as well.

As mentioned above, parameterization of $\bar{k}_L^d$ is adopted after deorientation, in order to extract information from the scattering matrix more intuitively and independently. In Eq. (3-3.5) with $\sin 2a = 0$, one of the co-pol S_{vv} or S_{hh} becomes 0 and makes indeterminant of the phase difference between S_{vv} and S_{hh}. If $\sin c = 0$, both S_{vv} and S_{hh} becomes 0, leaving both a and b as indeterminant. Thus, we define a new set of parameter u, v, w:

$$\begin{cases} u = \sin c \cos 2a \\ v = \sin c \sin 2a \cos 2b \\ w = \cos c \end{cases} \tag{3-3.20}$$

where w indicates the amplitude ratio of co-pol and x-pol scatterings, u indicates the amplitude ratio of co-pol scatterings in case of little x-pol scattering ($\sin c \to 1$), and v denotes the phase difference of co-pol scatterings when both co-pols have comparable amplitudes ($\sin 2a \to 1$).

On the u, v, w space, a quarter of sphere is shown in Fig. 3-3.2. Scattering is described within the sphere. The parameters u, v, w correspond to the axes OA, B_1B_2, OC, respectively. Since the situations of $u \geq 0$ or $u < 0$ differs only on which co-pol, vv or hh, is higher and the scattering mechanism is actually the same. This ambiguity problem is also addressed by Cameron and Leung (1990). Hence, only the part of $u \geq 0$ is plotted in Fig. 3-3.2. It can be seen from Fig. 3-3.2 that scattering distribution tends to Point C as x-pol scattering becomes large, and tends to plane B_1B_2A as x-pol scattering becomes small. On plane B_1B_2A, the scattering distribution tends to Point A as relative difference of co-pol scattering magnitudes becomes large, and tends to Line B_1B_2 as both co-pol scattering components become comparable. On Line B_1B_2, the scattering distribution is shifted from Point B_1 to B_2 as the phase difference of two co-pol scatterings decrease.

To show the distribution of typical scatter targets in Fig. 3-3.2(a): the helix scatterer is near Point C, the dipole scatterer is near Point A, a spherical scatterer is near Point B_1, the dihedral scatterer is near Point B_2, and special material scatterer, artificial device or complicated multiple scattering is near Point O. Most of the simple scattering targets can be described by the parameters u, v on plane B_1B_2A. In this case, the condition of Eq. (3-3.6) is satisfied and $v = \sin 2a \cos 2b = \cos 2\alpha$. We can found that on plane B_1B_2A the parameter v is equivalent to α of Pauli parameterization.

The scattering distribution and parameter relationships are plotted in Fig. 3-3.2(b), which is consistent with Cloude's explanation of α [Cloude 1986]. It can be seen from Eq. (3-3.20) and Fig. 3-3.2(b) that the parameter v is determined by $\sin 2a$ and $\cos 2b$, while $\sin 2a$ limits the v range and $\cos 2b$ makes v varying within $-\sin 2a$ and $\sin 2a$. The set of u, v contains all of the information that α has, and can present more characteristic information of scattering mechanisms. It demonstrates the feasibility of more precise classification by using the set of parameters u, v.

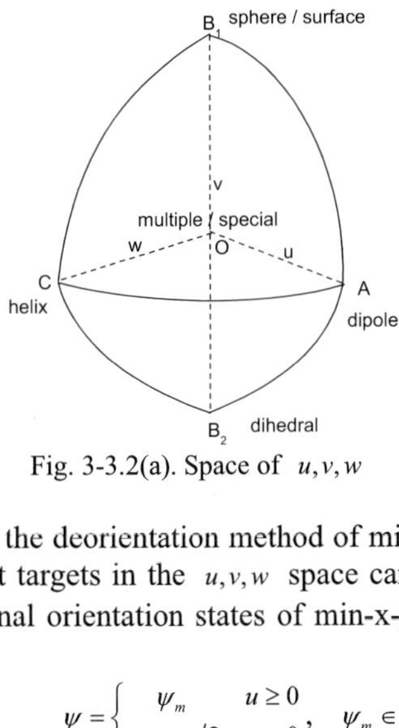

Fig. 3-3.2(a). Space of u, v, w

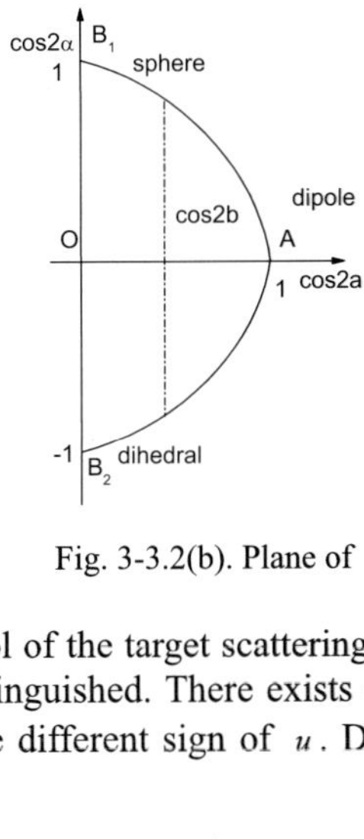

Fig. 3-3.2(b). Plane of u, v

Via the deorientation method of minimizing x-pol of the target scattering vectors, different targets in the u, v, w space can be well distinguished. There exists a pair of orthogonal orientation states of min-x-pol due to the different sign of u. Define ψ as

$$\psi = \begin{cases} \psi_m & u \geq 0 \\ \psi_m + \pi/2 & u < 0 \end{cases}, \quad \psi_m \in [0, \pi/2), \quad \psi \in [0, \pi), \tag{3-3.21}$$

As $w \to 1$, the target is a helix. As $w \to 0$ and $|u| \to 1$, it is a dipole. As $w \to 0, |u| \to 0$ and $v \to 1$, it is a sphere. As $w \to 0, |u| \to 0$ and $v \to -1$, it is a dihedral. And as $w \to 0$, $|u| \to 0$ and $v \to 0$, it means multiple scattering or special target scattering. The angle ψ denotes the target orientation deviation from the min-x-pol state.

D. Deorientation of non-deterministic targets

The Mueller matrix $[M]$, the coherency matrix $[T]$ and the covariance matrix $[C]$ can be obtained from the second moment of random targets scattering. The matrix $[M]$ is real, both of $[T]$ and $[C]$ are positive semi-definite Hermitian with positive real eigenvalues and orthogonal eigenvectors. Since T and C conserve similarity transformation, they have the same eigenvalues and eigenvectors [Cloude unitary and Pottier 1996] and are expressed as

$$[T]=\sum_{i=1}^{3}\lambda_i\bar{k}_{P,i}\cdot\bar{k}_{P,i}^{+},\quad [C]=\sum_{i=1}^{3}\lambda_i\bar{k}_{L,i}\cdot\bar{k}_{L,i}^{+} \tag{3-3.22}$$

where $\lambda_1\geq\lambda_2\geq\lambda_3\geq 0$ are the eigenvalues, and $\bar{k}_{P,i}$, $\bar{k}_{L,i}$ are the eigenvectors, correspondingly. Each eigenvector is an incoherent target scattering vector and can be parameterized into a set of target decomposition parameters. The entropy H is defined as [Cloude 1986, Cloude et al. 2001, Jin and Cloude 1994]

$$H=-\sum_{i=1}^{3}P_i\log_3 P_i \tag{3-3.23}$$

where $P_i=\lambda_i/\Sigma_{i=1}^{3}\lambda_i$ means the probability or priority of each incoherent scattering vector occurring among randomly distributed targets. In previous literatures, e.g. Cloude and Pottier (1997), the average parameters extracted from every eigenvectors are used to describe the average characteristics of randomly distributed targets as

$$\langle y\rangle=\sum_{i=1}^{3}p_i y_i \tag{3-3.24}$$

where y_i denotes one of the target decomposition parameters extracted from the eigenvector $\bar{k}_{P,i}$.

When the first eigenvalue is significantly larger than others, main properties of the targets are described by the first, i.e. principal, eigenvector. However, in the case of complex random targets with multiple scattering, e.g. multi-layered forest canopy with volumetric scattering of crown leaves, branches and trunks, surface scattering of underlying rough surface and interactions of canopy scatters and surface, the second and the third eigenvalues become noticeable. Each of them indicates a kind of scattering and contains its own information about the whole layer of random targets, which might be missed by the average method adopted by previous literatures.

Thus, we focus on the principal eigenvector and its parameterization.

Last but not least, two issues need to be addressed here. First, Lee et al. (2000) carried out a comparison of several methods for estimation of orientation from the coherency matrix in a non-deterministic situation. By using PolSAR data experiments of slope retrieval, Lee et al. (2000) conclude that directly using the ensemble averaged circular polarization to estimate the slope of ground is better than using the circular polarization derived from the principal eigenvector after eigen-analysis. However, the deorientation can only be implied to each single incoherent scattering vector, but rather than the whole coherency matrix, and the eigen-analysis method is suitable for the decomposition of a mixture of multiple types of targets.

Second, as presented by Schuler et al. (1998,1999) and Lee et al. (2000,2002), both contributions from targets in common sense e.g. vegetation, and surroundings e.g. underlying soil, are prominent to the orientation estimation, especially in L-band PolSAR data. It is regarded that the eigen-analysis will decompose the different types of scattering i.e. target and background, into different eigenvectors, and only the most significant scattering will be considered for deorientation and stand for the whole targets.

3-3.2 Numerical Simulation and Analysis

A. A single non-spherical particle

Consider scattering from a single non-spherical particle as shown in Fig. 3-3.3a. The local coordinate $(\hat{x}_b, \hat{y}_b, \hat{z}_b)$ of the particle is related to the principal coordinate $(\hat{x}, \hat{y}, \hat{z})$ by the Euler angles ω, ξ, η [Jin 1994]. Define the incident coordinate as $\left(\hat{h}, \hat{v}, -\hat{k}_i\right)$ shown in Fig. 3-3.3(b) and the Euler angles ω_0, ξ_0, η_0 of the local coordinate $\left(\hat{h}, \hat{v}, -\hat{k}_i\right)$.

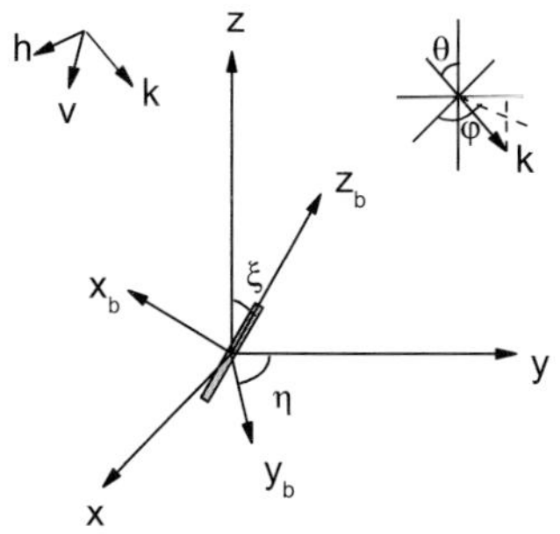

Fig. 3-3.3(a). In the principal coordinate

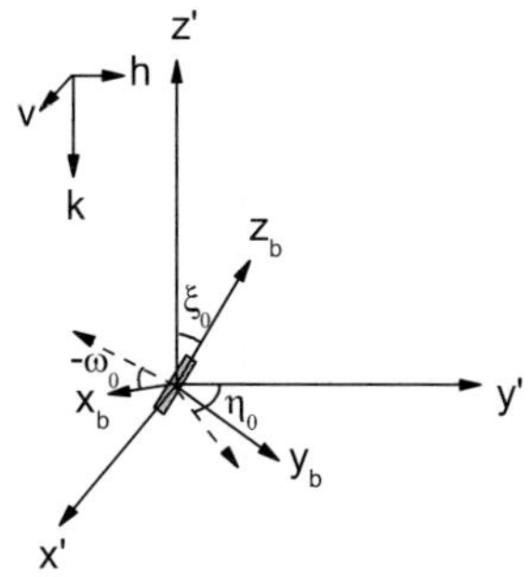

Fig. 3-3.3(b). In the incident coordinate

The relationship between ξ_0, η_0 and ξ, η can be derived as follows

$$\begin{cases} \cos\xi_0 = \sin\xi\cos\eta\sin\theta\cos\varphi - \sin\xi\sin\eta\sin\theta\sin\varphi + \cos\xi\cos\theta \\ \cos\eta_0\sin\xi_0 = \cos\xi\sin\theta - \sin\xi\cos\theta\cos(\eta+\varphi) \\ \sin\eta_0\sin\xi_0 = -\sin\xi\sin(\eta+\varphi) \end{cases} \tag{3-3.25}$$

Let $\varphi_i = 0^\circ$, Eq. (3-3.25) can be simplified as

$$\begin{cases} \cos\xi_0 = \sin\xi\cos\eta\sin\theta + \cos\xi\cos\theta \\ \cos\eta_0\sin\xi_0 = \cos\xi\sin\theta - \sin\xi\cos\theta\cos\eta \\ \sin\eta_0\sin\xi_0 = -\sin\xi\sin\eta \end{cases} \tag{3-3.26}$$

Note that in the incident coordinate, the incident angle is always 0°. Given that the incident angle is 0°, the scattering matrix of a small spheroid under Rayleigh approximation [Jin 1994] is written as

$$\begin{cases} S_{0_{hh}} = \dfrac{v_0 k^2}{4\pi}\left[G_a + (G_c - G_a)\sin^2\xi_0\sin^2\eta_0\right] \\ S_{0_{vv}} = \dfrac{v_0 k^2}{4\pi}\left[G_a + (G_c - G_a)\sin^2\xi_0\cos^2\eta_0\right] \\ S_{0_{hv}} = S_{0_{vh}} = \dfrac{v_0 k^2}{4\pi}(G_c - G_a)\sin^2\xi_0\sin\eta_0\cos\eta_0 \end{cases} \tag{3-3.27}$$

where $G_a = 1/\left[1/(\varepsilon_s - 1) + g_a\right]$, $G_c = 1/\left[1/(\varepsilon_s - 1) + g_c\right]$, and v_0 is the particle volume, k is the wave number; g_a and g_c are the functions related with the particle shape (three axes of a spheroid are assumed as $\ell_a = \ell_b, \ell_c$) [Jin 1994]. It has

$$2g_a + g_c = 1 \tag{3-3.28}$$

As $\ell_a = \ell_b > \ell_c$ for oblate (disk-like) spheroid, g_c is given as

$$g_c = \frac{1}{1-\rho^2}\left[1 - \frac{\rho}{\sqrt{1-\rho^2}}\tan^{-1}\frac{\sqrt{1-\rho^2}}{\rho}\right] \tag{3-3.29}$$

As $\ell_a = \ell_b < \ell_c$ for prolate (needle-like) spheroid, g_c is given as

$$g_c = \frac{1}{\rho^2 - 1}\left[\frac{\rho}{2\sqrt{\rho^2-1}}\ln\frac{\rho+\sqrt{\rho^2-1}}{\rho-\sqrt{\rho^2-1}} - 1\right] \tag{3-3.30}$$

where $\rho = \ell_c/\ell_a$. It can be seen that g_c and ρ conserves a monotonic relation. Hence, g_c can be used as a description of the particle shape property.

Following fom Eqs. (3-3.2a, 3-3.27), the target scattering vector $\bar{k}_P$ is derived as

$$\bar{k}_P = \frac{v_0 k^2}{4\pi}\begin{bmatrix} 2G_a + (G_c - G_a)\sin^2\xi_0 \\ -(G_c - G_a)\cos 2\eta_0 \sin^2\xi_0 \\ (G_c - G_a)\sin 2\eta_0 \sin^2\xi_0 \end{bmatrix} \tag{3-3.31}$$

Further from Eq. (3-3.4), the target decomposition parameters are derived as

$$\begin{cases} \alpha = \tan^{-1}\left[\dfrac{|G_c - G_a|\sin^2\xi_0}{|2G_a + (G_c - G_a)\sin^2\xi_0|}\right] \\ \beta = \cos^{-1}|\cos 2\eta_0| \\ \phi_1 = \arg\left[2G_a + (G_c - G_a)\sin^2\xi_0\right] \\ \phi_2 = \arg\left[-(G_c - G_a)\cos 2\eta_0\right] \\ \phi_3 = \arg\left[(G_c - G_a)\sin 2\eta_0\right] \end{cases} \tag{3-3.32}$$

It is seen that $\phi_3 - \phi_2 = 0, \pi$. From Eq. (3-3.18), the rotation angle of min-x-pol is

$$\psi_m = \begin{cases} \beta/2 & \phi_2 - \phi_3 = 0 \\ (\pi - \beta)/2 & \phi_2 - \phi_3 = \pi \end{cases} \tag{3-3.33}$$

and further from Eqs. (3-3.32, 3-3.33), it yields

$$\psi_m = \pi/2 - [\eta_0]_{\frac{\pi}{2}} \tag{3-3.34}$$

Substituting Eqs. (3-3.32~34) into Eq. (3-3.19), the deorientationed target scattering vector $\bar{k}_P^d$ can be written as

$$\bar{k}_P^d = \frac{v_0 k^2}{4\pi}\begin{bmatrix} 2G_a + (G_c - G_a)\sin^2\xi_0 \\ \pm(G_c - G_a)\sin^2\xi_0 \\ 0 \end{bmatrix} \tag{3-3.35}$$

where the sign $\pm$ is positive if $\eta_0 \in [0, \pi/2)$, and negative if $\eta_0 \in (\pi/2, 0)$. Following Eq. (3-3.2a), the deoriented $\bar{k}_L^d$ is written as

$$\bar{k}_L^d = \frac{v_0 k^2}{4\pi}\begin{bmatrix} G_a + (G_c - G_a)\sin^2\xi_0 \\ 0 \\ G_a \end{bmatrix} \text{ or } \frac{v_0 k^2}{4\pi}\begin{bmatrix} G_a \\ 0 \\ G_a + (G_c - G_a)\sin^2\xi_0 \end{bmatrix} \tag{3-3.36}$$

and by Eq. (3-3.5), the parameters are derived as

$$\begin{cases} a = \tan^{-1}\left|\dfrac{G_a}{G_a + (G_c - G_a)\sin^2\xi_0}\right| \text{ or } \tan^{-1}\left|\dfrac{G_a + (G_c - G_a)\sin^2\xi_0}{G_a}\right| \\ b = \pm\dfrac{1}{2}\arg\left[\dfrac{G_a}{G_a + (G_c - G_a)\sin^2\xi_0}\right] \\ c = \pi/2 \end{cases} \tag{3-3.37}$$

Further by Eqs. (3-3.20, 3-3.21), it yields the set of u, v, w, ψ as

$$\begin{cases} w = 0 \\ u = \pm \dfrac{|G_a|^2 - |G_a + (G_c - G_a)\sin^2\xi_0|^2}{|G_a|^2 + |G_a + (G_c - G_a)\sin^2\xi_0|^2} \\ v = 1 - \dfrac{|(G_c - G_a)\sin^2\xi_0|^2}{|G_a|^2 + |G_a + (G_c - G_a)\sin^2\xi_0|^2} \\ \psi = \eta_0 \end{cases} \qquad (3\text{-}3.40)$$

Eq. (3-3.40) gives the relationship between the particle's parameters (e.g. $\ell_a, \ell_c, \varepsilon_s$) and the target decomposition (e.g. u, v).

Fig. 3-3.4(a) shows the relationship between $|u|$ and the orientation ξ_0, the shape g_c of a particle. As the particle shape varies from needle-like to spherical, g_c changes from $0 \to 1/3$, and $|u|$ decreases monotoically. As the particle shape varies from sphere to disk-like, g_c changes from $1/3 \to 1$ and $|u|$ increases monotonically. As the angle ξ_0 varies from $0 \to \pi/2$, $|u|$ increases monotonically in the case of a needle-like or a disk-like particle, and remains constant in the case of a spherical particle. It demonstrates that $|u|$ indicates the non-symmetry of the particle's projection along with the sight line.

Fig. 3-3.4(b) shows the relationship between v and ξ_0, g_c. It shows that v is insensitive to either orientation or shape of the particle.

Fig. 3-3.4(c) gives the relationship between $|u|$, v and $|\varepsilon_s|$ of the complex dielectric constant $\varepsilon_s = \varepsilon_s' + i\varepsilon_s''$ of the needle-like particle. As $|\varepsilon_s|$ increases, $|u|$ increases monotonically and v decreases in opposite way, as shown in Eq. (3-3.20). However, as $\tan^{-1}(\varepsilon_s''/\varepsilon_s')$ increases, $|u|$ varies very slowly and v decreases monotonically. This means that $|\varepsilon_s|$ affects $|u|$, and $\varepsilon_s''/\varepsilon_s'$ affects v.

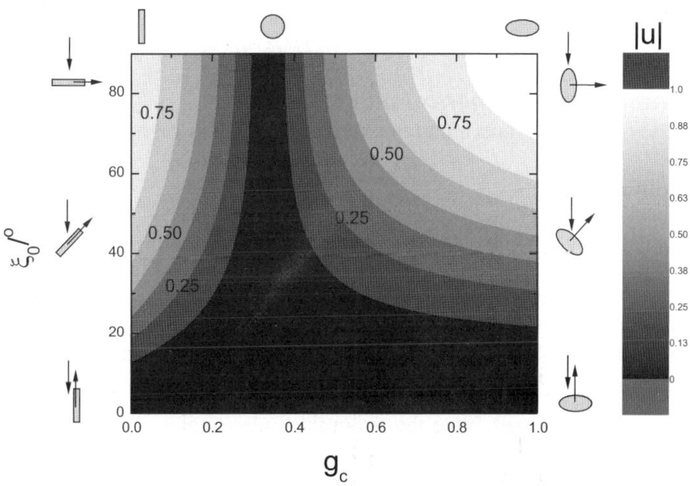

Fig. 3-3.4(a). Relationship between $|u|$ and ξ_0, g_c ($\varepsilon_s = 6.5 + 2.5i$)

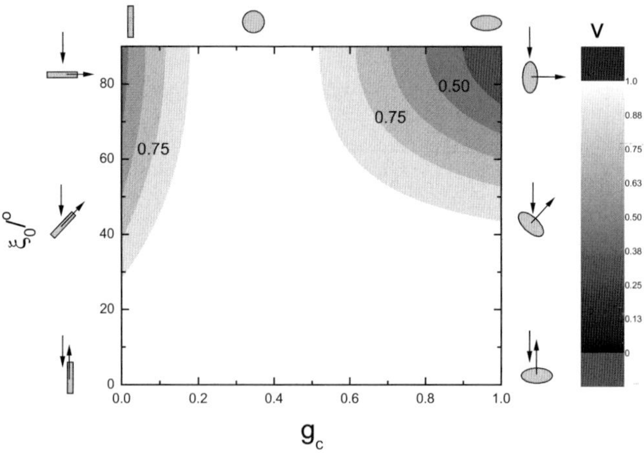

Fig. 3-3.4(b). Relationship between v and ξ_0, g_c ($\varepsilon_s = 6.5 + 2.5i$)

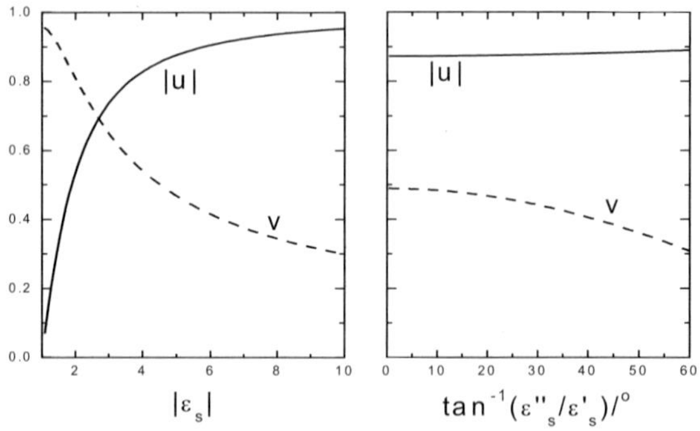

Fig. 3-3.4(c). Relationship between $|u|, v$ and ε_s
($\xi_0 = \pi/2$, $g_c = 0.9$, $\tan^{-1}(\varepsilon_s''/\varepsilon_s') = 20^\circ$ or $|\varepsilon_s| = 5$)

B. A layer of random non-spherical particles

The coherency matrix of random non-spherical particles is calculated by

$$\overline{\overline{T}} = \int_{x_1}\int_{x_2}\cdots\int_{x_n} \overline{k}_P \cdot \overline{k}_P^+ P_n(x_n)dx_n \ldots P_2(x_2)dx_2 P_1(x_1)dx_1 \tag{3-3.41}$$

where $x_1, x_2, \ldots, x_n$ are the parameters of random scatter properties, $P_1(x_1), P_2(x_2), \ldots, P_n(x_n)$ are the corresponding probability distribution function. If the target scattering vector is $\overline{k}_L$ in Eq. (3-3.41), the covariance matrix $\overline{\overline{C}}$ is then obtained.

We consider a layer of random needle-like particles whose random orientation is described by normal distribution of the Euler angles ξ, η with the means and variances $\langle\xi\rangle, \Delta\xi^2, \langle\eta\rangle, \Delta\eta^2$, respectively. Making deorientation of the principal target scattering vector, a set of $\psi, |u|, v, w$ is obtained. Note that from the Euler angles $\langle\xi\rangle, \langle\eta\rangle$, the corresponding Euler angles $\langle\xi_0\rangle, \langle\eta_0\rangle$ in the incident coordinate are obtained according to Eq. (3-3.26).

Fig. 3-3.5(a) gives the relationship between $\langle\eta_0\rangle$ and ψ on a different range $\Delta\eta$. Since ψ is derived from the principal scattering vector, it keeps a linear relationship between ψ and $\langle\eta_0\rangle$. Fig. 3-3.5(b) gives the relationship between the averaged $\langle\psi\rangle$ and $\langle\eta_0\rangle$. Here $\langle\psi\rangle$ is calculated by Eq. (3-3.24). It can be seen that the linearity between $\langle\psi\rangle$ and $\langle\eta_0\rangle$ becomes degraded, which has been discussed by Cloude et al. (1999).

Simulations also show that other parameters, $|u|, v, w$ and H, are insensitive to varying $\langle\eta_0\rangle$. However, as $\Delta\eta$ increases, it causes H to increase and $|u|$ to decrease. It is known that H indicates randomness and $|u|$ denotes the symmetry of targets; larger range of $\Delta\eta$ produces overall symmetry and uniformity of the layer of random targets.

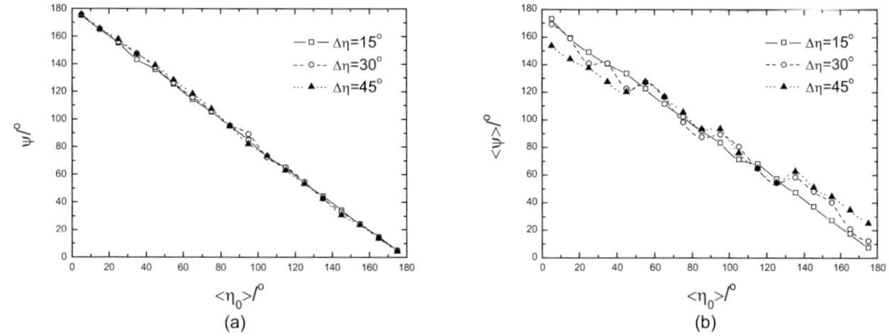

Fig. 3-3.5. Relationship between $\psi, \langle\psi\rangle$ and $\langle\eta_0\rangle$ with $\Delta\eta$ ($\xi_0 = 90^\circ, \Delta\xi = 15^\circ, \theta_i = 40^\circ, \varphi_i = 0^\circ, g_c = 0.9, \varepsilon_s = 6.5 + 2.5i$)

Fig. 3-3.6(a) gives the relationship between $|u|$ and $\sin^2\xi_0$. It shows a good linearity between $\sin^2\xi_0$ and $|u|$. Fig. 3-3.6(b) shows ψ versus $\langle\xi_0\rangle$ with $\eta_0 = 45^\circ$. It can be seen that as $\langle\xi_0\rangle$ increases, the projection of the particle along the sight line becomes asymmetrical, and ψ tends to be $180^\circ - \langle\eta_0\rangle$. This indicates no-orientation issue for symmetrical target.

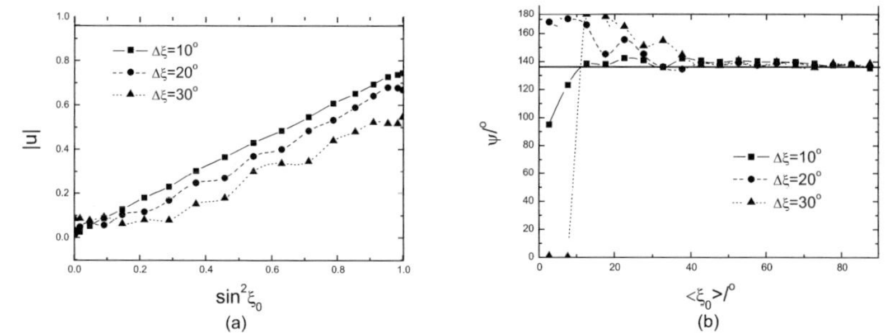

Fig. 3-3.6. Relationship between $\sin^2\xi_0$ and $|u|$, $\langle\xi_0\rangle$ and ψ with different $\Delta\xi$ ($\eta_0 = 45^\circ, \Delta\eta = 30^\circ, \theta_i = 40^\circ, \varphi_i = 0^\circ, g_c = 0.9, \varepsilon_s = 6.5 + 2.5i$)

Let g_c be a normal distribution with the mean $<g_c>$ and variance Δg_c^2. Fig. 3-3.7 shows $|u|$ versus g_c with different Δg_c. It shows that $|u|$ may also describe the averaged shape of a layer of random particles.

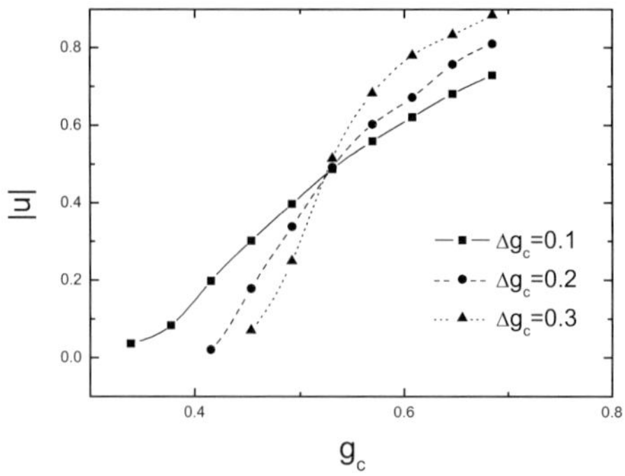

Fig. 3-3.7. Relationship between $|u|$ and g_c with different Δg_c
($\xi_0 = 90^\circ, \Delta\xi = 15^\circ, \eta_0 = 45^\circ, \Delta\eta = 30^\circ, \theta_i = 40^\circ, \varphi_i = 0^\circ, \varepsilon_s = 6.5 + 2.5i$)

C. A layer of random non-spherical particles above a rough surface

A model of a layer of random non-spherical particles above a rough surface is applied to numerical simulation of *u, v, H* at incidence 30° and L band 1.25GHz, as indicated in Table 3-3.1.

Table 3-3.1. Simulation model and parameters for some typical terrain surfaces

Type	Model and Parameters
Soil Land	An IEM rough surface without the canopy [Fung 1994], with *rms* height $\delta = 2 \sim 4cm$, correlation length $L = 8cm$ and dielectric constant $\varepsilon_2 = 15 + 2i$.
Sea Surface	Similar to the soil land but $\varepsilon_2 = 70 + 40i$,
Vegetation Type1	A layer of non-spherical particles above a rough surface. The scatter particles: $\ell_a = 0.1cm, \ell_c = 5cm$, $\varepsilon_s = 6.5 + 2.5i$, fractional volume $f_s = 0.005$, layer depth $d = 150cm$, Euler angle η distribution over $[0^\circ, 30^\circ] \sim [0^\circ, 90^\circ]$, ξ distribution over $[0^\circ, 360^\circ]$, and underlying rough surface: $\delta = 0.3cm$, $L = 10cm$, $\varepsilon_2 = 15 + 2i$.
Vegetation Type2	Similar to the vegetation type 1, but $\varepsilon_s = 22.5 + 7.5i$, $\ell_a = 0.1cm, \ell_c = 3cm$.
Forest	A layer of hybrid scatterers (spheroids for leaves and cylinders for trunks) above a rough surface. The spheroids parameters are: $\ell_a = 0.1cm,\ \ell_c = 3cm,\ \varepsilon_s = 6.5 + 2.5i,\ d = 300cm,\ \eta \in [0^\circ, 60^\circ],\ \xi \in [0^\circ, 360^\circ]$; the cylinders: radius $r = 15cm$, hight $h = 500cm$, dielectrical constant $\varepsilon_s' = 10 + 2i$; the underlying rough surface: $\delta = 0.3cm$, $L = 10cm$, $\varepsilon_2 = 15 + 2i$. Note that the fractional volume of trunks is $f_s' = 0.0005$, and the fractional volume of leaves varies as $f_s = 0.0001 \sim 0.005$ for sparse or dense forests.

Fig. 3-3.8(a) presents the distribution of some terrain surfaces on the $|u|-H$ plane. It can be seen that weak and single surface scatterings of the soil land and oceanic surface cause small H. In the case of vegetation canopy or forests, more scattering causes H increasing, and reach high value for thick (multi-layering) and dense forests. As the scatters of the dense forests become uniform, especially dominated by homogeneous crown leaves, H is then decreased.

Different dielectric properties of the soil land and oceanic surface also makes u identifiable on the $|u|$ axis.

Uniformity and symmetry of the random scatters makes $|u|$ decrease. The type 1 and 2 of the vegetation canopy can be identified in Fig. 3-3.8(a).

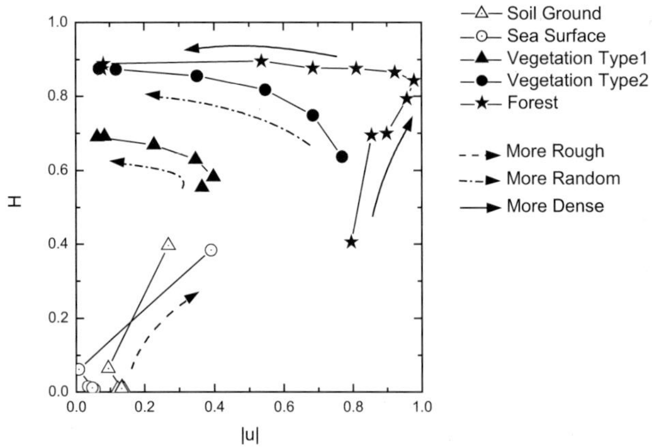

Fig. 3-3.8(a). Distribution of some typical terrain surfaces on the $|u|-H$ plane

Scattering in the Mueller matrix of the vegetation canopy model includes the contributions from: the underlying rough surface M_0, the scatter trunks and leaves M_1, double bounce of the underlying surface and trunks or leaves M_2, triple bounce of the underlying surface and trunks or leaves M_3, and so on. Fig. 3-3.8(b) presents the distribution of these scattering terms on the plane $|u|-v$. It can be seen that single scatterings locate on the area $v>0$, double scatterings locate on the area $v<0$, high-order triple scatterings locate on the area around $v=0$ because of the minimum x-pol enhanced by high order scattering. Multi-path scattering makes large x-pol and v approaches to zero.

Scattering contributed by different terms $M_i (i=1,2,3)$ and total scattering M, as the canopy changes from sparse to dense scatters, are shown on the $|u|-v$ plane of Fig. 3-3.8(b). It can be seen that dominant scattering is gradually shifted from trunks M_2 to crown M_1.

Briefly summarizing, H indicates the canopy randomness, u is useful for distinguishing different terrain targets and v is helpful to take account of scattering mechanisms of different-orders.

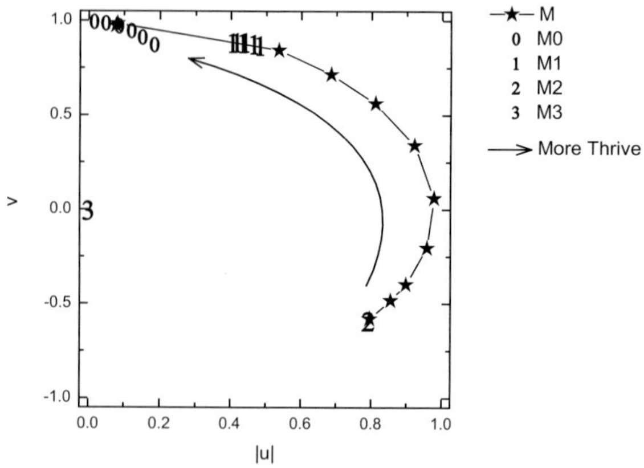

Fig. 3-3.8(b) Distribution of different scattering mechanisms on the $|u|-v$ plane

3-3.3 Applications toTerrain Surfaces Classification

A. Classification based on deorientation

The flow-chart of the classification method of PolSAR Data based on deorientation is shown in Fig. 3-3.9.

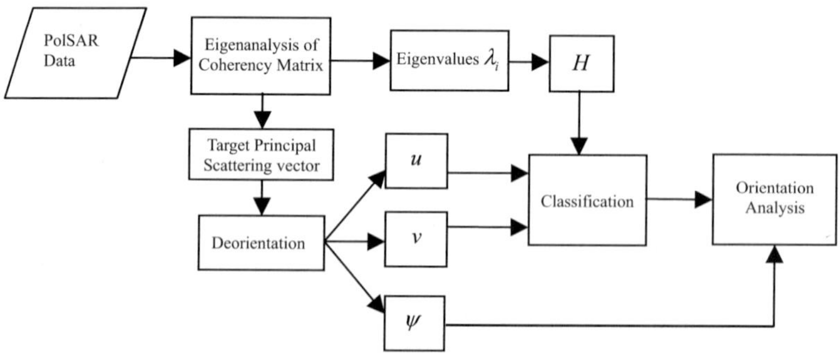

Fig. 3-3.9. Flow-chart of classification based on deorientation

We classify the terrain surfaces into 19 classes by u, v, H. According to different v, define single scattering as $(v > 0.2)$, double scattering as $(v < -0.2)$ and multiple scattering as $(-0.2 \le v \le 0.2)$. Further, single scattering and double scattering are divided into 9 classes as shown in Fig. 3-3.10, respectively. It is believed that triple or higher-order scatterings are not significant in most terrain surfaces at AIRSAR or SIR-C frequencies. The classification spectrum by single and double scatterings are shown in Fig. 3-3.10.

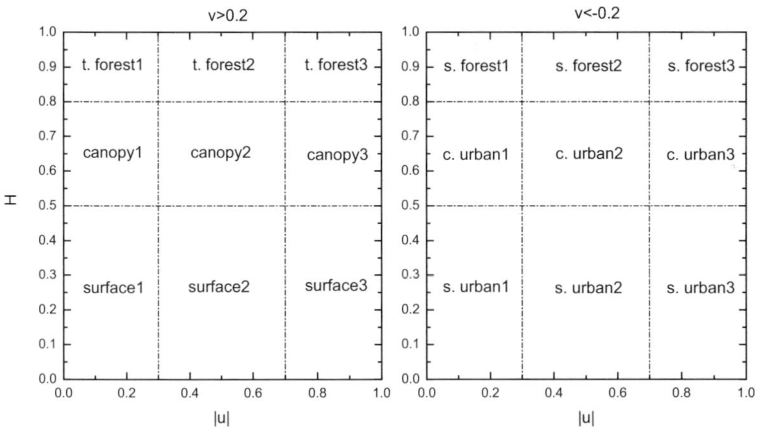

Fig. 3-3.10. Classification of single and double scatterings

We classify the single scattering as the surfaces-type if $H < 0.5$, as the canopy-type if $0.5 \le H \le 0.8$, and as the forests-type if $H > 0.8$. Further, each class of single scatterings is divided into 3 classes as $|u| < 0.3$, $0.3 \le |u| \le 0.7$ and $|u| > 0.7$.

Double scatterings are classified into simple urban area (s. urban) if $H < 0.5$, complex urban area (c. urban) if $0.5 \le H \le 0.8$ and forests-type if $H > 0.8$, where double bounce is due to the interactions between sparse trunks and ground (s. forest). Similarly, each class of double bounce is further divided into 3 classes as $|u| < 0.3$, $0.3 \le |u| \le 0.7$ and $|u| > 0.7$.

Note that those thresholds above are tunable in any practical applications.

B. Application to SIR-C data

As an example, the data of NASA/JPL SIR-C over China Guangdong Huiyang district in 1994 is tested for classification. Fig. 3-3.11 is the image of the total power of SIR-C data at L band.

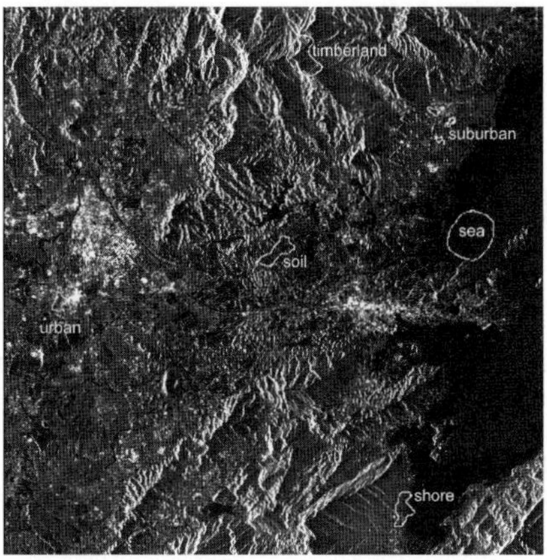

Fig. 3-3.11. Total power of SIR-C data at Huiyang, L band

In Fig. 3-3.11, five regions of soil land, timberland, urban area, suburban, shore and ocean are chosen according to the geographic map of this area.

The u,v,H distribution map of these terrain surfaces are shown in Fig. 3-3.12. It can be seen that ocean is mainly located on the zone of surface 2, soil land on the zone of surface 1, timberland on the zone of forest, urban area on the c. urban 3 and s. urban 3, suburban on the c. urban 1 and s. urban 1, and shore on the zone of canopy 3. This result proves the validity and effectiveness of the u,v,H classification approach.

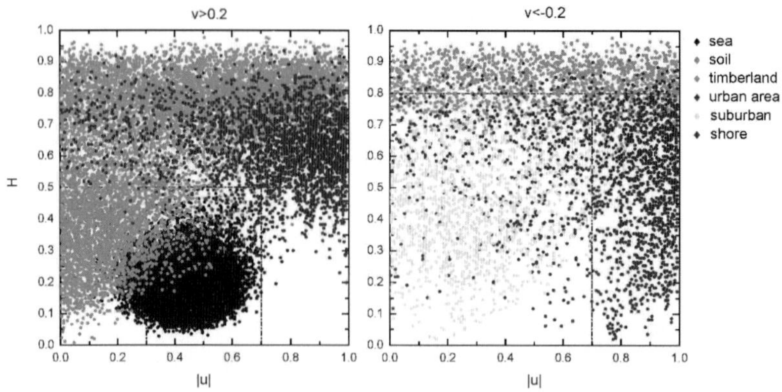

Fig. 3-3.12. Distributions of different terrain surfaces on the u,v,H spectrums

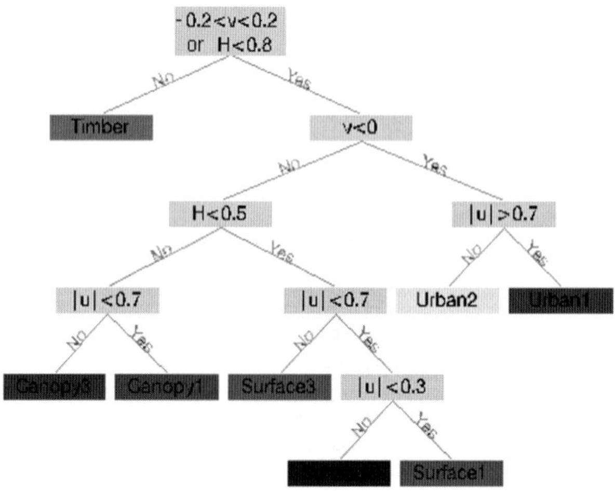

Fig. 3-3.13. Decision-tree

According to the distributions on Fig. 3-3.12, the terrain surfaces are divided into 8 classes following the decision-tree of Fig. 3-3.13: Timber, Urban 2, Urban 1, Canopy 2, Canopy 1, Surface 3, Surface 2 and Surface 1. After the u,v,H classification, a post-processing i.e. majority-rule analysis within a 5×5 window, is applied in order to make the classes distributing have continuity.

Comparing the result to geographic data, the terrain surfaces classified are well matched with the ground truth. The final classifications using u,v,H are shown in Fig. 3-3.14. The main types of land and site names in this district are labeled in Fig. 3-3.14 for comparing.

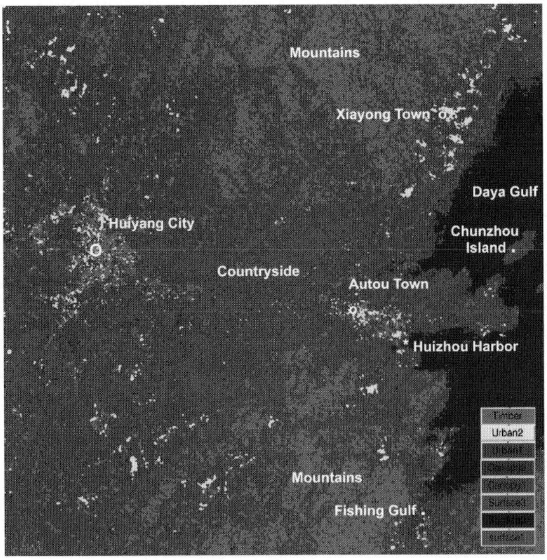

Fig. 3-3.14. The u,v,H classification of SIR-C data over Huiyang

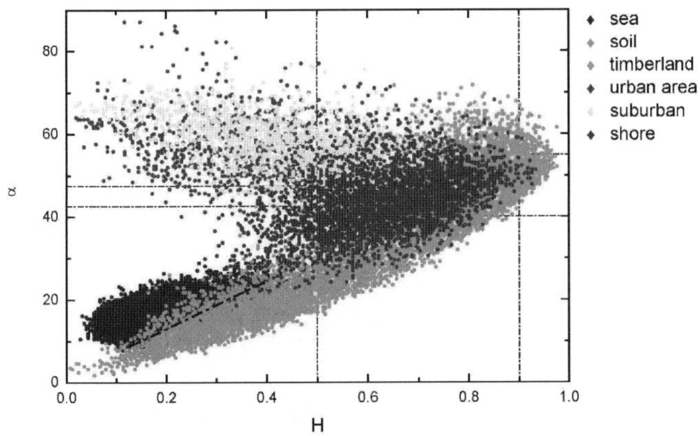

Fig. 3-3.15(a). Distributions of different terrain surfaces on the $\alpha-H$ plane

Comparing with the $\alpha-H$ method [Cloude and Pottier 1997], Fig. 3-3.15(a) gives the distributions of 5 regions on the α,H plane as selected in Fig. 3-3.11. It can be seen that ocean and soil land are intermixed in the Zone 9 and are not well separated. A dot-dashed line for possible separation is tilted on the $\alpha-H$ plane and cannot be determined by a single α. Moreover, the urban and suburban areas, and the shore and timber areas are also more or less confused.

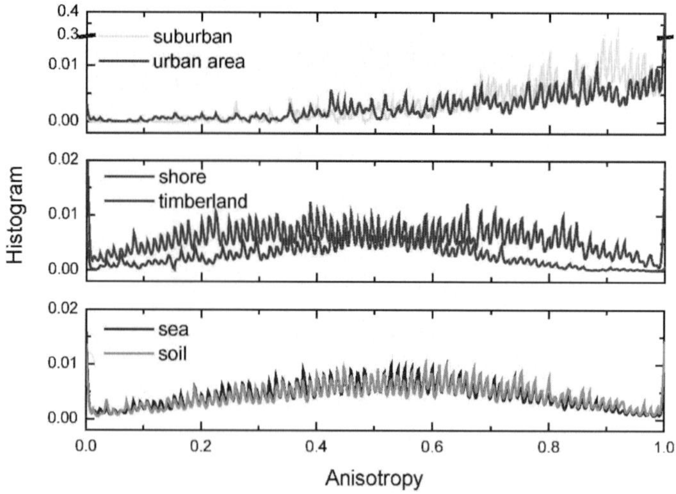

Fig. 3-3.15(b). Anisotropy histogram of different terrain surfaces

As an equal comparison, a 3-D parameter classification algorithm i.e. $\alpha - H - A$ [Pottier and Lee 1999] is carried out comparing to u, v, H 3-D parameter method. Here, anisotropy A is a parameter derived from eigenvalues, and its classification ability is also tested by those confused pairs of terrain surfaces in $\alpha - H$ plane, namely suburban and urban area, shore and timberland, sea and soil. The normalized histogram of each pair of terrain surfaces is plotted in Fig. 3-3.15(b). It can be seen that, if further divided by a extra threshold of $A > 0.5$, only the timberland and shore can be further classified, while sea and soil, suburban and urban area are still confused.

By the condition $A > 0.5$ or $A \leq 0.5$, each class in $\alpha - H$ plane is further divided into 2 classes. Thus, the final classification of 18 classes Z1-1, Z1-2, …, Z9-1, Z9-2 based on the H-A-α spectrum [Pottier and Lee 1999] is shown in Fig. 3-3.16, as a comparison of Fig. 3-3.14. Note that the 5×5 majority-rule analysis as a post-processing is also applied after classification.

As a primary and simple classification method, the set of u, v, H is better than single $\alpha - H$ or H-A-α for classifications of certain terrain surfaces. The set of u, v strengthens the capability to distinguish different targets. Another reason for the defects in $\alpha - H$ is due to the averaged parameters over all eigenvectors of the coherency matrix and leading to tilted separation lines on the $\alpha - H$ plane. As H becomes large, the parameter α is restricted within the valid zone [Cloude and Pottier 1997], i.e. high entropy leads to compression of the dynamic range of α. It makes the separation line tilted and degrades the classification capability.

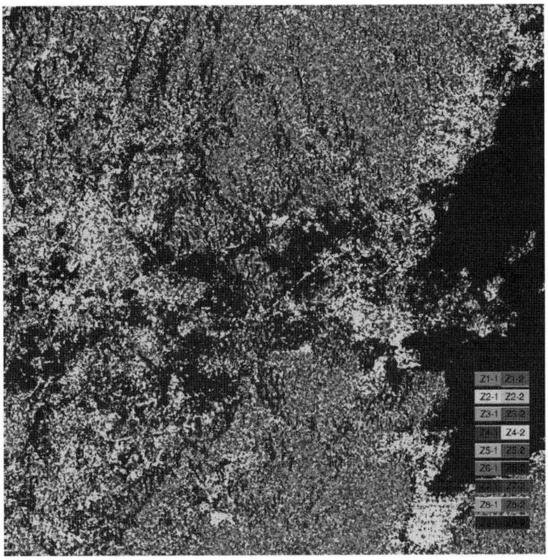

Fig. 3-3.16. The $\alpha - H$ classification as a comparison of Fig. 3-3.14

C. Orientation analysis

An AIRSAR data at L band of the Boreal area in Canada with rich resources of vegetation and an underlying flat ground is chosen for orientation analysis. The main kind of vegetation canopy on this area is the needle-leaf forests [Plummer and Curran 1998].

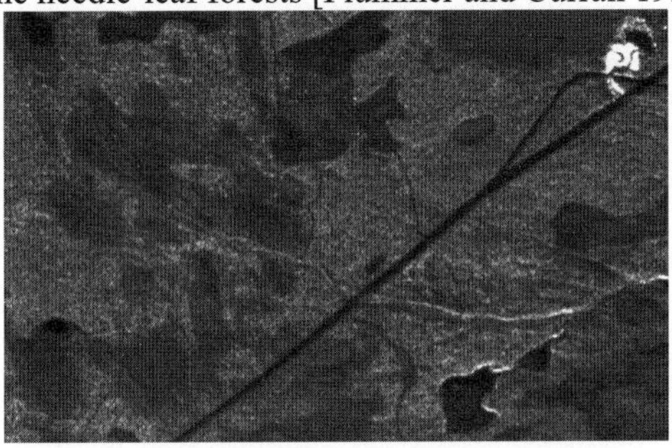

Fig. 3-3.17. Total power of AIRSAR data, Boreal area

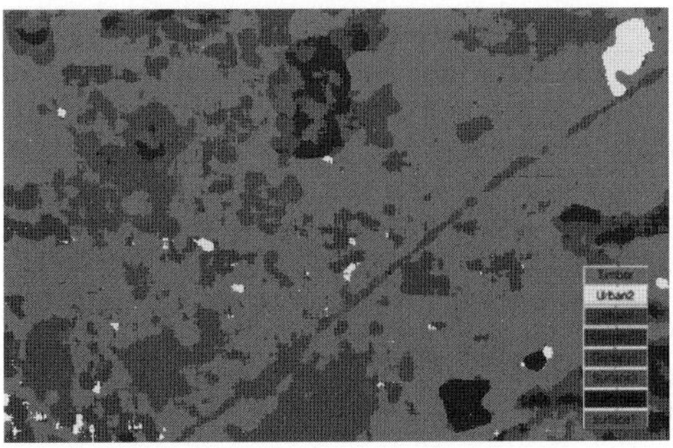

Fig. 3-3.18. Classification of AIRSAR data, Boreal area

Figs. 3-3.17 and 3-3.18 are the total power image and its deorientation classification over this area, respectively. The classification method is the same as Fig. 3-3.13, and a 9×9 window majority analysis is adopted. Note that classifications in Fig. 3-3.18 show that some vegetation areas are classified as Urban 1 (the red area). This is probably due to the sparse vegetation with mainly double scatterings, i.e. $v<0$ and lower H. Thus, more orientation analysis is necessary to confirm the class or resolve some confusions.

Fig. 3-3.19 presents the orientation distribution of 8 classes of the terrain surface targets. Note that the short segment in each pixel displays the orientation angle of the parameter ψ of this pixel. Only the center pixel from each 4×4 window is displayed to avoid overlapped confusion from neighbor pixels. It can be seen from Fig. 3-3.19 that

1. Most areas of Timber show random orientation e.g. the region A. Only is there few part with uniform orientation, e.g. the region B with almost vertical orientations.
2. There are almost uniform orientations of vertical or horizontal directions in the Urban 2, e.g. the region C, due to the regular features of artificial constructions.
3. There are uniformly vertical orientations in the Urban 1 (sparse forest of vertical trunks, instead of artificial constructions).
4. Random orientation in the Canopy 1 means that the vegetation canopy in this area might be the disordered bush. Note that the region D should be the roads inside the forest, whose randomness might be confused by the bush vegetation on the roadsides.
5. The Surface 2 can be divided into two subclasses by the orientation distribution, i.e. uniformly horizontal oriented area, e.g. the region E, and randomly oriented area, e.g. the region F. The horizontal oriented area is the water surface, while the randomly oriented area is disordered-random grass land.
6. The Surface 1 can also be divided into uniformly and horizontally oriented scatter targets as water surface, e.g. region G, and uniformly and vertically oriented targets as ordered grass land, e.g. region H.
7. There are not many surfaces as classed as Canopy 2 and Surface 3.

Following the above orientation analysis, the terrain surfaces are further classified into the subclasses and are identified by their types.

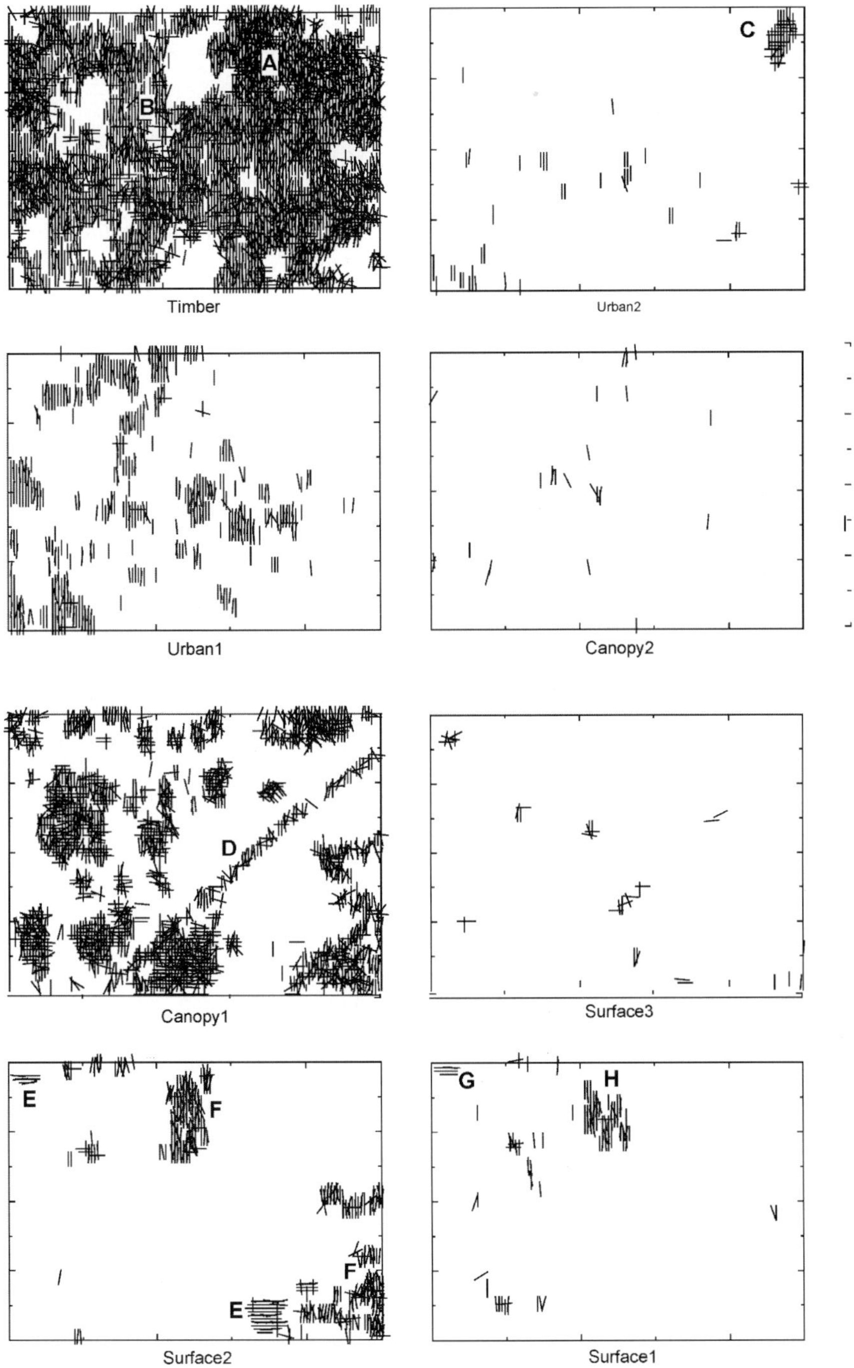

Fig. 3-3.19. Orientation distribution of 8 classes of Fig. 3-3.18

Appendix 3-3A

Cameron's conditions of the maximum symmetry target component are [Cameron and Leung 1990]

$$\begin{cases} \tan 2\vartheta = \dfrac{k_{P,2}k_{P,3}^* + k_{P,2}^* k_{P,3}}{|k_{P,2}|^2 - |k_{P,3}|^2}, & |k_{p,2}|^2 \neq |k_{p,3}|^2 \\ (k_{P,2}k_{P,3}^* + k_{P,2}^* k_{P,3}) \sin 2\vartheta > (|k_{P,3}|^2 - |k_{P,2}|^2) \cos 2\vartheta \end{cases} \quad (3\text{-}3A.1)$$

where ϑ is the angle used to construct the symmetry basis vector. Substituting Eq. (3-3.4a) into Eq. (3-3A.1) yields

$$\begin{cases} \tan 2\vartheta = \tan 2\beta \cos(\phi_2 - \phi_3), & \cos 2\beta \neq 0 \\ \sin 2\beta \cos(\phi_2 - \phi_3) \sin 2\vartheta + \cos 2\beta \cos 2\vartheta > 0 \end{cases}. \quad (3\text{-}3A.2)$$

Substituting the above first equation into (3-3A.2) into the second, it yields

$$\begin{cases} \tan 2\vartheta = \tan 2\beta \cos(\phi_2 - \phi_3), & \cos 2\beta \neq 0 \\ \cos 2\beta \cos 2\vartheta > 0 \end{cases} \quad (3\text{-}3A.3)$$

Comparing to Eq. (3-3.16), it has the same forms and we can obtain $\vartheta = 2\psi$.

The slope inversion algorithm using circular polarizations can be expressed as [Lee et al. 2000]

$$4\Theta = \left[\arg\langle S_{RR} S_{LL}^* \rangle + \pi \right], \quad (3\text{-}3A.4)$$

where

$$\arg\langle S_{RR} S_{LL}^* \rangle = \tan^{-1} \frac{-4\,\mathrm{Re}\langle (S_{hh} - S_{vv}) S_x^* \rangle}{-\langle |S_{hh} - S_{vv}|^2 \rangle + 4\langle |S_x|^2 \rangle}; \quad (3\text{-}3A.5)$$

S_{RR}, S_{LL} are circularly right and left co-pol; Θ is the defined slope orientation angle.

In the case of a deterministic situation, the assemble average $\langle \cdot \rangle$ in Eqs. (3-3A.4) and (3-3A.5) can be omitted. Hence, Eq. (3-3A.5) can be rewritten as

$$\arg\langle S_{RR} S_{LL}^* \rangle = \tan^{-1} \frac{(S_{hh} - S_{vv}) \cdot 2S_x^* + (S_{hh} - S_{vv})^* \cdot 2S_x}{|S_{hh} - S_{vv}|^2 - |2S_x|^2} \quad (3\text{-}3A.6)$$

According to Eq. (3-3.2a), Eq. (3-3A.6) can be further expressed using $\bar{k}_P$ elements as

$$\arg\langle S_{RR} S_{LL}^* \rangle = \tan^{-1} \frac{k_{P,2}k_{P,3}^* + k_{P,2}^* k_{P,3}}{|k_{P,2}|^2 - |k_{P,3}|^2} \quad (3\text{-}3A.6)$$

Referring to Eqs. (3-3A.1) and (3-3A.4), we can obtain

$$\tan(4\Theta - \pi) = \tan 2\vartheta \quad (3\text{-}3A.7)$$

Obviously, ψ, ϑ and Θ derived by different authors are in fact consistent with each other.

3-4 An Algorithm for Ship Wake Detection from the SAR Images

The great advance in SAR imaging technology has promoted image data study from all-whether space-borne observations. It has become known that a moving ship may generate surface perturbation to form so-called V-shaped Kelvin arms and turbulence wakes, and it can be sensed from SAR observations. Bright or dark streaks of radar echo (backscattering) in SAR image can be identified as a detection of moving ship. Information of the ship location, moving direction, and even its velocity might become available [Hennings et al. 1999, Rey et al. 1990, Copeland et al. 1995, Lin et al. 1996, Eldhuset 1996]. One method is first to identify the bright region in the SAR image as the possible location of the ship head, then further to find the ship wakes around this region [Rey et al. 1990, Copeland et al. 1995, Lin et al. 1996, Eldhuset 1996]. However, it seems difficult in many cases to isolate the bright region, especially in a strong noisy background. Another method is to employ a small square window moving over the image to detect linear textures [Copeland et al. 1995]. An automatic computer service based on the algorithms for linear wake detection has also been discussed [Lin et al. 1996]. However, the efficiency and practice of wake detection remain to be further studied.

The variation in the radar echo (bright or dark streaks, i.e. enhanced or reduced backscattering) modulated by ship wakes depends on the SAR wavelength, polarization, spatial resolution, SAR flight and observation directions, as well as the ocean environment such as sea surface winds and currents. However, owing to the linear textures of ship wakes, it is desirable to identify it from very large number of pixels in an SAR image.

The application of the Radon transform to digital image processing is actually an integration of all pixel values along a certain line. However, the Radon transform was usually carried out over the whole image; thus it does not seem to be effective, especially for wakes on a short scale [Copeland et al. 1995]. To improve the detection accuracy, it should distinguish ship wakes from other possible perturbations, e.g. those caused by sea surface winds or underwater bottom topography. It is usually to assign a threshold to obtain a binary image for detection. Accuracy is certainly affected by this empirical threshold.

In this session, the approach called as SWDRM (ship wakes detection based on Radon transform and morphological image processing) is developed [Jin and Wang 2004]. The SAR image is first divided into many sub-images by moving a small working window. Each sub-image is mapped into the Radon space with an adaptive threshold. The dilation and erosion operations are performed. Finally, an inverse Radon transform produces the grey-level and binary images for manual or automatic computer detection. This method can quickly and accurately detect ship wakes in the SAR image. The SWDRM algorithm is, essentially, constructing an inverted grey-level image with high signal-to-noise (SNR). Analyzing the intense, length, width and correlation of linear features, full information about moving ship can be generated. As an example, it is applied to a SEASAT SAR image with enhanced noise. The results are favorably demonstrated.

3-4.1 The SWDRM Algorithm

A ship wake in oceanic clutter is a good indicator of a moving ship due to its linearity and lengthy scale [Rey et al. 1990]. Because the retrieval technique usually assumes no motion of objects in SAR scene, it might not yield accurate location for moving ship appeared in an SAR image. Ship wake detection can give a better indication of the ship location and other information. The SWDRM algorithm, setting adaptive threshold in the

Radon space and carrying out erosion threshold, reduces clutter noise. The gray-level and binary images for detection can be inverted.

The Radon transform maps a straight line $\rho = x\cos\theta + y\sin\theta$ in the (x, y) image plane into a point (ρ, θ) in the Radon plane, essentially similar to the Hough transform in the computer visualization domain. The Radon transform is defined as

$$R(\rho,\theta) = \iint_D f(x,y)\delta(\rho - x\cos\theta - y\sin\theta)dxdy \quad (3\text{-}4.1)$$

where $f(x,y)$ is the image intensity in the (x, y) plane, D is the entire image, δ is the Dirac delta function. Eq.(3-4.1) transforms a linear wake in the (x,y) plane into a specified point in (ρ, θ) space. It can also be reversed as follows:

$$y = (\rho/\sin\theta) - x\cot\theta \quad (3\text{-}4.2)$$

The peaks or troughs in Radon space correspond to the bright or dark streaks respectively of radar signatures in the (x,y) plane. Setting the threshold in Radon space, clutter noise can be reduced and a high signal to noise ratio (SNR) can be obtained.

Morphological image processing provides the structure of the entire image. At each pixel, the specified logical operations, namely erosion and dilation, are performed. It is defined as Castleman (1996).

Dilation is given by

$$E = B \oplus S = \{x,y|S_{xy} \cap B \neq \Phi\} \quad (3\text{-}4.3)$$

and erosion is given by

$$E = B \otimes S = \{x,y|S_{xy} \subseteq B\} \quad (3\text{-}4.4)$$

where E is an image obtained from dilation and erosion of the image B by structuring the element S. Generally, the wake width takes several pixels.

The dilation combines some neighboring peaks. The dilation is then used to eliminate a single peak. Thus, the number of most unlikely non-wake peaks and troughs is reduced. Certainly, some linear

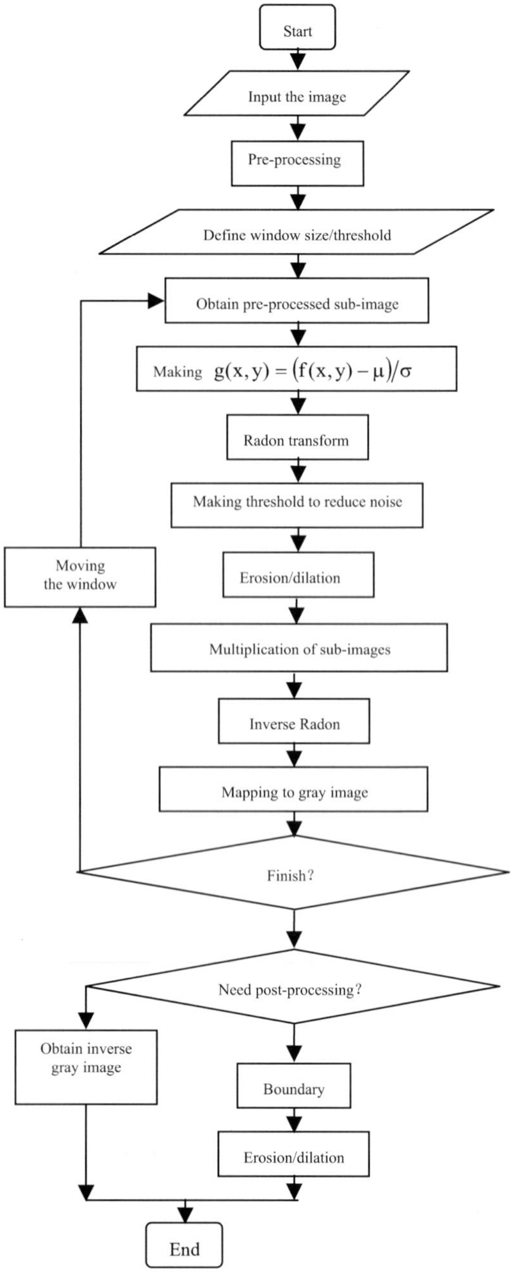

Fig. 3-4.1. Flow chart of the algorithm

textures, e.g. that perturbed by underwater bottom topography, might remain for further clarification.

The SWDRM algorithm contains the following steps:

(1) A pre-image process is carried out to classify the ocean region and median filtering over the entire image to reduce single speckle noise.

(2) A working window is moved so as to divide the image into many sub-images. It is calculated in each pixel as follows:

$$g(x,y) = \left(f(x,y) - \mu\right)/\sigma \tag{3-4.5}$$

where μ and σ are the mean value and the standard deviation respectively of a sub-image. It is possible that the Radon transform for an image with the same grey-level have sinusoidal peak lines. This second step can reduce mosaic pattern for reconstruction of the grey-level image.

(3) The Radon transform is applied with two thresholds for peaks and troughs, respectively to the sub-images. All pixels with the grey level between the thresholds are defined as zero:

$$R(\rho,\theta) = 0 \tag{3-4.6}$$

as $-\mu_R - K_1\sigma_R \le R(\rho,\theta) \le \mu_R + K_2\sigma_R$,where μR and σR are the local mean value and standard deviation respectively of the sub-image in the Radon space. The thresholds K_1 and K_2 have been assumed to correspond to the brightest peaks and darkest troughs respectively in Radon space. Then, using the morphological image process, i.e. erosion and dilation operations, single peak is eliminated.

(4) Finally, the inverse Radon transform is performed to obtain a gray-level image.

Moving the working window over the entire image and repeating these steps, an inverted grey-level image is constructed. If necessary, a binary image can be further obtained. Fig. 3-4.1 presents a flow chart of the algorithm.

3-4.2 Application of SWDRM to SEASAT SAR Image

Fig. 3-4.2 is a part of the SEASAT image over the southern bight of the North Sea from orbit 1473 with the figure center at (51°10'21"*N*, 1°50'15"*E*). It has 850×750 pixels and actually contain two ship wakes [Hennings et al. 1999]. The angle between ship moving and radar observation is 228°. Typical Kelvin arms can be seen in the figure.

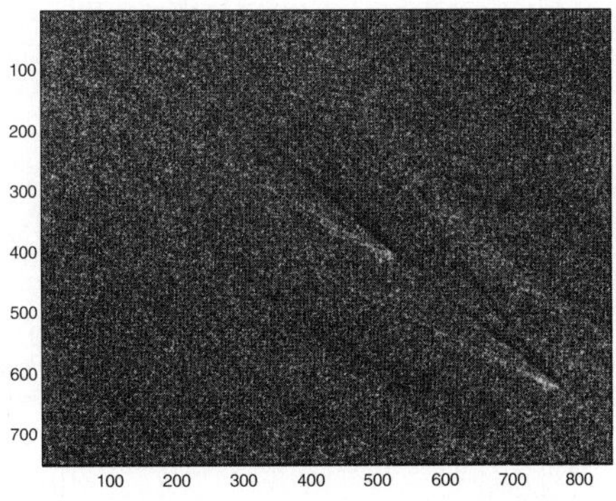

Fig.3-4.2. An image of SEASAT SAR

Fig. 3-4.3 presents the SWDRM results. The working window has 100×100 pixels. A test area with 250×200 pixels is chosen from the center region in Fig. 3-4.2 (see Fig.

3-4.3(a)). Using the median filtering, it becomes Fig. 3-4.3(b). The Radon transform without dilation and erosion yields Fig. 3-4.3(c). It can be seen that there are many random lines in addition to real ship wakes. Using dilate and erode process, the Radon transform yields Fig. 3-4.3(d), where the dilation length is 4 pixels, and erosion length is 5 pixel. It can be seen that random lines are significantly reduced. Detection can become much accurate. Here, the thresholds $K_1 = 3.4$ and $K_2 = 4.5$ are taken. Certainly, the threshold values affect inverted figure and detection. However, we have found that this algorithm can always yield the satisfactory results as $K_1 \in (2.5, 4)$ and $K_2 \in (3, 4.8)$.

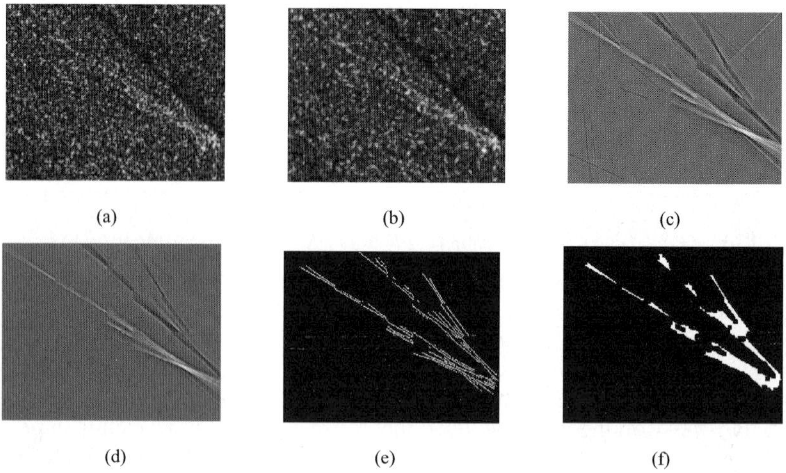

Fig. 3-4.3. A testing image and processed image

Using post-image process, a binary image can also be obtained. As an example, Fig. 3-4.3(e) is obtained by a boundary test on Fig. 3-4.3(d). Using iterative algorithm of dilation and erosion process, a much better figure is obtained as shown in Fig. 3-4.3(f).

The grey-level and binary images for the whole of Fig. 3-4.2 of the SEASAT SAR image are given in Figs. 3-4.4(a, b). Two moving ships are easily demonstrated.

Fig. 3-4.4. A processed gray-level image and a binary image

Speckle noises [Ulaby et al. 1986], e.g. due to oceanic clutter, affect the image quality employed for ship detection. Using integration or summation of the pixel values along a specified linear segment, the Radon transform can actually smooth noisy fluctuation to clarify the ship wakes.

Another part of the SEASAT SAR image of the same orbit 1473 with the figure center at (51°28'17"*N*, 2°05'26"*E*) is tested. The noise variance is manually increased to double the previous value. In Fig. 3-4.5(a), it seems now not easy to visualize the ship wakes. Using the SWDRM with the same working window and thresholds in the previous example, it yields an inverse grey-level image as shown in Fig. 3-4.5(b). In the upper left corner, the ship wake, ship location and moving direction are easily recognized. On the right-hand side, the Kelvin arms can be also visible. It demonstrates that the SWDRM algorithm is robust in presence of strong noise.

If the thresholds $K_1 = 2.7$, $K_2 = 3.5$ are taken, the inverted image is shown in Fig. 3-4.5(c). Two moving ships, Kelvin arms and even the darker turbulence wake become visible compared with Fig. 3-4.5(b). Additionally, if the thresholds are changed, more clutter lines appear in the inverse image; analysis of the correlation of random lines does not misinterpret the wakes and noise lines.

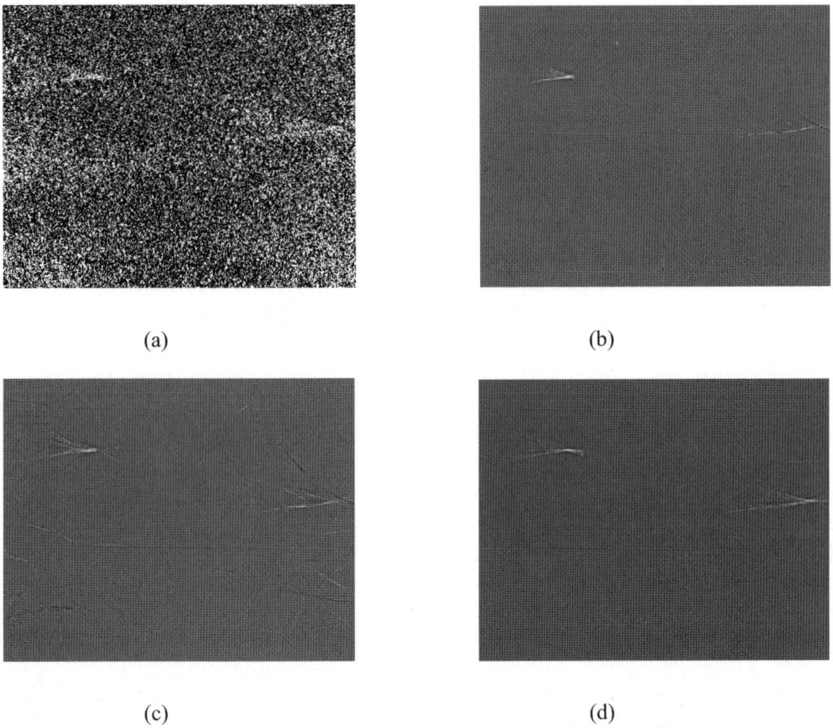

(a) (b) (c) (d)

Fig. 3-4.5. SAR image corrupted by noise and its inverted images

In the above examples, a smaller window is utilized. In fact, if an appropriate working window is chosen, the result might become better. For example, if a working window of 150×150 pixels is taken, the thresholds K_1 and K_2 are kept as before, a better result is obtained as shown in Fig. 3-4.5(d). In practice, the size of the working window can be chosen from 100×100 to 200×200 pixels. The SWDRM show that it is easily able to detect ship wakes, even when the noise is increased.

3-5 An Improved Method of the Minimum Entropy for Refocusing the Moving Target Image

Utilizing the synthetic technology of the pulse duration and Doppler shift [Hovanessian 1980], a high resolution image has been achieved by space- or airborne SAR observation.

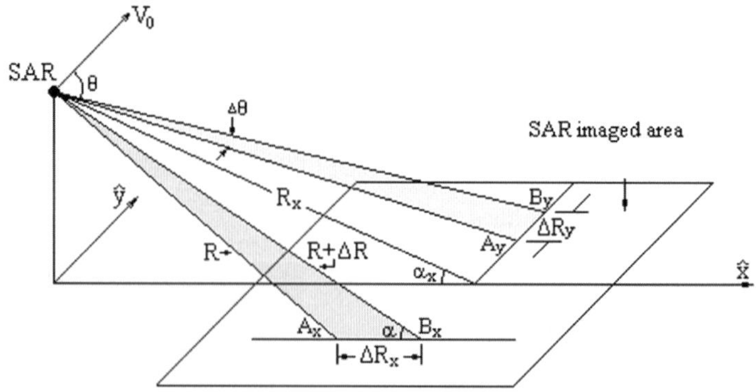

Fig. 3-5.1. SAR imagery and spatial resolution

If a fast-moving target appears in the imaged area, its image might be blurred by unexpected DP shift, which will seriously degrade image quality, especially for purposes of target detection and classification. Some theoretical models and algorithms for moving target detection such as the Doppler centroid tracking method [Werness et al. 1994], the phase gradient autofocus technique [Li et al. 1999], the prominent point process [Kasilingam et al. 2000], etc. have been developed. However, all of these approaches need some prior conditions, e.g. the centroidor or prominent points of the target, etc. In most practices for target detection, these prior conditions are not always available.

The minimum entropy method (MEM) [Itoh et al. 1996, Wahl et al. 1994] is an optimized method with no need for additional prior parameters. The entropy as a measure of uncertainty is defined by the pixel values of the SAR image. Based upon the principles of SAR imagery and defocusing phenomena, the MEM utilizes the entropy as a cost function, which is to be optimized for the phase regulation and refocusing. In this session, an improved MEM is studied to speed up optimization and phase regulation and is applied to fast refocusing of a simulated SAR image of a moving aircraft [Jin and Chen 2001].

3-5.1 Spatial Resolution of SAR Image

Making use of the time difference and DP shift of the received echoes, a high resolution SAR image can be achieved [Hovanessian 1980]. As described in Fig. 3-5.1, the SAR is flying along the direction $\hat{y}$ and its side-looking direction is $\hat{x}$. Because all points along the direction $\hat{x}$ on the imaged area keep the same relative velocity with the flying SAR, the resolution at the $\hat{x}$ direction only depends on the pulse duration of the SAR. The arrival time of the echo is measured for determining the distance between the SAR radar and target.

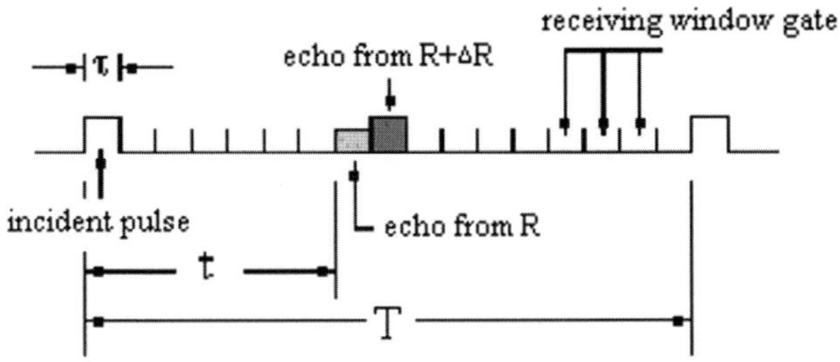

Fig. 3-5.2. Echoes received by SAR at $\hat{x}$ direction

As illustrated in Fig. 3-5.2, the pulse duration is τ and interpulse period is T. The receiving window gate is also simply assumed as τ (τ is inversely proportional to the sampling frequency). As shown in Fig. 3-5.1, the receiving times of the echoes from $A_x(R)$ and $B_x(R+\Delta R)$ are respectively

$$t = \frac{2R}{c}, \quad t+\tau = \frac{2(R+\Delta R)}{c} \tag{3-5.1}$$

where c is light velocity. Thus, it simply gives

$$\tau = \frac{2\Delta R}{c} \tag{3-5.2}$$

From Fig. 3-5.1, it yields

$$\Delta R_x = \frac{\Delta R}{\cos\alpha} = \frac{c\tau}{2\cos\alpha} \tag{3-5.3}$$

where ΔR_x is the spatial resolution along the side-looking direction $\hat{x}$. It only depends on the pulse duration τ and SAR observation angle α. Therefore, the distance ΔR_x between A_x and B_x, or target moving along $\hat{x}$ can be well identified from the receiving signals.

However, the spatial resolution in the $\hat{y}$ direction is achieved using different DP shifts from different points, such as A_y and B_y, of the imaged area, also as illustrated in Fig. 3-5.1.

Suppose the incident electromagnetic wave of the SAR

$$S_t = \sin 2\pi f_0 t \tag{3-5.4}$$

where f_0 is the carrier frequency. The echo received by the SAR is

$$S_r = \sin[2\pi f_0 (t - \frac{2R_0 + 2R't}{c})] \tag{3-5.5}$$

where R_0 is the initial distance between the imaged area and SAR, R' is the velocity of SAR relative to the stationary imaged area. From Eq. (3-5.5), the frequency shift and phase of the radar echoes are, respectively, written as

$$f_d = -\frac{2R'f_0}{c} = -\frac{2R'}{\lambda} \tag{3-5.6a}$$

$$\phi_r = -2\pi\frac{2R_0}{\lambda} \tag{3-5.6b}$$

where λ is the wavelength of incident wave ($\lambda = c/f_0$).

From Fig. 3-5.1, the frequency shifts at the positions A_y and B_y are respectively

$$f_{dAy} = \frac{2V}{\lambda}\cos(\theta + \Delta\theta), \quad f_{dBy} = \frac{2V}{\lambda}\cos\theta \tag{3-5.7}$$

Their difference is

$$\Delta f_d = f_{dBy} - f_{dAy} = \frac{2V_0}{\lambda}\Delta\theta \sin\theta \tag{3-5.8}$$

where V_0 is the SAR flight velocity. Using $\Delta\theta = \Delta R_y / R$, the SAR resolution at the $\hat{y}$ direction is

$$\Delta R_y = \frac{\lambda R \Delta f_d}{2V_0 \sin\theta} \tag{3-5.9}$$

It can be seen that the SAR resolution in the $\hat{y}$ direction is closely related to the SAR flight velocity. If a moving target appears along the $\hat{y}$ direction in the imaged area, the DP shift would be much different from the stationary imaged area. Because the sampling and filtering of the SAR was designed only to compensate for the DP shift due to the relative motion between the SAR and the stationary area, this additional DP difference would smear the output, i.e. defocusing.

3-5.2 Defocused Target in the SAR Image

Suppose a moving target with velocity V_t appearing in the imaged area as shown in Fig. 3-5.3.

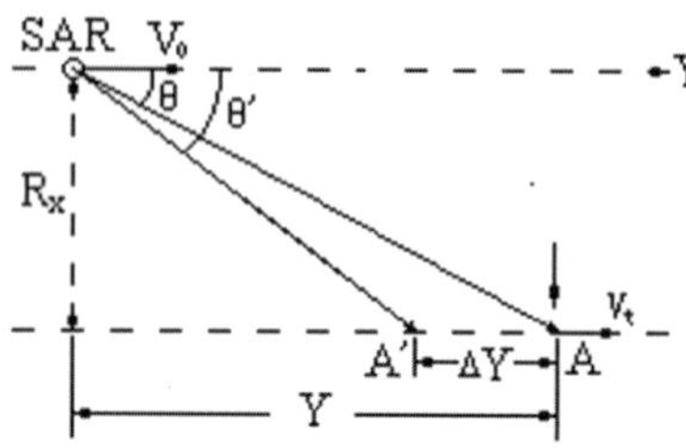

Fig. 3-5.3. SAR observation of a moving target

From Eq. (3-5.8), if $V_t = 0$, the DP shift of the echo is proportional to $(2V_0/c)\cos\theta$. However, if $V_t \neq 0$, the DP shift becomes proportional to $[2(V_0 - V_t)/c]\cos\theta$. Because the signal processing of the SAR sampler and filter was designed for the SAR velocity V_0, the target position at A is then misunderstood as the position A'. It yields

$$\frac{2V_0}{c}\cos\theta' = \frac{2(V_0 - V_t)}{c}\cos\theta \tag{3-5.10}$$

and

$$\frac{R_x}{Y} = \tan\theta, \quad \frac{R_x}{Y - \Delta Y} = \tan\theta' \tag{3-5.11}$$

where $R_x = h/\sin\alpha_x$, θ' is the misunderstood observation angle. Therefore, it yields

$$\Delta Y = (\text{ctan}\theta - \text{ctan}\theta')R_x = \{\text{ctan}\theta - \text{ctan}[\cos^{-1}(\frac{V_0 - V_t}{V_0}\cos\theta)]\} \tag{3-5.12}$$

This means that the real position of the target Y is shifted to Y-ΔY in the SAR image. As the target is moving, its position in the SAR image yields the variation of

$$Y'(t) = Y_0 + V_t \cdot t - \Delta Y(t) \tag{3-5.13}$$

where Y_0 is the initial position of the observed target.

It is noted that, even though there appears to be no time variable in Eq. (3-5.12), the side-looking angle θ of the SAR actually has time dependence. Assuming the initial observation angle as θ_0, it can be obtained that at any time t

$$\frac{R_x}{\tan\theta} - \frac{R_x}{\tan\theta_0} = -(V_0 - V_t)t \tag{3-5.14}$$

which yields

$$\theta = \mathrm{ctan}^{-1}(\mathrm{ctan}\theta_0 - \frac{V_0 - V_t}{R_x}t) \tag{3-5.15}$$

From Eqs. (3-5.12, 3-5.15), the position of a moving target shown in the SAR image with time dependence can be obtained.

Since the SAR image is produced by the signal echoes averaged over a time period, a point of the moving target would appear in the image as a smeared shifted line, instead of a simply shifted point. The phase of Eq. (3-5.6b) that is due to motion in the $\hat{y}$ direction causes the error response in the SAR filtering. Thus, the image of fast-moving target can be smeared or defocused.

If a target is moving in any other direction, only the component of its velocity in the $\hat{y}$ direction, taken as V_t of Eqs. (3-5.10~15), can cause image smearing and it is then processed by the MEM.

3-5.3 The Minimum Entropy Method (MEM)

The DP shift of the moving target can be seen as a source of the phase error. It is desirable to find a way to compensate for the phase error and then refocus the image. The concept of entropy in the information theory, as a measure of uncertainty, was first presented by Shannon and has been extensively studied in information technology.

The MEM utilizes the entropy as a cost function, which is to be optimized for phase regulation and finally erase the phase errors [Itoh et al. 1996].

Suppose a set of discrete signals in time-domain ($f_0, f_1, f_2, \cdots, f_{N-1}$) with some phase errors. Define the entropy as

$$H = -\sum_{n=0}^{N-1} f_n \cdot \ln f_n \tag{3-5.16}$$

Making discrete Fourier transformation (FT) (all FT is the fast FT (FFT) in our discussion), the signals in frequency domain are obtained as ($F_0, F_1, F_2, \cdots, F_{N-1}$). The first step is to regulate the phase of F_0. The initial regime for regulation is [$-\pi$, 0, π]. It yields

$$\begin{cases} F_0(-\pi) = F_0 \cdot e^{-j\pi} \\ F_0(0) = F_0 \cdot e^{j0} = F_0 \\ F_0(\pi) = F_0 \cdot e^{j\pi} \end{cases} \tag{3-5.17}$$

Substituting $F_0(p), p = -\pi, 0, \pi$, into the original series of ($F_0, F_1, F_2, \cdots, F_{N-1}$), a new series is obtained

$$\begin{cases} F'(-\pi) = (F_0(-\pi), F_1, F_2, \cdots, F_{N-1}) \\ F'(0) = (F_0(0), F_1, F_2, \cdots, F_{N-1}) \\ F'(\pi) = (F_0(\pi), F_1, F_2, \cdots, F_{N-1}) \end{cases} \tag{3-5.18}$$

Making inverse FFT (IFFT) to Eq.(3-5.18), a new series in time domain $f'(-\pi), f'(0)$ and $f'(\pi)$ can be obtained. Their entropies are $H_0(-\pi), H_0(0)$ and $H_0(\pi)$. Take the phase with the minimum entropy as the result of this regulation, i.e.

$$\phi = [\underset{\phi=[-\pi,0,\pi]}{\arg\min}\ (H_0), 0, 0, ..., 0] \tag{3-5.19}$$

Based on the regulation of F_0, the next step is to regulate the phase of F_1. The signal series for entropy computation is now as $(f_0', f_1', f_2, \cdots, f_{N-1})$ in order to be sure that the entropy of the next phase regulation will be smaller than the previous one.

Repeating this algorithm, the phase regulation of the signal series is finally obtained as

$$\begin{aligned} \phi &= [\underset{\phi=[-\pi,0,\pi]}{\arg\min}\ (H_0),\ \underset{\phi=[-\pi,0,\pi]}{\arg\min}\ (H_1),\ \underset{\phi=[-\pi,0,\pi]}{\arg\min}\ (H_2), ...,\ \underset{\phi=[-\pi,0,\pi]}{\arg\min}\ (H_{N-1})] \\ f_\phi &= \mathrm{IFFT}[F \cdot \exp(j\phi)] \end{aligned} \tag{3-5.20}$$

If the entropy of the series f_ϕ is still larger than a prescribed allowance E_1, comparing with the initial series f, this series needs regulation again within the regime [-π,0,π]. Returning to F_0, regulate another round until the difference between two rounds is smaller than E_1. Now the regime of phase regulation of [-π,0,π] is reduced as [-π/2,0,π/2].

Repeat all operations from Eq. (3-5.17). The first regime of phase regulation is [-π,0,π], the second one is [-π/2,0,π/2], … , the m-th is $[-\pi/2^{m-1}, 0, \pi/2^{m-1}]$, and so on. Of course, it is not infinite. As the entropy difference of the regulation with the previous one is smaller than the prescribed allowance E_2, the regulation is finished and signals in the time domain with erased phase error are obtained.

A. Two-dimensional (2-D) MEM for SAR image

Define the entropy of the SAR image as

$$H = -\sum x \ln x \tag{3-5.21}$$

where x is the value of the image pixel.

It is known from previous discussion that only movement along the $\hat{y}$ direction causes defocusing. If the target is not rotating, all components of the target actually have the same velocity along the $\hat{y}$ direction. It can be assumed that all phase errors of the whole target body are caused in the same way. Therefore, by regulating the lines along the $\hat{y}$ direction in the same way, refocusing can be finished for the whole image. This is the 2-D MEM for the SAR image [Itoh et al. 1996].

Take a moving aircraft as shown in Fig. 3-5.4(a). A simulation of defocused aircraft image is presented in Fig. 3-5.4(b).

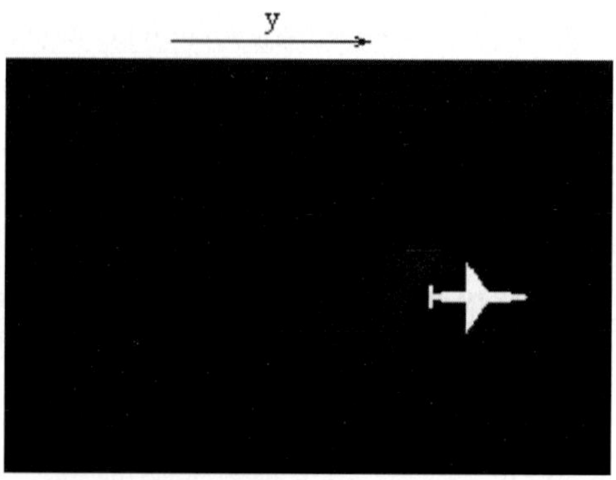

Fig. 3-5.4(a). A moving aircraft in the SAR image

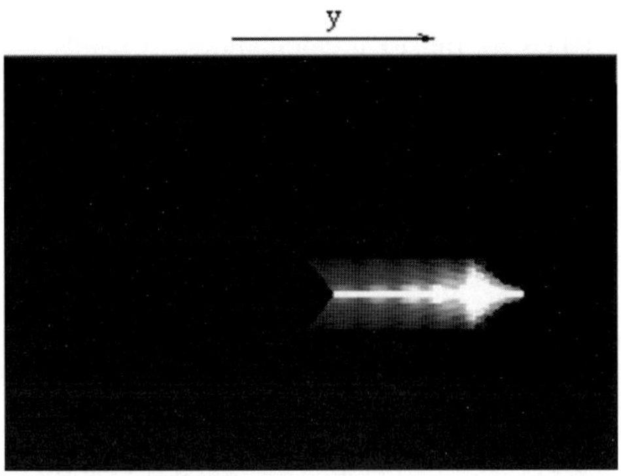

Fig. 3-5.4(b). Defocusing image of a moving aircraft

This simulation of the defocused image is made using following parameters: the aircraft velocity V_t=500 *m/s* along the $\hat{y}$ direction, the pulse repetition frequency (PRF) of SAR [Oliver and Quegan 1998] is 2 *kHz*, 1000 pulse echoes on average (i.e. SAR look number=2), which means one observation takes 0.5 *s*. During a 0.5 *s*, the flight distance of the aircraft is 250 *m*. Suppose the distance between SAR and the center of the imaged area as 50 *km*, the SAR resolution 5 *m*. The initial observation angle θ_0=88°. From Eqs. (3-5.12, 3-5.15), when t =0 *s*, ΔY= -436.86 *m*, and when t =0.5 *s*, ΔY= -249.07 *m*. During this time, flight distance is 250 *m*, which yields a smeared length of the moving aircraft image about 440 *m*.

The phase errors of Fig. 3-5.4(b) are due to the DP shift, moving status as well as the background clutter. Since the phase errors in all lines are assumed to be of the same nature, the FFT and IFFT regulations are carried out for all lines.

Fig. 3-5.5 show the final result, where both E_1 and E_2 are taken as 10^{-3}. If the allowances E_1 and E_2 were smaller, the result would be better and but take much larger time to obtain. The work shown in Fig. 3-5.5 was carried on the Pentium-III CPU of 800 *mHz*, 128*M*, windows 2000 and took 468 *s* (almost 8 *min*).

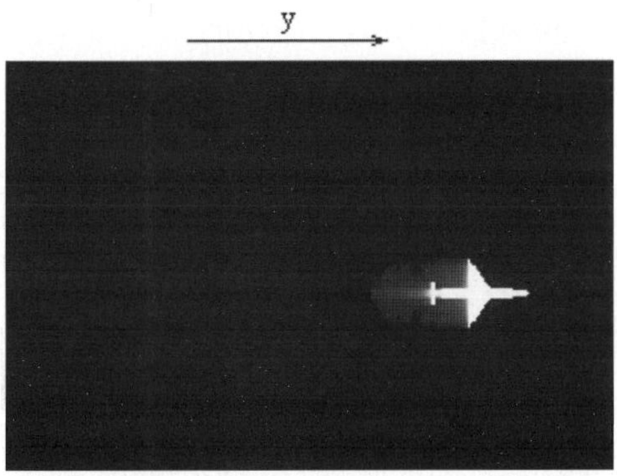

Fig. 3-5.5. Refocused SAR image of the moving aircraft

B. An improved MEM

The time is mainly taken by the FFT calculations. It needs 2N FFT to complete one entropy regulation for a series of N signals. To satisfy allowance of E_1 and E_2, the FFTs are very time consuming.

Suppose a series of N signals $(f_0, f_1, f_2, \cdots, f_{N-1})$ in time domain. Its discrete FFT is defined as

$$F(k) = \sum_{n=0}^{N-1} f_n \cdot \exp(-j2\pi \frac{kn}{N}), \quad N=0, 1, 2, \ldots, N\text{-}1 \tag{3-5.22}$$

Regulating the phase to the signal, e.g. n=2, of $F(k)$, it is obtained that

$$\begin{aligned} F'(0) &= \sum_{n=0}^{N-1} f_n \cdot \exp(-j2\pi \frac{0 \cdot n}{N}) = F(0) \\ F'(1) &= \sum_{n=0}^{N-1} f_n \cdot \exp(-j2\pi \frac{1 \cdot n}{N}) = F(1) \\ F'(1) &= \{\sum_{n=0}^{N-1} f_n \cdot \exp(-j2\pi \frac{2 \cdot n}{N})\} \cdot \exp(j\phi) \qquad \phi = \pm\phi, 0 \\ &\ldots \\ F'(N-1) &= \sum_{n=0}^{N-1} f_n \cdot \exp(-j2\pi \frac{(N-1) \cdot n}{N}) = F(N-1) \end{aligned} \tag{3-5.23}$$

Making IFFT to three new series in frequency domain, three new series in time domain can be obtained. Having chosen the series with the minimum entropy as the phase regulation, carry out next round of phase regulation.

Actually, using the following simple algorithm, three new series in time domain can be easily obtained. Since

$$f(n) = \sum_{k=0}^{N-1} F(k) \exp(j2\pi \frac{kn}{N}) \tag{3-5.24}$$

it follows that

$$\begin{aligned} f'(n) &= \sum_{k=0}^{N-1} F'(k) \exp(j2\pi \frac{kn}{N}) \\ &= F'(0) \cdot W_N^0 + F'(1) \cdot W_N^1 + F'(2) \cdot W_N^2 + \ldots F'(0) \cdot W_N^{N-1} \\ &= F(0) \cdot W_N^0 + F(1) \cdot W_N^1 + F(2) \cdot W_N^2 \cdot \exp(j\phi) + \ldots F'(0) \cdot W_N^{N-1} \\ &= f(n) + F(2) \cdot W_N^2 \cdot \exp(j\phi) - F(2) \cdot W_N^2 \\ &= f(n) + F(2) \cdot W_N^2 \cdot [\exp(j\phi) - 1] \end{aligned} \tag{3-5.25}$$

Therefore, three IFFT can be saved for one phase regulation. Similarly, for all subsequent rounds, the FFT and IFFT are not repeated again and again. It means that one FFT can complete the entire process of the MEM.

As in Fig. 3-5.4, our method of MEM takes 6 per cent of the conventional MEM, i.e. less than 0.5 *min*.

3-6 References

Besag, J. (1974), "Spatial interaction and the statistical analysis of lattice systems (with discussion)", *Journal of Royal Statistical Society, series B*, 36(2): 192-326.

Besag, J. (1986), "On the statistic analysis of dirty pictures", *Journal of Royal Statistical Society, series B*, 48(3): 259-302.

Bruzzone, L. and D.F. Prieto (2000), "Automatic analysis of the difference image for unsupervised change detection", *IEEE Transactions on Geoscience and Remote Sensing*, 38(3): 1171-1182.

Castleman, K.R. (1996), *Digital Image Processing*, New York: Printice Hall.

China Integrative Atlas (1990), Beijing: China Map Publishing Company, 151.

Cloude, S.R. (1986), "Group theory and polarization algebra", *Optik*, 75(1): 26-36.

Cloude, S.R. and E. Pottier (1996), "A review of target decomposition theorems in radar polarimetry", *IEEE Transactions on Geoscience and Remote Sensing*, 34(2): 498-518.

Cloude, S.R. and E. Pottier (1997), "An entropy based classification scheme for land application of polarimetric SAR", *IEEE Transactions on Geoscience and Remote Sensing*, 35(1): 68-78.

Cloude, S.R. (1999), "Wide-band polarimetric radar inversion studies for vegetation layers," *IEEE Transactions on Geoscience and Remote Sensing*, 37(5): 2430-2441.

Cloude, S.R., K.P. Papathanassiou and E. Pottier (2001), "Radar polarimetry and polarimetric interferometry", *IEICE Transactions on Electronics*, E84-C (12): 1814-1822.

Copeland, A.C., G. Ravichandran and M.M. Trivedi (1995), "Localized Radon transform-based detection of ship wakes in SAR images", *IEEE Transactions on Geoscience and Remote Sensing*, 33(1): 35-45.

Demmel, J. (1997), *Applied Numerical Linear Algebra*, Society for Industrial and Applied Mathematics.

Dempster, A.P., N.M. Laird and D.B. Rubin (1977), "Maximum linkelihood from incomplete data via the EM algorithm", *Journal of Royal Statistical Society*, 39(1): 1-38.

Eldhuset, K. (1996), "An automatic ship and ship wake detection system for spaceborne SAR images in coastal region", *IEEE Transactions on Geoscience and Remote Sensing*, 34(4): 1010-1019

Chan, F.H.Y., F.K. Lam and H. Zhu (1998), "Adaptive thresholding by variational method", *IEEE Transaction on Geoscience Remote Sensing*, 7(3): 468-473.

Freeman, A. and S.L. Durden (1998), "A three-component scattering model for polarimetric SAR data", *IEEE Transactions on Geoscience and Remote Sensing*, 36(3): 963-973.

Fukunaga, K. (1990), *Introduction to Statistical Pattern Recognition*, 2nd ed. London: Academic.

Fung, A.K. (1994), "*Microwave scattering and emission models and their applications*," Boston: Artech House.

Ghiglia, D.C. and L.A. Romero (1994), "Robust two-dimensional weighted and unweighted phase unwrapping that uses fast transforms and iterative methods", *Journal of Optical Society of American A*, 11(1): 107-117.

Hennings, R. et al. (1999), "Radar imaging of Kelvin arms of ship wakes", *International Journal of Remote Sensing*, 20(13): 2519-2543

Hovanessian, S.A. (1980), *Introduction to Synthetic Array and Imaging Radars*, Mass: Artech House.

Huynen, J.R. (1970), "*Phenomenological Theory of Radar Targets*", PhD. thesis, University of Technology, Delft, Netherlands.

Itoh, T., H. Sueda and T. Wantanabe (1996), "Motion compensation for ISAR via centroid tracking", *IEEE Transactions on Aerospace and Electronic Systems*, 32(3): 1191-1197.

Jin, Y.Q. (1994), *Electromagnetic Scattering Modeling for Quantitative Remote Sensing*, Singapore: World Scientific.

Jin, Y.Q. and S.R. Cloude (1994), "Numerical eigenanalysis of the coherency matrix for a layer of random nonspherical scatterers", *IEEE Transactions on Geoscience and Remote Sensing*, 32(6): 1179-1185.

Jin, Y.Q. (2000), *Information of Electromagnetic Scattering and Radiative Transfer in Natural Media*, Beijing: Science Press.

Jin, Y.Q. and S. Wang (2001), "An algorithm for ship detection from SAR image using the radon transform and topographical image processing", *The Imaging Science Journal*, 48(4): 159-163.

Jin, Y.Q. and Z. Li (2002), "Numerical simulation for radar surveillance of a ship target in oceanic clutters", *Chinese Science Bulletin,* 47(21): 1766-1771.

Jin, Y.Q. and Y. Chen (2002), "An improved method of the minimum entropy for refocusing moving target image in the SAR observation", *The Imaging Science Journal*, 50(6): 147-152.

Jin, Y.Q. and L. Luo (2004), "Terrain topographic inversion from single-pass polarimetric SAR image data by using polarimetric stokes parameters and morphological algorithm", *Science in China, (F)*, 47(4): 490-500.

Jin, Y.Q., F. Chen and L. Luo (2004), "Automatic analysis of change detection of multi-temporal ERS-2 SAR images by using two-thresholds EM and MRF algorithms", *The Imaging Science Journal*, 52: 234-241.

Kasetkasem, T. and P. Varshney (2002), "An image change detection algorithm based on Markov random field model", *IEEE Transactions on Geoscience and Remote Sensing*, 40(8): 1815-1823.

Kasilingam, D. et al. (2000), Focusing of synthetic aperture radar images of moving targets using minimum entropy adaptive filters", *International Geoscience and Remote Sensing Symposium* (IGARSS'00): 74-76.

Kong, J.A. et al. (1990), "Classification of earth terrain using polarimetric synthetic aperture radar images," *PIER (Progress in Electromagnetic Research)*, ed. by J. A. Kong, New York: Elsevier, 3(6).

Lee, J.S., R.W. Jansen and D.L. Schuler (1998), "Polarimetric analysis and modeling of multi-frequency SAR signatures from gulf stream fronts", *IEEE Journal of Oceanic Engineering*, 23(4): 322-333.

Lee, J.S. et al. (1999), "Unsupervised classification using polarimetric decomposition and the complex Wishart classifier", *IEEE Transactions on Geoscience and Remote Sensing*, 37(5): 2249-2258.

Lee, J.S., D.L. Schuler and T.L. Answorth (2000), "Polarimetric SAR data compression for terrain azimuthal slope variation", *IEEE Transaction on Geoscience Remote Sensing*, 38(5): 2153-2163.

Lee, J.S. et al. (2004), "Unsupervised terrain classification preserving polarimetric scattering characteristics," *IEEE Transactions on Geoscience and Remote Sensing*, 42(4): 722-731.

Li, X. et al. (2000), "Auto-focusing of ISAR images based on entropy minimization", *IEEE Transactions on Aerospace and Electronic System*, 35(4): 1240-1251.

Lin, I.I. et al. (1996), "Ship and ship wake detection in the ERS SAR imagery using computer-based algorithm", *International Geoscience and Remote Sensing Symposium*

(IGARSS'96): 151-153

Ridd, M.K. and J.J. Liu (1998), "A comparison of four algorithms for change detection in an urban environment", *Remote Sensing of Environment*, 63(2), 95-100.

Moon, T.K. (1996), "The expectation-maximization algorithm", *Signal Processing Magazine*, 13(6): 47-60.

Oliver, C. and S. Quegan (1998), *Understanding Synthetic Aperture RadarImages*, Mass: Artech House.

Plummer, S. E., and P. J. Curran (1998), "BOREAS RSS-04 1994 Jack pine leaf biochemistry and modeled spectra in the SSA, data set", Available on-line [http://www.daac.ornl.gov] from Oak Ridge National Laboratory Distributed Active Archive Center, Oak Ridge, Tennessee, U.S.A.

Pritt, M.D. and J.S. Shipman (1994), "Least-squares two-dimensional phase unwrapping using FFT's", *IEEE Transaction on Geoscience Remote Sensing*, 32(3): 706-708.

Pritt, M.D. (1996), "Phase unwrapping by means of multi-grid techniques for interferometric SAR", *IEEE Transactions on Geoscience Remote Sensing*, 34(3): 728-738.

Render, A.P. and H.F. Walker (1984), "Mixture densities maximum likelihood and the EM algorithm", *SIAM Review*, 26(2): 195-239.

Rey, M.T. et al. (1990), "Application of Radon transform techniques to wake detection in Seasat-A SAR images", *IEEE Transactions on Geoscience and Remote Sensing*, 28(4): 553-560

Rignot, E.J.M. and J.J. Van Zyl (1993), "Change detection techniques for ERS-1 SAR data", *IEEE Transactions on Geoscience and Remote Sensing*, 31(4): 896-906.

Schuler, D.L., J.S. Lee, T.L. Answorth and M.R. Grunes (2000), "Terrain topography measurements using multipass polarimetric synthetic aperture radar data", *Radio Science*, 35: 813-832.

Schuler, D.L., T.L. Answorth and J.S. Lee (1998), "Topographic mapping using polarimetric SAR data", *International Journal of Remote Sensing*, 19(1): 141-160.

Shahshahani, B.M. and D.A. Landgrebe (1994), "The effect of unlabeled samples in reducing the small size problem and mitigating the Hughes phenomenon", *IEEE Transactions on Geoscience and Remote Sensing*, 32(5): 1087-1095.

Singh, A. (1989), "Digital change detection techniques using remotely-sensed data", *International Journal of Remote Sensing*, 10(6): 989-1003.

Takajo, H. and T. Takahashi (1998), "Least-squares phase estimation from phase difference", *Journal of Optical Society of American (A)*, 5: 416-425.

Ulaby, F.T. et al. (1986), "Textural information in SAR images" , *IEEE Transactions on Geoscience and Remote Sensing* , 20(2): 35-45.

Van Zyl, J.J. (1989), "Unsupervised classification of scattering behavior using radar polarimetry data", *IEEE Transactions on Geoscience and Remote Sensing*, 27(1): 36-45.

Wahl, D.E. et al. (1994), "Phase gradient auto-focusing — A robust tool for high resolution SAR phase correction". *IEEE Transactions on Aerospace and Electronic Systems*, 30(3): 827-835.

Ward, K.D. and S. Watts (1985), "Radar sea clutter", *Microwave Journal*, 28(6): 109-121

Werness, S. et al. (1990), "Moving target imaging algorithm for SAR data". *IEEE Transactions on Aerospace and Electronic Systems*, 21(1): 57-67.

Xu, F. and Y.Q. Jin (2005), "Deorientation theory of polarimetric scattering targets and application of terrain surface classification", *IEEE Transactions on Geoscience and Remote Sensing*, 43(10):1 - 16.

Zebker, H.A. et al. (1991), "Calibrated imaging radar Polarimetry: Techniques examples and applications", *IEEE Transactions on Geoscience and Remote Sensing*, 29(6): 942-961.

Zhang, Q. et al. (2002), "Urban built-up land change detection with road density and spectral information from multi-temporal Landsat TM data", *International Journal of Remote Sensing*, 23(15): 3057-3078.

Chapter 4

Remote Sensing Data Validation and Applications 151

4-1 Terrain Surface Moisture Mapping Based on Multi-Year Passive Microwave Remote Sensing

Terrain surface moisture in regional and continental scales is a key factor for studying land-atmosphere interaction and hydrology, improving weather and climate forecast, surveying crop growth, and monitoring drought or floods, etc. Assimilation of terrain surface moisture remotely sensed in different scales to hydrologic modeling can greatly improve the global study.

Scientific mission of space-borne microwave radiometer was started from the Nimbus Series in 1970s, and multi-channels at 6.63, 10.69, 18, 21 and 37 *GHz* (all dual polarization) were employed in Nimbus 7. A new channel at 85 *GHz* was utilized mainly for rainfall remote sensing in the DMSP (the Defense Meteorological Satellite Program) SSM/I (Special Sensor Microwave/Imager) launched in 1989 [Hollinger 1989]. The SSM/I has 7 channels from 19.35 to 85 *GHz* and has been in operation for over 10 years. Multi-temporal and multi-channels imagery data have been available for applications of monitoring terrain surface moisture, snow, sea ice, sea surface wind, atmospheric vapor and rainfall etc. [Ferraro et al. 1996]. The approaches and algorithms are largely based on empirical regression of multi-channels data. The AMSR-E (Advanced Microwave Scanning Radiometer) of the Earth Observing System (EOS) with the frequency channels at 6.9, 10.7, 18.7, 23.8, 36.5 and 89 *GHz* (all dual polarization) [Njoku et al. 2003] was launched on May 2002. Its lower frequency channels, such as 6.9 and 10.7 *GHz*, might minimize influence from canopy scattering and are capable of retrieving the terrain surface moisture.

Generally, there are two difficulties to information retrieval of the terrain surface moisture. Firstly, the observed brightness temperature of the satellite-borne radiometer contains contributions from the soil land surface, vegetation canopy, land surface roughness, and atmosphere (temperature and vapor). It is difficult to identify radiance variation caused by which one and how much one is. Secondly, it is due to large regional and continental mapping. Large territory contains so much heterogeneity, even over a single pixel of the radiometer resolution within 30-70 km^2. Survey of land surface conditions in different scales is always laborious and difficult work. There have been a lot of studies [Paloscia et al. 2001, Jackson et al. 1999, 2001, Njoku and Li 1999] on the theoretical models of land-canopy scattering-emission and fields experiments made at local regions. However, it is difficult to extend the conclusion from a local region at a certain time to another region and another time, and to obtain consistent and quantitative evaluation of terrain surface moisture in large regional and continental scales, as operational service often requires.

In this session, the polarization index (PI) is derived from the radiative transfer model of half-space of canopy-soil land. It can be seen that this index well indicates the polarized radiance effects, respectively, from the terrain moisture of canopy-soil surface, and other interference due to surface roughness and atmosphere. We do not pretend to be able to clearly separate the contributions from vegetation canopy and underlying land surface. It always keeps as a question for further study as many people are doing. But, we admit that variations in soil land moisture are positively correlated with variations in vegetation water content. At this point, it seems not critical to identify which one, vegetation canopy (especially sparsely distributed canopy) or soil land, is prominent to contribute to change of the terrain surface moisture. We are alert on the areas with large change of surface moisture, especially suffering draught or floods within the imagery pixels.

A spatial distribution of the average $< PI >$ over large scale, e.g. over China vast and heterogeneous territory, during a month period is statistically obtained from 5 or 6 years

(1997-2002) DMSP SSM/I data at lowest 19.35 *GHz,* as available to us. Such average $<PI>$ is taken as a normal reference. The deviation $\Delta_n PI$ $(\equiv (PI-<PI>)/<PI>)$ at the same time period might indicate characteristic variation of terrain surface moisture, based on such knowledge that no seasonal cycle is assumed, i.e. all the weather, surface roughness, vegetation canopy and so on at the same time period annually are almost kept the same. Actually, we admit that nothing in nature can be fixed. However, as operational service of one or two channel brightness temperature observation required, those seasonal cycles of vegetation and atmosphere are minimized, and the cost is the accuracy of our approach. In this sense, the mapping $\Delta_n PI$ can detect the change of terrain surface moisture in regional and continental heterogeneous scales.

Particularly, significant anomaly of $\Delta_n PI$ can indicate the extreme cases of possible drought and floods. To quantitatively define the seriousness, the *PIs* at each imagery pixel are categorized into 9 levels based on occurrence probability. Bigger deviation from the middle level $<PI>$ indicates more possibility of drought or floods. The mapping of $\Delta_n PI$ and its anomaly over continental scale such as over China vast territory is designed to provide operational service for monitoring terrain surface moisture. To minimize the scattering effect of vegetation canopy and optimize the model, our approach is recommended to lower frequency channels such as the AMSR-E and China's microwave imager radiometer of Fengyun Series in the future [Yan and Jin 2004].

Additionally, if the $<PI>$ and ground truth of average soil volumetric moisture $<m_v>$ are correctly evaluated, it is possible to retrieve the soil land moisture m_v by using the *PI* data and $<PI>$, $<m_v>$. As a test, some data of *PI*, $<PI>$ and $<m_v>$ at 10.7 and 19.3 *GHz* from TRMM TMI (Tropical Rainfall Measuring Mission, Microwave Imager) are collected from Njoku et al. (2003). The retrieval of m_v seems very favorable.

4-1.1 Radiative Transfer and *PI* Index

In clear atmosphere, scattering of atmospheric molecular and aerosol in microwave VRT is negligible. With the profiles of atmospheric temperature and moisture, the VRT equation is written as

$$\cos\theta \frac{d}{dz} I_p = -\kappa_a(z) I_p + \kappa_a C T_a(z), \qquad p = v, h \tag{4-1.1}$$

and the boundary conditions are

$$I_p(\pi-\theta, z=\infty) \approx 0, \qquad I_p(\theta, z=0) = e_{sp}(\theta) C T_s + r_{sp}(\theta) I_p(\pi-\theta, z=0) \tag{4-1.2}$$

where $I_p(\pi-\theta, z=\infty) \approx 0$ is the downward radiation from up-space, $e_{sp}(\theta), r_{sp}(\theta)$ are the emissivity and reflectivity of the land surface, respectively, and $r_{sp}(\theta) = 1 - e_{sp}(\theta)$, T_s and $T_a(z)$ are the physical temperature of the land surface and atmosphere, respectively.

It can be obtained that the brightness temperature at $z = d(\to\infty)$, $I_p(\theta, z=d)$ or $T_{Bp}(\theta, z=d) = I_p(\theta, z=d)/C$ is

$$T_{Bp}(\theta, z=d) = e_{sp}(\theta) T_s e^{-\int_0^d \kappa_a(z)\sec\theta dz} + r_{sp}(\theta) e^{-\int_0^d \kappa_a(z)\sec\theta dz} \int_0^d T_a(z)\kappa_a(z)\sec\theta\, e^{-\int_0^z \kappa_a(z')\sec\theta dz'} dz$$
$$+ \int_0^d T_a(z)\kappa_a(z)\sec\theta\, e^{-\int_z^d \kappa_a(z')\sec\theta dz'} dz \tag{4-1.3}$$

where $\kappa_a(z)$ is the absorption coefficient of atmospheric oxygen and water vapor, and the atmospheric opacity $\tau(\theta)=\sec\theta\int_0^d \kappa_a(z)dz$. Simply assuming $T_a(z)=T_{a0}$ as a constant, Eq. (4-1.3) becomes

$$T_{Bp}(\theta)=(1-r_{sp})T_s e^{-\tau}+(1-e^{-\tau})(1+r_{sp}e^{-\tau})T_a \tag{4-1.4}$$

To consider the vegetation canopy overlaying the land surface, assuming isotropic scattering of canopy point scatters, the phase function in the scalar RT equation is written as

$$P=\frac{\kappa_{sc}}{4\pi}=\kappa_{ec}\frac{\widetilde{\omega}_c}{4\pi} \tag{4-1.5}$$

where $\widetilde{\omega}_c=\kappa_{sc}/\kappa_{ec}$ is the albedo of the canopy (the subscript c), the VRT solution or the brightness temperature observed is written as [Kerr and Njoku 1990]

$$\begin{aligned}T_{Bp}(\theta)&=(1-r_{sp})T_s e^{-\tau_c-\tau_a}+(1-\widetilde{\omega}_c)(1-e^{-\tau_c})e^{-\tau_a}T_c+(1-\widetilde{\omega}_c)(1-e^{-\tau_c})r_{sp}e^{-\tau_c-\tau_a}T_c\\&+T_a^{\uparrow}+r_{sp}(T_a^{\downarrow}+T_\infty^{\downarrow}e^{-\tau_a})e^{-2\tau_c-\tau_a}\end{aligned} \tag{4-1.6}$$

where the first term is the emission of the land surface through vegetation canopy and atmosphere, the second term is the upward emission of vegetation canopy through atmosphere, the third term is the downward and then reflected emission of vegetation canopy by the land surface through canopy and atmosphere, the fourth term is the upward emission of atmosphere $T_a^{\uparrow}$, and the fifth term is the reflected downward emission of atmosphere $T_a^{\downarrow}$ (including $T_\infty^{\downarrow}$ from up-space) by the land surface through canopy and atmosphere. τ_c is the canopy opacity and $\tau_c=k'(\varepsilon_c''/\varepsilon_c')d_1\sec\theta$. $\varepsilon_c'+i\,\varepsilon_c''$ is the dielectric constant of the vegetation canopy. It is usually assumed that $T_a^{\uparrow}=T_a^{\downarrow}=(1-e^{-\tau_a})T_a$, $T_\infty^{\downarrow}=0$.

As $e^{-\tau_c}\approx 1, \widetilde{\omega}_c\approx 0$ at low microwave frequency, Eq. (4-1.6) is reduced to (4-1.4).

Scattering of vegetation canopy as an integrated part of terrain surface made another issue in the RT model [Kerr and Njoku 1990]. In the model of Eq. (4-1.4), the vegetation canopy is now not specifically treated as a layer. This simplicity might perturb our final results, especially in the case of thick vegetation canopy within the imagery pixels and at high microwave frequency channel, but is tractable to yield prominence to underlying land surface moisture, especially applicable to sparsely-distributed vegetation canopy and future lower frequency channels, e.g. in AMSR-E etc. We call the terrain surface to take account of the canopy effect.

Emissivity of the terrain surface $e_p=1-r_p$ is a function of soil land moisture m_v, interactions of scattering and emission of vegetation canopy and surface roughness. T_a and T_s are, respectively, the average physical temperature of atmosphere and terrain surface of half space (the physical temperatures of vegetation canopy and underlying land surface are not individually separated in this model) . We simply assume $T_a=(1-\delta_T)T_s$, $0\le\delta_T<<1$.

The polarization index $PI=(T_{Bv}-T_{Bh})/(T_{Bv}+T_{Bh})$ was defined to mainly show the polarized emission from underlying terrain surface and eliminate the physical temperature terms. From Eq. (4-1.4), we derive that

$$PI=\frac{T_{Bv}-T_{Bh}}{T_{Bv}+T_{Bh}}=\frac{(r_h-r_v)e^{-2\tau}[1+\delta_T(e^{\tau}-1)]}{2-(r_v+r_h)e^{-2\tau}-\delta_T(1-e^{-\tau})[2+(r_v+r_h)e^{-\tau}]} \tag{4-1.7}$$

Additionally, to take account of depolarization due to surface roughness and other effect, an empirical model of the surface reflectivity [Dobson et al. 1985] takes

$$r_p=[(1-Q)r_{p0}+Qr_{q0}]e^{-\sigma} \tag{4-1.8}$$

where r_{p0} and r_{q0} $(p,q=v,h)$ are the p,q-polarized reflectivities of the smooth surface, and Q,σ are the roughness parameters with $0 \le Q < 0.5$. Eq.(4-1.7) yields

$$PI \approx \frac{R_{(-)}(1-2Q)(1+\delta_T)}{2e^{2\tau}e^{\sigma}(1-\delta_T)-R_{(+)}} \tag{4-1.9}$$

where $R_{(-)} \equiv r_{h0} - r_{v0}$, $R_{(+)} \equiv r_{h0} + r_{v0}$. Eq. (4-1.9) separates some relevant factors in an explicit way.

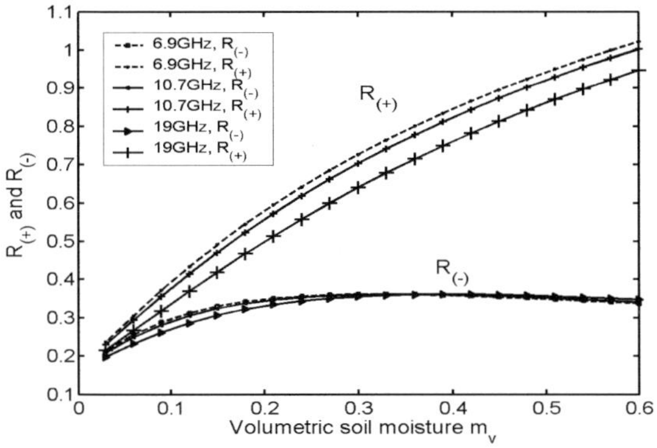

Fig. 4-1.1. $R_{(-)}$ and $R_{(+)}$ versus m_v

Let's first see how the variation of m_v affects the surface emissivity e_p or reflectivity r_p. Variation of m_v changes the dielectric constant of soil $\varepsilon(m_v)$, and further r_{h0}, r_{v0} and $R_{(-)}$, $R_{(+)}$. From the empirical formulation of $\varepsilon(m_v)$ given by Dobson et al. (1985), the parameter $\varepsilon(m_v)$ and functions r_{h0}, r_{v0}, $R_{(-)}$ and $R_{(+)}$ at 6.9, 10.7, 19*GHz* are calculated. Fig. 4-1.1 shows $R_{(-)}$ and $R_{(+)}$ versus m_v. It can be seen that $R_{(-)}$ is almost invariable for different m_v, but $R_{(+)}$ is significantly increasing when m_v rises. This is a main reason why *PI* in Eq. (4-1.9) increases when m_v rises. Existence of vegetation canopy won't significantly change such functional behavior of $R_{(-)}$ and $R_{(+)}$ upon m_v.

The merit of Eq. (4-1.9)'s *PI* is that different contributions by the soil moisture m_v via $R_{(-)}$ and $R_{(+)}$, depolarization factors of surface roughness Q and $e^{-\sigma}$, atmospheric opacity τ, and atmospheric parameter δ_T are physically and functionally separated.

For example, if τ, Q and σ, or $(1-\delta_T)$ increase, the polarization difference from underlying terrain surface is darken and *PI* decreases. On the other hand, increasing m_v yields high $R_{(+)}$. Thus, from Eq. (4-1.9), *PI* is increased and the polarization difference becomes significant. As mentioned in Introduction that no seasonal cycle is assumed, the parameters, τ, Q, σ and δ_T, and the status of vegetation canopy are almost kept as similar in the same time period annually, variation of *PI* is dominantly governed by varying surface moisture m_v.

4-1.2 The *PI* Mapping at SSM/I 19.35 *GHz*

The *PI* data shows considerable variations in different regions and different months.

Taking the *PI* data of SSM/I descending F14 at 19.35 *GHz* channel in May of 5 years (1998-2002), the monthly average $<PI>$ in the Shanxi area ($\sim 110.0^\circ - 115.0^\circ E$, $35.0^\circ - 40.0^\circ N$) of China is obtained. The location is indicated by a top square in Fig. 4-1.2. Its $<PI>$ of 5 years May is shown in Fig. 4-1.3(a). The rectangular size of each pixel (longitude 1/3°×latitude 1/3°) is about 30 *km*. Multiple passes data are co-registered into each rectangular pixel by averaging the data of this pixel recorded over the satellite footprints.

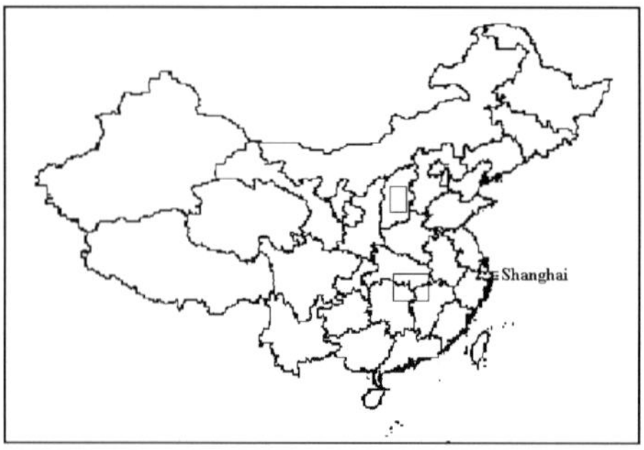

Fig. 4-1.2. The drought and flooding areas in China as examples

In order to describe the PI's deviation from the mean $<PI>$ in the same time (e.g. a month) period, the normalized average deviation is defined as

$$\Delta_n PI = (PI - <PI>)/<PI> \qquad (4\text{-}1.10)$$

To admit that the parameters of τ, Q, σ and δ_T, and the status of vegetation canopy in the same region and the same month period are kept as usual as the averages. The $\Delta_n PI$ of Eq. (4-1.10) then individually demonstrates variation of PI governed by m_v, i.e. $\Delta_n PI > 0$ indicates increased m_v, and $\Delta_n PI < 0$ indicates decreased m_v.

Of course, as we have mentioned, in case of thick vegetation canopy, variation of $\Delta_n PI$ indicates change of total water content embedded in canopy and underlying land surface, and especially indicates canopy moisture at high frequency channel. It is also functional upon terrain surface moisture. We do not extend this discussion in our RT model of Eq. (4-1.4), and focus on moisture mapping of large scale terrain surface. Figs. 4-1.3(b) and 3(c) show the monthly PI and $\Delta_n PI$ of May 2001, respectively, when the region was undergoing a serious spring drought (more warm color in the figure). It indicates the drought ($\Delta_n PI < 0$, see the red and brown color) around the center and northern part of the region.

As a comparison, Fig. 4-1.3(d) presents $\Delta_n PI$ of May 2002 in the same region. It can be seen that $\Delta_n PI \approx 0$ in most areas (more green and blue color) indicates no serious drought as reported indeed. Its top left corner is close to drying semi-desert.

Figs. 4-1.4(a-d) show a flooding example. Fig. 4-1.4(a) presents the 6 years $<PI>$ of August in the highly moistured region around Dongtinghu Lake and Poyanghu Lake ($\sim 26.0^\circ - 33.0^\circ N$, $111.0^\circ - 118.0^\circ E$, see the location of the low square in Fig.4-1.2) during August 1997-2002. Two lakes and a part of Yangtze River are roughly profiled.

Figs. 4-1.4(b,c,d) show the PI and $<PI>$ in August 1998, and $\Delta_n PI$ in August 2002, respectively. A serious flooding caused by heavy precipitation took place in the Yangtze River area during July-August 1998.

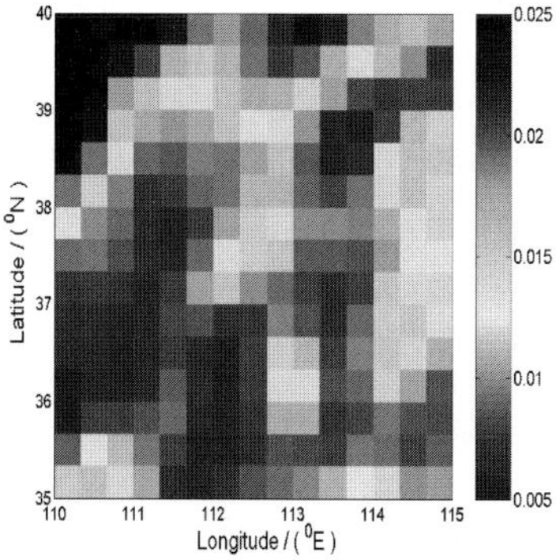

(a) 5 years $<PI>$ in May

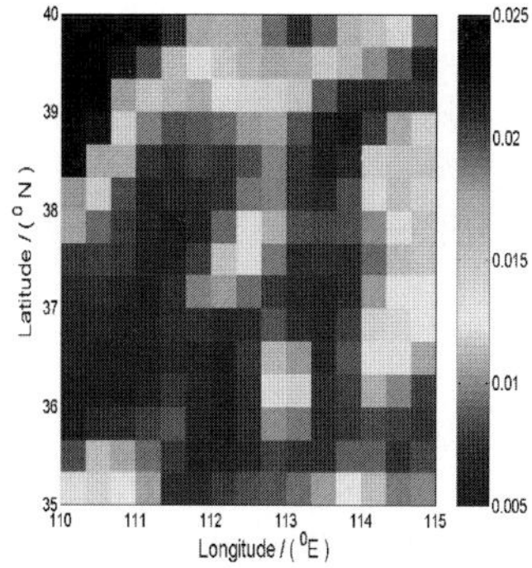

(b) PI in May 2001

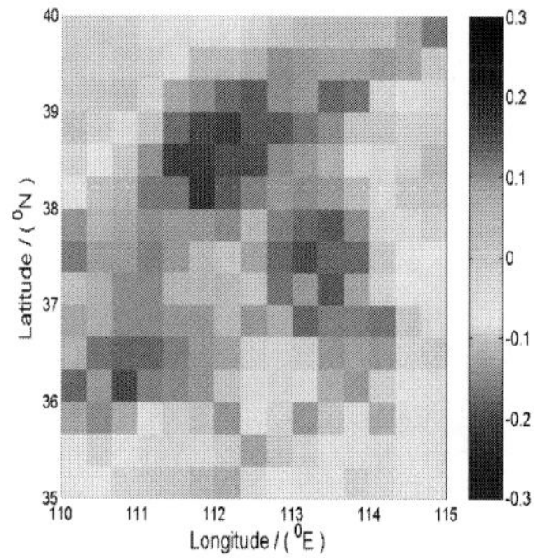

(c) $\Delta_n PI$ in May 2001

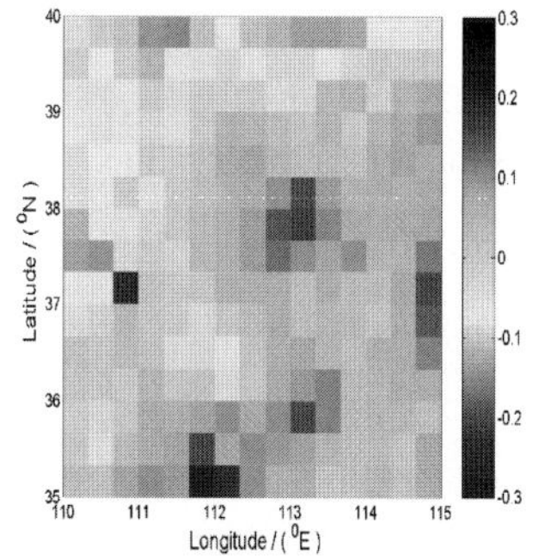

(d) $\Delta_n PI$ in May 2002

Fig. 4-1.3. $<PI>$, PI and $\Delta_n PI$ in the Shanxi area, China

Fig. 4-1.4(c) shows significant increase of $\Delta_n PI$ (more green and blue color) during the flooding, especially around two lakes. As a comparison, Fig. 4-1.4(d) shows $\Delta_n PI \approx 0$ when the region becomes normal in August 2002.

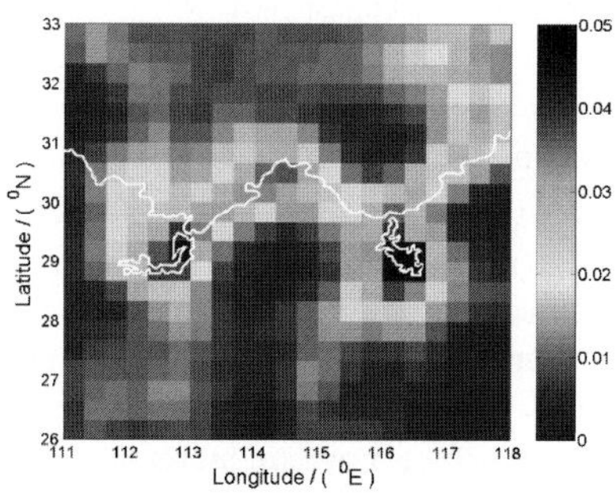

(a) 6 years $<PI>$

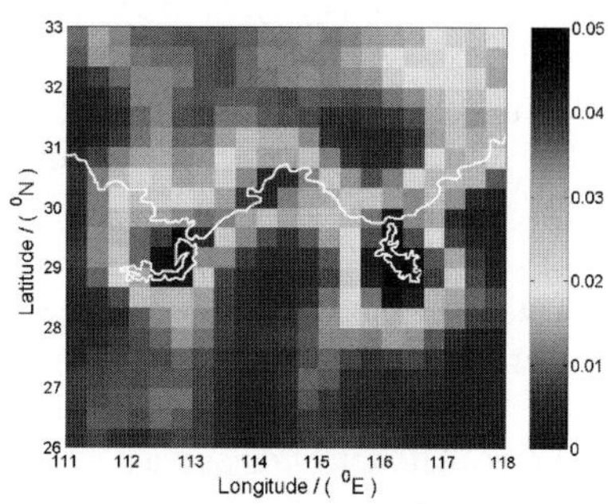

(b) PI in August 1998

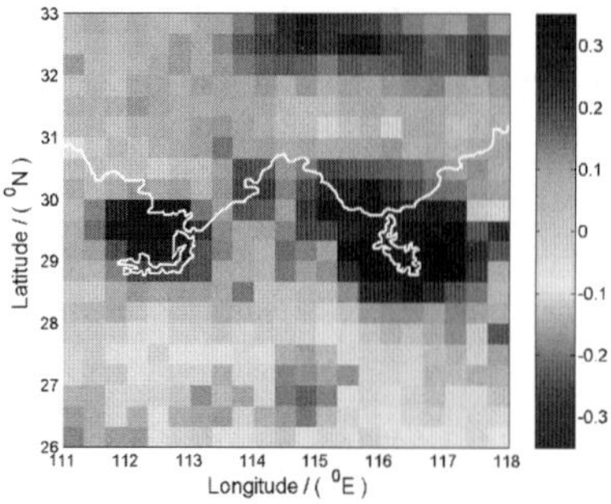

(c) $\Delta_n PI$ in August 1998

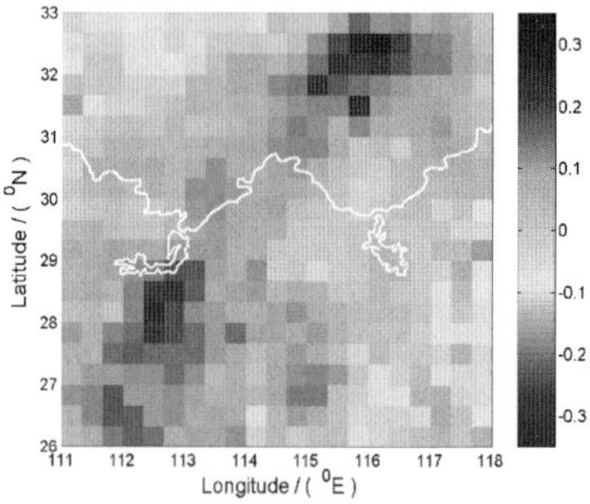

(d) $\Delta_n PI$ in August 2002

Fig. 4-1.4. 6 years $< PI >$ and $\Delta_n PI$ near the Dongtinghu Lake and Poyanghu Lake in August 1998 and 2002

Fig. 4-1.5 presents the $\Delta_n PI$ mapping over China territory in August 2002 to show spatial distribution of the terrain surface moisture.

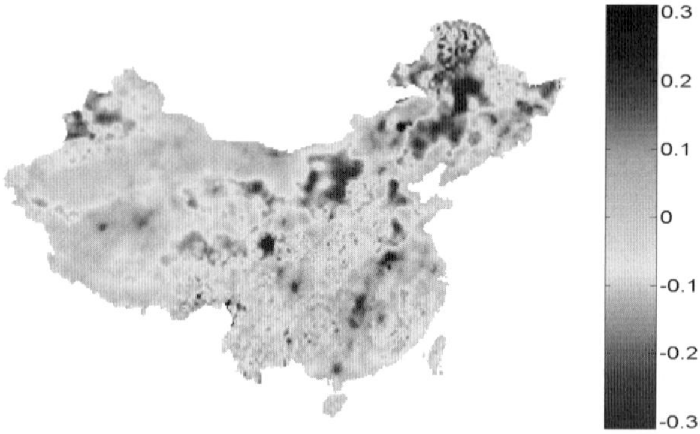

Fig. 4-1.5. $\Delta_n PI$ mapping as m_v variation in August 2002

4-1.3 Precipitation Causes Flooding

The scattering index *SI* was usually defined as [Jin 1999]

$$SI \equiv \frac{T_{B22v} - T_{B85v}}{T_{B22v} + T_{B85v}} \tag{4-1.11}$$

to indicate scattering caused by precipitation or other scattering effect. Stronger scattering of rainfall at 85*GHz*, comparing with 22*GHz*, darkens emission from underlying terrain surface. It yields $T_{B85v} << T_{B22v}$ and significantly increases *SI*. Thus index *SI* is usually taken as an index for rainfall detection and an antecedent index for surface moisture increase.

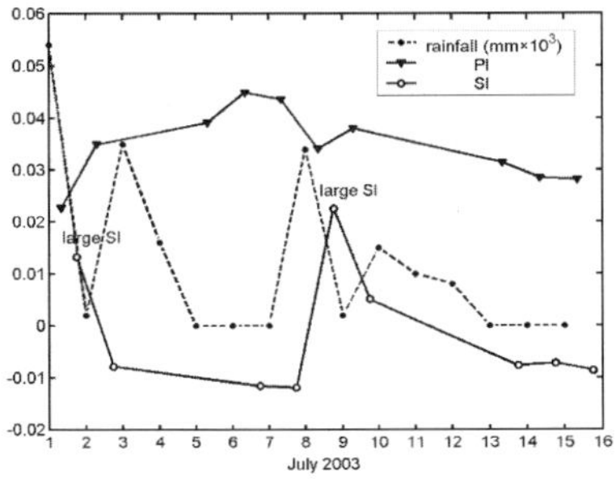

Fig. 4-1.6(a). *PI, SI*, and precipitation records at beginning July 2003 in Fuyang, Anhui.

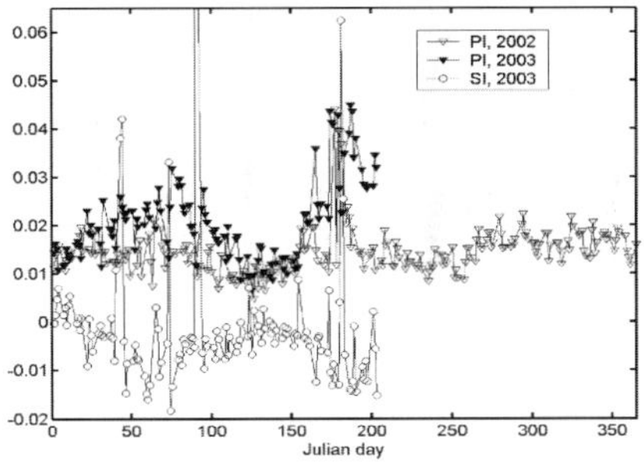

Fig. 4-1.6(b). *PI*, *SI* in the years 2002 and 2003.

Fig. 4-1.6(a) shows the indexes *PI, SI* data of SSM/I F14 and local ground records of precipitation ($mm \times 10^3$, in dashed line) in Fuyang, Anhui Province, China (about $115.8^\circ E$, $32.9^\circ N$) from 1 to 16 July 2003. There is a little time difference between the local ground record and F14 flight. It can be seen that the rainfalls at 1 and 8 July recorded by the solid circles indicated significant increases of *SI*, and darkens *PI* (Note that second rainfall peak was missed in the *SI* data because no ground record was reported). The precipitation record can be found from *www.t7online.com*. Due to those precipitations, it then drives *PI* increasing during the days after 2 and 9 July. The size of the area studied for Fig. 4-1.6 is one pixel with 1/3°×1/3° covering about 1000 km^2. Flooding detection by using our *PI* index over this large size should be well identified from paddy fields with possible water irrigation.

Fig. 4-1.6(b) presents indices *PI* and *SI* during January-July 2003 in Fuyang, Anhui. As a comparison, it also shows the *PI* index of 2002 considered as a normal year. It can be seen that index *PI* of 2003 keeps higher levels than 2002, especially during February-April and July. It identifies higher terrain surface moisture during these months when heavy precipitations happened as indicated by the *SI* peaks.

Making multi-year average $< PI >$ during 21 June to 20 July from 1997 to 2003 and the data of SSM/I F13, F14 and F15 descending flights in Huaihe River area on 1, 3, 4 and 6 July 2003, the $\Delta_n PI$ distribution is obtained as shown in Figs. 4-1.7(a~d). Take account of variation in data noise, empirically define $SI = 0.005$ as a threshold for rainfall detection. The white triangles in Figs. 4-1.7(a-d) indicate those locations of $SI > 0.005$ as raining regions.

From Figs. 4-1.7(c,d), the northern Huaihe River area on 4 and 6 July has large $\Delta_n PI$ distribution which indicates great soil moisture resulting from the antecedent precipitations on 1 and 3 July. The PI on 1 and 3 July is relatively small due to darkening of raindrops scattering which can be identified by the triangle covered area in Figs. 4-1.7(a,b). After rainfall, PI is then significantly increased as seen from Fig. 4-1.7(c). As a conclusion, the characteristic indices PI, SI and $\Delta_n PI$ are capable of providing spatial mapping for soil moisture variations and flooding evolution.

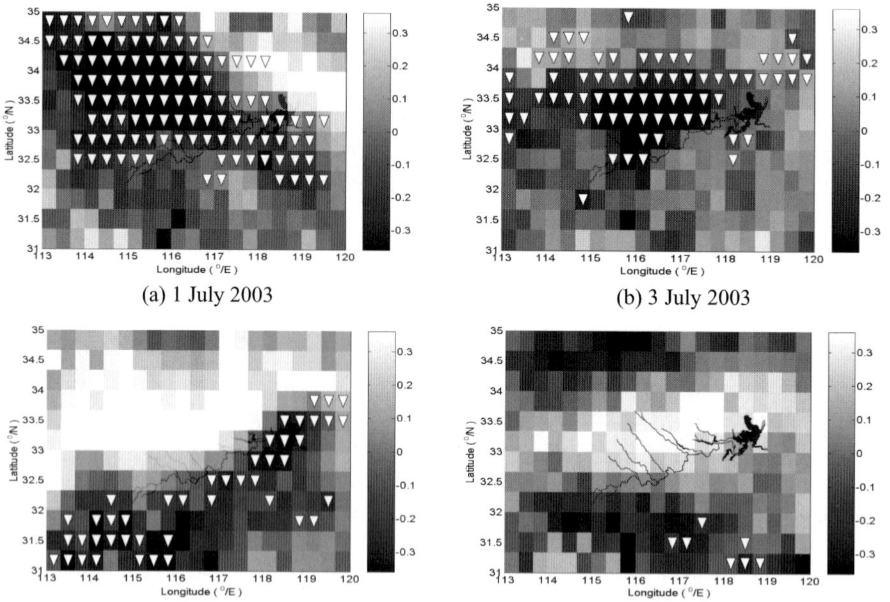

Fig. 4-1.7. $\Delta_n PI$ variation on 1-6 July 2003 (Huaihe River and Hongzehu Lake are marked)

Fig. 4-1.8 presents the $\Delta_n PI$ averaged from 1 to 15 July 2003. During this time period, PI over the northern side of Huaihe River shows extremely high soil moisture resulting from floods over that region. This spatial distribution is consistent with the results obtained from RADARSAT-1 SAR image [Xu 2003]. However, our result is more applicable for continuously monitoring soil moisture variation and its saturation distribution.

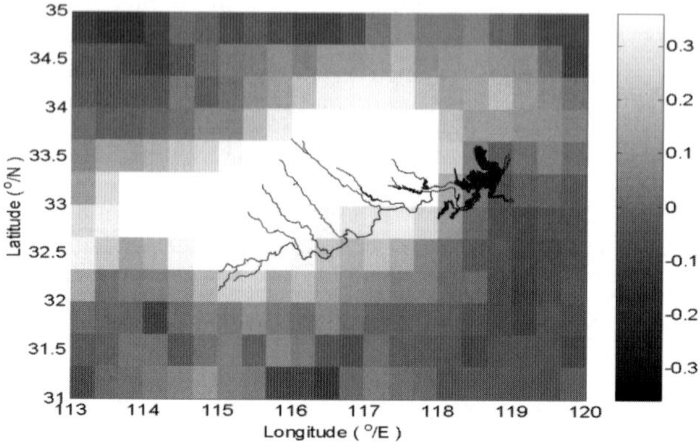

Fig. 4-1.8. Average $\Delta_n PI$ distribution in 1-15 July 2003

4-1.4 The *PI* Average Deviation and the Anomaly Index *K* of $\Delta_n PI$

The variation of *PI*, namely ΔPI, is related to the $< PI >$ in that region. The same ΔPI of two regions does not mean the same variation of m_v. Though the normalized $\Delta_n PI$

represents the relative variation of PI versus $<PI>$ in the region, comparison between different or adjacent regions, which one is more serious remains uncertain.

Suppose there are M times of observations during a certain period and divide all those PI data at each pixel region into 9 levels in descending order, sequentially. Each level takes $0.1M$ times data except the middle 5-th level with $0.2M$ times as most normal PI range in this region. Denote the anomaly index K $(K=1,2,\cdots,9)$ to each level, i.e. $K<5$ means the drying PI less than normal $<PI>$, and $K>5$ the wetting PI higher than $<PI>$. Large $|K-5|$ means serious anomaly. For example, if the anomaly index K in Region A is low as 9, and $K=6$ in Region B, then the situation in Region A is more serious as wet. Therefore, the mapping of the anomaly index K is applicable to the situation evaluation of terrain surface moisture over regional and continental scales in a consistent manner.

Dividing the PI data (in our case, $M\approx 200$) of July 1-15 from year 1997 to 2003 into 9 levels, Figs. 4-1.9(a,b) show the spatial distribution of the anomaly index K around the Huaihe River on 1 July and 1-15 July 2003, respectively. Comparing with Fig. 4-1.9(a), Figs.4-1.9(a,b) quantitatively classify the flooding regions into different levels of anomalies. It might be said that the pixels with $K\geq 8$ are identified as seriously flooding area.

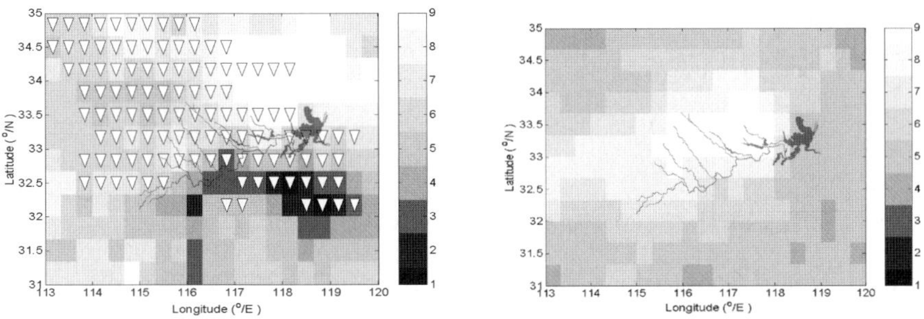

(a) A anomaly index K on 1 July 2003 (b) Anomaly index K in 1-15 July 2003

Fig. 4-1.9. Spatial distribution of the anomaly index K

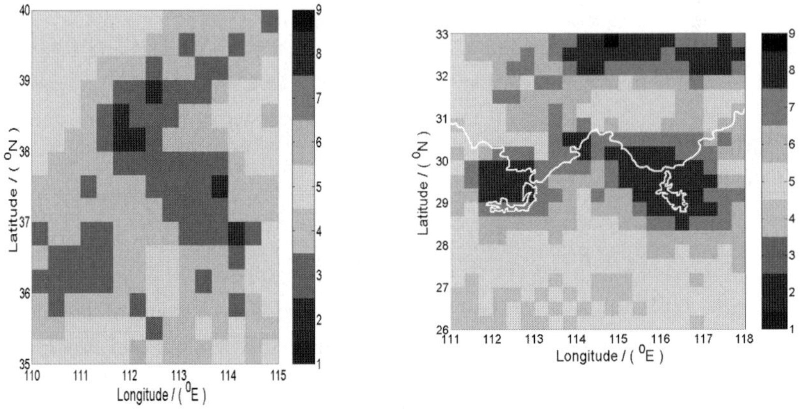

(a) The anomaly index in the Shanxi May 2001 (b) The anomaly index around the Yangtze River, August 2002

Fig. 4-1.10. The mapping of anomaly index for drought and flood situation

Figs. 4-1.10(a,b) show the K mapping over the Shanxi area in drying May 2001 and the Yangtze River area in flooding August 2002, respectively. It won't be surprised that there is similarity between the K and $\Delta_n PI$ mapping. But, the K mapping presents a simple way to make conclusion where the situation with large $\Delta_n PI$ is most serious.

Fig. 4-1.11 presents the K mapping over China in August 2002. It describes where the situation of the land surface moisture becomes more serious or noticeable over such vast and heterogeneous territory.

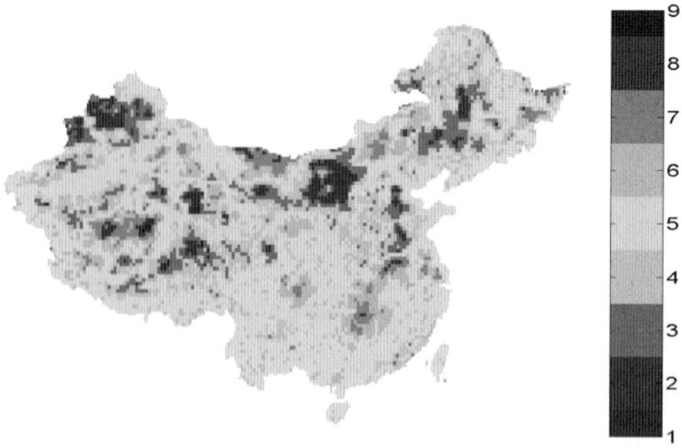

Fig. 4-1.11. The mapping of anomaly index over China in August 2002

4-1.5 Retrieval of Soil Surface Moisture at 10.7 and 19.3*GHz* Channels

As mentioned in Introduction, scattering from vegetation canopy certainly affects the accuracy of our final results at such frequency 19.35 *GHz* of SSM/I. Taking lower frequency channels (such as the AMSR-E's) to minimize scattering from vegetation canopy and opacity variation of atmosphere is more appropriate for retrieval of land surface moisture. Our model shows such feasibility and is recommended to the AMSR-E lower-frequency channels, as long as multi-years AMSR-E data become easily available.

Let's admit that the parameters of atmosphere, surface roughness and canopy, τ, Q, σ and δ_T, in the same time period (e.g. a month) of each year are almost not much different (i.e. within this month of each year, the atmospheric temperature and vapor are similar, surface roughness is not significantly changed, growth of crop canopy or biomass are similar, and so on). From Eq. (4-1.3), we obtain

$$\frac{PI}{<PI>} \approx \frac{2e^{2\tau}e^{\sigma}(1-\delta_T) - R_{(+)}(<m_v>)}{2e^{2\tau}e^{\sigma}(1-\delta_T) - R_{(+)}(m_v)} \equiv \frac{a - R_{(+)}(<m_v>)}{a - R_{(+)}(m_v)} \tag{4-1.12}$$

where $a \equiv 2e^{2\tau}e^{\sigma}(1-\delta_T)$. If a, $<m_v>$ and $<PI>$ are known from the ground truth, the function $R_{(+)}(m_v)$ can be inverted from the observation of PI via Eq. (4-1.12). Further, from Fig. 4-1.1, the soil surface moisture m_v can be retrieved, even the parameters τ, σ and δ_T have not been specifically known.

Njoku et al. (2002) have discussed potential application of the soil surface moisture retrieval by using the AMSR-E data. As an example of the methodology, we collected the PI

data of TRMM TMI and the ground truth of the soil surface moisture in Lonok, Arkansas, USA ($91.8667^{\circ}W$, $34.8333^{\circ}N$) from 11 December 1999 to 1 February 2000 listed in Njoku et al. Firstly, two days data of m_v and *PI* at 11 December 1999 and 18 January 2000 are chosen as the known and are used to invert the unknown *a*. Then, the m_v at other 6 days can be retrieved by using the respective *PI* via Eq. (4-1.12). Fig. 4-1.12 shows comparison between the measured m_v of Njoku et al. and our inverted m_v at 10.7 and 19.35 *GHz*. It can be seen that the inversion is well validated.

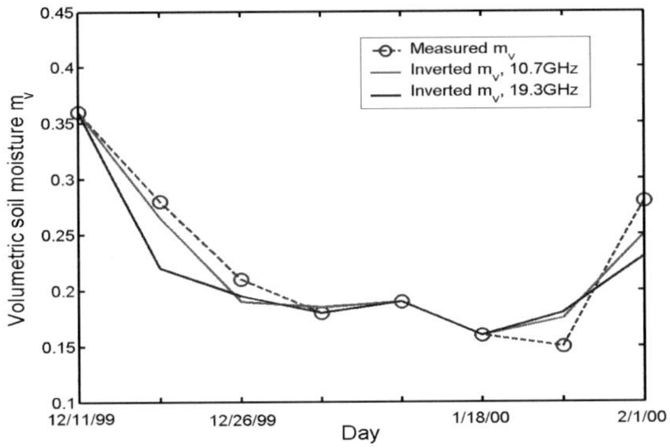

Fig. 4-1.12. Comparison of inverted and ground truth m_v

4-2 Monitoring the Sandstorm Using the SSM/I Data and Getis Statistics

In recent years serious sandstorms in different scale have been taking place in the north of China, especially during March and April each year. A massive sandstorm, enveloping most north China from west to east on 19-20 March 2002, reported as the most significant one since 1990s [Luo 2002]. Monitoring sandstorms and their temporal and spatial variations has become a critical issue for environment surveillance and protection in China. Desertification in north China has also turned into a serious problem that the desert area keeps increasing annually, totaling 9.4% of Chinese land area [Chen 1998]. As origins of sandstorms, Badain Jaran Desert and Tengger Desert tend to merge and expand. As a result, vegetation canopy in near region is diminishing, and even more seriously, entire region becomes gradually desertified. For example, the tree canopy around Badain Jaran Desert has lost an area of nearly 2000 km^2 in the past 40 years. Additionally, owing to grassland degeneration and windy weather, the local desert area is constantly expanding and the extent of desertification becomes more serious.

The DMSP SSM/I currently in operation since 1989 has 7 channels (19, 37, 85*GHz* with dual polarization and 22*GHz* with vertical polarization only). The local time of descending (DS) and ascending (AS) flights of the satellite F13 is about 6*am* and 6*pm*, respectively, whereas the DS and AS time for F14 and F15 is about 9:30*am* and 9:30*pm*. The spatial resolutions are correspondingly 15*km* (85*GHz*), 30*km* (37*GHz*) and 50*km* (19*GHz*). The satellite-borne microwave radiometer has become one of the most important tools for monitoring terrain surface because of its ability for all-whether operation to penetrate through precipitation and provide continuous service for many years. Observation of DMSP SSM/I has been applied to invert physical, hydrological, biological and other scientific information from atmosphere, sea surface and land covers. However, how to monitor sandstorm and land desertification by use of satellite-borne microwave remote sensing has not been well studied.

In this session, by using the VRT equation, sandstorm influence over the radiative properties is discussed. Employing multi-channel SSM/I data and VRT, we propose a sandstorm index *SDI* to monitor the occurrence and evolution of sandstorm. Application of this index is demonstrated by using the SSM/I observations of 19-21 March 2002 over Hunshandake Desert, Inner Mongolia and nearby regions when a serious sandstorm happened. A polarization index, PI_{37}, namely desertification index *DI*, defined by the SSM/I data at 37*GHz* channel, is employed for verification of spatial desertification, and two-year *DI* data of 1997 and 2001 over the Badain Jaran Desert and Tengger Desert are discussed. In order to evaluate the spatial correlation of regional radiative properties, the Getis spatial auto-correlation statistics are presented for *DI* index of the desert region to quantitatively describe desertification evolution at spatial scales.

4-2.1 Sanstorm index *SDI* defined by SSM/I data

The VRT equation for clear atmosphere is as Eq. (4-1.4):

$$T_{Bp} = (1-r_p)e^{-\tau}T_s + (1-e^{-\tau})(1+r_p e^{-\tau})T_a, \quad p = v,h \tag{4-2.1}$$

where r_p is p-polarized reflectivity of underlying surface, T_s is the physical temperature of the land surface, T_a is the averaged physical temperature of atmosphere and τ is the atmospheric opacity at this channel. For simplicity in our modeling, suppose $T_s = T_a \equiv T_o$. We can have

$$T_{Bp} = (1 - r_p e^{-2\tau})T_o \tag{4-2.2}$$

Because the sand particles are very small with radius a about $0.15mm$ unequally, under the Rayleigh approximation ($k\,a << 1$ at low frequency, here k denotes the wave number), the scattering, absorption and extinction coefficients κ_s, κ_a, κ_e of the sand particles can be written as [Jin 1994]

$$\kappa_e = \kappa_s + \kappa_a = 2\,fk^4 a^3 \left|\frac{\varepsilon_{sr} - 1}{\varepsilon_{sr} + 2}\right|^2 + fk\frac{\varepsilon''_{sr}}{\varepsilon'_{sr}}\left|\frac{3}{\varepsilon_{sr} + 2}\right|^2 \tag{4-2.3}$$

where $\varepsilon_{rs} = \varepsilon'_{sr} + i\varepsilon''_{sr}$ is the relative dielectric constant of sand particles, f is the particles' fraction volume per unit volume in the sandstorm layer.

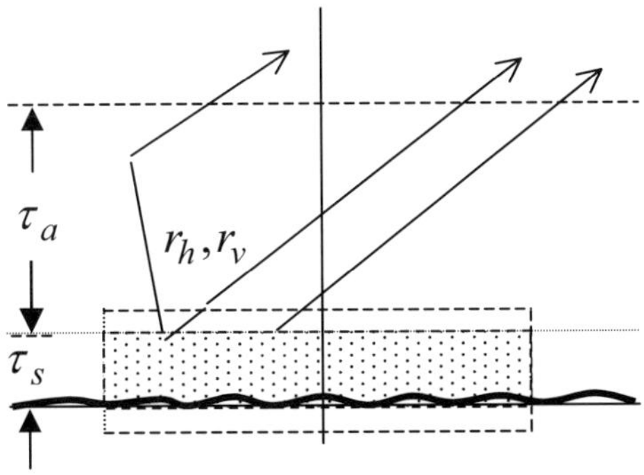

Fig.4-2.1. Geometry of radiative transfer in stand stormy atmosphere

According to the report of March 2002 by the surveillance network of the China State Environment Protection Administration (SPEA), we take the sandstorm parameters as follows: a=0.15mm, fractional volume $f = m/\rho$=5.5×10^{-9}, where total weight per unit cubage, M=0.011 g/m^3 and the sand density ρ=2×10^6 g/m^3 were measured. Assuming the thickness of sandstorm layer be d=3 km, and relative dielectric constant ε_{rs}=5.0+i0.25 [Goldhirsh 2001], the opacity of sandstorm layer is calculated as

$$\tau_s = \kappa_e d \sec\theta \tag{4-2.4}$$

At the 85GHz channel and SSM/I observation angle θ=53°, Eq. (4-2.4) yields $\tau_{s85} \approx 1 \times 10^{-3}$. It is known that the opacity of standard atmosphere at 85GHz is about $\tau_{a85} \approx 0.22$ or more [Jin 1994]. It can be seen that the sandstorm opacity would not be able to significantly affect whole atmospheric opacity. In other words, one cannot identify any significant change of the opacity from microwave radiometers in satellite-borne remote sensing.

Sandstorm existence would affect radiative transfer from the underlying surface and most likely darken the difference between vertically and horizontally polarized reflectivities, as shown in Fig. 4-2.1. Polarization index $PI_{37} \equiv T_{B37v} - T_{B37h}$ has been used to characterize desert surface [Jin 1997], and scattering index $SI_{2285v} \equiv T_{B22v} - T_{B85v}$ has been used to describe the scattering difference between high (85GHz) and low (22GHz) SSM/I channels. It is noted that SI_{2285v} is good indicator to detect rainfall. In our cases, there is no rainfall. Following Eq. (4-2.2), it is derived as

$$PI_{37} = (r_{37h} - r_{37v})e^{-2\tau_{37}}T_o \tag{4-2.5}$$

$$SI_{2285} = (r_{85v}e^{-2\tau_{85}} - r_{22v}e^{-2\tau_{22}})T_o \quad (4\text{-}2.6)$$

Both PI_{37} and $SI_{2285\,v}$ are the function of T_o and surface reflectivity r_p .

We can simply assume $r_{85v} \approx \alpha\, r_{37v}$, $r_{22v} \approx \beta\, r_{37v}$ for dry surface, where α , β are actually very close to 1. To minimize the variation from T_o, we define the sandstorm index *SDI* as

$$SDI = \frac{SI_{2285}}{PI_{37}} = \frac{1}{(r_h / r_v) - 1} \cdot \frac{\alpha \cdot e^{-2\tau_{85}} - \beta \cdot e^{-2\tau_{22}}}{e^{-2\tau_{37}}} \quad (4\text{-}2.7)$$

where r_h , r_v are the *h,v*-polarized reflectivities at 37*GHz*, which take into account the sandstorm and underlying land surface indicated by the dashed line in Fig. 4-2.1. It always has $r_v < r_h$.

It can be seen that a sandstorm makes r_v approach r_h, and increases *SDI*. The *SDI* from SSM/I measurements at 22, 37 and 85*GHz* is seen as a criterion for detecting sandstorm.

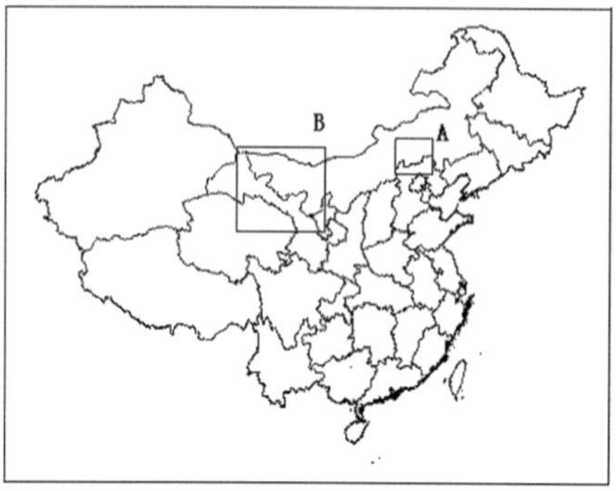

Fig. 4-2.2. Location of our study regions

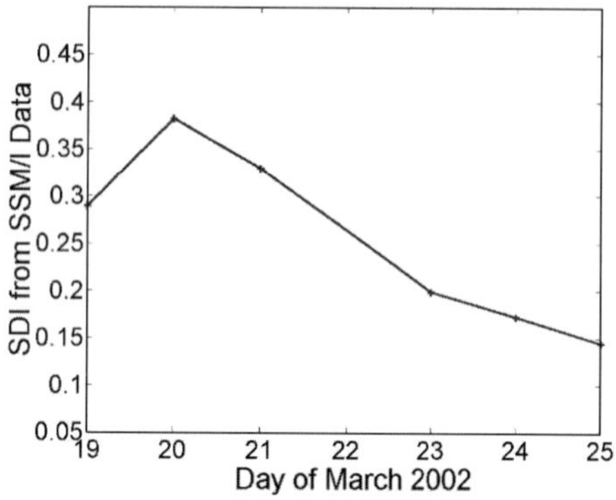

Fig. 4-2.3. Averaged *SDI* variations in Region A before and after the sandstorm

Region A in Fig. 4-2.2 underwent sandstorm for many times during 19-21 March 2002. Fig. 4-2.3 shows the averaged *SDI* from F13 DS data in Region A during 19-25 March 2002. It can be seen that the rise of *SDI* on 20-21 March indicates sandstorm occurrence.

As a base for comparison, Fig. 4-2.4(a) gives the averaged *SDI* of no sandstorm in Region A during 23-25 March 2002. The brightness in Fig. 4-2.4(a) indicates high-value *SDI*, and dull blackness means low-value *SDI*.

Fig. 4-2.4(b) shows the *SDI* distribution in Region A observed by DS-F13 on the morning 21 March 2002 when a sandstorm occurred. The difference between Fig. 4-2.4(b) and Fig. 4-2.4(a) is obtained as shown in Fig.4-2.4(c), where the brightness is now due to the sandstorm occurrence of 21 March 2002.

Fig. 4-2.4(d) was released to the Internet by China State Satellite Meteorology Center (SSMC) as a visualization of sandstorm occurrence at 12:00 on 21 March 2002 from observation of China Meteorological Satellite FY1-C. Fig. 4-2.4(c) is well validated by Fig. 4-2.4(d) if time difference between these two observations is taken into account.

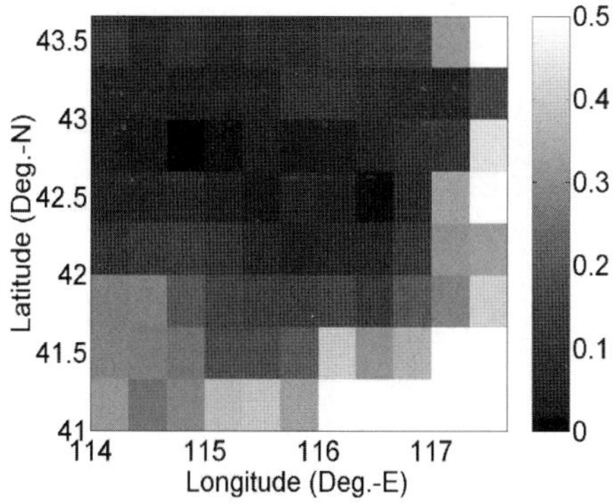

(a) *SDI* distribution without sandstorm

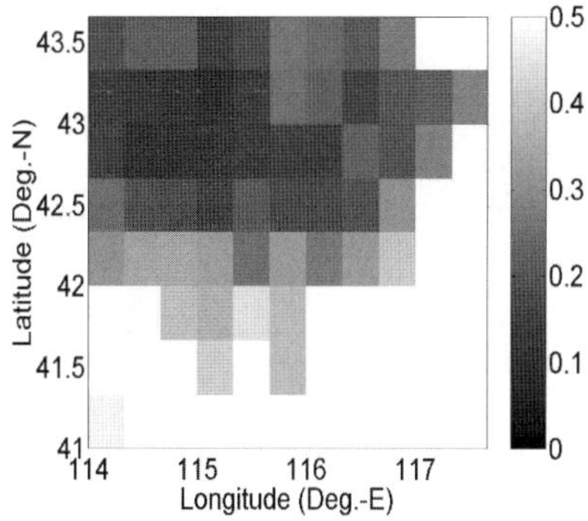

(b) *SDI* at 9:30am on 21 March with a sandstorm

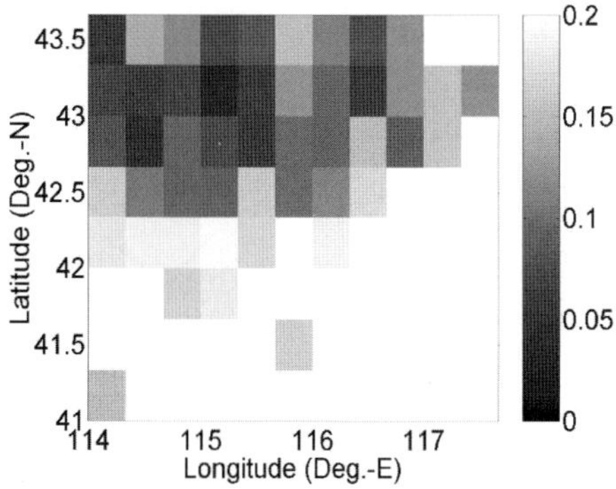

(c) *SDI* difference between (b) and (c) indicating the sand stormy region

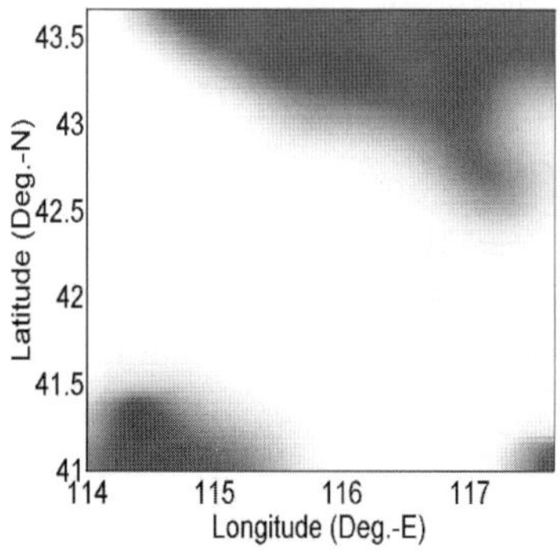

(d) The sand stormy region at 12:00 on 21 March issued by China SSMC

Fig. 4-2.4. The index *SDI* distribution

4-2.2 Desert Index *DI* for Monitoring Desertification

Desertification is characterized by canopy disappearance, soil exposure and its conversion into sand. Dry sand surface drives smaller difference between *v*, *h*-polarized reflectivities of land surface. We have defined the polarization index PI_{37} as a characteristic parameter for desert surface [Jin 1997]. We might as well name the desertification index as

$$DI = PI_{37} = (r_h - r_v)e^{-2\tau}T_o \tag{4-2.8}$$

Figs. 4-2.5(a, b) present the averaged *DI* distribution derived from SSM/I F13 over China's Badanin Jaran Desert and Tengger Desert during March-August in 1997 and 2001, respectively. Brightness in the figures depicts the desertification extent of two deserts.

It is noted that the regions with $DI > 20K$ in Fig. 4-2.5(b) of 2001 have been expanded and the boundary between two deserts seems to be merged. Fig.4-2.5(c) shows the difference between Fig. 4-2.5(b) and Fig. 4-2.5(a) to display the merging and expansion of two deserts. Fig. 4-2.5(d), especially, presents the *DI* variation at the Yabulai town, Inner Mongolia (103.0° *E*, 39.6° *N*), between March-August 1997 and 2001. It demonstrates a significant increase of *DI* during past 4 years. It also can be seen that possible canopy during summer season decreases *DI*.

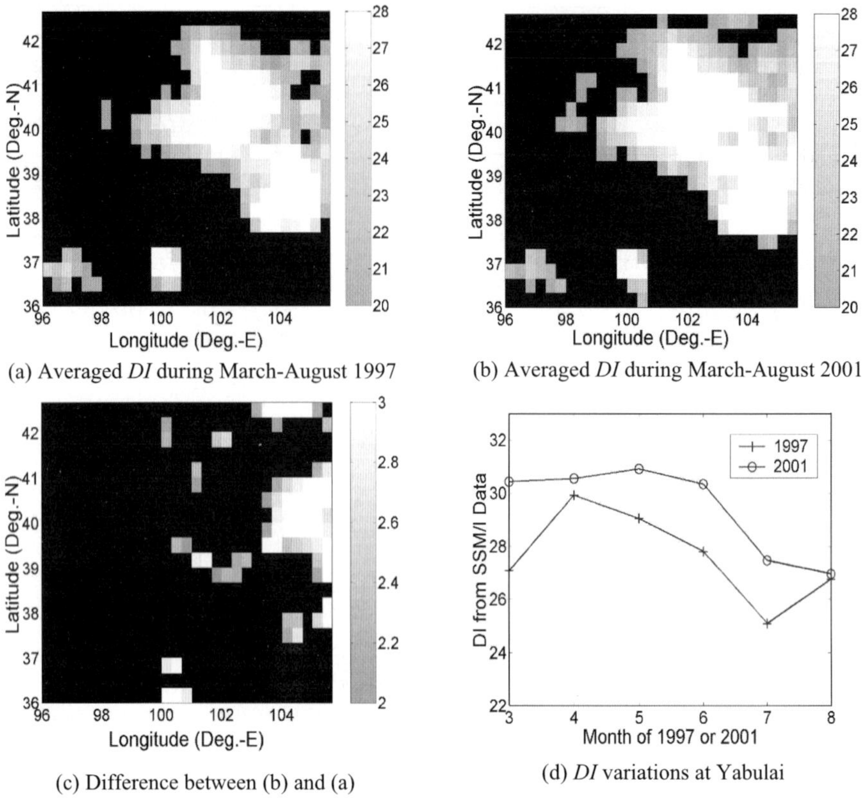

(a) Averaged *DI* during March-August 1997

(b) Averaged *DI* during March-August 2001

(c) Difference between (b) and (a)

(d) *DI* variations at Yabulai

Fig.4-2.5. Desertification index *DI* distribution

4-2.3 The Getis Statistics of Spatial Auto-Correlation

Statistics of spatial auto-correlation quantitatively evaluates the correlation of characteristic variable or parameter x over the whole region. In conventional statistics, all variables at different locations are always assumed as independent of each other. In environmental remote sensing, local radiative properties usually demonstrate highly spatial correlation, e.g., canopy growth or disappearance, drought or flooding in one location certainly affects the environment in another nearby location. The Getis statistics [Getis and Ord 1992, Wulder and Boots 1998, Bhandari 2000] are studied for the range and clustering pattern of the variable x. Looking into an n-pixel image, express the sum of weighted variable values within a specified distance d of a particular observation pixel (including the i-th pixel) as a proportion of the sum of the variable values for entire study area:

$$G_i^*(d) = \sum_{j=1}^{n} W_{ij}(d)x_j \Big/ \sum_{j=1}^{n} x_j \tag{4-2.9}$$

where W_{ij} is a spatial weights matrix, and set to 1 as the pixel j is within distance d, otherwise to 0. The normalized Z-score is calculated as

$$Z_i(d) = \frac{[G_i^*(d) - E(G_i^*(d))]}{\sqrt{Var(G_i^*(d))}} \tag{4-2.10}$$

Eq. (4-2.10) describes the clustering pattern of regional variable x. High positive Z values computed from Eq. (4-2.10) indicate clustering of high pixel values, while high negative ones denote clustering of low pixel values. Having *DI* as variable, we may examine the spatial correlation and clustering pattern of desertification. If a pixel has great correlation with another region identified as desert, the similarity in radiative properties between this pixel and the desert region reflects this kind of correlation, thus providing a way to validate the extent of local desertification. With regards to this point, we cannot reach it by mere use of above-mentioned difference of *DI*.

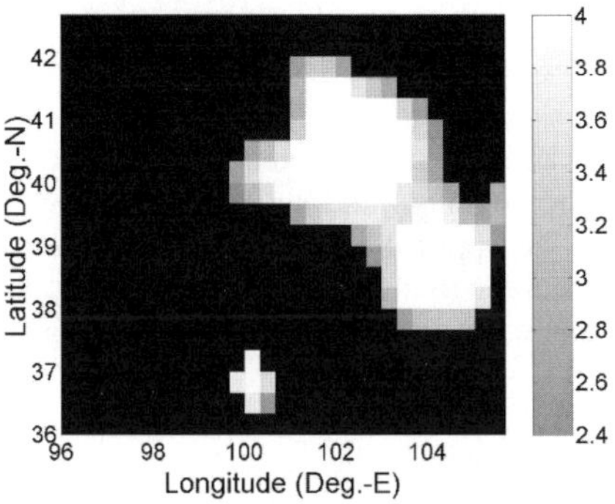

(a) Getis distribution of *DI* in 1997

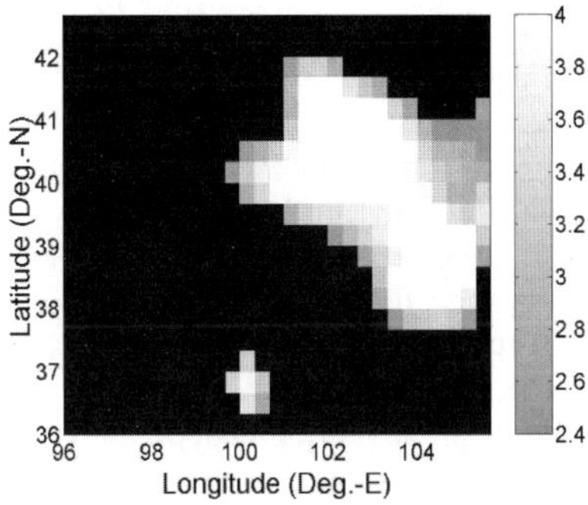

(b) Getis distribution of *DI* in 2001

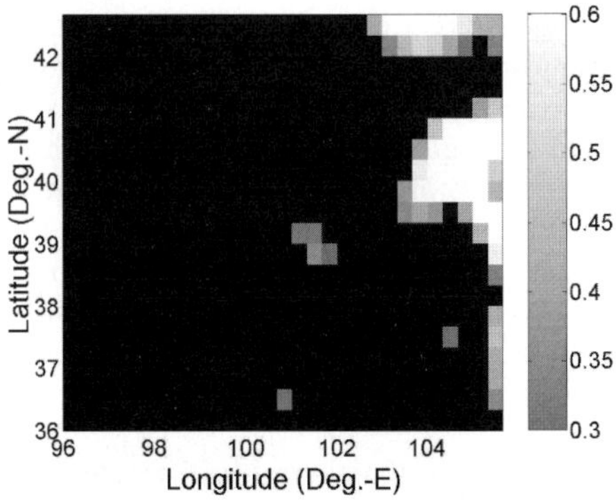

(c) The Getis spatial correlation variations between 1997 and 2001

Fig.4-2.6. The Getis spatial correlation

Figs. 4-2.6(a, b) present the distribution of normalized Z value using the Getis statistics during March-August 1997 and 2001, respectively. A merging tendency can be seen at the boundary of the two deserts in 2001. Meanwhile, the top-right corner of Fig. 4-2.6(c) shows a possible desertification happening to the northeast of Tengger Desert, where spatial correlation is identified.

The Getis statistics demonstrates spatial auto-correlation, and will not be affected by those variations in the whole region, e.g. increases of atmospheric temperature and opacity, etc. We can also find that some high *DI* values in Fig. 4-2.5(c) have disappeared in the Getis distribution of Fig. 4-2.6(c), which has low correlation. The Getis statistics has rich physical insight into spatial variation of the whole region, and can provide a quantitative description for desertification information and its analysis.

4-3 A Genetic Algorithm to Simultaneously Retrieve Land Surface Roughness and Soil Wetness

One of the most important purposes for land remote sensing is to retrieve the land wetness information. Variation of land wetness changes soil permittivity, and affects observation of the backscattering coefficient σ and the brightness temperature T_B. Because the land surfaces in nature are always rough, the roughness, as well as wetness, also affects the observation of σ and T_B. It is difficult to identify which one, wetness, roughness or both, causes variation of σ and T_B. How to invert the wetness and roughness from remote sensing has been one of the most interesting problems to be resolved.

A mesh graph method [Huang and Jin 1995] and the artificial neural network (ANN) method [Jin and Liu 1997] to retrieve the surface roughness and soil wetness have been proposed. These approaches have been proved to be applicable. Of course, for the mesh graph method the soil structure must be known in order to simulate and draw many mesh lines. The ANN method always takes a large amount of data and a long time to train the network.

In recent years, the genetic algorithm (GA), which is based on the principle of optimal selection in a natural evolutionary process, has been widely applied to optimal problems in engineering, such as automatic design, pattern recognition, robotics control and synthesis of neural networks, etc. [Haupt 1994, 1995; Sarabandi 1997]. The gene is the basic genetic building block. A set of genes is a chromosome, which represents a kind of creature. A certain number of chromosomes combination composes a population. Only organisms adapted to environments can survive, mate, and develop so that optimal descendants are produced. The population undergoes genetic operations such as natural selection, cloning, mutation and mating to eventually arrive at the most optimal status.

In this session, we develop a genetic algorithm to invert the most optimal multi-parameters of wetness and roughness from angular measurements of backscattering coefficients. The cost, or objective, function is calculated by using a two-scale rough surface model. The retrieved wetness and roughness of bare soil surfaces are compared well with the ground data measurements available [Jin and Wang 2001].

4-3.1 The Genetic Algorithm to Retrieve Multi-Parameters

The genetic algorithm (GA) was first presented by Holland (1992) based on the concept of the optimal selection of the natural evolution process. The GA algorithm is a globally iterative, numerical optimization method. Due to the rapid development of computer technology in recent decades, the GA algorithm has been widely applied in many engineering areas for optimized selection and simulation. The GA algorithm, as well as the ANN method, shows how computer programs that evolve in ways that resemble natural selection can solve complex problems that even their creators do not fully understand [Holland 1992].

In the GA algorithm, each parameter is encoded into a gene, which is simply a binary representation. A trial solution of a set of genes composes a chromosome. A number of different chromosomes mean a population. The initial population is generated randomly. Then the population undergoes through the natural selection, i.e. the chromosomes are tested for their fitness to the cost function. All of the chromosomes with higher fitness survive as the parents for the next generation, while all of the chromosomes with lower fitness are discarded. Cloning of the surviving parents improves the population quality. Mating parents produce new offspring, which inherit their parents' fitness. During this operation, the size of the population remains constant. The binary-encoded parents can mate with each other by swapping their binary bits with their partner. To maintain the population diversity and avoid the algorithm being trapped at a local minimum, a random operation of mutation or new blood is introduced into the population. To keep the parents fitness, the ratio of mutation and new blood should not be too large, otherwise it would damage the good chromosomes, and even may cause the algorithm to undertake random-searching. The mutation to take a small change is very important for optimization when parameters are continuous. A good mutation operator will speed up convergence.

As a new population is obtained, the process above is iterated through natural selection, cloning, mating, mutation and introducing new blood, and the most suitable population is acquired eventually. A good tutorial introduction can be found in Haupt (1995).

In comparison with traditional optimization techniques based on gradient calculation, the GA algorithm is superior for global optimization, without complicated math calculations. And it is especially effective when the parameter number is very large.

In this session, the GA algorithm is developed to simultaneously retrieve multi-parameters of land surface roughness and soil wetness from angular backscattering observation. Parameters of soil wetness and surface roughness are encoded into genes. The cost function is constructed by numerical calculation of backscattering of the two-scale rough surface model [Wu and Fung 1972]. By using the GA algorithm, the optimized parameters with the best fitness to the cost are obtained.

4-3.2 Two-Scale Backscattering Model of Rough Surface

The volumetric wetness m_v is defined as the water content per unit volume. Variation of water content is a main factor affecting soil permittivity ε $(=\varepsilon' + i\varepsilon'')$. The relationship between m_v and ε at microwave region has been formulated empirically by Wang and Schmugge (1980), and Dobson et al. (1984).

Surface roughness is generally described by the standard deviation σ and the surface correlation length ℓ of surface height fluctuation. As the surface roughness meets the conditions

$$k\ell > 6 \quad \text{and} \quad \ell^2 > 2.76\sigma\lambda \tag{4-3.1}$$

where k is the wave number and λ is the wavelength, the surface is randomly perturbed in a large-scale. The backscattering coefficient may be calculated by the Kirchhoff approximation and stationary phase method [Wu and Fung 1972]. Co-polarized (*vv, hh*) backscattering coefficients are written as

$$\sigma_{hh}^{KA}(\theta) = \frac{|R_h(0)|^2}{2s^2\cos^4\theta}\exp(-\frac{\tan^2\theta}{2s^2}) \tag{4-3.2}$$

$$\sigma_{vv}^{KA}(\theta) = \frac{|R_v(0)|^2}{2s^2\cos^4\theta}\exp(-\frac{\tan^2\theta}{2s^2}) \tag{4-3.3}$$

where θ is the incident angle, s is the RMS slope of the surface, here $s=\sqrt{2}\sigma/\ell$ for a Gaussian rough surface. $R_h(0)$, $R_v(0)$ are the h,v-polarized Fresnel reflection coefficient of a flat surface at θ=0, respectively.

As the surface roughness satisfies small perturbation condition

$$k\sigma_1 < 0.3 \qquad \text{and} \qquad \sqrt{2}\sigma_1/\ell_1 < 0.3 \tag{4-3.4}$$

where σ_1, ℓ_1 are the standard deviation and the surface correlation length of small perturbation, respectively, co-polarized (*hh, vv*) backscattering coefficient is calculated by the small perturbation approximation (SPA) [Ulaby et al. 1982].

$$\sigma_{hh}^{SPA}(\theta) = 8k^4\sigma_1^2\cos^4\theta\,|\,a_h\,|^2\,W(2k\sin\theta,0)$$
$$\sigma_{vv}^{SPA}(\theta) = 8k^4\sigma_1^2\cos^4\theta\,|\,a_v\,|^2\,W(2k\sin\theta,0) \quad (4\text{-}3.5)$$

where

$$a_h = \frac{\cos\theta - \sqrt{\varepsilon_r - \sin^2\theta}}{\cos\theta + \sqrt{\varepsilon_r - \sin^2\theta}}$$
$$a_v = (\varepsilon_r - 1)\frac{\sin^2(\theta) - \varepsilon_r(1+\sin^2\theta)}{[\varepsilon_r\cos\theta + (\varepsilon_r - \sin^2\theta)^{1/2}]^2} \quad (4\text{-}3.6)$$

here $W(k_\perp = 2k\sin\theta, k_z = 0)$ is the normalized Fourier transform of the surface correlation function, i.e. the spectrum of the small perturbation.

The two-scale rough surface takes account of both large and small-scale perturbation. The first-order SPA solution is averaged over the slop distribution of the large-scale perturbation to account for the tilting effect of the large slope [Wu and Fung 1972]. The co-polarized backscattering coefficient is written as

$$\sigma_{pp}(\theta) = \sigma_{pp}^{KA}(\theta) + <\sigma_{pp}^{SPA}(\theta)> \qquad pp=vv,\ hh \quad (4\text{-}3.7)$$

where the second term of the RHS of Eq.(4-3.7) is written as

$$<\sigma_{pp}^{SPA}(\theta)> = \frac{1}{2\pi}\int_{\pi/2}^{-\pi/2}\{\sigma_{pp}^{SPA}(\theta')|_{\theta_n=\tan^{-1}(s)} + \sigma_{pp}^{SPA}(\theta')|_{\theta_n=-\tan^{-1}(s)}\}d\varphi_n \quad (4\text{-}3.8)$$

here $\theta' = \cos^{-1}[\cos\theta_n\cos\theta + \sin\theta_n\cos\varphi_n\sin\theta]$.

As the retrieved parameters are obtained in the GA algorithm, they are used to calculate the backscattering coefficient by using the two-scale model.

4-3.3 The GA Algorithm to Retrieve Multi-Parameters

Five parameters of $k\sigma_1, k\ell_1, s$ and ε', ε'' are encoded into genes to be optimized. A set of $[k\sigma_1, k\ell_1, s, (\varepsilon', \varepsilon'')]$ composes a chromosome. Co-horizontally polarized backscattering coefficient $\sigma_{hh}^{T}(\theta)$ (the superscript *T* denotes "true data") at five observation angles, 30°, 40°, 50°, 60° and 70°, are used to retrieve the parameters. From the trial solutions of chromosomes in the GA algorithm, the backscattering coefficients $\sigma_{hh}^{R}(\theta)$ in two-scale model of Eq. (4-3.7) are calculated (the superscript *R* denotes "retrieved data"). The cost function is constructed as

$$C = \sum_{\theta}\left|\sigma_{hh}^{T}(\theta) - \sigma_{hh}^{R}(\theta)\right|^2 \quad (dB) \quad (4\text{-}3.9)$$

Fitness, how ever good the retrieval is, is measured by the cost function *C*. The lower the value of the cost function, the better the fitness is. As the cost function is less than a certain value, e.g. 3 *dB* in this work, retrieval is accomplished.

A population of size *N* (*N* kinds of $[k\sigma_1, k\ell_1, s, (\varepsilon', \varepsilon'')]$, in this work *N*=8) is generated randomly. The value of each parameter can vary within the range restricted by its physical sense, i.e. $k\sigma_1 \in [0,1]$, $k\ell_1 \in [0,10]$, $s \in [0,1]$, $\varepsilon' \in [3,30]$, $\varepsilon'' \in [0,10]$. The cost function is calculated to compare $\sigma_{hh}^{R}(\theta)$ with $\sigma_{hh}^{T}(\theta)$ for each trial solution of chromosomes of a population from the GA algorithm. The chromosomes are then ranked from the best-fit to the least-fit according to their respective cost values, and *N*/2 chromosomes of the least fit are discarded.

Of surviving $N/2$ chromosomes, the two best ones are chosen to mate with each other as parents. $N/2$ pairs of mating coefficients are generated and are imposed on these two best parents to produce $N/2$ offspring. Thus, the size of population N is maintained.

The cost function of the new population is calculated again to perform the natural selection operation. Then m least-fitting chromosomes (in this case m=2) are replaced with m new chromosomes (that is, new blood) generated randomly. Note that the ratio of replacement should not be too big, otherwise the genetic algorithm would deteriorate to a random-search method.

Mutation is a random, small change of the gene. When the parameters are continuous, the value of the gene is increased or decreased appropriately. If this variation is too big, the suitable value to keep the parents fitness might be missed. If this variation is too small, it might slow the rate of convergence. And a good mutation should be subject to the evolutionary process, i.e. if the population fitness is low, the mutation variation should become larger, and vice verse. For example, the mutation of $k\ell_1$ is designed as

$$k\ell_1(i+1) = k\ell_1(i) \pm 10 \times \frac{1}{20 + \dfrac{1}{R^{C_0/t(i)}}} \tag{4-3.10}$$

where i is the iteration number, 10 is the range of $k\ell_1$ The fig. 20 is intended to limit the largest variation of mutation. C_0 is the constant value of the cost function when the program is terminated. $t(i)$ is the value of cost function at the generation i with $k\ell_1(i)$. R is the random value between 0 and 1. Eq. (4-3.10) satisfies the requirements for mutation. It has been proved very efficient to speed up the GA convergence.

If the cost function Eq. (4-3.9) is less then $3dB$, the optimal chromosome is obtained and the GA algorithm to optimize or invert parameters is finished.

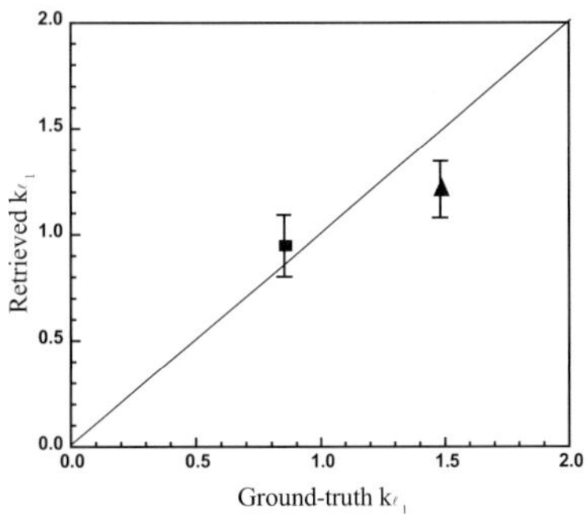

Fig. 4-3.1. Retrieved $k\ell_1$ and ground-truth data

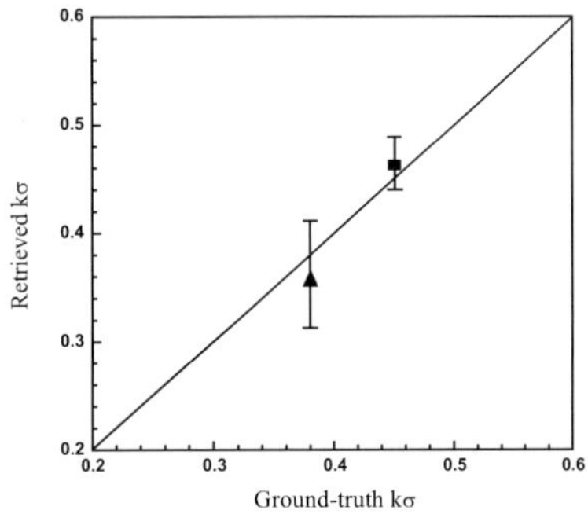

Fig. 4-3.2. Retrieved $k\sigma$ and ground-truth data

An active microwave signature catalogue (2-12 GHz) has been developed by Wegmuller and Matzler (1993). Because the two-scale model is used in the cost function, we have to choose those backscattering data whose angular variation can be well described by this model. Two groups of experimental measurements are used for the GA retrievals. One is the bare land (indicated as BA2) of Marfeldingen, Switzerland in 1 June, 1986 at the frequency 3.1GHz and the other is the bare land (BA3) of Niederried, Switzerland in 14 and 15 August 1986 at the frequency 4.6 GHz. In these two experiments, the parameters $k\ell_1$, $k\sigma$ and m_v had been measured. Our retrieved results $(k\sigma_1, k\ell_1, s, m_v)$ $(s = \sqrt{2}k\sigma / k\ell)$ are well compared with the ground truth measurements, as shown in Figs. 4-3.1~3, respectively. The square points represent BA2 case and the triangles indicate the BA3 case. The length of line represents the standard deviation of retrieved results. Each solution of the GA algorithm takes about few min. on 586 PC.

Fig. 4-3.1 and Fig. 4-3.2 show the comparison of the retrieved results with ground data for the roughness $k\ell_1$ and $k\sigma$, respectively.

Fig. 4-3.3 shows the comparison of the retrieved results with ground data for the wetness of m_v (two wetness for BA3 in 14, 15 August). The wetness m_v was inverted by the retrieved ($\varepsilon', \varepsilon''$) following the empirical relationship between m_v and ε.

To show the angular dependence of backscattering, Fig. 4-3.4 gives the σ_{hh}^R calculated by using five retrieved parameters, and comparison with the discrete points of experimental measurement of BA2.

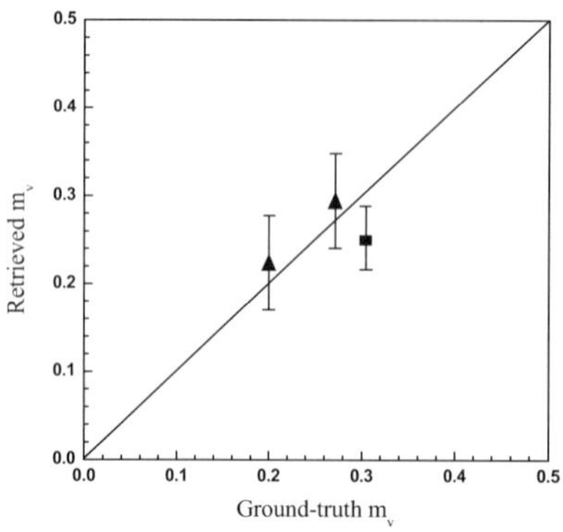

Fig.4-3.3. Retrieved m_v and ground-truth data

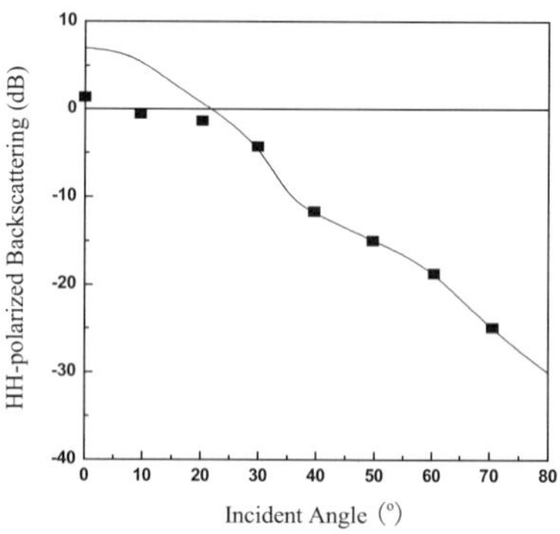

Fig.4-3.4. Backscattering measurements and model

Since the two-scale model is employed for calculation of $\sigma_{hh}^R(\theta)$ and the cost function C, the ground truth of the land surfaces have to satisfy the model conditions adequately. In other words, if the surface does not match the model conditions very well, the cost value has to be released and the retrievals would be less accurate. The two-scale model is not a unique choice. As long as an appropriate model can be well matched with the real surface situations, the GA algorithm will be applicable to multi-parameters retrievals. The GA algorithm with the cost function links the simulation with inversion. This approach is also applicable to the inversion of surface canopy over the underlying land, if good model of surface canopy is applied.

4-4 Correlation of the ERS and SSM/I Observations over Snowpack and Numerical Simulation

Among current advances of satellite-borne microwave remote sensing in the 1990's global surveys is ERS (European Remote Sensing Satellite) and DMSP (Defense Meteorological Space Program). The active C-band (5.3 *GHz*) scatterometer aboard the ERS-1 and ERS-2 operates at vertical polarization (*vv*) and provides backscattering measurement over the Earth surfaces while their three antennae look 45° forwards, sideways, and backwards with respect to the satellite flight direction, the incidence angle ranges from 18° to 57°. The passive SSM/I (Special Sensor Microwave/Imager) of DMSP operating at seven channels (19, 37, 85 *GHz* with dual polarization, and 22 *GHz* with vertical polarization only) and at an observation angle 54° has been in operation since 1987. It has become a primary source for microwave data interpretation and derived hydrological products. However, the correlative characteristics of backscattering coefficient (σ) and brightness temperature (T_B) measured by active scatterometer and passive radiometers, respectively, on the same area have not been closely studied. It remains to answer whether the measurements by active ERS and passive SSM/I are mutually consistent and supplementary, and how both active and passive measurements can resolve ambiguity in data interpretation and inversion.

Snow is dense scatterer media (fraction of ice grains is always greater than 0.2), or is seen as strongly-fluctuating media (dielectric fluctuation by random ice grains). Scattering and emission from snowpack depend on geometrical and dielectric properties such as ice grain size, fraction, wetness, stratigraphy, etc. and changes with frequency. In this session, some data collected from ERS-1 and DMSP SSM/I observations on Greenland, Siberia etc. are presented for analysis of active/passive correlation and comparison of model simulation. It is demonstrated that multichannel analysis by both active and passive sensors is very helpful for monitoring complex and highly variable forms of terrain surfaces such as snow covers. By using the dyadic Green's function of stratified media and the Born approximation, backscattering from multi-layered, strongly fluctuating random media is calculated for data simulation of the ERS C-band scatterometer. The radiative transfer equations of thermal emission from multi-layer media of dense scatterers is solved for SSM/I data simulation. Theoretical results of backscattering and emission, and their functional dependence on snow properties are discussed.

4-4.1 The Active ERS-1 and Passive SSM/I Observations

Fig. 4-4.1 gives the temporal variation of vertically polarized brightness temperature (T_{Bv}) of SSM/I AS-F13 at 22, 37 and 85 *GHz* over Greenland (68.16°*N*, 41.16°*W*) during 1995-1996. It can be seen that there was a rapid decline of T_{B85v} of the 85 *GHz* channel in July 1995. A small chart in Fig. 4-4.1 [Wismann et al. 1997] shows backscattering σ_{vv} of 5.3 *GHz* at incidence angle 40° measured by ERS-1 at the same location during the same period. The interruption episode in July 1995 can be seen from both SSM/I and ERS data. Backscattering decrease was almost linear with fluctuation less than 1 *dB* during the entire time span, except a rapid jump by 5 *dB* in the July episode. It lasted less than 2 weeks before σ_{vv} again decreased similarly. However, multichannel T_B of SSM/I seems to give more information about temporal variation.

The 85 *GHz* channel has been especially studied for detecting snow and precipitation because of strong scattering sensitivity. As frequency is high, 85 *GHz*, scattering becomes much stronger and leads to darkening thermal emission and brightness temperature of underlying media. A scattering index SI is usually defined as $SI_{19-37} = T_{B19v} - T_{B37v}$ or $SI_{22-85} = T_{B22v} - T_{B85v}$, and a rule $SI_{22-85} > 0$ as a scattering signature is used for identification of scattering media. Fresh snow cover always shows $SI_{22-85} > 0$.

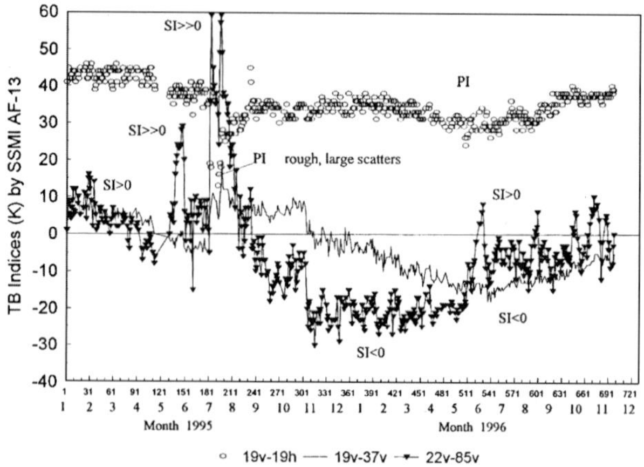

Fig. 4-4.1. Temporal variation of SSM/I data over Greenland in 1995-96 with a comparison of ERS-1 data

However, it has been revealed [Jin 1997] that from stratified snowpack with highly reflective ice-crust sheets, scattering signature anomalously shows $SI_{22-85} < 0$. Stratigraphy is due to so-called melt-related metaphorphism of snow into firn. As another factor, surface roughness and melted-refrozen large ice grains in the top layer can enhance scattering and diminish the polarization difference. A surface polarization index PI is usually defined as $PI_{19} = T_{B19v} - T_{B19h}$ to show variation of v,h-polarized reflectivity's difference of the terrain surface.

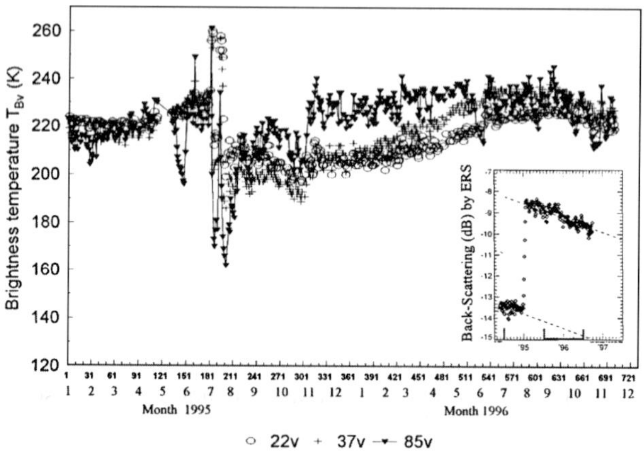

Fig. 4-4.2. The T_B indices (SI_{22-85}, SI_{19-37} and PI_{19}) of Fig. 4-4.1 SSM/I data

Fig. 4-4.2 shows the T_B indices SI_{22-85}, SI_{19-37} and PI_{19} of Fig. 4-4.1 SSM/I data. Temporal evolution of the snowpack seems more evident if following daily variation of both

T_B indices and backscattering measurements in Fig. 4-4.1. Fresh snow causes $SI > 0$ and σ_{vv} at ERS 5.3 *GHz* to slowly increase with snow depth. Wet snow causes emissivity to increase, and SI and σ_{vv} to decrease. Because of crust sheets formed by the melting and refrozing process, the snow firn shows anomalously $SI < 0$. Surface roughness and coarser ice grains formed during melting-refrozing process in the top layer increase scattering σ_{vv} and SI, and decrease PI.

Fig. 4-4.3 gives another example of temporal variation of SI_{22-85} from SSM/I AS (ascending)-F13 data over Siberia (70.83°*N*, 76.5°*E*) during 1995-1996. A small chart [Boehnka and Wismann 1997] in Fig. 4-4.3 shows backscattering σ_{vv} measured by the ERS-1 during March-July 1995. It can be seen that, from 1 April, σ_{vv} gradually increased over 5 *dB* due to melting-refreezing, since stronger scattering was caused by recrystallized coarser ice grains. Since the middle of June (Julian Day 166), σ_{vv} increases due to underlying thawed and rough land. The SI_{22-85} data of Fig. 4-4.3 shows consistent comparison during surface variations of fresh dry snow, melting-refreezing snow, and thawed rough land surface. Fluctuation of SI_{22-85} and σ_{vv} during melting-refreezing period is especially significant in Figs. 4-4.1~3.

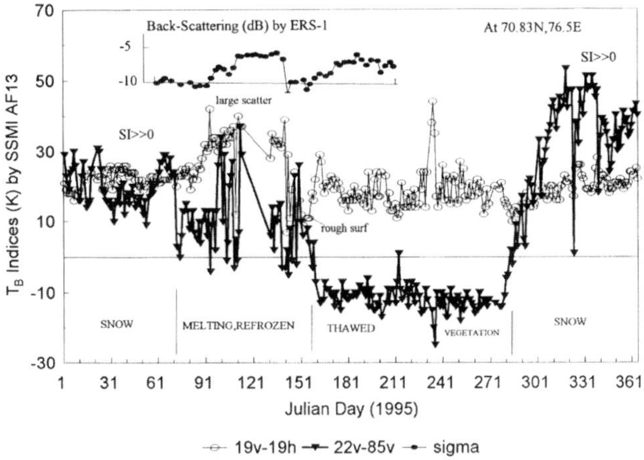

Fig. 4-4.3. Temporal variation of SSM/I T_B indices with comparison of ERS-1 data over Siberia

4-4.2 Scattering and Emission Modeling of Multi-Layer Random Media

Snowpack consists of ice grains randomly embedded in the air background medium. It may be seen as a strongly fluctuating random medium since the fraction of ice grains is dense. A multilayer model of strongly fluctuating random media is employed for complex-layer snow covers. Dielectric constant $\varepsilon_m(\vec{r})$ at any point $\vec{r}$ in the *m*-th medium, here m denotes the *m*-th medium, is random number fluctuating with the randomly embedded ice-grains. Single scattering is taken into account for ERS observation at 5.3 *GHz*. By using the mean dyadic Green's function of stratified media and the Born approximation of internal fields in the media, co-polarized (*pp*=*vv* or *hh*) backscattering coefficient from the multi *M*-layer random media was derived by Zuniga et al. (1979) as

$$\sigma_{pp}(\theta_{0i}) = \sum_{m=1}^{M} \delta_m (k'_m)^4 [\, R_{pp}^{(m)} (d_m - d_{m-1})\, \Phi_m(2\bar{k}_{\rho i},0) + \mathrm{S}_{\mathrm{pp}}^{(\mathrm{m})} \frac{\Phi_m(2\bar{k}_{\rho i}, 2\bar{k}_{mzi})}{k''_{mzi}} \,] \qquad (4\text{-}4.1)$$

where there are M layers in total over a underlying land medium. The interface is located at $z=-d_m$ beneath the (m-1)-th layer, so $d_m - d_{m-1}$ is the thickness of the m-th layer. $\Phi_m(\bar{k}_\rho, k_{mz})$ is the three-dimensional Fourier transform of the correlation function of random dielectric fluctuation in the m-th layer. The mean dielectric constant ε_{gm} of the strongly fluctuating m-th medium is calculated by [Jin 1994, Tsang et al. 1985]

$$f_{sm}\frac{\varepsilon_{sm}-\varepsilon_{gm}}{\varepsilon_{sm}+2\varepsilon_{gm}}+(1-f_{sm})\frac{\varepsilon_b-\varepsilon_{gm}}{\varepsilon_b+2\varepsilon_{gm}}=0 \qquad (4\text{-}4.2)$$

where f_{sm}, ε_{sm} are, respectively, the fractional volume and dielectric constant of scatterers (ice grains) in the m-th layer; ε_b is the dielectric constant of background medium, i.e. $\varepsilon_b=\varepsilon_0$ of air in our models, and $k'_m=\text{Real}(\omega\sqrt{\mu_0\varepsilon_{gm}})$. The correlation function is usually assumed as

$$C(\bar{r}-\bar{r}')=<\xi(\bar{r})\xi^*(\bar{r}')>=\delta_m(k_m)^4\exp(-|\bar{\rho}-\bar{\rho}'|^2/\ell_{\rho m}^2-|z-z|/\ell_z) \qquad (4\text{-}4.3)$$

where random fluctuation $\xi(\bar{r})\equiv 3(\varepsilon_{gm}/\varepsilon_0)[\varepsilon_m(\bar{r})-\varepsilon_{gm}]/[\varepsilon_m(\bar{r})+2\varepsilon_{gm}]$, and $\varepsilon_m(\bar{r})$ takes $\varepsilon_{sm}, \varepsilon_b$, randomly. In Eqs. (4-4.1,3) δ_m is the fluctuation variance of $<\xi\xi^*>$, and ℓ_ρ, ℓ_z are the correlation length in the transverse $(\hat{\rho})$ and vertical direction $(\hat{z})$, respectively. $R_{pp}^{(m)}, S_{pp}^{(m)}$ in Eq. (4-4.1) are the functions of multi-layer reflectivities, which are listed in the Appendix 4-4A as a reference with a typo correction [Zuniga et al.1979].

In the case of dry snow, the dielectric constant of an ice particle, ε_s, is usually taken as a constant without frequency dependence, e.g. $\varepsilon_s=(3.2+i0.0032)\varepsilon_0$. In the case of wet snow, ε_s can be taken as an effective dielectric constant ε_{ws} of ice spheres with water-coating [Bohren and Huffman 1983] as follows

$$\frac{\varepsilon_{ws}-\varepsilon_0}{\varepsilon_{ws}+2\varepsilon_0}=\frac{(\varepsilon_2-\varepsilon_0)(\varepsilon_1+2\varepsilon_2)+S(\varepsilon_1-\varepsilon_2)(\varepsilon_0+2\varepsilon_2)}{(\varepsilon_2+2\varepsilon_0)(\varepsilon_1+2\varepsilon_2)+2S(\varepsilon_2-\varepsilon_0)(\varepsilon_1-\varepsilon_2)} \qquad (4\text{-}4.4)$$

where $\varepsilon_1, \varepsilon_2$ are the dielectric constant of ice and water, respectively. The dielectric constant of water is strongly dependent upon frequency. The parameter S is related with water-coating thickness and may be expressed by the fractional volumes of total ice particles f_s and water coating f_w such that $S=(1-f_w/f_s)^{1/3}$. Empirical formula of mean dielectric constant of dry and wet snowpack as given in Matzler et al. (1997) can be seen as our ε_{gm} in calculations.

For SSM/I observation at 19, 22, 37, 85 *GHz* channels, the snowpack is modeled as a multi-layer media of dense spherical scatterers. The VRT equation of the n-th layer, to take account of multiple scattering, is written as

$$\cos\theta_n\frac{d}{dz}\bar{T}_{Bn}(\theta_n,z)=-\kappa_{en}\bar{T}_{Bn}(\theta_n,z)+\int_0^\pi d\theta'_n\,\sin\theta'_n\bar{\bar{P}}_n(\theta_n,\theta'_n)\cdot\bar{T}_{Bn}(\theta'_n,z)+\kappa_{an}\bar{T}_n \qquad (4\text{-}4.5)$$

where $\bar{T}_{Bn}$ is the polarized (v, h) brightness temperature and $\bar{T}_n$ is the physical temperature of the n-th layer (n = 1, ..., M). The extinction coefficient κ_{en} of the n-th layer is calculated by the effective propagation constant of dense scatterers with Percus-Yevick distribution [Tsang et al. 1985]. The coupled VRT equations of multilayer media with the boundary conditions can be numerically solved by the discrete ordinate and eigenanalysis method [Jin 1994, 1997]. The p-polarized (p=v,h) brightness temperature observed by satellite-borne SSM/I to take account of atmosphere is obtained as

$$T_{Bp}^{SSMI}(\theta_0) = e_p T_s e^{-\tau_a} + (1 - e^{-\tau_a})(1 + r_p e^{-\tau_a}) T_a \tag{4-4.6}$$

where $e_p T_s$ ($\equiv T_{Bp}(\theta_0, z=0)$) of Eqs. (4-4.4, 4-4.5) is the solution of multi-layer VRT equations, T_s is the physical temperature of the top surface of snowpack. The surface reflectivity r_p is $r_p = 1 - e_p$; and τ_a , T_a are atmospheric opacity and mean physical temperature, respectively. Atmospheric opacity, τ_a, is a function of frequency and is chosen by the USA standard atmosphere [Waters 1976] in calculations. Effect of atmospheric opacity on C-band σ_{vv} is simply negligible.

4-4.3 Numerical Simulation of Scattering and Emission from Multi-Layer Snowpack

Eq. (4-4.1) is calculated for several cases of snow covers such as fresh dry snow, layering snow, wet snow, ice crust cover etc. Fig. 4-4.4 shows numerical simulation of backscattering σ_{vv} at 5.3 *GHz* and 40° incidence angle for seven cases of snow covers, indicated by the filled triangles. As a comparison, the scatterometer measurements by Strozzi et al. (1997) on the Swiss Alps area are, correspondingly, given by the open triangles in Fig. 4-4.4. Depth and layering in theoretical simulations follow the ground data given by Strozzi et al. (1977), and other parameters such as correlation lengths (related with ice grain sizes) are assumed and are listed under the figure title. Different sizes of the filled triangles correspond to different depth of the lowest snow layer for cases 1-5 and 7, except case 6. For example, small triangle with *d*50 means the result with the depth of the lowest layer equal to 50 *cm*, and so on. In case 6, different triangles correspond to variation of the top crust thickness as $d_1 = 5, 10, 15, 20\,cm$, respectively. It can be seen that fresh dry snow increases σ_{vv} in cases 1 and 2, whilst wet snow significantly decreases σ_{vv} in cases 3 and 5. Scattering from layered snow depends on layer structures, especially on the top layers. Large ice grains or large correlation length at the loose snow can enhance scattering as shown in case 4. However, the top ice crust scattering is darkened by the underlying snowpack, as shown in case 6. In case 7, the grass canopy is modeled as a layer of tenuous random medium.

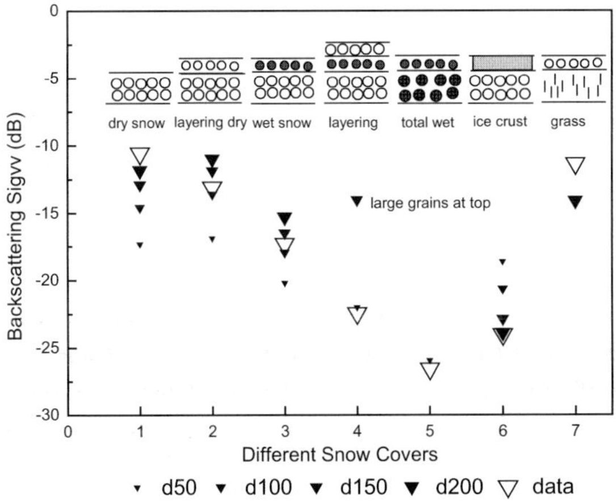

Fig. 4-4.4. Backscattering coefficients σ_{vv} at 5.3 *GHz* for different snow covers

Land: $\varepsilon_{land} = (6 + i0.6)\varepsilon_0$;

(1). Fresh dry snow: $d_1 = 50, 100, 150, 200\ cm,\ f_1 = 0.35,\ \varepsilon_{s1} = (3.2 + i0.0032)\varepsilon_0,\ \ell_\rho = \ell_Z \equiv \ell = 0.08\ cm$;

(2). Layering dry snow: $d_1 = 10cm,\ f_1 = 0.35,\ \ell = 0.06cm,\ \varepsilon_{s1} = (3.2 + i0.0032)\varepsilon_0$;

$d_2 - d_1 = 40, 90, 140, 190cm,\ f_2 = 0.4,\ \ell = 0.08cm,\ \varepsilon_{s2} = (3.2 + i0.0032)\varepsilon_0$;

(3). Wet snow: $d_1 = 3cm,\ f_w = 0.04,\ f_1 = 0.35,\ \ell = 0.06cm,$ ice grains with water coating;

$d_2 - d_1 = 50,100,150,200cm,\ f_2 = 0.4,\ \ell = 0.08cm,\ \varepsilon_{s2} = (3.2 + i0.01)\varepsilon_0$;

(4). Multi-layer snow: $d_1 = 30cm,\ f_1 = 0.35,\ \ell = 0.05cm,\ \varepsilon_{s1} = (3.2 + i0.0032)\varepsilon_0$, ;

$d_2 - d_1 = 25cm,\ f_2 = 0.85,\ \ell = 0.06cm,\ \varepsilon_{s2} = (3.2 + i0.1)\varepsilon_0$;

$d_3 - d_2 = 45cm,\ f_3 = 0.4,\ \ell = 0.08cm,\ \varepsilon_{s3} = (3.2 + i0.01)\varepsilon_0$

In the case of large grains at top layer, $\ell = 0.15\ cm$.

(5). Total wet snow: $d_1 = 3cm,\ f_{w1} = 0.04,\ f_1 = 0.35,\ \ell = 0.06cm,$ ice grains with water coating;

$d_2 - d_1 = 50cm,\ f_{w2} = 0.02,\ f_2 = 0.4,\ \ell = 0.08cm$;

(6). Crust cover: $d_1 = 5,10,15,20cm,\ f_1 = 0.85,\ \ell = 0.06cm,\ \varepsilon_{s1} = (3.2 + i0.1)\varepsilon_0$;

$d_2 - d_1 = 50cm,\ f_2 = 0.4,\ \ell = 0.08cm,\ \varepsilon_{s2} = (3.2 + i0.01)\varepsilon_0$;

(7). Grass and snow cover: $d_1 = 5cm,\ f_1 = 0.3,\ \ell = 0.06cm,\ \varepsilon_{s1} = (3.2 + i0.0032)\varepsilon_0$;

$d_2 - d_1 = 30cm,\ f_2 = 0.0055,\ \ell_\rho = 0.5cm,\ \ell_z = 1.0cm,\ \varepsilon_{s2} = (10 + i2)\varepsilon_0$.

For SSM/I data simulation, Fig. 4-4.5 shows SI_{22-85} at observation angle 54° calculated by Eqs. (4-4.5,6) for different snow covers. Parameters such as depth, correlation lengths, and layering are assumed and are listed under the figure title.

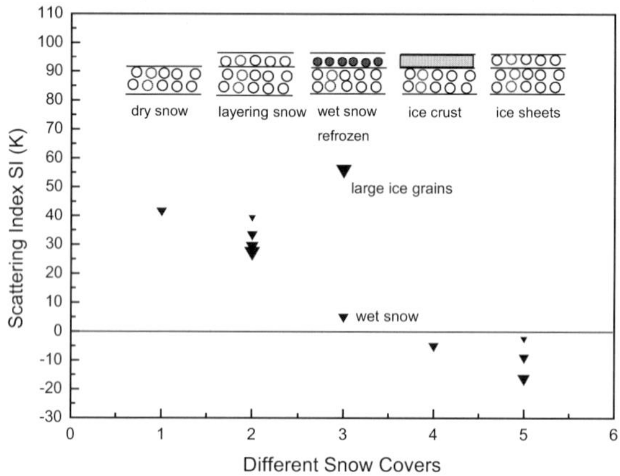

Fig. 4-4.5. Scattering index SI of SSM/I channels for different snow covers

Land: $\varepsilon_{land} = (5 + i0.5)\varepsilon_0;\ T_a = 278K,\ T_s = 263K,$

Atmosphere: $\tau_a = 0.06(19GHz), 0.12(22GHz), 0.08(37GHz), 0.25(85GHz)$.

(1). Fresh dry snow: $d_1 = 50cm,\ f_1 = 0.35,\ \varepsilon_{s1} = (3.2 + i0.0032)\varepsilon_0$, ice radius a=0.03cm;

(2). Dry layering snow: $d_1 = 20cm,\ f_1 = 0.35,\ a = 0.03cm,\ \varepsilon_{s1} = (3.2 + i0.0032)\varepsilon_0$;

$d_2 - d_1 = 10,20,30,50,60cm,\ f_2 = 0.4,\ a = 0.08cm,\ \varepsilon_{s2} = (3.2 + i0.01)\varepsilon_0$;

(3). Wet snow: $d_1 = 3cm,\ f_w = 0.02,\ f_1 = 0.35,\ a = 0.03cm,$ ice grains with water coating;

$d_2 - d_1 = 50cm,\ f_2 = 0.4,\ a = 0.08cm,\ \varepsilon_{s2} = (3.2 + i0.01)\varepsilon_0$;

Refrozen snow with coarser grains in top layer: $d_1 = 3cm,\ f_1 = 0.4,\ a = 0.06cm$;

(4). Crust cover: $d_1 = 3cm,\ f_1 = 0.85,\ a = 0.08cm,\ \varepsilon_{s1} = (3.2 + i0.01)\varepsilon_0$;

$d_2 - d_1 = 50cm, f_2 = 0.4, a = 0.08cm, \varepsilon_{s2} = (3.2 + i0.01)\varepsilon_0$;

(5). Layering snow with an ice sheet:

$d_1 = 5,\ 3,\ 1\ cm, f_1 = 0.35,\ a = 0.03cm,\ \varepsilon_{s1} = (3.2 + i0.0032)\varepsilon_0$;

$d_2 - d_1 = 1cm, f_2 = 1,\ \varepsilon_{s2} = (3.2 + i0.1)\varepsilon_0$; $d_3 - d_2 = 50cm,\ f_3 = 0.4,\ a = 0.08cm,\ \varepsilon_{s3} = (3.2 + i0.01)\varepsilon_0$.

It can be seen that presence of snowpack in cases 1-3 shows the scattering signature as positive SI_{22-85}. Scattering from fine ice grains at the top layer may increase SI_{22-85} since strong scattering at high 85 *GHz* cools emission from underlying media. But, coarser grains and increasing depth of underlying snow layers decrease SI_{22-85} since scattering at 85 *GHz* might be less sensitive to these changes of underlying media. This is shown in case 2. In wet snow case, case 3, SI_{22-85} can be further decreased. However, due to recrystallized coarser ice-grains in the top layer, scattering at 22 and 85 *GHz* channels may enhance the scattering signature SI_{22-85}. In case 4, the top crust can totally darken scattering and SI_{22-85} becomes negative as an emission signature. Case 5 shows anomalous $SI_{22-85} << 0$ due to ice sheets, as shown in Jin (1997).

It can be seen from Fig. 4-4 and Fig. 4-5 that scattering and emission from snow covers are functionally dependent on geometric and dielectric properties of snow covers. The effects due to temporal and spatial variations of these parameters can be simulated by our theoretical models. The observations of ERS and SSM/I have correlated characteristics. Due to the highly variable forms of terrain surfaces and complexity of its geometric and dielectric parameters, consistent observations by multichannel active/passive sensors may resolve ambiguous data interpretation.

Appendix 4-4A

In Eq. (4-4.1), the functions in the *m*-th layer derived by Zuniga et al. (1979) are as follows:

$$R_{hh}^{(m)} = 2\pi^2[|\beta_{mi}^h A_{mi}|^2 + |\alpha_{mi}^h B_{mi}|^2 + 2\,\mathrm{Re}(A_{mi}\beta_{mi}^h B_{mi}^* \alpha_{mi}^{h*})]$$

$$S_{hh}^{(m)} = \frac{\pi^2}{2}[|\alpha_{mi}^h A_{mi}|^2 + |\beta_{mi}^h B_{mi}|^2 \cdot \exp(-4k_{mzi}(d_m + d_{m-1}))](\exp(4k_{mzi}d_m) - \exp(4k_{mzi}d_{m-1}))$$

$$R_{vv}^{(m)} = 2\pi^2[|\beta_{mi}^v C_{mi}|^2 + |\alpha_{mi}^v D_{mi}|^2 + 2\,\mathrm{Re}(C_{mi}\beta_{mi}^v D_{mi}^* \alpha_{mi}^{v*})]\frac{|k_{\rho i}^2 - k_{mzi}^2|^2}{k_m^4}$$

$$S_{vv}^{(m)} = \frac{\pi^2}{2}[|\alpha_{mi}^v C_{mi}|^2 + |\beta_{mi}^v D_{mi}|^2 \cdot \exp(-4k_{mzi}(d_m + d_{m-1}))](\exp(4k_{mzi}d_m) - \exp(4k_{mzi}d_{m-1}))$$

The amplitudes A_m, B_m, α_m^h, and β_m^h are determined through the propagation matrix formalism of *hh*-polarization as

$$\begin{bmatrix} \xi_m \cdot \exp(-ik_{mz}d_m) \\ \eta_m \cdot \exp(ik_{mz}d_m) \end{bmatrix} = \overline{\overline{V}}_{m,m-1}^h \begin{bmatrix} \xi_{m-1} \cdot \exp(-ik_{(m-1)z}d_{m-1}) \\ \eta_{m-1} \cdot \exp(ik_{(m-1)z}d_{m-1}) \end{bmatrix}$$

where

$$\overline{\overline{V}}_{m,m-1}^h = \frac{\Gamma}{X_{m,m-1}} \begin{bmatrix} \exp(-ik_{mz}(d_m - d_{m-1})) & R_{m,m-1}\exp(-ik_{mz}(d_m - d_{m-1})) \\ R_{m,m-1}\exp(ik_{mz}(d_m - d_{m-1})) & \exp(ik_{mz}(d_m - d_{m-1})) \end{bmatrix}$$

and $X_{m,m-1} = 1 + R_{m,m-1}$. Here, $R_{m,m-1}$ and $X_{m,m-1}$ are the reflection and transmission coefficients between the interface of the *m*-th and (*m*-1)-th layers, respectively. The unknown ξ_m represents A_m or α_m^h, and η_m represents B_m or β_m^h while $\xi_0 = R^h$, $\eta_0 = 1$, and $\Gamma = 1$.

The reflection coefficient of multi-layer media with *hh*-polarization with a typo corrected is as follows:

$$R^h = \frac{1}{R_{01}} + \frac{[1-(1/R_{01}^2)]}{(1/R_{01})} + \frac{e^{i2k_{1z}d_1}}{R_{12}} + \frac{[1-(1/R_{12}^2)]e^{i2(k_{2z}+k_{1z})d_1}}{(1/R_{12})e^{i2k_{2z}d_1}}$$
$$+\cdots+\frac{e^{i2k_{(n-1)z}d_{n-1}}}{R_{(n-1)n}} + \frac{[1-(1/R_{(n-1)n}^2)]e^{i2(k_{nz}+k_{(n-1)z})d_{n-1}}}{(1/R_{(n-1)n})e^{i2k_{nz}d_{n-1}}} + R_{nt}e^{i2k_{nz}d_n}$$

The amplitudes $C_m, D_m, \alpha_m^v, \beta_m^v$ and R^v for *vv*-polarization are obtained from the above results be letting $\Gamma = k_{m-1}/k_m$ and by making the replacements $R_{m,m-1} \to S_{m,m-1}$ and $X_{m.m-1} \to Y_{m,m-1}$, here $S_{m,m-1}$ and $Y_{m,m-1}$ are the reflection and transmission coefficients of the *vv*-polarization, respectively. In this case ξ_m represents C_m or α_m^v and η_m represents D_m or β_m^v.

4-5 Correlated Observations by the ERS-1 and SSM/I over Ocean and Numerical Simulation

One of the most current advances of satellite-borne microwave remote sensing in the global surveys of the 1990's is the ERS (European Remote Sensing Satellite) and DMSP (Defense Meteorological Space Program). The active C-band (5.3 *GHz*) scatterometer aboard the ERS-1 and ERS-2 operates at vertical polarization (*vv*) and provides backscattering measurement over the Earth surfaces. The passive SSM/I with 7 channels (19, 37, 85 *GHz* with dual polarization, and 22 *GHz* with vertical polarization only) has been in operation since 1987. It has become a primary source for microwave data interpretation and derived hydrological products.

Scattering and emission from a randomly rough sea surface driven by and depending on wind have been of great importance in ocean remote sensing. It has also been observed that a sea surface can be covered by foam and whitecaps even over a third of its area if the wind speed is as high as 25 *m/s*. In this session, a composite model of a two-scale randomly rough surface covered by a layer of the Rayleigh particles is employed [Jin 1994, Huang and Jin 1995]. Numerical simulations of backscattering and thermal emission from rough sea surface are obtained, and are compared with the ERS and SSM/I observations over China ocean. Correlation of the ERS and SSM/I observations is discussed by both simulation and measurements.

4-5.1 The Active ERS-1 and Passive SSM/I Observations over China Ocean

Surface roughness can reduce the polarization difference of surface reflectivity and emissivity. In this session a polarization index of brightness temperature (T_B), *PI*, is mainly defined as $PI_{37} = T_{B37v} - T_{B37h}$ at the 37 *GHz* channel, which has better spatial resolution (~30 *km*) than 19GHz and less atmospheric sensitivity than 85 *GHz*. Increase in the index PI_{37} may indicate high sea surface speeds.

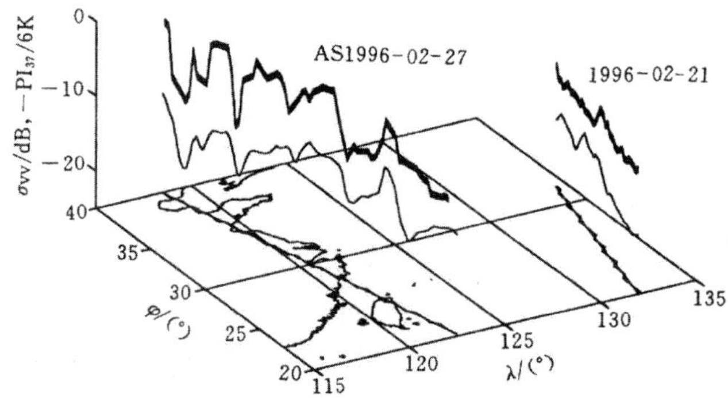

Fig. 4-5.1. Data of σ_{vv} and the T_B index $-PI_{37}/6$ (*K*) on February 21 and 27 1996

Some data of the backscattering σ_{vv} (at incidence angle 40°) and the T_B index $-PI_{37}/6$ measured by the ERS and SSM/I, separately, on 21 and 27 February 1996 over the China sea areas are presented in Fig. 4-5.1. The bold lines denote $-PI_{37}/6$ (*K*), and thin lines are σ_{vv} (*dB*). The data are collected from a DS (descending) flight of the ERS and SSM/I AF-10 on 21 February and another AS (ascending) flight of the ERS and SSM/I

AF-13 on 27 February. The satellite footprints are also, respectively, indicated in Fig. 4-5.1. Difference of overpass time of the SSM/I and ERS is about 4 hours. Fig. 4-5.1 demonstrates that σ_{vv} (thin curves) and $-PI_{37}/6$ (bold curves) have close correlation even they were separately measured by active ERS and passive SSM/I.

Fig. 4-5.2 gives more data of $\sigma_{vv}(40^\circ)$ and T_B indices $PI_{37}, PI_{19}, SI_{22v-85v}$ (scattering index $\equiv T_{B22v} - T_{B85v}$) at different latitude along the flight line from the ERS and SSM/I AS-F13 observations on 27 February 1996. The incidence angle 40° (in the incidence range from 18° to 57°) was specially chosen only for convenience. The letters W and L denote the water and land areas, respectively. It shall be interesting to see variation of σ_{vv} and T_B indices as crossing sea and land coastlines. It can be seen that the plain areas such as L_3 and L_4 are not clearly distinguished from water areas if only using σ_{vv}. However, the SSM/I indices *SI* and *PI* may well identify the plain surfaces from water area.

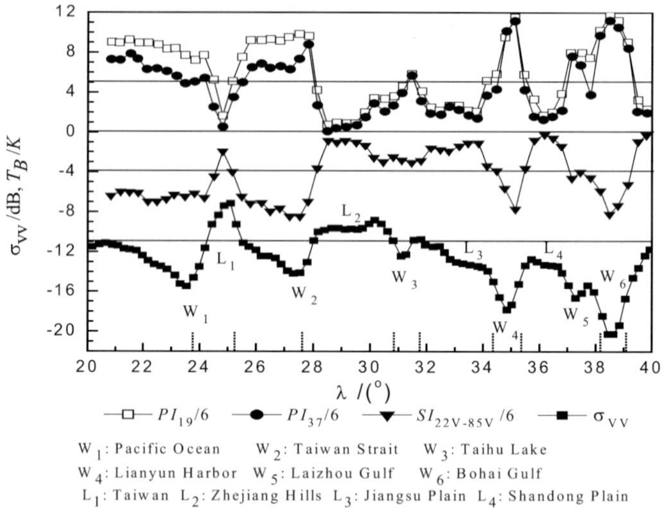

Fig. 4-5.2. Data of $\sigma_{vv}(40^\circ)$ and T_B indices along the footprints of ERS and SSM/I observations (27/2/1996)

Fig. 4-5.3(a) graphically classifies the variation of σ_{vv} and T_B indices. As $\sigma_{vv} < -11dB$ and $PI_{37} > 30K$, it is water area with no rains (denotes by *W*, *NR*). Sea surface winds cause PI_{37} decreasing (high speed *U*, or small speed *u*). If $0 < PI_{37} < 30K$ (PI_{37} always >0), the scattering index $SI_{22v-85v}$ is employed. As $SI_{22v-85v} > 0$, it is a scattering signature, and rains over sea or land surfaces, or snowpack are possible (thick snow *S*, high rain rate *R* or thin snow *s*, light rain rate *r*). If $SI_{22v-85v} \to 0$ with high σ_{vv}, a land surface with vegetation (*V*) canopy is there. As $SI_{22v-85v}$ is negative as an emission signature such as <-25*K*, it is water area with no rains (cold underground and no rain scattering). If $-25K < SI_{22v-85v} < 0$, it is an emission signature from the land surfaces. Generally, σ_{vv} (at 5.3*GHz*) is always <0 *dB*. High σ_{vv}>0 *dB* is most likely coming from special targets or environments.

Fig. 4-5.3(b) used the data of 27 February of the ERS and SSM/I as from Fig. 4-5.3(a) to validate the classification of Fig. 4-5.3(a). The index $SI_{22v-85v}$ was never greater than zero, since there was no rain at that date along the flight. The decision lines such as -25*K*, 30*K* etc. in Figs. 4-5.3(a, b), are not very restrictive, depending atmosphere and ground truth.

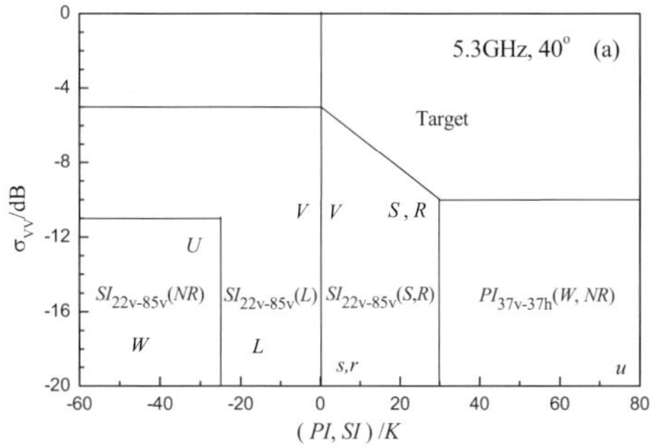

Fig. 4-5.3(a). Classification of σ_{vv} and T_B indices

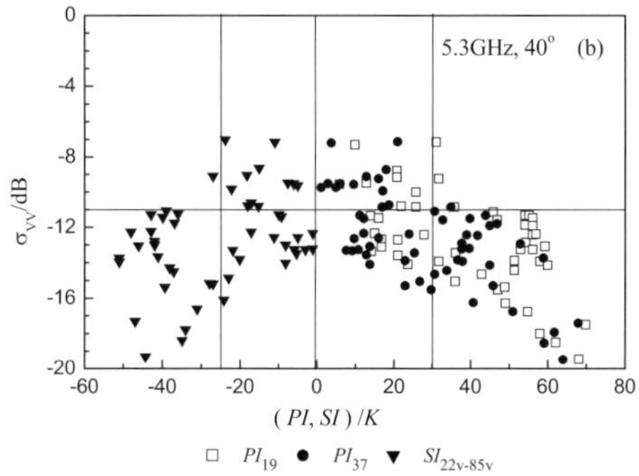

Fig.4-5.3(b). Validation by the data of Fig. 4-2

4-5.2 A Composite Model for Scattering and Emission from Sea Surface

To simulate variation of σ_{vv} and T_B indices over sea surfaces, a composite model of randomly rough sea surface has been developed as shown in Fig. 4-5.4 [Jin 1994, Huang and Jin 1995]. The observed bistatic scattering coefficient from a rough sea surface, which is partially covered by a layer of foam scatterers, is written as

$$\gamma_{pq}^{OBS}(\theta_s,\theta_i;\phi_s,\phi_i)=C_w\gamma_{pq}^{COM}+(1-C_w)\gamma_{pq}^{TS}\ ,\qquad p,q=v,h \tag{4-5.1}$$

where C_w is the percent coverage of foam and is empirically formulated by matching observations as a function of sea surface wind speed u as

$$C_w=11.12e^{0.063u}-16.23\quad (\text{as}\ u\geq 7\ m/s) \tag{4-5.2}$$

The coverage C_w takes account of functional dependence of foam cover upon surface wind speeds. The bistatic scattering γ_{pq}^{COM} of our composite (*COM*) model is contributed by scattering from two-scale rough surface and foam scatterers. By using iterative method to the VRT equation of a layer of foam scatterers over a two-scale randomly rough surface, the bistatic scattering is obtained in the zero-th and first-order solutions as

$$\gamma_{pq}^{COM} = \gamma_{pq}^{(0)} + \gamma_{pq}^{(1)} \tag{4-5.3}$$

where $\gamma_{pq}^{(0)} = \gamma_{pq}^{TS} e^{-\kappa_e d(\sec\theta_i + \sec\theta_s)}$ accounts for the scattering from a two-scale (*TS*) rough surface with attenuation through a layer of random foam particles; $\gamma_{pq}^{(1)}$ accounts for single scattering and multi-reflections from foam particles and underlying rough surface. The formulation of $\gamma_{pq}^{(1)}$ can be found in Jin (1994), and Huang and Jin (1995).

Foam particles are modeled as spherical water droplets under the Rayleigh scattering approximation. Since the scattering and extinction coefficients κ_s, κ_e are proportional to the fractional volume of particles f_s, both factors $\kappa_s d$ and $\kappa_e d$ are proportional to $f_s d$, which is actually the number of particles per unit area. Since there are no real measurement available for the fraction, size of foam particles and layer thickness etc., the product $f_s d$ as a model was simply proposed as a constant 0.1, which has been well tested for SMMR data matching at different sea surface speeds for different frequency [Jin 1994, Huang and Jin 1995].

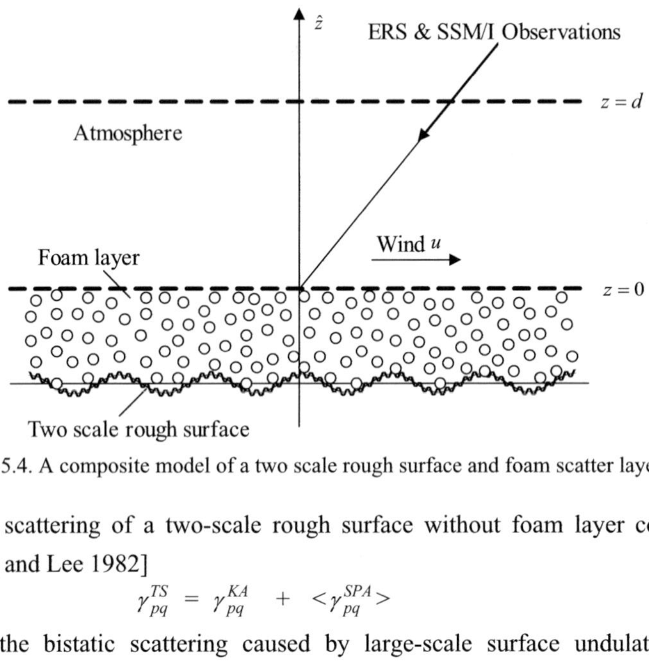

Fig. 4-5.4. A composite model of a two scale rough surface and foam scatter layer

The bistatic scattering of a two-scale rough surface without foam layer cover, γ_{pq}^{TS}, is written as [Fung and Lee 1982]

$$\gamma_{pq}^{TS} = \gamma_{pq}^{KA} + \langle \gamma_{pq}^{SPA} \rangle \tag{4-5.4}$$

where γ_{pq}^{KA} is the bistatic scattering caused by large-scale surface undulations in the Kirchhoff approximation (*KA*) as follows

$$\gamma_{pq}^{KA}(\theta_s, \varphi_s; \theta_i, \varphi_i) = \frac{\pi \, | \bar{k}_i - \bar{k}_s |^4}{\cos\theta_i \, | \hat{k}_i \times \hat{k}_s |^4 \, (k_{iz} - k_{sz})^4} F_{pq} P(\xi_{x0}, \xi_{y0}) \tag{4-5.5}$$

where $\bar{k}_i, \bar{k}_s$ are incident and scattering wave vectors, respectively; $P(\xi_{x0}, \xi_{y0})$ is the probability density function at the stationary phase points of (ξ_{x0}, ξ_{y0}). The stationary phase points are determined by the vectors $\bar{k}_i, \bar{k}_s$. The Cox-Munk probability density function is used in our model [Cox and Munk 1954, Jin 1994]; the functions of F_{pq} are related with the reflection coefficients of fluctuating surface. All formulations can be found in Jin (1994).

The second term at RHS of Eq. (4-5.4), $<\gamma_{pq}^{SPA}>$, is the bistatic scattering of small irregularities in the small perturbation approximation (*SPA*) averaged over the distribution of large undulations. It is written as

$$<\gamma_{pq}^{SPA}(\theta',\phi';\theta_i,\phi_i)>=\iint d\xi_{x'}d\xi_{y'}[1+(\xi_x\cos\phi_i+\xi_x\sin\phi_i)\tan\theta_i]P(\xi_{x'},\xi_{y'})\gamma_{pq}^{SPA}(\theta',\phi';\theta_i,\phi_i) \quad (4\text{-}5.6)$$

where $\xi(x,y)$ is random surface height and at $<\xi>=0$ at z=0 (here, the slope of randomly fluctuating surface are written as $\xi_x=\partial\xi/\partial x, \xi_y=\partial\xi/\partial y$). Note that the slope of ξ_x, ξ_y takes account of transition of the local coordinates. The formulation of γ_{pq}^{SPA}, proportional to the sea surface spectrum, can be found in Jin (1994). The semi-empirical spectrum $W(\kappa,\phi)$ [Fung and Lee 1982] is employed to calculate γ_{pq}^{SPA}. Functional dependence of γ_{pq}^{TS} on surface wind speed u and direction ϕ have been included in $W(\kappa,\phi)$ and $P(\xi_x,\xi_y)$.

From Eq. (4-5.1), backscattering coefficient is obtained as

$$\sigma_{pq}(\theta_0)=\gamma_{pq}^{OBS}(\theta_0,\pi-\phi_0,\theta_0,\phi_0)\cos\theta_0 \quad (4\text{-}5.7)$$

By using the reciprocity, the p-polarized emissivity from this composite surface model is obtained as

$$e_p(\theta_0)=1-\frac{1}{4\pi}\int_0^{2\pi}d\varphi\int_0^{2\pi}d\theta\sin\theta\;[\gamma_{vp}^{OBS}(\theta,\varphi;\theta_0)+\gamma_{hp}^{OBS}(\theta,\varphi;\theta_0)]\;,\quad p=v,h \quad (4\text{-}5.8)$$

To take into account atmospheric emission, the brightness temperature observed by space-borne radiometer SSM/I is written as

$$T_{Bp}^{SSM/I}(\theta_0)=e_p(\theta_0)T_s e^{-\tau}+(1-e^{-\tau})[1+r_p(\theta_0)e^{-\tau}]T_a \quad (4\text{-}5.9)$$

where the reflectivity $r_p=1-e_p, p=v,h$; τ is the total atmospheric opacity, which has strong frequency dependence and is chosen from the US Standard Atmosphere given by Waters (1976); T_s is the physical temperature of sea surface and T_a is the average physical temperature of atmosphere.

Backscattering coefficients observed by the ERS are

$$\sigma_{pq}^{ERS}(\theta_0)=\sigma_{pq}(\theta_0)e^{-\tau} \quad (4\text{-}5.10)$$

Actually, at the C band of ERS, $e^{-\tau}$ is negligible. Thus, by using Eqs. (4-5.9,10), $T_{Bp}^{SSMI}(\theta_0)$ at SSM/I channels and $\sigma_{vv}^{ERS}(\theta)$ at ERS channel can be numerically simulated.

4-5.3 Numerical Simulation of $T_{Bp}^{SSMI}(\theta_0)$ and $\sigma_{vv}^{ERS}(\theta)$

Fig. 4-5.5 gives numerical simulation of $T_{Bp}^{SSMI}(\theta_0=53^\circ)$, p=v,h, at 37 *GHz* vs sea surface wind speeds u (*m/s*). Solid lines are obtained from Eqs. (4-5.1,5,6), including emission from two-scale rough surface, foam layer and atmosphere. Parameters in calculations are listed under the figure title. Dotted lines are obtained without contribution from foam layer (i.e. $C_w=0$). It can be seen that as sea surface speed increases, foam emission becomes significant, especially for horizontally polarized T_{Bh}. In other words, detection of foam existence and its sensitivity indicate increasing surface winds. The polarization difference of surface emission also becomes smaller.

As a comparison, Fig. 4-5.6 shows numerical simulation of $\sigma_{pp}^{ERS}(\theta=40^\circ)$, pp=vv, hh, at ERS 5.3 *GHz* versus sea surface wind u (*m/s*). Dotted lines (without foam scattering) cannot be seen and are almost totally merged to the solid lines (including foam scattering). It means that low frequency channel such as 5.3 *GHz* is not sensitive to foam scattering since Rayleigh scattering is proportional to fourth order of frequency. Therefore, $\sigma_{pp}^{ERS}(\theta)$ would be not sensitive to high sea surface speeds when sea surface is covered by foam. To combine correlated active and passive measurements should be of helpful to retrieve sea surface wind speeds, especially high wind speeds.

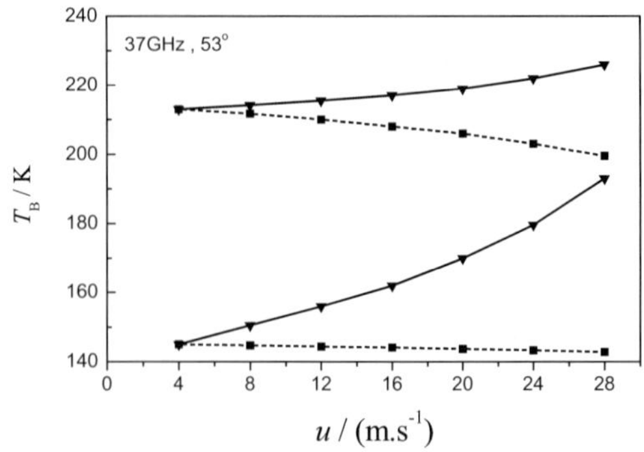

Fig. 4-5.5. Simulation of T_{Bp}^{SSMI} vs. wind speed u

($\tau_{37} = 0.16$, $\tau_{5.3} = 0.004$, $T_s = 288K, T_a = 288K$)

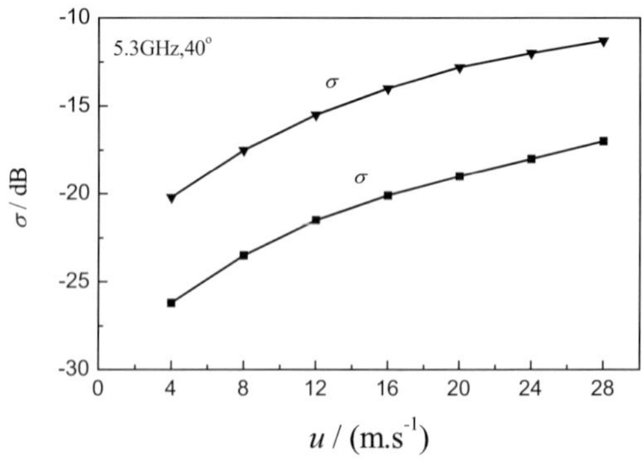

Fig. 4-5.6. Simulation of $\sigma_{pp}^{ERS}(40^{\circ})$ vs. wind speed u

Fig. 4-5.7 collected some data over vast areas of China sea from the ERS and SSM/I observations during the days 21, 22, 23, 26, 27 February 1996. Discrete points in Fig. 4-5.7 show correlation of SSM/I index PI_{37} and ERS $\sigma_{vv}^{ERS}(40^{\circ})$. The location and date when the observations were made are listed under the figure.

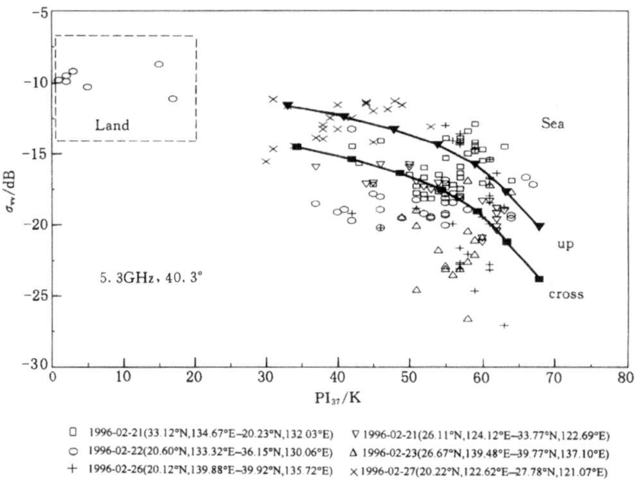

Fig. 4-5.7. Correlation of SSM/I and ERS Data and numerical simulations atmospheric opacity $\tau_{37} = 0.16, \tau_{5.3} = 0.004$, $T_s = 288K, T_a = 288K$.

By using numerical simulation of our composite model, two solid lines give theoretical results of PI_{37}^{SSMI} versus σ_{vv}^{ERS} at different sea surface wind speeds and up-wind and cross-wind direction, respectively. It has been found that the effects from the atmosphere (water vapor, clouds etc.) and real sea surface spectrum are significant. That is probably one of the reason why the collected data in Fig. 4-5.7 spread. Variation of the opacity τ can show the sensitivity from the atmospheric clouds and water vapor. The temperatures T_s, T_a are not sensitive parameters to affect the final results. Our numerical results only intend to show a simulation of possible cases. To combine active σ_{pp} and passive T_{Bp}, and other collateral and available satellite data sources, ground observations etc. should greatly improve the data interpretation and resolve ambiguity in inversion.

4-6 Data Fusion of RADARSAT SAR and DMSP SSM/I for Monitoring Sea Ice of China's Bohai Sea

During recent decades, a significant advance of satellite-borne microwave remote sensing has been the continuous monitoring of global physical and hydrological parameters. Active and passive microwave remote sensing, such as the synthetic aperture radar (SAR) and radiometry technologies, has greatly promoted the sea ice study [Lythe et al. 1997, Sandven et al. 2001, Bertoria 2001, Shokr et al. 1999, Carlstrom et al 1997]. The detection and specification of sea ice is of great importance to studies of global change, the radiation budget of the earth's hydrosphere, polar climate, and other characteristics. The main issues currently involve understanding and quantitatively retrieving the physical and hydrological information from those multi-channels, multi-polarized, and multi-sensor measurements.

For example, the Canadian RADARSAT SAR launched in November 1995 presented measurements of the co-horizontally polarized backscattering coefficient (σ_{hh}) at the C band. One of most successful programs of passive microwave remote sensing since 1989 is the DMSP SSM/I operated at seven channels to measure the brightness temperatures (T_B). Analysis of both active and passive observations of sea ice have been also discussed [e.g. Kwok et al. 1998, Lythe et al. 1999, Johannessen et al. 1996]. Monitoring regional sea ice is of special interest to operations involved in the exploration of mineral and petroleum resources as well as in navigation. However, most sea ice research has focused on the polar region at high latitudes, where most sea ice is multi-year (MY) ice. Shore ice at the middle latitudes, such as at China's Bohai Sea or Huang Hai Sea, lasts only a few months during the winter season. Its physical constitution and high temporal and spatial variability are significantly different from ice at the polar regions. Therefore, it is especially necessary to collect data of both active and passive remote sensing and develop analysis and algorithms for shore ice at these regions.

By using both active and passive data from the RADARSAT SAR and DMSP SSM/I observations over the Bohai Sea ice on the same day (January 23, 1999 as a representation), we present a data fusion analysis for sea ice detection and classification. Sea ice is usually modeled as a component medium of ice/brine background embedded randomly by air bubbles. By using numerical simulations of scattering and thermal radiative transfer of a sea ice layer, and comparing these with valid data from the RADARSAT and SSM/I observations, we can study the characteristic variation of the Bohai Sea shore ice. The scattering index (*SI*), polarization index (*PI*), and polarization ratio (*PR*) from passive T_B observations are defined to categorize the existence of sea ice. Further, by using active σ_{hh} measurements from the RADARSAT high-resolution SAR, we can classify the sea ice of the Bohai Sea. Temporal and spatial variations of these indexes can be effectively used to monitor the sea ice of the Bohai Sea.

4-6.1 Scattering and Emission of Sea Ice

Sea ice is usually classified and young ice, new ice, first-year ice (FY), multi-year (MY) ice, and so forth [Ulaby et al. 1985]. The ice thickness might be listed, accordingly, in a similar order, from a few centimeters to one meter or more. The sea ice of China's Bohai Sea usually lasts from the end of November to the beginning of March of the next year. The ice thickness is usually from 30 or 40 *cm* (shore ice) to a few centimeters towards the ocean. The temporal and spatial variability of the shore ice is quite high compared to that of MY ice in the polar region. The dielectric constant ε_b of the ice/brine medium, as a background body of sea ice, has been discussed in the literature [Ulaby et al. 1985, Jin, 1994], and is taken as $\varepsilon_b = (3.2 + i0.2)\varepsilon_0$ in our calculations. The subscript b denotes the sea ice background. The fractional volume of air bubbles, f_a, is generally less than 0.05 for FY ice. It yields a specific gravity of about $0.87\, g/cm^3$. In contract, MY ice is usually multi-layered with f_a of about 0.2, and a specific gravity of $0.7\, g/cm^3$.

Sea ice can be modeled as a strongly fluctuating random medium [Jin 1994]. The dielectric fluctuation is described by the variables of

$$\xi_a = 3\frac{\varepsilon_g}{\varepsilon_0}\left(\frac{\varepsilon_b - \varepsilon_g}{\varepsilon_b + 2\varepsilon_g}\right) \text{ and } \xi_a = 3\frac{\varepsilon_g}{\varepsilon_0}\left(\frac{\varepsilon_a - \varepsilon_g}{\varepsilon_a + 2\varepsilon_g}\right) \tag{4-6.1}$$

where the effective permittivity ε_g can be determined by solving

$$f_a \xi_a + (1 - f_a)\xi_b = 0 \tag{4-6.2}$$

Note that the dielectric constants for air bubbles, $\varepsilon_a = \varepsilon_0$, and for ice/brine, ε_b, are almost constant with no frequency dependence.

As a model of one-layer sea ice, the co-polarized *pp* (=*vv* co-vertically, and *hh* co-horizontally)-backscattering coefficient has three contributions:

$$\sigma_{pp}(\theta) = \sigma_{ICE}(\theta) + \sigma_{SUR1}(\theta) + \sigma_{SUR2}(\theta)\exp(-2k_g'' \, d \sec\theta) \tag{4-6.3}$$

where the first term on the right-hand side (RHS) is the volumetric scattering due to air bubbles in the ice medium, and the second and third terms are the surface scattering from the top and underlying interfaces of the ice layer, respectively. Attenuation through the ice layer with thickness d is taken into account in the exponential term, where $k_g'' = \text{Imag}(k_g)$.

According to the strong fluctuation theory [Jin 1994], the first term of the RHS in *hh*-polarization is written as

$$\sigma_{ICE,hh}(\theta) = k_0^4 W \left|\frac{X_{01i}}{D_{2i}}\right|^2 \Phi_{hh} \tag{4-6.4}$$

where k_0 is the wavenumber of free space, $|X_{01i}/D_{2i}|^2$ takes into account multi- reflections between the top and underlying interfaces,

$$W = \frac{1}{3}a^3(f_a|\xi_a|^2 + f_b|\xi_b|^2) \tag{4-6.5}$$

and

$$\Phi_{hh} = \frac{1}{4k_{gzi}''}(1 - e^{-4k_{gzi}''d})(1 + |R_{12i}|^4 e^{-4k_{gzi}''d}) + 4d|R_{12i}|^2 e^{-4k_{gzi}''d} \tag{4-6.6}$$

and where a in Eq. (4-6.5) is the radius of the spherical air bubble scatter, R_{12i} is the h-polarized reflection coefficient of the underlying smooth interface, and

$$k_{gzi}'' = \text{Imag}(k_{gzi}), \text{and } k_{gzi} = \sqrt{k_g^2 - k_0^2 \sin^2\theta}, \quad k_g = k_0\sqrt{\varepsilon_g/\varepsilon_0}.$$

The rough surface scattering is calculated by using the two-scale model [Ulaby et al., 1985] as

$$\sigma_{SURn} = \sigma_n^{KA} + \langle \sigma_n^{SPA} \rangle \tag{4-6.7}$$

where n = 1 (top interface) or 2 (bottom interface), and *KA* and *SPA* indicate the Kirchhoff (large-scale) and small perturbation (small-scale) approximations, respectively. Specifically,

$$\sigma_{pp}^{KA} = \frac{|R_p(0)|^2}{2s^2\cos^4\theta}\exp(-\frac{\tan^2\theta}{2s^2}), \quad p=h \text{ or } v \tag{4-6.8}$$

where $R_p(0)$ is the p-polarized reflection coefficient of smooth interface at $\theta = 0$, and the Gaussian surface slope $s = \sqrt{2}\delta/\ell$, where δ and ℓ are, respectively, the square root of the surface variance and correlation length in large scale. Further,

$$\sigma_{pp}^{SPA}(\theta) = 8k_0^4\delta_1^2\cos^4\theta\,|A_p|^2\,W(2k_0\sin\theta,0) \tag{4-6.9}$$

where $W(k_\perp,k_z)$ is the Fourier transform of the correlation function of surface height with the square root of the variance δ_1 and correlation length ℓ_1, and the function A_p is related to the p-polarized reflection coefficient of the smooth interface at θ. Eq. (4-6.9) is averaged over the large-scale undulation as [Ulaby et al. 1985]

$$<\sigma_{pp}^{SPA}(\theta)>=\frac{1}{2\pi}\int_{\pi/2}^{-\pi/2} d\phi_n[\sigma_{pp}^{SPA}(\theta')|_{\theta_n=\tan^{-1}s}+\sigma_{pp}^{SPA}(\theta')|_{\theta_n=-\tan^{-1}s}] \tag{4-6.10}$$

where $\theta' = \cos^{-1}(\cos\theta_n\cos\theta + \sin\theta_n\sin\theta\cos\phi_n)$.

Fig. 4-6.1 shows simulated σ_{hh} at the C band and incident angle of 35° for increasing ice thickness d. Because the air bubble's f_a in FY ice is always small, scattering enhancement is mainly contributed by surface scattering of σ_{SUR1} and σ_{SUR2} (because d very small). As d keeps increasing, σ_{SUR2} gradually darkens due to attenuation through the ice layer, and scattering is dominated by σ_{SUR1}. Sea surface wind can drive a free sea surface to rougher and also enhance scattering. Thus, without any prior knowledge, it is not appropriate to use a single instance of increasing σ_{hh} to classify sea ice existence.

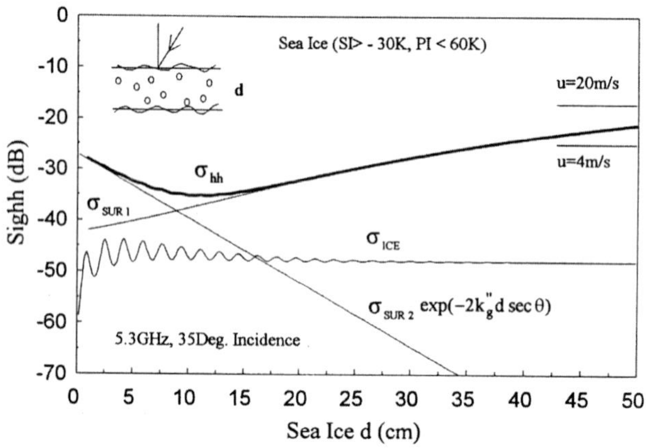

Fig.4-6.1. Backscattering σ_{hh} versus sea ice thickness d at C band and incident angle 35° air bubbles: $f_a = 0.05$, radius $a = 0.025cm$; $d<50$ cm.

Roughness of the top surface: $\delta = 2 + (d-1)/49$ cm, $\ell = 15$ cm, $\delta_1 = 0.1$ cm, $\ell_1 = 5$ cm.

Roughness of the bottom surface: $\delta = 2$ cm, $\ell = 20$ cm, $\delta_1 = 0.1$ cm, $\ell_1 = 5$ cm.

The VRT equation for calculation of $\overline{T}_B(=T_{Bv},T_{Bh})$ is written as

$$\cos\theta\frac{d}{dz}\overline{T}_B(\theta,z) = -\kappa_a(\theta)\overline{T}_B(\theta,z) + \int_0^\pi d\theta'\sin\theta'\overline{\overline{P}}(\theta,\theta')\cdot\overline{T}_B(\theta',z) + \kappa_a\overline{T} \tag{4-6.11}$$

By solving VRT Eq. (4-6.11) with the boundary conditions, we can calculate the brightness temperature $T_{Bp}(p=v,h)$ at SSM/I channels. The extinction coefficient κ_e, absorption coefficient κ_a, and phase matrix $\overline{\overline{P}}$ of small air bubble scatters, and the numerical approach of the VRT equation is given by Jin (1994).

Due to the existence of sea ice, the T_B of all SSM/I channels is increased, and the difference between different channels becomes smaller. Because atmospheric absorptions at 22 *GHz* and 85 *GHz* channels are fairly close, compared to other channels, the scattering index is defined as $SI = T_{B22v} - T_{B85v}$ to indicate a reduced T_B difference in high- and low-frequency channels. *SI* is sensitive to the existence of sea ice. When no sea ice exists, it has $SI << 0$, e.g. $SI < -40K$. When sea ice appears, the index *SI* gradually increases (> -30*K*). Meanwhile, the T_B difference between vertical and horizontal polarization also becomes smaller. The polarization index is defined as $PI = T_{B19v} - T_{B19h}$.

Fig. 4-6.2 shows numerical simulations of *SI-PI* for different situations of increasing ice thickness from 0 to 15 *cm* obtained by solving the VRT equation (4-6.11). The solid points and line indicate the numerical results, and the circle points are some typical SSM/I data selected from observations over the Bohai Sea during October 1995-April 1996. Fig. 2 roughly indicates a rule of ice detection as *SI* > -30 *K*, *PI* < 60 *K* [Jin 1998].

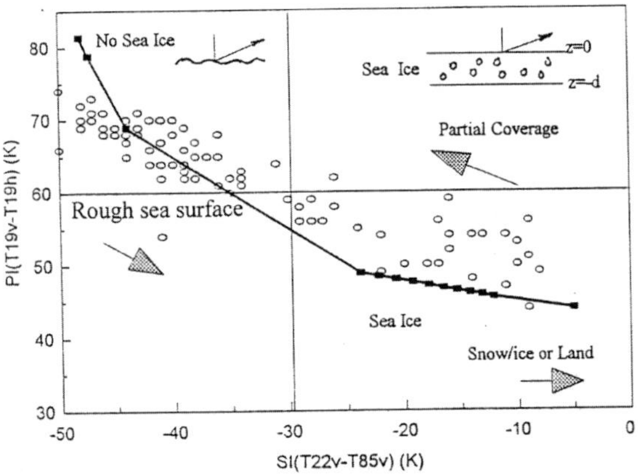

Fig. 4-6.2. The *SI-PI* graph and VRT results for detection of sea ice
Ocean: $T_2 = 273\,K$; Sea ice: $T_1 = 268\,K$; Air bubble: *a*=0.02 *cm*, $f_a = 0.08$.
Atmosphere: $T_a = 278\,K$, Atmospheric opacity: $\tau(19) = 0.06$, $\tau(22) = 0.12$, $\tau(37) = 0.08$, $\tau(85) = 0.22$.

4-6.2 Data of SSM/I and Radarsat SAR over the Bohai Sea

Fig. 4-6.3 is an image of the brightness index (*DN*) from observation of RADARSAT SAR data on January 23, 1999, over the Bohai Sea. Following the instructions of the RADARSAT Data Products Specification (1997), the backscattering coefficient σ_{hh} (*dB*) at the *j*-th pixel can be obtained from the *DN* (*dB*) value as

$$\sigma_{hh}(j) = \beta(j) + 10\log_{10}(\sin\theta_j) \tag{4-6.12a}$$

$$\beta(j) = 10\log_{10}[DN(j)^2 + A_3 / A_2(j)] \tag{4-6.12b}$$

where $\beta(j)$ is the radar brightness of the j-th pixel, $A_2(j)$ is the scaling gain value and A_3 is a fixed offset, and $\theta(j)$ is the local incidence angle to the j-th pixel, determined by the satellite altitude, the earth radius, and the slant range from the satellite to the j-th pixel.

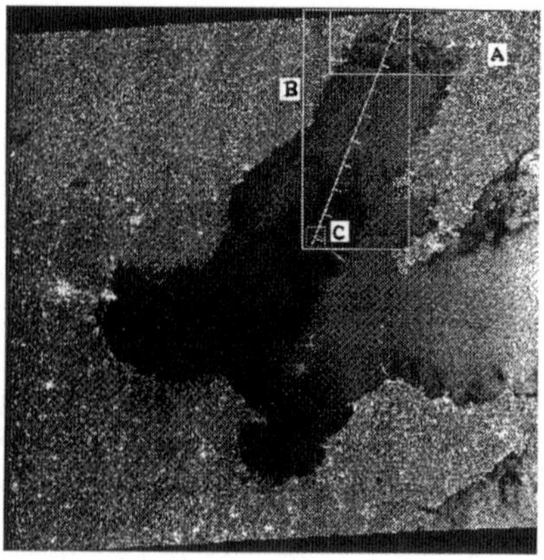

RADARSAT SAR DN Image

Fig. 4-6.3. Image of *DN* values from RADARSAT SAR

Figs. 4-6.4(a, b) show the σ_{hh} values of the areas circled by frames A and B in Fig. 4-6.3, respectively. Fig. 4-6.4(c) shows the variation of σ_{hh} along line C in Fig. 4-6.3. The numbers on the horizontal axis indicate the pixel location from the north (value 0) to the south.

σ_{HH} Image of Region A

Fig. 4-6.4(a). σ_{hh} for frame A of Fig. 4-6.3

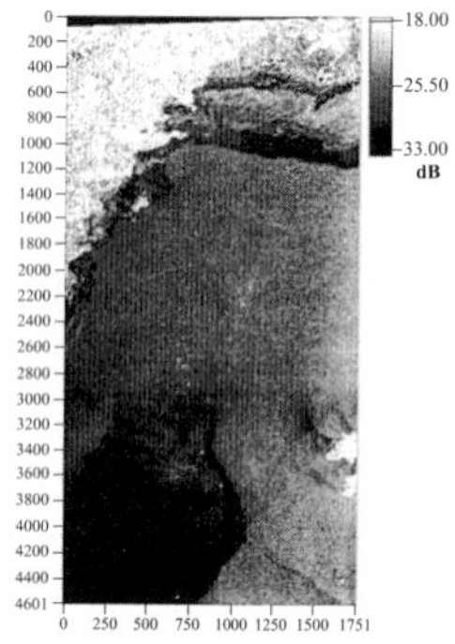

σ_{HH} Image of Region B

Fig. 4-6.4(b). σ_{hh} for frame B of Fig. 4-6.3

It can be seen from Fig. 4-6.4c that σ_{hh} is, in comparison with the sea water area, significantly increased (from a low of -30 to –18 *dB*), due to the north shore ice of the North Gulf of (locations 0 to 500. The large fluctuation between locations 300 and 1000 corresponds to the heterogeneous distribution of ice type and thickness. Location 1000 seems to be the border between sea ice and seawater. After location 1000, the value of σ_{hh} remains within -35*dB* to -25*dB*; these locations have no ice, and scattering is coming from the rough sea surface. Without a priori knowledge or the variations of the spatial correlation of σ_{hh}, it would not be easy to distinguish the sea ice from the water areas. After location 3500, σ_{hh} decreases as a typical value of flat- surface sea water e.g. less than –40 *dB*.

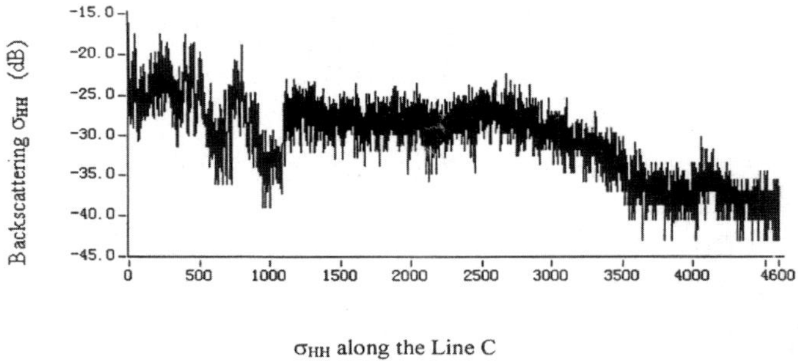

Fig. 4-6.4(c). Variation of σ_{hh} along line C of Fig. 4-6.3

Figs. 4-6.5(a, b) present, respectively, the images of the *SI* and *PI* indexes from the SSM/I T_B observations on the same day and in the same area as Figs. 4-6.3 and 4-6.4. The rule of ice detection using *SI* and *PI* as discussed above seems direct for finding the shore ice, especially near Ying Kou City.

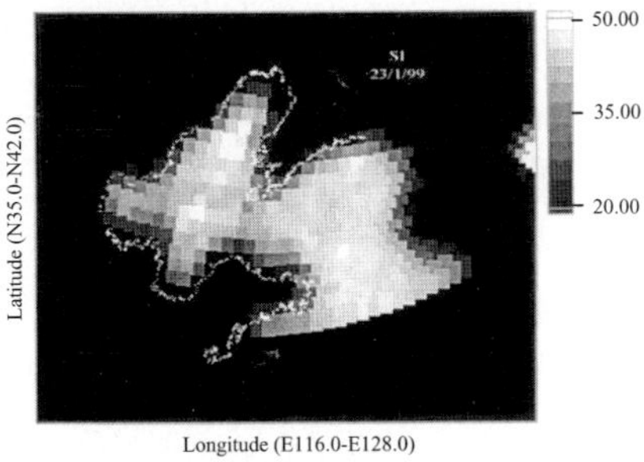

Fig. 4-6.5(a). *SI* from SSM/I T_B data

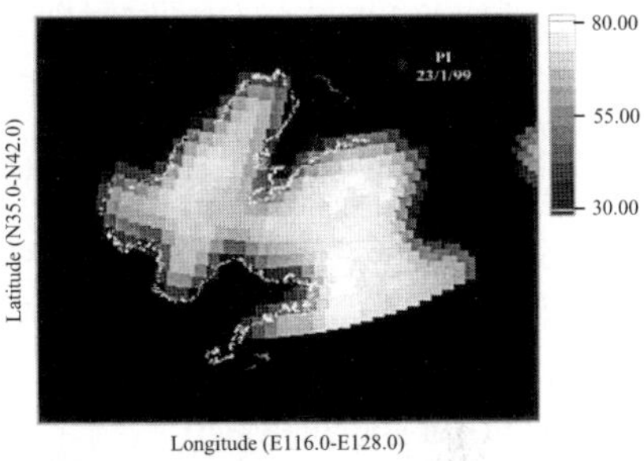

Fig. 4-6.5(b). *PI* from SSM/I T_B data

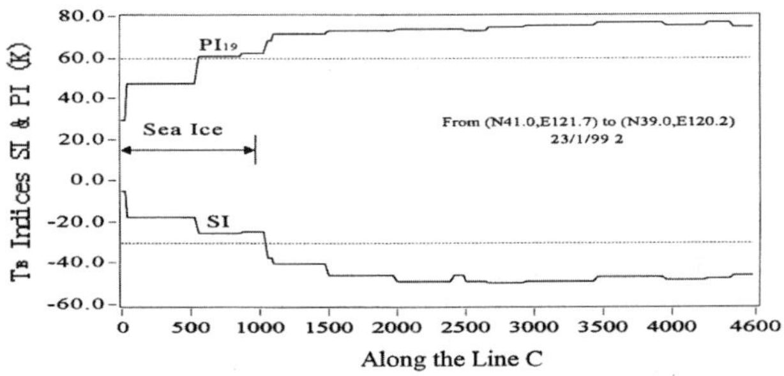

Fig. 4-6.5(c). *SI* and *PI* indexes along line C

Fig. 4-6.5(c) gives the *SI* and *PI* values along line C of Fig. 4-6.3. It can be seen that sea ice exists from location 0 to location 1000. Further, from the high resolution σ_{hh} images of SAR in Figs. 4-6.4(a, b), and its spatial variation, the physical status of sea ice, such as thickness, surface roughness, and snow cover, can then be detailed.

Due to the spatially heterogeneous distribution of sea ice and seawater, an index of the polarization ratio (*PR*) at the SSM/I 37*GHz* channel was proposed to describe ice coverage percentage (concentration) [Haggar et al. 1998]:

$$PR = \frac{T_{B37v} - T_{B37h}}{T_{B37v} + T_{B37h}} \tag{4-6.13}$$

Empirically, at *PR*=0.1851, the surface is 100% water, and at *PR*=0.0157, it is 100% ice. The usefulness of these empirical values has not yet been validated. Actually, the radiative transfer equation of the atmosphere and underlying sea ice/water can be simply written as

$$T_{Bp} = e_p T_s e^{-\tau} + (1-e^{-\tau})(1+re^{-\tau})T_a + s_p T_s e^{-\tau}, \quad p = v,h \tag{4-6.14}$$

where e_p and τ_p are, respectively, the $p(v,h)$-polarized surface emissivity and reflectivity, with $e_p = 1 - r_p$, τ is the atmospheric opacity; T_s and T_a are, respectively, the surface physical temperature and atmospheric effective temperature. On the RHS of Eq. (4-6.14), the first term represents the emission from sea ice to water. The second term is the atmospheric emission, and the third term is the additional contribution to T_{Bp} due to surface roughness, which is taken into account by using the empirical parameter s_p $(p = v,h)$. By letting $T_a \approx T_s$, we obtain

$$PR = \frac{e^{-\tau}[(r_h - r_v)e^{-\tau} + (s_v - s_h)]}{[2-(r_h + r_v)]e^{-\tau} + (1-e^{-\tau})[2+(r_h + r_v)e^{-\tau}] + (s_v + s_h)e^{-\tau}} \tag{4-6.15}$$

Thus, *PR* has a functional dependence on atmospheric opacity τ and surface reflectivity r_p. The reflectivity of an ice layer over water is calculated as

$$r_p = \left| \frac{R_{01p} + R_{12p}\exp(ik_{gz}d)}{1 + R_{01p}R_{12p}\exp(ik_{gz}d)} \right|^2 \tag{4-6.16}$$

where R_{01p} and R_{12p} are, respectively, the *p*-polarized reflection coefficients of the interfaces of air-ice and ice-water.

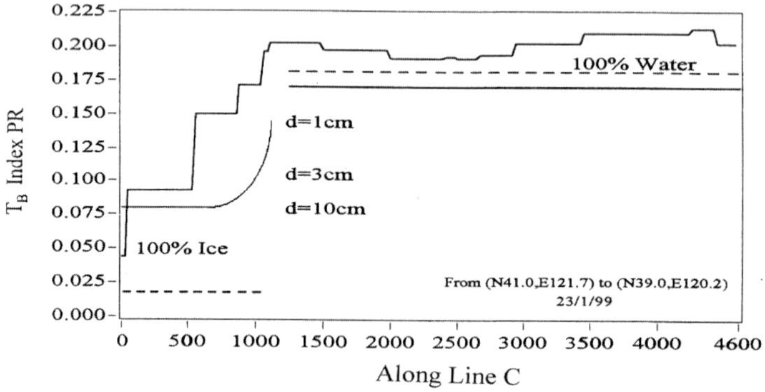

Fig. 4-6.6. Index *PR* along line C of Fig. 4-6.3 compared with Eq.(4-6.15) and empirical values $\tau = 0.05$, $s_v = 0.1$, $s_h = 0.01$.

Fig. 4-6.6 shows the *PR* values of Eq. (4-6.13) from SSM/I data along line C of Fig. 4-6.3, where the solid lines near the 100% indications for water and ice give the *PR* values from Eq. (4-6.15), and the dotted lines are the empirical values given by Haggar et al.(1998). Fig. 4-6.6 shows that the empirical value for a 100% ice surface is underestimated because it is even less than the value for land at location 0. Considering partial ice coverage within the spatial resolution, the detection rule of the *SI* and *PI* indexes should be empirically modulated.

We propose that the *SI*, *PI*, and *PR* indexes be employed to categorize the existence of sea ice, and the variation of σ_{hh} with high spatial resolution then be used to determine the sea ice status. Further collection of ground truth data should be of much help to improve validation and classification. A study of data fusion using the infrared, visible light and microwaves observations is further helpful.

4-7 Spatial Auto-correlation Using Optimal Multi-Layering Multi-Scale Getis Statistics

Spatial distribution and variation of characteristic parameters of the terrain surface are involved in complex way from observations of space-borne remote sensing. Measurements from the regions with similar topography should demonstrate the similarity. Each pixels of remotely sensed image are actually not independent each other, and are more or less spatially correlated. For example, the data values of the nearby pixels at the same channel might be similar to indicate correlation of a spatial construction. Temporal autocorrelation can also show temporal variation of the spatial construction. Autocorrelation analysis is based on comparison of the data of nearby pixels and won't be significantly influenced by other occasional factors in observation. Thus, spatial autocorrelation has become a powerful tool to study the spatial characteristic and classification of the terrain surface.

For example, Moran' I and Geary approaches [Goodchild 1986] and some others [Wulder and Boots 1998] studied spatial dependency and assessed the auto-correlated nature of pixels. As Wulder and Boots (1998) pointed, these approaches yield a single summary measure and unrepresentative if the spatial autocorrelation varies significantly over the image. The Getis statistic measures local variations within patterns of spatial dependence and illustrates local clustering of variables based on spatial scale [Getis and Ord 1992, Wulder and Boots 1998]. However, because the scale is prescribed for an image, the Getis statistic does not show spatial autocorrelation of multiple scales and are not sensitive to the variables clustering associated with medium values.

In this session, different scales are assigned to measure the Getis autocorrelation. As an example, spatial autocorrelation of different scales from the SSM/I observation is studied. Multi-layering Getis statistic is developed to evaluate the variables clustering associated with medium values. Some simulated images and ERS-2 SAR data are applied to analysis of spatial autocorrelation distribution and classification of characteristic parameters of the terrain surface.

4-7.1 Autocorrelation of Getis Statistic

The Getis statistic is defined as [Wulder and Boots 1998]

$$G_i^*(d) = \sum_{j=1}^{n} W_{ij}(d)x_j \Big/ \sum_{j=1}^{n} x_j \tag{4-7.1}$$

where W_{ij} is a spatial weights matrix centered at the point i. The points centered around i with the weight $W_{ij}\neq 0$ construct a window, generally as a square, i.e. $W_{ij}=1$ within d of the point i and zero otherwise.

Standardized Z value is defined as [Wulder and Boots 1998]

$$Z_i(d) = \frac{[G_i^*(d) - E(G_i^*(d))]}{\sqrt{Var(G_i^*(d))}} \tag{4-7.2}$$

which describes clustering of the variable x in the image. Positive Z indicates clustering of high x, and negative Z means clustering of low x.

Fig. 4-7.1(a) shows the desertification index DI ($\equiv T_{B37v} - T_{B37h}$, the difference of vertically and horizontally polarized brightness temperature) measured by passive microwave remote sensing DMSP SSM/I during March-August 2001 around China's Badain Jaran Desert and Tengger Desert (96°~106° E, 36°~43° N) [Jin and Yan 2004]. The index DI was defined as a characteristic parameter for classifying desertification.

Figs. 4-7.1(b~d) show the Z distributions, respectively, using $d=1(3\times3$ window)、$d=2(5\times5$ window) and $d=3(7\times7$ window), where the black areas indicate negative spatial autocorrelation as clustering of low *DI* values, and white areas are clustering of the high *DI* values indicating desert. It can be seen that even the values of some points between two deserts are not high, but, due to surrounding by high values of nearby points, the Z value of these points are moved to high value. Large window takes account of autocorrelation of large area and smooth transition of clustering of high and low values.

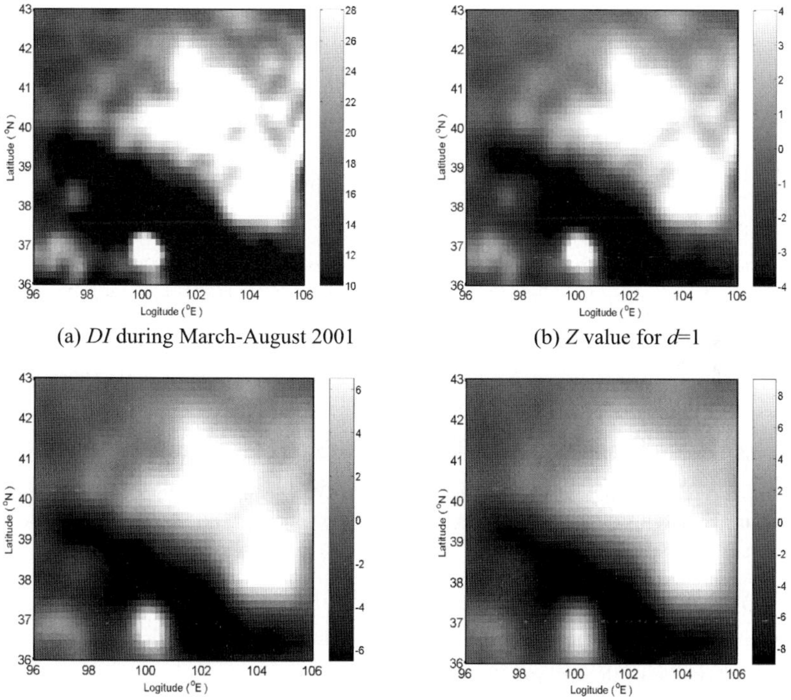

(a) *DI* during March-August 2001 (b) Z value for d=1

(c) Z value for d=2 (d) Z value for d=3

Fig. 4-7.1. Distribution of the *DI* and Getis Statistic

4-7.2 Scale Distribution of Spatial Autocorrelation

As described above, a scale of d is assigned for distribution of the standardized Z for spatial autocorrelation of the Getis statistic. Generally, a 3×3 or 5×5 window is taken as the spatial scale.

Let's suppose to have a matrix as

$$\begin{array}{cccccccccccccc}
m & m & m & m & m & m & m & m & m & m & m & m & m & m \\
m & A & A & A & A & A & m & m & B & B & B & B & B & m \\
m & A & A & A & A & A & m & m & B & B & B & B & B & m \\
m & A & A & A & A & A & m & m & B & B & B & B & B & m \\
m & A & A & A & A & A & m & m & B & B & B & B & B & m \\
m & A & A & A & A & A & m & m & B & B & B & B & B & m \\
m & m & m & m & m & m & m & m & m & m & m & m & m & m
\end{array}$$

where there are totally $7\times14=98$ elements. Let $A=m(1+c)$, $B=m(1-c)$, $0<c<1$, calculate $Z_i(d)$ centered at A, respectively, for $d=1$ and $d=2$ as shown in Table 4-7.1.

	$d=1$	$d=2$
G_i^*	$\frac{9(1+c)}{98}$	$\frac{25(1+c)}{98}$
$E(G_i^*)$	$\frac{9}{98}$	$\frac{25}{98}$
$Var(G_i^*)$	$4.39\times10^{-4}c^2$	$1.00\times10^{-3}c^2$
$Z(G_i^*)$	4.39	8.07

Table 4-7.1. The standardized Z for different window size

It can be seen from Table 4-7.1 that large d makes $Z(G_i^*)$ increased, e.g. the value of Z of d = 2 is almost double of that of d = 1. The spatial scale of the region centered at A should be 5×5, instead of 3×3.

Generally, there exists an optimal scale for the point i that

$$d_{op}=\arg\max(Z_i(d)),\ \ |Z(d_{op})|>T_i \tag{4-7.3}$$

where T_i is a control parameter to make different thresholds for different spatial scales. If Eq. (4-7.3) is not true, then take $d_{op}=0$. Suppose to have spatial autocorrelation for clustered variables within the regions (min x_a) and (max x_b), a coefficient α is defined as

$$\alpha=\int_{\min}^{x_a}p(x)dx=\int_{x_b}^{\max}p(x)dx \tag{4-7.4}$$

where $p(x)$ is the probability density function of the variable x. The coefficient α indicates how large clustering is seen as spatially correlated, and determines T_i directly from the pixels value of the image. T_i is then written as

$$T_{ia}=Z(d_i)\big|_{x\equiv x_a},\ \ \ T_{ib}=Z(d_i)\big|_{x\equiv x_b} \tag{4-7.5}$$

Eqs. (4-7.4, 4-7.5) are used to determine the thresholds. Under a certain condition, there exists an optimal scale d_{op} (which is not equal to zero) to indicate how far spatially it can reach the correlation.

Fig. 4-7.2(a) presents an image, which is randomly simulated with normal distribution (mean $\bar{x}=0$, variance $\sigma=10$). Its size is 100×100 pixels, where three big blocks with the scale $d=1$, $d=2$ and $d=3$, respectively, are artificially embedded. The pixel values within the blocks are either maximum or minimum.

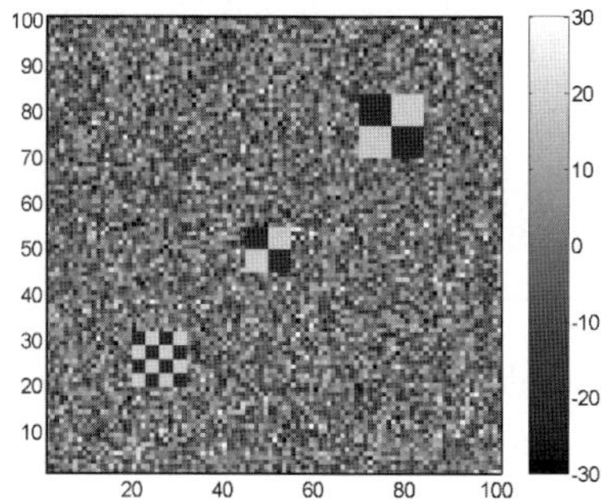

(a) A simulated image

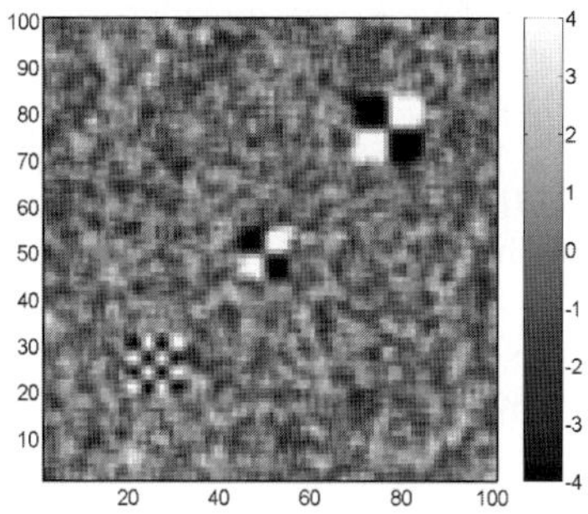

(b) Z of Getis statistic

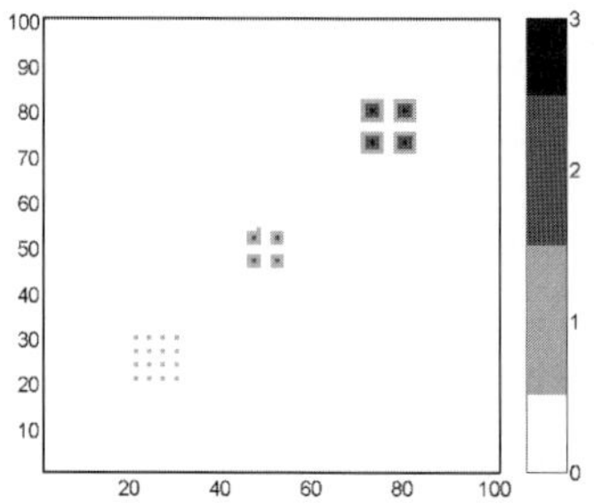

(c) Calculation of the scale d

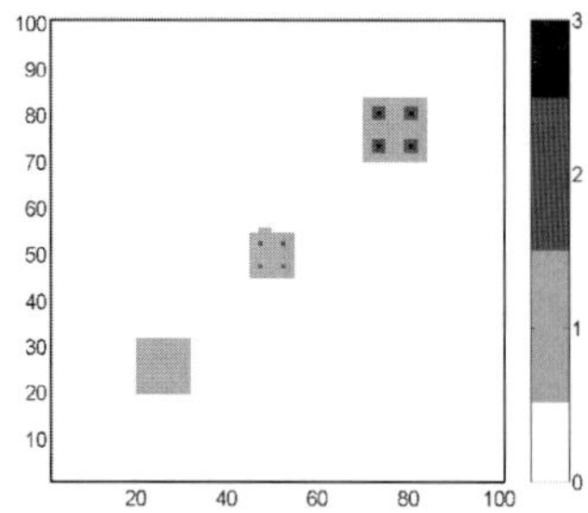

(d) Distribution of the scale d after filling

Fig. 4-7.2. Spatial scale distribution of the simulated image

Fig. 4-7.2(b) shows the Z distribution for $d = 1$. Fig. 4-7.2(c) shows the scale distribution at ($\alpha = 0.1$). It can be seen that the centers of each small square can be well identified. Based on the size of d of each block, further fill the scale within the centered region. It is shown in Fig. 4-7.2(d). Comparing Fig. 4-7.2(b) with Figs. 4-7.2(c, d), it can be seen that the relative value of Z distribution cannot well evaluate the autocorrelation and retrieve optimal scale information.

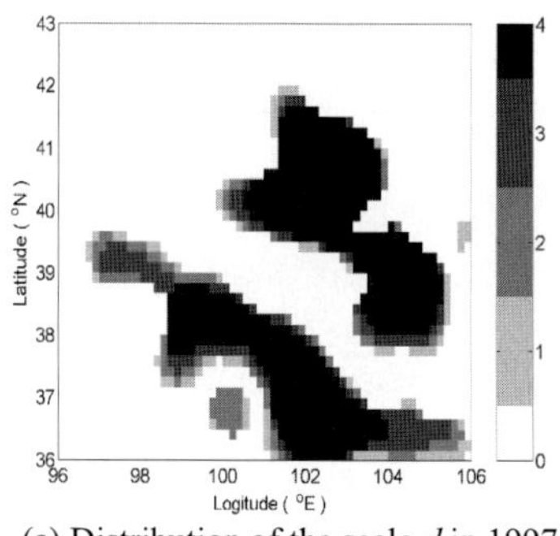

(a) Distribution of the scale d in 1997

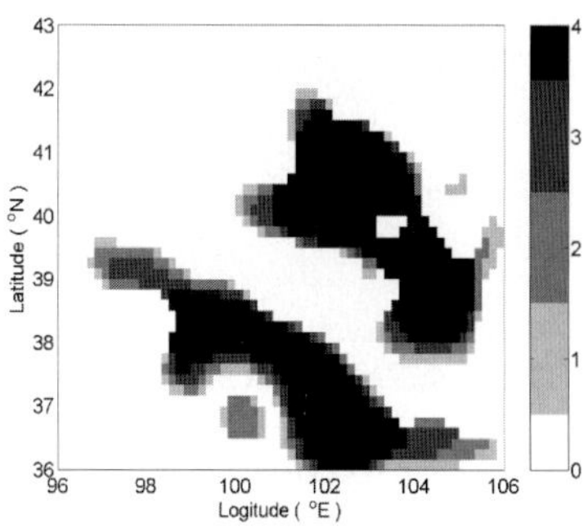

(b) Distribution of the scale d in 2001

Fig. 4-7.3. Distribution of spatial scale d in 1997 and 2001 at Badanjilin Desert and Tenggeli Desert.

Taking different scales d, Figs. 4-7.3(a, b) presents the distribution of the scales d from the DI observations of SSM/I around China's Badain Jaran Desert and Tengger Desert, respectively during March-August 1997 and 2001. It can be seen that two deserts seems merging as it is really happening. Various scales can show temporal variation of spatial characteristics.

4-7.3 Multi-Layering Getis Distribution

It has been pointed out [Wulder and Boots 1998] that the Getis statistic cannot well evaluate clustering of the points associated with medium value. Spatial autocorrelation of those pixels are often missed.

Suppose to have obtained the spatial d_{op} under the restriction of α, it is then defined as the 1-st layer. Calculate spatial distribution of d_{op} for rest points again using

$$G_i^{*\prime}(d) = \sum_{j=1}^{n} W'_{ij}(d) x_j \Big/ \sum_{j=1}^{n'} x_j \tag{4-7.6}$$

where n' are the number of those pixel points which was seen as spatially not correlated from the 1-st step or layer. The weights matrix $W'_{ij}(d)$ satisfies $\sum_j W'_{ij}(d) > N_T$, where N_T is the minimum of the pixel points associated with spatial correlation. If it is less than N_T, the points are seen as isolated and no spatial correlation are found. It then yields the 2-nd layer. Keeping the layering till the ℓ-th layer, those points with $d = 0$ are seen as no any spatial correlation.

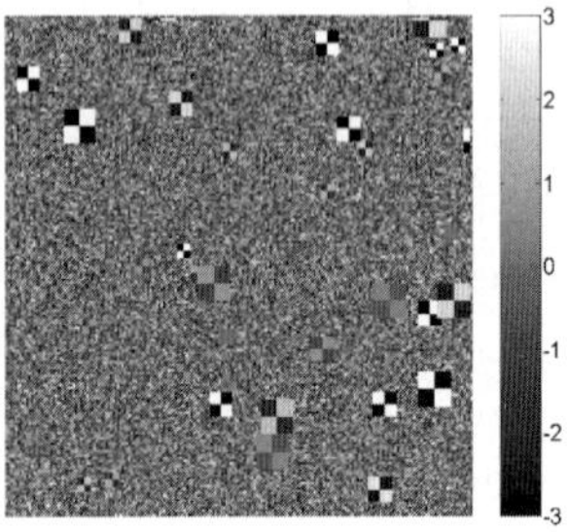

(a) A simulated image

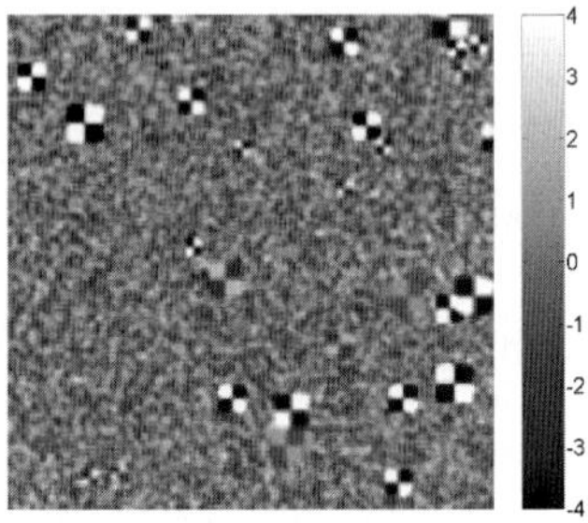

(b) Standardized Z for d=1

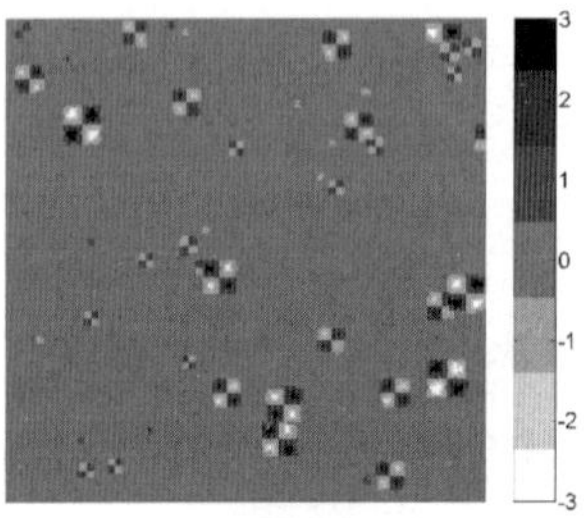

(c) Distribution of spatial scale d

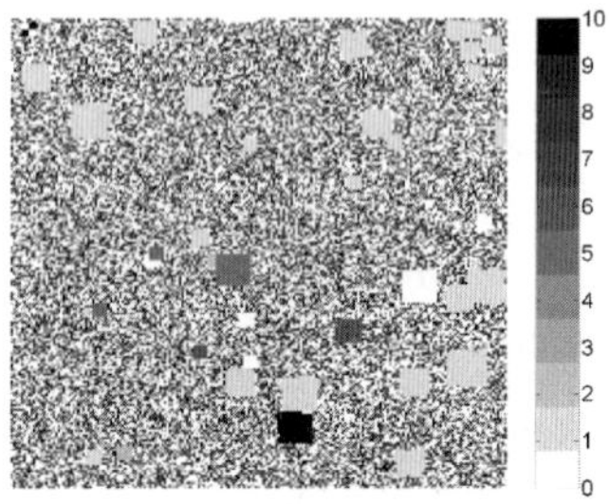

(d) Layering distribution（0 denotes no correlation）

Fig. 4-7. 4. Multi-layering spatial scales of the simulated image

After calculation of layering d_{op} is finished, to fill those correlated pixel points with the same statistic as their center. These points and their center have the same spatial autocorrelation belong to the same category.

Fig. 4-7.4(a) is a simulated image with normal distribution (mean value $\bar{x}=0$, variance $\sigma=1$). The image size is 200×200 and there are some data blocks with $d=1$、$d=2$ and $d=3$.The pixel values of the same small blocks are the same, and the values of different blocks are randomly chosen.

Fig. 4-7.4(b) is the standardized Z distribution for $d=1$, and cannot identify the blocks with the values neither high nor low.

Fig. 4-7.4(c) shows the result of multi-layering (10 layers) scale. Negative scale is also introduced. It can be seen that all blocks in Fig. 4-7.4(a) have been identified from Fig. 4-7.4(c), and the d_{op} values are correspondingly assigned (filling).

Fig. 4-7.4(d) presents the layer sequence. It can be seen that some blocks missed in the 1-st layer are now found. The layer number corresponds to the threshold value T_i to indicate the window size. It is also a kind of spatial information of the image.

Fig. 4-7.5 is the total power image of ERS-2 SAR at C band over Shanghai Hongqiao Airport in 9 April 2004. The image size is 250×450 pixels and spatial resolution is $12.5m\times12.5m$. The imaged region includes the airport (low left corner), residence area (north) and park (south).

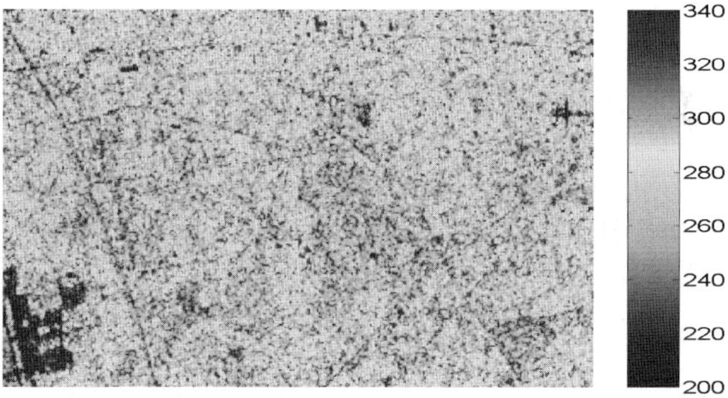

Fig. 4-7.5. ERS-2 SAR image over Hongqiao Airport in April 9, 2004 (=10×log(total power))

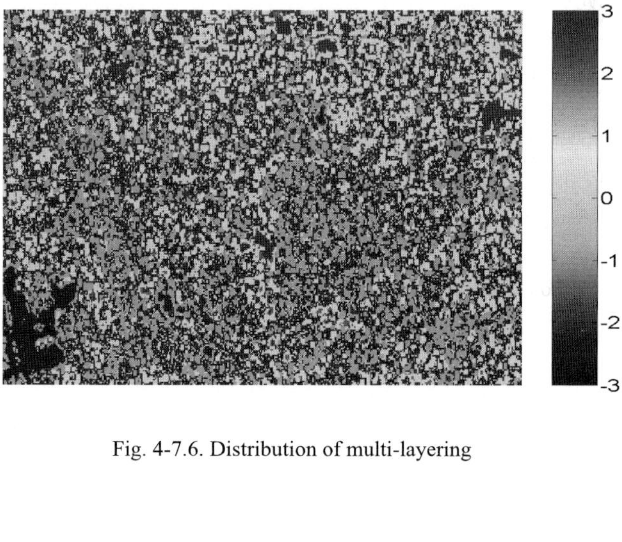

Fig. 4-7.6. Distribution of multi-layering

Suppose the maximum size of the window is 7×7 and $\alpha = 0.22$, the scale distribution of 20 layering is shown in Fig. 4-7.6. It can be seen that autocorrelation of the airport area is –3 as a low value clustering. Most of the scales of the north residence area are positive as a high value clustering. The south area (park etc.) shows many scales around -1.

Fig. 4-7.6 and the layer number can be seen as an input information for retrieval of characteristic parameters from the image. Positive autocorrelation most likely indicates the random constructions such as the building etc. with double or high order backscattering. Negative autocorrelation covers the areas such as grass field, highway, river etc. with less backscattering.

Based on Fig. 4-7.6, a binary black-white figure is sorted as Fig. 4-7.7 to classify characteristic distribution of the terrain surface around the Hongqiao area.

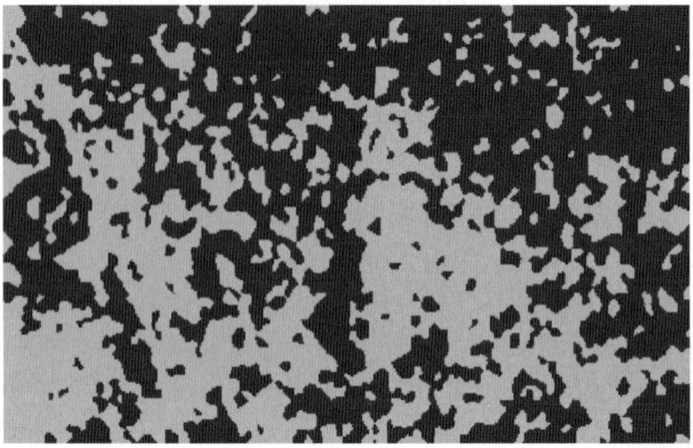

Fig. 4-7.7. Classification of positive and negative scales (black $d>0$, white $d<0$)

4-8 A Hybrid BP-ANN/GA Algorithm for Data Fusion of Landsat ETM+ and ERS-2 SAR in Classification of Urban Terrain Surfaces

Remote sensing of multiple sensors, e.g. at visible, infrared and microwave frequencies, presents different views on the geometries and spectrum reflectance of the terrain surfaces because of their different measuring physics. Information retrieval from a single sensor's data is always restrictive. Data fusion of multiple sensors is a new tool to retrieve more accurate and richer information in remote sensing applications. To make the terrain surface classification, the minimum distance method and Maximum Likelihood (ML) method based on the data statistics with the prior probability knowledge are usually employed [Devijver et al. 1982]. During recent years, the ANN (artificial neuron network) method, especially the most popular BP-ANN (back-propagation ANN), has been applied to study information retrieval of remote sensing [Yoshida et al. 1994, Jin et al. 1997, Hansen et al. 2001, Paola et al. 1995, and others]. However, it has been seen that slow convergence of the BP-ANN might be stuck in the local minima because of its gradient descending algorithm for the weights calculation. The GA (genetic approach), as a new optimization algorithm, can be helpful to improve the BP-ANN convergence.

Combining BP-ANN with GA, some studies have focused on evolution of BP-ANN's connection weights and its architecture on the learning rule [Yao 1993, Wei β 1994]. In our approach, the algorithm includes two steps: applying GA to optimization of the initial weights of BP-ANN under a defined generation, and selecting the initial weights of BP-ANN as input for network training. As the optimization only runs in given times, the total operation time is reduced.

After reducing the speckle noise of the SAR image and making the registration with the infrared images, the samples of surface classification are selected from each band of the infrared image and input to training the BP-ANN. According to the error of the BP-ANN output and its convergence speed, the GA is selected to initialize its weights and the BP-ANN/GA algorithm is created to make faster and more accurate classification. To take the good capability of the SAR image for detecting the man-made objects in complex urban areas, the infrared Landsat ETM+ and microwave ERS SAR images over the same study area are fused with the principal components analysis (PCA) method. Then, the samples of each surface class from the first three components are input into the training BP-ANN/GA to yield better surface classification than single sensor data [Cao and Jin 2005].

An example of the Shanghai Pudong area using the fused data of the Landsat ETM+ and ERS-2 SAR images well demonstrates the classification speed and accuracy of our BP-ANN/GA algorithm.

4-8.1 Data Preprocessing

The study area is located near the Century Park of Shanghai Pudong Destrict, China where there are river and lake, new buildings, flat grass fields etc. The infrared data was acquired by the Landsat ETM+ on 27 November, 2002. Its spatial resolution is $30m$. The ERS-2 SAR data was acquired on 9 April, 2002 at 5.3 *GHz*, *vv*-polarization and spatial resolution $12.5m$.

In urban areas, the existence of various man-made objects such as the buildings and roads with different materials, heights and orientations etc. attributes more randomness to the image noise character. As the first step, the Gamma adaptive filter with a 3×3 window is employed to reduce the image noise and keep the image main details [Lopbs et al. 1993].

In order to make data fusion, two different images are projected to the same geometry and grey space. The quadratic method or the cubic convolution method is employed (Jensen 1996). Then, the images of 400×400 pixels of Landsat ETM+ and ERS-2 SAR are ready as shown in Fig. 4-8.1 and Fig. 4-8.2, respectively.

Fig. 4-8.1. the Landsat ETM+ image at Band 4

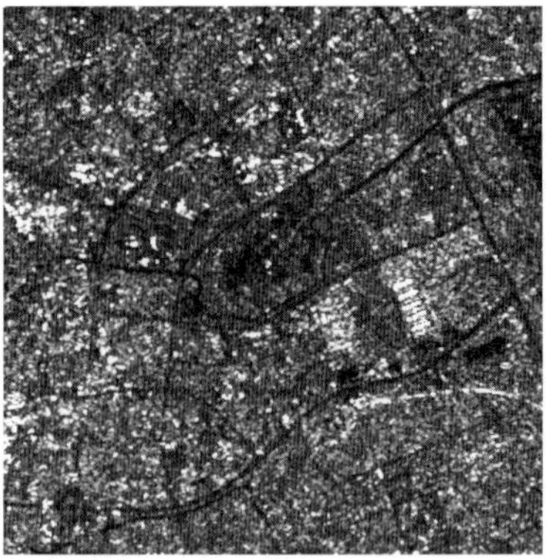

Fig. 4-8.2. the ERS-2 SAR image

Because the output of the activation function (Sigmoid function) is taken between 0 and 1, all input data are normalized in order to avoid the saturated area of the function as follows:

$$x^* = \frac{x - x_{\min}}{x_{\max} - x_{\min}} \tag{4-8.1}$$

where x is the original data, $x_{\min}$, $x_{\max}$ are the minimum and maximum, respectively.

4-8.2 The ANN Training

A. *BP-ANN training with the infrared data*

Solving non-linear problem by the BP-ANN approach [Klein 1999], the feature vectors extracted from the image are fed into the input layer, and the classifications are made as the output through the trained ANN.

Three layers (one input, one hidden and one output) of BP-ANN are constructed in this paper. The number of neurons of the input and output layers are equal to the features of the image (3) and the number of the classes (5), respectively. Twelve neurons in the hidden layer of the ANN are selected through trial and error, and the set of its initial candidates is decided empirically. The activation functions of both the hidden layer and output layer are the Sigmoid function:

$$f(x) = \frac{1}{1+e^{-x}} \tag{4-8.2}$$

The desired outputs for the 5 surface classes: water, grass, building, road and flat land-field are defined as (1 0 0 0 0), (0 1 0 0 0), (0 0 1 0 0), (0 0 0 1 0) and (0 0 0 0 1), respectively. If the ANN output is beyond the allowed error, the error between the real output and the desired output are evaluated as follows:

$$E = \frac{1}{2}\sum_{k=1}^{5}(t_k - r_k)^2 \tag{4-8.3}$$

where t_k is the real output of the k-th neuron of the output layer and r_k is its desired output. Then, the error feedback is propagating in the opposite direction to adjust the weights until the satisfied results are acquired. We first choose the training data for BP-ANN from the Landsat ETM+ image at the bands 4, 5, 7 and four samples are chosen for each class.

The real output and desired output are listed in Table 1, the error is shown in Fig. 4-8.3. The classification accuracy in Table 4-8.1 is the overall accuracy calculated by using the outputs of all samples of corresponding surface as follows:

$$CA = \frac{M - \sum_{i=1}^{M}\sum_{k=1}^{N}(t_{ik} - r_{ik})}{M} \tag{4-8.4}$$

where CA is the classification accuracy, M is the number of the samples ($M = 4$), N is the number of the neurons of the output layer ($N = 5$), t_{ik} is the real output of the i-th sample in the k-th neuron of the output layer, and r_{ik} is its desired output.

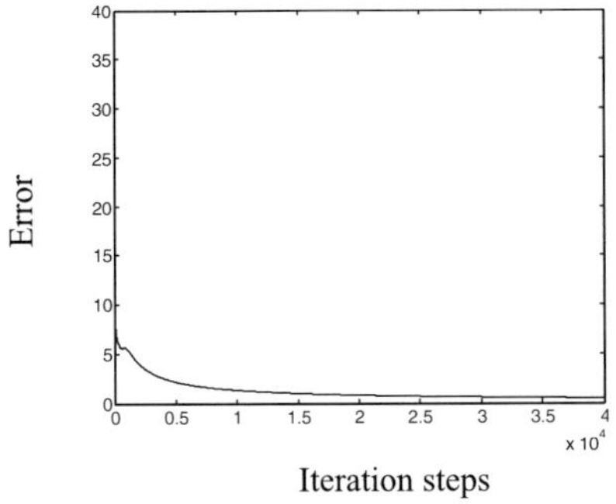

Fig. 4-8.3. The error of the BP-ANN with the infrared training data

Table 4-8.1. The real and desired outputs of the BP-ANN with the infrared training data

Output / Surface	Real output				Desired output	Classification accuracy (%)
	Sample 1	Sample 2	Sample 3	Sample 4		
Water	0.9410	0.9646	0.9298	0.9751	1	85.15
	0.0155	0.0167	0.0166	0.0181	0	
	0.1779	0.1521	0.1688	0.1395	0	
	0.0187	0.0214	0.0143	0.0433	0	
	0.0005	0.0003	0.0007	0.0001	0	
Grass	0.0131	0.0304	0.0005	0.0137	0	74.22
	0.9839	0.8678	0.6423	0.9543	1	
	0.0001	0.0009	0.0049	0.0002	0	
	0.0535	0.1546	0.0005	0.1663	0	
	0.0000	0.0000	0.0388	0.0000	0	
Building	0.0195	0.0247	0.0149	0.0035	0	60.28
	0.0075	0.0062	0.0098	0.0060	0	
	0.6579	0.6975	0.5947	0.7564	1	
	0.0386	0.0405	0.0329	0.0279	0	
	0.0100	0.0100	0.0107	0.0326	0	
Road	0.0281	0.0177	0.0241	0.0548	0	61.25
	0.1652	0.0248	0.0807	0.1583	0	
	0.0427	0.3407	0.1039	0.0489	0	
	0.9140	0.8979	0.9238	0.8041	1	
	0.0000	0.0000	0.0000	0.0000	0	
Flat field	0.0000	0.0000	0.0000	0.0000	0	71.28
	0.1160	0.0410	0.1797	0.0833	0	
	0.1134	0.3122	0.0625	0.1549	0	
	0.0000	0.0000	0.0000	0.0000	0	
	0.9974	0.9912	0.9627	0.9631	1	

Fig. 4-8.3 shows that the minimum and maximum errors between the real and desired outputs are 0.1964 and 36.6402, respectively, which indicate some misclassification there. It can be seen from Table 4-8.1 that the BP-ANN seems good for classification of water body and flat land-fields (the classification accuracy is 85.15%, 74.22%, respectively), but becomes poor for the types such as building, road and grass fields (especially, the classification accuracies for building and road are 60.28%, 61.25%, respectively). It can also be seen that the real output of the third sample of building is (0.0149 0.0098 0.5947 0.0329 0.0107), and the accuracy is very poor and much deviated from real value (0 0 1 0 0). The GA algorithm is used to initialize the weights of BP-ANN. With the help of the GA to optimize the weights, the training speed and classification accuracy are improved.

B. BP-ANN/GA training with the infrared data

The GA is an adaptive optimization algorithm to simulate the inheritance and evolution of the biology in the nature (Bäck 1996, Jin et al. 2001). Training BP-ANN with GA is divided into two parts: the first one is to optimize the weights with GA, and the second one is to train the BP-ANN with the result of GA.

Because the weights of BP-ANN are real, the chromosomes are real coded, different from conventional binary codes. Responding to the 96 weights in the BP-ANN, each chromosome of the population has 96 genes. The population scale designed is 60 and the initial population is selected randomly between 0 and 1. The whole evolution generation is 200. And the fitness function is defined as follows:

$$f(t,p)=\frac{err_{\max}(t)-err(t,p)}{err_{\max}(t)-err_{\min}(t,p)} \qquad (4\text{-}8.5)$$

where $f(t,p)$ is the fitness of the p-th population of the t-th generation, the errors $err_{\max}(t)$ and $err_{\min}(t)$ are the maximum and minimum errors in all the populations of the t-th generation, the error $err(t,p)$ is the output error of the p-th population in the t-th generation. The selection operation follows that the higher fitness and the more probabilities. Arithmetic crossover and Gaussian mutation operations are chosen according to the real genes. The crossover and mutation probabilities are 0.6 and 0.05, respectively.

At last, we choose the chromosome with the minimum error from the last generation of the GA algorithm as the initial weights of BP-ANN to make BP-ANN training with the infrared data.

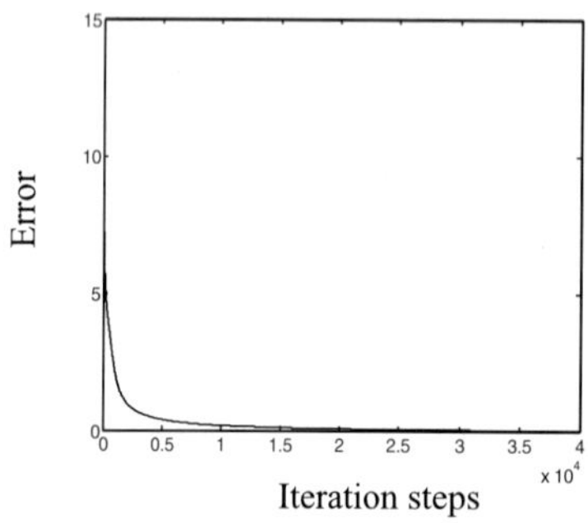

Fig. 4-8.4. The error of the BP-ANN/GA with the infrared training data

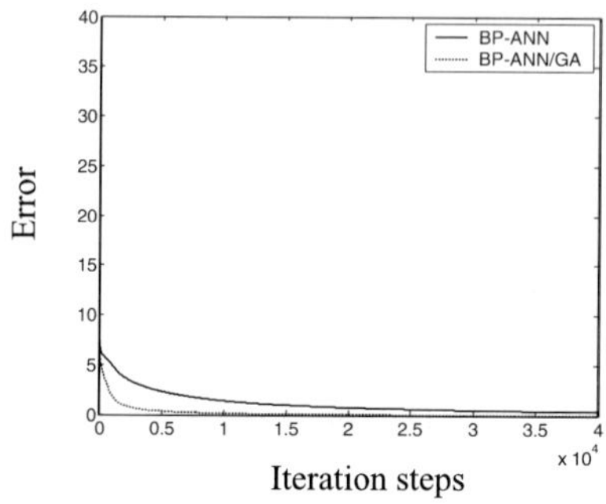

Fig. 4-8.5. The error comparison between BP-ANN/GA and BP-ANN with the infrared training data

Table 4-8.2. The real and desired outputs of the BP-ANN/GA with the infrared training data

Output / Surface	Real output				Desired output	Classification accuracy (%)
	Sample 1	Sample 2	Sample 3	Sample 4		
Water	0.9806	0.9885	0.9796	0.9885	1	92.81
	0.0100	0.0147	0.0115	0.0161	0	
	0.0583	0.0362	0.0489	0.0289	0	
	0.0000	0.0000	0.0000	0.0000	0	
	0.0001	0.0000	0.0001	0.0000	0	
Grass	0.0045	0.0079	0.0016	0.0015	0	93.75
	1.0000	0.9915	0.8944	0.9995	1	
	0.0000	0.0000	0.0000	0.0000	0	
	0.0027	0.0177	0.0002	0.0631	0	
	0.0000	0.0000	0.0372	0.0000	0	
Building	0.0186	0.0240	0.0148	0.0032	0	88.07
	0.0001	0.0001	0.0001	0.0000	0	
	0.9291	0.9566	0.8610	0.9678	1	
	0.0087	0.0071	0.0089	0.0501	0	
	0.0072	0.0069	0.0081	0.0329	0	
Road	0.0004	0.0004	0.0003	0.0026	0	86.74
	0.0504	0.0007	0.0093	0.0375	0	
	0.0002	0.0887	0.0019	0.0005	0	
	0.9009	0.9798	0.8430	0.9387	1	
	0.0000	0.0000	0.0000	0.0000	0	
Flat field	0.0002	0.0003	0.0005	0.0003	0	92.34
	0.0271	0.0022	0.0739	0.0109	0	
	0.0030	0.1009	0.0007	0.0101	0	
	0.0000	0.0001	0.0000	0.0002	0	
	0.9971	0.9898	0.9623	0.9645	1	

The real and desired outputs of BP-ANN/GA are listed in Table 4-8.2.

Fig. 4-8.4 shows that the minimum and maximum errors between the real and desired outputs of the BP-ANN/GA are 0.0155 and 14.0632, respectively, which are both better than the BP-ANN without GA.

Comparison between the error trends of the BP-ANN/GA and BP-ANN in Fig. 4-8.5 shows that the training speed of ANN in this case is speeded up. Total number of iterations for the BP-ANN/GA convergence is 3424 while it is 10610 for BP-ANN. Besides, the computing time of the BP-ANN/GA for a given error threshold 0.25 takes 445.125 seconds (431.01 seconds for BP-ANN training and 14.094 seconds for GA optimization) while it takes 542.86 seconds for the BP-ANN.

It can be seen from Table 4-8.2 that the classification accuracies for all surfaces are improved, especially for building, road and flat field, (the classification accuracy is 88.07%, 86.74% and 92.34% correspondingly, improving 27.79%, 25.49% and 21.06% over the accuracy of the BP-ANN). The accuracies for water and grass classifications are both better than 90%. It also can be seen that the building class takes a better output as (0.0148 0.0001 0.8610 0.0089 0.0081) with a little difference from (0 0 1 0 0).

C. *BP-ANN/GA training with the fusion of infrared and SAR images*

It is well known that the infrared image may show rich spectral information, but is lack of sensitivity to mistakes of different spectrum for the same object and the same spectral for different objects. On the other hand, the SAR image has spatially high resolution to make view on different geometric objects, but some measurements of back-scattering, such as for water body, flat fields and roads, are often not distinguishable. To take both advantages of infrared and microwave observations, the principal component analysis [PCA, Gonzalez 2003] is applied to fusing the infrared and microwave image data. PCA is an orthonormal linear transformation to derive the uncorrelated eigenvectors. It can decrease the data dimensions without much information lost. The eigenvalues of the images' covariance matrices are calculated and arranged in descending order at first. Then, a transformation matrix is constructed by the eigenvectors, correspondingly, with which we can transform the images into their components.

The first three components of the Landsat ETM+ infrared data and ERS-2 SAR data are presented in Figs. 4-8.6(a~c). The forth component is also obtained and ignored here due to little information. It can be seen that the figures of PCA1 and PCA2 contain most of images information (Fig. 4-8.6(a,b)) and the figure of PCA3 supplements some information for the surfaces such as water body, grass field and wide road (Fig. 4-8.6(c)). Taking 4 samples of each surface type from these three components as input into training BP-ANN/GA, the real output and desired output are listed in Table 4-8.3 and the error is shown in Fig. 4-8.7.

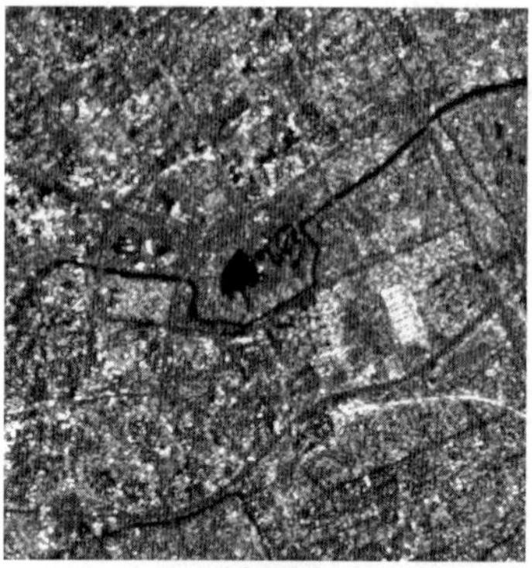

(a) PCA1

(b) PCA2

(c) PCA3

Fig. 4-8.6. Different components of PCA

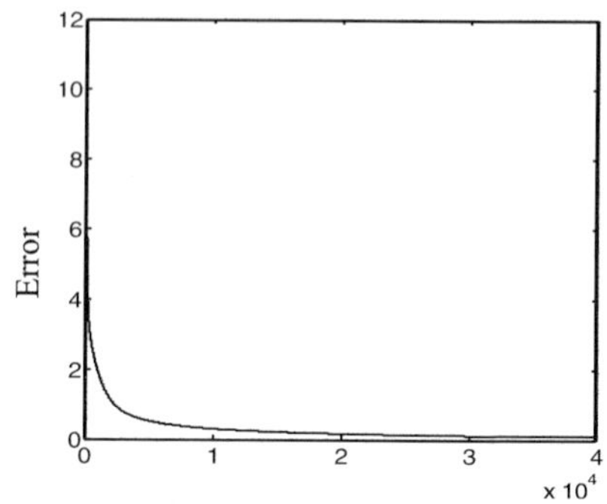

Fig. 4-8.7. The error with the fusion training data

Comparing Fig. 4-8.7 with Fig. 4-8.4, it can be seen that the minimum and maximum errors between the real output and the desired output in Fig. 4-8.7 become smaller, as 0.0154 and 11.0149. It means that fusing the infrared and SAR images improves the classification accuracy. Table 4-8.3 shows that the difference between the desired output and the real output for all of the surface types are very trivial. Classification accuracies for all surfaces are better than 90%.

Table 4-8.3. The real and desired outputs of the BP-ANN/GA with the fusion data

Output / Surfaces	Real output				Desired output	Classification accuracy (%)
	Sample 1	Sample 2	Sample 3	Sample 4		
Water	0.9925	0.9391	1.0000	1.0000	1	93.53
	0.0000	0.0000	0.0000	0.0000	0	
	0.0000	0.0000	0.0172	0.0266	0	
	0.0591	0.0877	0.0000	0.0000	0	
	0.0000	0.0000	0.0000	0.0000	0	
Grass	0.0003	0.0005	0.0006	0.0005	0	92.01
	0.9830	0.9678	0.9396	0.9848	1	
	0.0000	0.0000	0.0000	0.0000	0	
	0.0010	0.0440	0.1198	0.0055	0	
	0.0144	0.0024	0.0020	0.0038	0	
Building	0.0076	0.0194	0.0024	0.0013	0	94.62
	0.0000	0.0000	0.0000	0.0000	0	
	0.9980	0.9338	0.9817	0.9832	1	
	0.0011	0.0002	0.0014	0.0017	0	
	0.0057	0.0015	0.0255	0.0440	0	
Road	0.0004	0.0597	0.0034	0.0054	0	93.57
	0.0001	0.0015	0.0062	0.0732	0	
	0.0000	0.0000	0.0000	0.0000	0	
	0.9974	0.9981	0.9986	0.9155	1	
	0.0163	0.0000	0.0004	0.0001	0	
Flat field	0.0000	0.0000	0.0000	0.0000	0	91.88
	0.0001	0.0001	0.0001	0.0000	0	
	0.0075	0.0124	0.0128	0.0614	0	
	0.0240	0.0926	0.0180	0.0076	0	
	0.9967	0.9962	0.9582	0.9606	1	

4-8.3 The ANN Studying

After ANN training, three cases of the ANN studying and surface classifications are made: the infrared image of Landsat ETM+ 457 as input into the BP-ANN and the BP-ANN/GA, respectively, and the fused infrared and SAR images as input into the BP-ANN/GA. The classification results are, respectively, shown in Fig.8, Fig.9 and Fig.10. In these classification images, five colors refer to five classes: blue color standing for water body (river and lake), green color for grass field, orange color for building area, yellow color for flat land-field and brown color for road. The classifications of grass and flat land-fields in Figs.8 and 9 are similar, but the classification of water body seems poor. The BP-ANN/GA is better than the BP-ANN, including the continuous roads. At the up-right corner of Fig.9, many small paths are also detected.

However, many rectangular areas of building are largely mixed with roads. It might be due to the fact that the materials of the buildings and roads are similar in view of the infrared spectral reflectance.

Fig. 4-8.10 employs data fusion and BP-ANN/GA, and shows the best classifications of all 5 surfaces. It is interesting to compare Fig. 4-8.10 with the Shanghai tourist map of Fig. 4-8.11. High spatial resolution and good sensitivity of backscattering from geometrically different buildings and roads offer good capability to the SAR image. Because two images of Landsat ETM+ and ERS-2 SAR in our case are acquired at different time, it might be another reason that seasonal cycle might affect final accuracy of fusion classification.

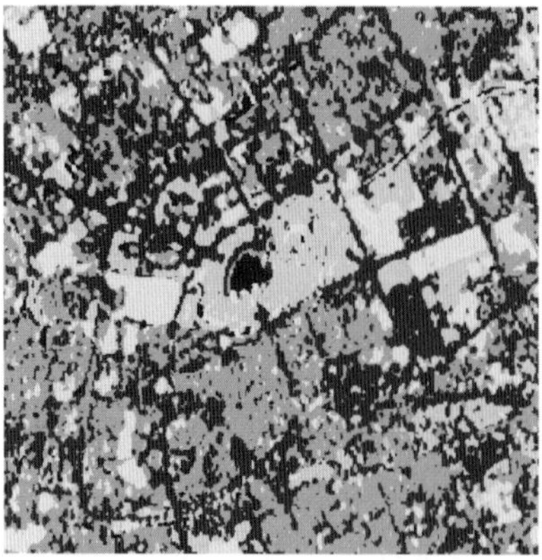

Fig. 4-8.8. BP-ANN classification of ETM+457

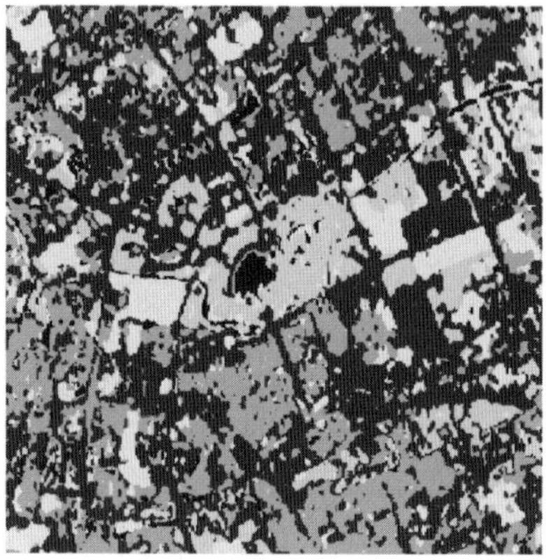

Fig. 4-8.9. BP–ANN/GA classification of ETM+457

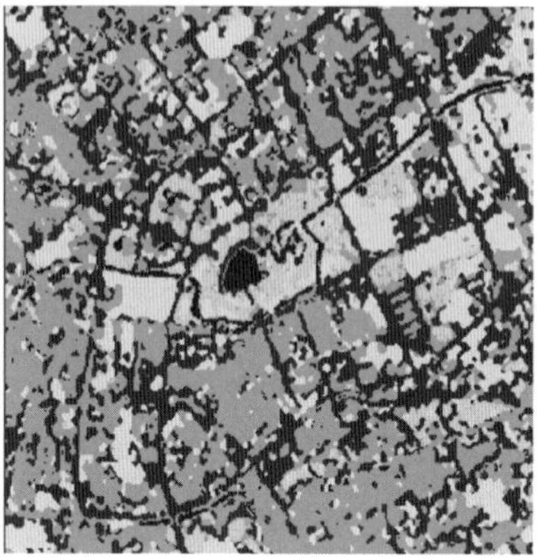

Fig. 4-8.10. BP–ANN/GA classification of the fusion of ETM+457 and SAR

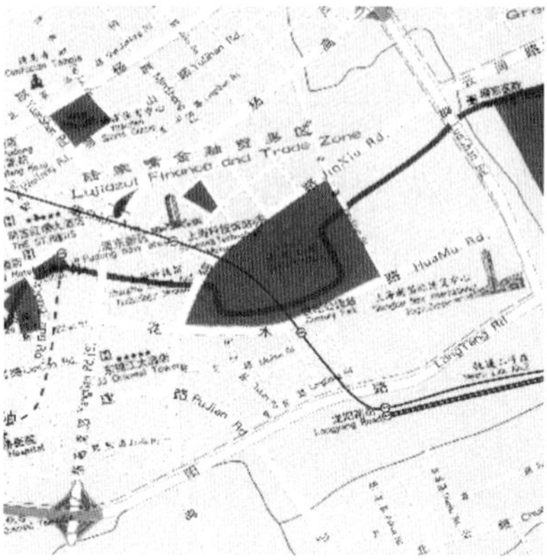

Fig. 4-8.11. The tourist map of the studying area

4-9 References

Bäck, T. (1996), *Evolutionary Algorithms in Theory and Practice*, New York: Oxford University Press.

Bertoria, S.V. (2001), "Special issue-stuudy of multi-polarization C-band backscattering signatures for Arctic Sea ice mapping with future satellite SAR", *Canadian Journal of Remote Sensing*, 27(5): 387-402.

Bhandari, S.M. (2000), "Comments on 'monitoring of brightness temperature over India and adjoining regions using SSM/I data' by Kumar et al. (1999)", *International Journal of Remote Sensing*, 21(16): 3187-3189.

Boehnke, K. and V. Wismann (1997), "Thawing processes during Siberian spring observed by ERS scatterometer and SAR", *International Geoscience and Remote Sensing Symposium Proceedings* (IGARSS'97 Singapore), 4: 1826-1828.

Bohren, C.F. and D.R. Huffman (1983), *Absorption and Scattering of Light by Small Particles*, New York: Wiley.

Cao, G.Z. and Y.Q. Jin (2005), "A hybrid algorithm of BP-ANN/GA for data fusion of Landsat ETM+ and ERS-2 SAR in urban terrain classification", *International Journal of Remote Sensing*.

Carlstrom, A. (1997), "A microwave backscattering model for deformed first-year sea ice and comparisons with SAR data", *IEEE Transactions on Geoscience and Remote Sensing*, 35(2): 378-391.

Chen, S. (1998), *The Earth System Science (in Chinese)*, Beijing: China Science and Technology Press.

Choudhury, B.J., Y.H. Kerr, E.G. Njoku and P. Pampaloni (1995), *Passive Microwave Memote Sensing of Land-atmosphere Interactions*, Netherlands: VSP International Science Publishers.

Cox, C. and W. Munk (1954), "Measurement of roughness of the sea surface from photographs of the sun's glitter", *Journal of Optical Society America*, 44(11): 838-850.

Devijver, P.A. and J. Kittler (1982), *Pattern Recognition: A Statistical Approach*, New York: Prentice-Hall.

Dobson, M.C., F. Kouyate and F.T. Ulaby (1984), "An reexamination of soil textural effects", *IEEE Transactions on Geoscience and Remote Sensing*, 22(6): 530-535.

Dobson, M.C., F.T. Ulaby, M.T. Hallikainen and M.A. El-Rayes (1985), "Microwave dielectric behavior of wet soil-part 2:dielectric mixing models", *IEEE Transactions on Geoscience and Remote Sensing*, 23(1): 35-45.

Ferraro, et al. (1996), An eight-year (1987-1994) time series of rainfall, clouds, water vapor, snow cover and sea ice derived from SSM/I measurements. *Bulletin of the American Meteorological Society*, 77(5): 891-905.

Fung, A.K. and K.K. Lee (1982), "A semi-empirical sea-spectrum model for scattering coefficient estimation", *IEEE Journal of Oceanic Engineering*, OE7(4): 166-176.

Getis, A. and J.K. Ord (1992), "The analysis of spatial association by use of distance statistics", *Geographical Analysis*, 24(3): 189-206.

Goldhirsh, J. (2001), "Attenuation and backscatter from a derived two-dimensional duststorm model", *IEEE Transactions on Antennas and Propagation*, 49(12): 1703-1711.

Gonzalez, R.C. and R.E. Woods (2003), *Digital Image Processing* (second edition). New York: Prentice-Hall.

Goodchild, M. (1986), *Spatial autocorrelation, Concepts and Techniques in Modern Geography*, 47 Norwich: Geo Books: 3-6.

Haggar, C.E., C. Garrity and R.O. Ramseier (1998), "The modeling of sea ice melt-water ponds for the high Arctic using an airborne line scan camera and applied to SSM/I", *International Journal of Remote Sensing*, 18(12): 2373-2394.

Hansen, J.V. and J.B. McDonald (2001), "Some experimental evidence on the performance of GA-designed neural networks", *Journal of Experimental and Theoretical Artificial Intelligence*, 13(3): 307-321.

Haupt, R.L. (1994), "Thinned arrays using Genetic Algorithm", *IEEE Transactions on Antennas Propagation*, 42(7): 993-999.

Haupt, R.L. (1995), "An introduction to Genetic Algorithms for electromagnetics", *IEEE Antennas Propagation Magazine*, 37(2): 7-14.

Haverkamp, D., L.K. Soh and C. Tsatsoulis (1995), "Automated approach to determining sea ice thickness from SAR data", *IEEE Transactions on Geoscience and Remote Sensing*, 33(1): 46-50.

Hollinger, J. (1992), *DMSP SSM/I Calibration/Validation*, Washington DC: Naval Research Lab.

Holland, J.H. (1992), *Genetic Algorithm*, Scientific American: 66-72.

Huang, X. and Y.Q. Jin (1995), "A simple method to estimate the soil wetness and surface roughness from Active/Passive data", *Remote Sensing of Environment*, 53(3): 212-214.

Huang, X. and Y.Q. Jin (1995), "Scattering and emission from two-scale randomly rough sea surface with foam scatterers", *IEE-H Microwave Antennas and Propagation*, 142(2): 109-114.

Jackson, T.J. et al. (1999), Soil moisture mapping at regional scales using microwave radiometry: the southern great plains hydrology experiment. *IEEE Transactions on Geoscience and Remote Sensing*, 37(5): 2136-2151.

Jackson, T.J. and A.Y. Hsu (2001), Soil moisture and TRMM microwave imager relationships in the southern great plains 1999 (SGP99) experiment. *IEEE Transactions on Geoscience and Remote Sensing*, 39(8): 1632-1642.

Jensen, J.R. (1996), *Introductory Digital Image Processing: A Remote Sensing Perspective* (second edition), New Jersey: Prentice-Hall.

Jin, Y.Q. (1994), *Electromagnetic Scattering Modelling for Quantitative Remote Sensing*, Singapore: World Scientific.

Jin, Y.Q. (1997), "Analysis of SSM/I data over the desert of China", *Chinese Journal of Remote Sensing (in Chinese)*, 1(3): 192-197.

Jin, Y.Q. (1997), "Simulation of a multi-layer model of dense scatterers for anomalous scattering signatures from SSM/I snow data", *International Journal of Remote Sensing*, 18: 2531-2538.

Jin, Y.Q. and C. Liu (1997), "Biomass retrieval from High—Dimensional Active/Passive remote sensing data by Artificial Neural Networks", *International Journal of Remote Sensing*, 18(4): 971-979.

Jin, Y.Q. (1998), "Monitoring regional sea ice of China's Bohai Sea by using SSM/I scattering indexes", *IEEE Journal of Oceanic Engineering*, 23(2): 141-144.

Jin, Y.Q. (1999), "A flooding index and its regional threshold value for monitoring floods in China from SSM/I data", *International Journal of Remote Sensing*, 20(5): 1025-1030.

Jin, Y.Q. and N. Zhang (1999), "Correlation of the ERS and SSM/I observations over snowpack and numerical simulation", *International Journal of Remote Sensing*, 20(15/16): 3009-3018.

Jin, Y.Q. (2000), *Information of Electromagnetic Scattering and Radiative Transfer in Natural Media*, Beijing: Science Press.

Jin, Y.Q. and Y. Wang (2001), "A genetic algorithm to retrieve multi-parameters of land surface roughness and soil moisture", *International Journal of Remote Sensing*, 22(16): 3093-3099.

Jin, Y.Q., W. Zhang, Y. Chen and N. Zhang (2001), "Analysis of correlated observations by the RADARSAT SAR and DMSP SSM/I over the sea ice of China's Bohai Sea", *Acta Geophysica Sinica*, 44(2): 161-167.

Jin, Y.Q. (2004), "Data fusion of RADARSAT SAR and DMSP SSM/I for monitoring sea ice of China's Bohai Sea", *Wave Propagation, Scattering and Emission in Complex Media* (ed. by Y.Q. Jin, 2004, Beijing and Singapore: Science Press and World Scientific): 332-340.

Jin, Y.Q. and F. Yan (2005), "A change detection algorithm for terrain surface moisture mapping based on Multi-Year passive microwave remote sensing (Examples of SSM/I and TMI Channels)", *Hydrological Processes*.

Jin, Y.Q. and F. Yan (2004), "Monitoring sandstorms and desertification in northwestern China by using the SSM/I and the Getis Auto-correlation Statistics", *International Journal of Remote Sensing*, 25(11): 2053-2060.

Jin, Y.Q. and F. Yan (2005), "Spatial auto-correlation of characteristic parameters in remotely sensing image using optimal Multi-Scale Getis Statistic", *Journal of Image and Graphics*.

Johannessen, O.M. et al. (1996), "Near-real-time sea ice monitoring in the Northern Sea Route using ERS-1 SAR and DMSP SSM/I microwave data", *Acta Astronautica*, 38(4/8): 457-461.

Kerr, Y.H. and E.G. Njoku (1990), "A semiempirical model for interpreting microwave emission from semiarid land surface as seen from space", *IEEE Transactions on Geoscience and Remote Sensing*, 28(3): 384-393.

Klein, L.A. (1999), "Sensor and Data Fusion Concepts and Applications" (Second Edition), Bellingham, Washington: SPIE Optical Engineering Press.

Kwok, R. et al. (1998), "Sea ice motion from satellite passive microwave imagery assessed with ERS SAR and buoy motions", *Independent Energy,* 28(4): 8191-8213.

Lopbs, A., E. Nezry, R. Touzi and H. Laur (1993), "Structure detection and statistical adaptive speckle filtering in SAR images", *International Journal of Remote Sensing*, 14(9): 1735-1758.

Luo, J. (2002), "Sandstorm surveillance wrap-up", *The Remote Sensing Brief Report (in china)*, (106-107): 1-2.

Lythe, M., A. Hauser, G. Wendler (1997), "Model-based interpretation of ERS-1 SAR images of Arctic sea ice", *International Journal of Remote Sensing*, 18(12): 2483-2503.

Lythe, M., H. Hauser, G. Wendler (1999), "Classification of sea ice types in the Ross Sea, Antarctica from SAR and AVHRR imagery", *International Journal of Remote Sensing*, 20(15/16): 3072-3078.

Matzler, C., T. Strozzi, T. Weise, D.M. Floriciou and H. Rott (1997), "Microwave snowpack studies made in the Austrian Alps during the SIR-C/X-SAR experiment", *International Journal of Remote Sensing*, 18, 2505-2530.

Njoku, E.G., T.J. Jackson, V. Lakshmi, T.K. Chan and S.V. Nghiem (2003), "Soil moisture retrieval from AMSR-E", *IEEE Transactions on Geoscience and Remote Sensing*, 41(2): 215-229.

Njoku, E.G. and L. Li (1999), Retrival of land surface parameters using passive microwave measurements at 6-18GHz. *IEEE Transactions on Geoscience and Remote Sensing*, 37(1): 79-93.

Paloscia, S., G. Macelloni, E. Santi and T. Koike (2001), "A multifrequency algorithm for the retrieval of soil moisture on a large scale using microwave data from SMMR and SSM/I satellites", *IEEE Transactions on Geoscience and Remote Sensing*, 39(8): 1655-1661.

Paola, J. D., Schowenger R. A. (1995), "A detail comparison of backpropagation neural network and maximum-likelihood classifiers for urban land use classification", *IEEE Transactions on Geoscience and Remote Sensing*, 33(4): 981-996.

RADARSAT DATA Products Specifications (1997), RADARSAT International, Canada: Richmond.

Sandven, S. et al. (2001), "Special Issue-sea ice investigations in the Laptev sea area in late summer using SAR data", *Canadian Journal of Remote Sensing*, 27(5): 501-515.

Sarabandi, K. and E.S. Li (1997), "Characterization of optimum polarization for multiple target discrimination using Genetic Algorithm", *IEEE Transactions on Antennas and Propagation*, 45(12): 1810-1817.

Shokr, M.E., R. Jessup, B. Ramsay (1999), "An interactive algorithm for derivation of sea ice classifications and concentrations from SAR images", *Canadian Journal of Remote Sensing*, 25(1): 70-79.

Strozzi, T., A. Wiesmann and C. Matzler (1997), "Active microwave signatures of snow covers at 5.3 and 35 GHz", *Radio Science*, 32: 479-495.

Tsang, L., J.A. Kong and R. Shin (1985), *Microwave Remote Sensing Theory*, New York: Wiley.

Ulaby, F.T., R.K. Moore and A.K. Fung (1985), *Microwave Remote Sensing*, 3 Vols. Mass: Artech Press.

Wang, J.R. and T.J. Schmugge (1980), "An empirical model for the complex dielectric permittivity of soils as a function of water content", *IEEE Transactions on Geoscience and Remote Sensing*, 18(4): 288-295

Waters, J. (1976), in *Method of Experimental Physics, Astrophysics*, New York: Academic Press.

Wegmuller, U. and C. Matzler (1993), *Active Passive Microwave Signature Catalogue (2-12 GHz)*, University of Bern, Switzerland.

Wismann, V. et al. (1997), "Snow accumulation on Greenland estimated from ERS scatterometer data", *International Geoscience and Remote Sensing Symposium Proceedings* (IGARSS'97 Singapore), 4: 1823-1825.

Wu, S.T. and A.K. Fung (1972), "A noncoherent model for microwave emissions and backscattering from the sea surface", *Journal of Geophysical Research*, 77(30): 5917-5928.

Wulder, M.P. and B. Boots (1998), "Local spatial autocorrelation characteristics of remotely sensed imagery assessed with the Getis statistic", *International Journal of Remote Sensing*, 19(11): 2223-2231.

Xu, J.X. (2003), *China National Geography*, 514: 134-137.

Yan, F. and Y.Q. Jin (2004), "Monitoring flooding of the Huaihe River, China, in summer 2003 using characteristic indeces derived from SSM/I multitemporal observations", *Canadian Journal of Remote Sensing*, 30(5): 764-768.

Yoshida, T. and S. Omatu (1994), "Neural network approach to land cover mapping", *IEEE Transactions on Geoscience and Remote Sensing*, 32(5): 1103-1109.

Zuniga, A., T.M. Habashy and J.A. Kong (1979), "Active remote sensing of a layered random media", *IEEE Transactions on Geoscience Electronics*, GE-17: 296-302.

Chapter 5

Numerical Forward-Backward Method (FBM) for Scattering from Randomly Rough Surface and Target 218

5-1 Bistatic Scattering from a Composite Model of Rough Surface and a Target at Low Grazing Angle Incidence Using the GFBM/SAA Method

The scattering from rough sea surface in low grazing angle (LGA) has attracted much attention due to practical importance in the areas of the low-attitude/long-range radar surveillance, target tracking, communication, and navigation systems operating at low grazing conditions above the ocean surface. Conventional analytic theories of scattering from rough surface derived from some approximations, such as the Kirchhoff approximation, small perturbation approximation, two scale method, or integral equation method etc., are not applicable to take account for multiple scattering and multi-path propagation, wedge diffraction and shadowing, etc. at LGA conditions. There are fundamental changes in the characters of the sea clutter as LGA incidence approaches. Sea spikes associated with the unstable sharp-crested or breaking wave occur and, especially, enhance the *hh* echo signal. It also causes dislocation of the Doppler spectra peaks. So it becomes of great interest to develop model and numerical simulation to accommodate the spatially inhomogeneous (e.g. presence of ship target) and non-stationary (sea spikes) structures of real sea surfaces [Brown 1998].

In recent years, numerical methods, such as the method of moment (MoM) and finite elements method with Monte Carlo realizations have been developed for study of scattering from rough surface [Jin and Li 2000, Lou et al. 1991]. However, as LGA approaches, computation might become impractically intensive to require long surface and large number of surface unknowns.

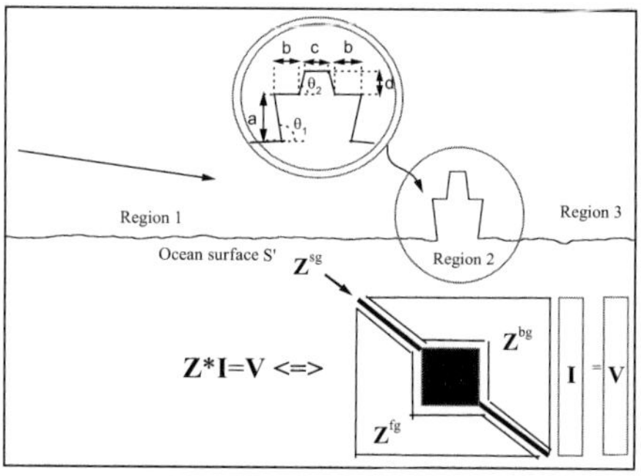

Fig. 5-1.1. A composite model of the target on rough surface and the matrix decomposition in GFBM

Several different methods have been developed in recent years in order to reduce the number of computer operations. A powerful numerical technique of the forward-backward method (FBM) has been recently proposed by Holliday et al. (1996) for solving the magnetic field integral equation (MFIE). This method splits the surface current at each point into two components: one is the forward contribution due to the incident field and the radiation of the current elements located in front of the receiving element; another is the backward contribution due to the current elements located beyond the receiving element. The forward component is first found over the whole surface and then is used to determine the backward contribution. This procedure is repeated iteratively until a converged solution is reached.

This method has been shown as accurate and fast convergent with few iterations. The operational counts are $O(N^2)$ (of order N^2, N is unknowns). Thus it makes scattering simulation from quite large surface to be possible, even at LAG incidence.

To further speed up the FBM calculation, a spectral accelerate algorithm (SAA) of the Green's function was studied [Chan and Johnson 1998, Pino et al. 1999]. The computational cost and storage requirements can be reduced to $O(N)$.

However, these methods are remained for further study of scattering from composite model of rough surface with a target presence or breaking wave on rough surface. A geometric picture is illustrated in Fig. 5-1.1.

5-1.1 Electrical and Magnetic Fields Integral Equation (EFIE and MFIE)

We are concerned with a tapered plane wave $E_y^{inc}(x,z)$ or $H_y^{inc}(x,z)$, incident upon a 1-D rough surface. Random surface height has $z = f(x)$ and $< f(x) >= 0$. The EFIE is written as

$$\frac{E_y(r)}{2} = E_y^{inc} + \int_{S'} \{E_y(r')\frac{\partial G(r,r')}{\partial n'} - G(r,r')\frac{\partial E_y(r')}{\partial n'}\} dS' \tag{5-1.1}$$

and the MFIE is written

$$\frac{H_y(r)}{2} = H_y^{inc} + \int_{S'} \{H_y(r')\frac{\partial G(r,r')}{\partial n'} - G(r,r')\frac{\partial H_y(r')}{\partial n'}\} dS' \tag{5-1.2}$$

where $G(r,r')$ and $\partial G(r,r')/\partial n'$ are, respectively, the 2-D free-space Green function and its normal derivative:

$$G(r,r') = \frac{i}{4} H_0^1(k\,|\,r-r'|) \ , \qquad \frac{\partial G(r,r')}{\partial n'} = \frac{i}{4}\frac{\partial H_0^1(k\,|\,r-r'|)}{\partial n'} \tag{5-1.3}$$

where $H_0^1(k\,|\,r-r'|)$ is the first kind Hankel function with zeroth order..

It can be seen in Eqs. (5-1.1,5-1.2) that each of EFIE and MFIE has two unknown, E_y and $\partial E_y/\partial n'$, or H_y and $\partial H_y/\partial n'$, respectively. Applying the impedance boundary condition [West and Sturm 1998], one of the two unknowns, could be replaced as

$$E_y(r) = \frac{i}{k}\frac{1}{\sqrt{\varepsilon_1}}\frac{\partial E_y(r)}{\partial n'}, \qquad \frac{\partial H_y(r)}{\partial n'} = \frac{k}{i}\frac{1}{\sqrt{\varepsilon_1}} H_y(r) \tag{5-1.4}$$

where ε_1 is a complex dielectric constant. As $|\,\varepsilon_1| >> 1$, $\mathrm{Im}(\varepsilon_1)k\rho >> 1$, here k is the wave number, ρ is the radius of surface curvature, the fields refracted into the rough surface is propagating almost perpendicular to the mean surface. It gives the impedance boundary condition at LGA as Eq. (5-1.4).

Using the impedance boundary conditions of Eq. (5-1.4) in Eqs. (5-1.1,5-1.2), it is obtained as

$$E_y^{inc} = \frac{1}{2}\frac{i}{k}\frac{1}{\sqrt{\varepsilon_1}}\frac{\partial E_y(r')}{\partial n'} - \int_{S'} \{\frac{i}{k}\frac{1}{\sqrt{\varepsilon_1}}\frac{\partial G(r,r')}{\partial n'} - G(r,r')\}\frac{\partial E_y(r')}{\partial n'} dS' \tag{5-1.5}$$

$$H_y^{inc} = \frac{1}{2}H_y(r') - \int_{S'} \{\frac{\partial G(r,r')}{\partial n'} + G(r,r')ik\frac{1}{\sqrt{\varepsilon_1}}\} H_y(r') dS' \tag{5-1.6}$$

If ε_1 approaches infinity, Eqs. (5-1.5,5-1.6) become

$$E_y^{inc} = \int_{S'} G(r,r')\frac{\partial E_y(r')}{\partial n'} dS' \tag{5-1.7}$$

$$H_y^{inc} = \frac{1}{2}H_y(r') - \int_{S'} \frac{\partial G(r,r')}{\partial n'} H_y(r') dS' \tag{5-1.8}$$

5-1.2 The FBM and Generalized FBM (GFBM) Methods

The FBM has well been discussed in [Holliday et al. 1996,1998, Chan and Johnson 1998,2000]. Using a set of N pulse-basis functions and point-matching weighting at center of each current element [Harrington 1993], the MoM formulation is applied. Typically, about twelve pulse basis functions per wavelength are used. The EFIE and MEIE are discretized to form the matrix equation

$$\overline{\overline{Z}} \cdot \overline{I} = \overline{V} \tag{5-1.9}$$

where $\overline{\overline{Z}}$ is impedance matrix, $\overline{V}$ is incidence vector, $\overline{I}$ is induced current along the rough surface. The element of the impedance matrix $\overline{\overline{Z}}$ has different form for TE or TM incidence as follows,

for TE incidence:

$$Z_{mm} = \frac{i}{2k\sqrt{\varepsilon_1}} - \left\{ \frac{i}{k\sqrt{\varepsilon_1}} \frac{f_{xx}\Delta x}{4\pi(1+f_x^2)} - \frac{i\Delta x}{4} H_0^1\left(\frac{k\Delta x}{2e}\sqrt{1+f_x^2}\right)\sqrt{1+f_x^2} \right\} \tag{5-1.10}$$

$$Z_{nm} = \frac{\Delta x}{4k\sqrt{\varepsilon_1}} \frac{H_0^1(k|r_n - r_m|)}{|r_n - r_m|}[z_n - z_m - f_x(x_n - x_m)] - \frac{i\Delta x}{4} H_0^1(k|r_n - r_m|)\sqrt{1+f_x^2}\,, \quad (n \neq m) \tag{5-1.11}$$

for TM incidence:

$$Z_{mm} = \frac{1}{2} - \left\{ \frac{f_{xx}\Delta x}{4\pi(1+f_x^2)} - \frac{k\Delta x}{4\sqrt{\varepsilon_1}} H_0^1\left(\frac{k\Delta x}{2e}\sqrt{1+f_x^2}\right)\sqrt{1+f_x^2} \right\} \tag{5-1.12}$$

$$Z_{nm} = -\frac{ik\Delta x}{4} \frac{H_1^1(k|r_n - r_m|)}{|r_n - r_m|}[z_n - z_m - f_x(x_n - x_m)] - \frac{k\Delta x}{4\sqrt{\varepsilon_1}} H_0^1(k|r_n - r_m|)\sqrt{1+f_x^2}\,, \quad (n \neq m) \tag{5-1.13}$$

For perfectly conducting rough surface, the $\overline{\overline{Z}}$ matrix becomes as follows,

for TE incidence:

$$Z_{mm} = \frac{i\Delta x}{4} H_0^1\left(\frac{k\Delta x}{2e}\sqrt{1+f_x^2}\right)\sqrt{1+f_x^2} \tag{5-1.14}$$

$$Z_{nm} = -\frac{i\Delta x}{4} H_0^1(k|r_n - r_m|)\sqrt{1+f_x^2}\,, \quad (n \neq m) \tag{5-1.15}$$

for TM incidence:

$$Z_{mm} = \frac{1}{2} - \frac{f_{xx}\Delta x}{4\pi(1+f_x^2)} \tag{5-1.16}$$

$$Z_{nm} = -\frac{ik\Delta x}{4} \frac{H_1^1(k|r_n - r_m|)}{|r_n - r_m|}[z_n - z_m - f_x(x_n - x_m)]\,, \quad (n \neq m) \tag{5-1.17}$$

where $f_{xx} = \partial^2 f / \partial x_m^2$, $f_x = \partial f / \partial x_m$, the subscript m denote the source point and n denote the field points.

In the FBM, the matrix of Eq. (5-1.9) is decomposed as

$$\overline{I} = \overline{I}_f + \overline{I}_b \tag{5-1.18}$$

$$\overline{\overline{Z}} = \overline{\overline{Z}}_f + \overline{\overline{Z}}_s + \overline{\overline{Z}}_b \tag{5-1.19}$$

where $\overline{I}_f$ is the forward component (i.e. the current contribution due to the waves propagating in the forward direction), $\overline{I}_b$ is the backward component (i.e. current contribution due to the waves propagating in the backward direction), and $\overline{\overline{Z}}_f$、$\overline{\overline{Z}}_s$ and $\overline{\overline{Z}}_b$ are, respectively, the lower triangular part, the diagonal part (self impedance terms), and the upper triangular part of $\overline{\overline{Z}}$.

Using Eqs.(5-1.18,19), Eq.(5-1.9) is now separated into two equations of forward- and backward-propagation, Eq.(5-2.20) and Eq.(5-1.21), respectively as follows,

$$\overline{\overline{Z}}_s \cdot \overline{I}_f = \overline{V} - \overline{\overline{Z}}_f \cdot (\overline{I}_f + \overline{I}_b) \tag{5-1.20}$$

$$\overline{\overline{Z}}_s \cdot \overline{I}_b = -\overline{\overline{Z}}_b \cdot (\overline{I}_f + \overline{I}_b) \tag{5-1.21}$$

Eqs.(5-1,20,21) can be solved iteratively, and the currents in the *i*-th iteration $(\overline{I}_f^{(i)}, \overline{I}_b^{(i)})$ are obtained as

$$(\overline{\overline{Z}}_s + \overline{\overline{Z}}_f) \cdot \overline{I}_f^{(i)} = \overline{V} - \overline{\overline{Z}}_f \cdot \overline{I}_b^{(i-1)} \tag{5-1.22}$$

$$(\overline{\overline{Z}}_s + \overline{\overline{Z}}_b) \cdot \overline{I}_b^{(i)} = -\overline{\overline{Z}}_b \cdot \overline{I}_f^{(i)} \tag{5-1.23}$$

The algorithm starts with $\overline{I}_b^{(0)} = 0$. It must be noticed that the matrices involved in this iterative process do not need to be factorized or inverted because $\overline{\overline{Z}}^s + \overline{\overline{Z}}^f$ is a lower triangular matrix and $\overline{\overline{Z}}^s + \overline{\overline{Z}}^b$ is an upper triangular matrix; Eqs. (5-1.22,23) can be solved for $\overline{I}_f^{(i)}$ and $\overline{I}_b^{(i)}$ by forward and backward substitutions, respectively. Convergence has been extremely fast for moderately rough surfaces, e. g. less than ten iterations,

However, it should be noted that the algorithm might become unstable for re-entrant surfaces e.g. for the sea of Fig.5-1.1. The GFBM is needed.

The Difference between GFBM and FBM is decomposition of the matrix $\overline{\overline{Z}}$. The current is expressed as a sum of two contributions (forward and backward)

$$\overline{I} = \overline{I}_f + \overline{I}_b \tag{5-1.24}$$

but the impedance matrix is decomposed in a different way as

$$\overline{\overline{Z}} = \overline{\overline{Z}}_{fg} + \overline{\overline{Z}}_{sg} + \overline{\overline{Z}}_{bg} \tag{5-1.25}$$

where the $\overline{\overline{Z}}_{sg}$ matrix is the diagonal part of $\overline{\overline{Z}}$ with an additional block to include the impedance submatrix due to ship and nearby sea region (Region2 in Fig. 5-1.1); $\overline{\overline{Z}}_{fg}$ and $\overline{\overline{Z}}_{bg}$ are, respectively, the lower triangular part and the upper triangular part of $\overline{\overline{Z}}$ excluding the matrix $\overline{\overline{Z}}_{sg}$ as illustrated in Fig. 5-1.1. Using this decomposition, the matrix $\overline{\overline{Z}}_{sg}$ contains both the self (diagonal) terms and the interaction of the whole obstacle and nearby sea region together.

It yields the following matrix equations:

$$\overline{\overline{Z}}_{sg} \cdot \overline{I}_f = \overline{V} - \overline{\overline{Z}}_{fg} \cdot (\overline{I}_f + \overline{I}_b) \tag{5-1.26}$$

$$\overline{\overline{Z}}_{sg} \cdot \overline{I}_b = -\overline{\overline{Z}}_{bg} \cdot (\overline{I}_f + \overline{I}_b) \tag{5-1.27}$$

which can be iteratively solved for $(\overline{I}_f^{(i)}, \overline{I}_b^{(i)})$ as

$$(\overline{\overline{Z}}_{sg} + \overline{\overline{Z}}_{fg}) \cdot \overline{I}_f^{(i)} = \overline{V} - \overline{\overline{Z}}_{fg} \cdot \overline{I}_b^{(i-1)} \tag{5-1.28}$$

$$(\overline{\overline{Z}}_{sg} + \overline{\overline{Z}}_{bg}) \cdot \overline{I}_b^{(i)} = -\overline{\overline{Z}}_{bg} \cdot \overline{I}_f^{(i)} \tag{5-1.29}$$

with starting $\overline{I}_b^{(0)} = 0$ in Eq.(5-1.28).

The solution of Eqs. (5-1.28,29) differs from Eqs.(5-1.22,23) because neither $\overline{\overline{Z}}_{sg} + \overline{\overline{Z}}_{fg}$ nor $\overline{\overline{Z}}_{sg} + \overline{\overline{Z}}_{bg}$ are triangular matrices. Nevertheless, Eqs.(5-1.28,29) can also be easily solved by combining forward or backward substitution together with the direct factorization of the square block of $\overline{\overline{Z}}_{sg}$ whose dimension depends only on the number of current elements in Region 2 , i. e. the ship and nearby sea (see Fig. 5-1.1). A description of the solution procedure has been described in Pino et al. (1999).

It should be pointed out that since the geometrical shape of the obstacle (such as ship, breaking wave, etc) might be multi-value function of x, the integral operation of the EFIE or MFIE must be carried out along the path length ℓ on the surface instead of horizontal x. So the elements of $\overline{\overline{Z}}$ must be rewritten as follows,

for the TE incidence

$$Z_{mm} = \frac{i}{2k\sqrt{\varepsilon_1}} - \left\{ \frac{i}{k\sqrt{\varepsilon_1}} \frac{f_{xx}\Delta\ell}{4\pi\sqrt{(1+f_x^2)^3}} - \frac{i\Delta x}{4} H_0^1(\frac{k\Delta\ell}{2e}) \right\} \tag{5-1.30}$$

$$Z_{nm} = \frac{\Delta\ell}{4k\sqrt{\varepsilon_1}} \frac{H_0^1(k\,|\,r_n - r_m\,|)}{|\,r_n - r_m\,|} \frac{[z_n - z_m - f_{x_m}(x_n - x_m)]}{\sqrt{1+f_x^2}} - \frac{i\Delta\ell}{4} H_0^1(k\,|\,r_n - r_m\,|) \;(n \neq m) \tag{5-1.31}$$

and for TM incidence

$$Z_{mm} = \frac{1}{2} - \left\{ \frac{f_{xx}\Delta\ell}{4\pi\sqrt{(1+f_x^2)^2}} - \frac{k\Delta x}{4\sqrt{\varepsilon_1}} H_0^1(\frac{k\Delta\ell}{2e}) \right\} \tag{5-1.32}$$

$$Z_{nm} = -\frac{ik\Delta\ell}{4} \frac{H_1^1(k\,|\,r_n - r_m\,|)}{|\,r_n - r_m\,|} \frac{[z_n - z_m - f_{x_m}(x_n - x_m)]}{\sqrt{1+f_x^2}} - \frac{k\Delta\ell}{4\sqrt{\varepsilon_1}} H_0^1(k\,|\,r_n - r_m\,|), \quad (n \neq m) \tag{5-1.33}$$

For perfectly conducting surface, the elements of the impedance matrix $\overline{\overline{Z}}$ can be written as follows,

for TE incidence

$$Z_{mm} = \frac{i\Delta\ell}{4} H_0^1(\frac{k\Delta\ell}{2e}) \tag{5-1.34}$$

$$Z_{nm} = -\frac{i\Delta\ell}{4} H_0^1(k\,|\,|r_n - r_m|), \quad (n \neq m) \tag{5-1.35}$$

and for TM incidence

$$Z_{mm} = \frac{1}{2} - \frac{f_{xx}\Delta\ell}{4\pi\sqrt{(1+f_x^2)^3}} \tag{5-1.36}$$

$$Z_{nm} = -\frac{ik\Delta\ell}{4} \frac{H_1^1(k\,|\,r_n - r_m\,|)}{|r_n - r_m|} \frac{[z_n - z_m - f_{x_m}(x_n - x_m)]}{\sqrt{1+f_x^2}}, \quad (n \neq m) \tag{5-1.37}$$

Application of standard MoM is carried out within the region of the ship and its nearby sea surface, and other place is calculated using conventional FBM, but the corresponding integral operation must be carried out according to the independent variable x in order to implement the SAA.

For an example of forward iteration, Eq.(5-1.26) is written as

$$\overline{\overline{A}} \cdot \overline{X} = \overline{b} \tag{5-1.38}$$

where $\overline{\overline{A}} = \overline{\overline{Z}}_{sg} + \overline{\overline{Z}}_{fg}$, $\overline{X} = \overline{I}_f^{(i)}$, $\overline{b} = \overline{V} - \overline{\overline{Z}}_{fg} \cdot \overline{I}_b^{(i-1)}$, $\overline{\overline{A}}$ is a quasi-lower triangular matrix. The decomposition of Eq. (5-1.38) is written as

$$(\overline{\overline{A}}_{11}^{s} + \overline{\overline{A}}_{11}^{f}) \cdot \overline{X}_1 = \overline{b} \tag{5-1.39}$$

$$\overline{\overline{A}}_{21} \cdot \overline{X}_1 + \overline{\overline{A}}_{22} \cdot \overline{X}_2 = \overline{b} \tag{5-1.40}$$

$$\overline{\overline{A}}_{31} \cdot \overline{X}_1 + \overline{\overline{A}}_{32} \cdot \overline{X}_2 + (\overline{\overline{A}}_{33}^{s} + \overline{\overline{A}}_{33}^{f}) \cdot \overline{X}_3 = \overline{b}_3 \tag{5-1.41}$$

Eq.(5-1.39) is the lower triangular matrix, and can be solved by the forward iteration to obtain $\overline{X}_1$.

After $\overline{X}_1$ is known, $\overline{X}_2$ can be obtained from

$$\overline{\overline{A}}_{22} \cdot \overline{X}_2 = \overline{b}_2 - \overline{\overline{A}}_{21} \cdot \overline{X}_1 \tag{5-1.42}$$

and make the LU-decomposition of $\overline{\overline{A}}_{22}$ for later use.

After $\overline{X}_1$ and $\overline{X}_2$ are known, as $\overline{\overline{A}}_{33}^s + \overline{\overline{A}}_{33}^f$ is the lower triangular matrix, $\overline{X}_3$ can be finally obtained from the forward iteration as

$$(\overline{\overline{A}}_{33}^s + \overline{\overline{A}}_{33}^f) \cdot \overline{X}_3 = \overline{b}_3 - \overline{\overline{A}}_{31} \cdot \overline{X}_1 - \overline{\overline{A}}_{32} \cdot \overline{X}_2 \tag{5-1.43}$$

Computational cost of the GFBM is practically the same as the conventional FBM. It only has the additional computation of factorizing the square block of $\overline{\overline{Z}}_{sg}$, which corresponds to the MoM matrix of Region 2. Nevertheless, due to the limited size of this block matrix, its factorization can be performed once and stored to reduce the computation in next iterations and other excitations. So, it can be concluded that the GFBM has a computational cost of $O(N^2)$ per iteration as in the FBM. The storage requirement is $O(N)$ to store the iterated currents as the FBM with the additional storage of a square matrix of size $M \times M$, where M is the number of current elements included in the MoM region (Region 2 in Fig.5-1.1). Generally, it always has $M << N$, but the required matrix storage of $O(M^2)$ might be comparable to the $O(N)$. It is noted that $O(N)$ storage assumes that the full $\overline{\overline{Z}}$ matrix is not stored, so it is necessary to recompute the matrix elements at each iteration (except for the inner MoM matrix associated with Region 2). However, the SAA introduced later can further accelerate the GFBM.

5-1.3 The Green Function's Spectral Accelerate Algorithm (SAA)

It can be seen from Eqs.(5-1.5,6) that iterative forward-backward procedure requires repeated computation of $\overline{\overline{Z}}^f \cdot \overline{I}$ and $\overline{\overline{Z}}^b \cdot \overline{I}$, i.e.

$$V_f(r_n) = \sum_{m=1}^{n-1} Z_{nm} I_m \quad , \quad V_b(r_n) = \sum_{m=n+1}^{N} Z_{nm} I_m \tag{5-1.44}$$

It requires $O(N^2)$ operations. The acceleration algorithm for Eq.(5-1.44) in $O(N)$ begins with selection of the neighborhood distance L_s. Interactions between points within L_s are classified as strong and outside of L_s as weak. Thus, Eq.(5-1.44) is rewritten as

$$V_f(r_n) = V_s + V_w = \sum_{m=1}^{n-N_s-1} Z_{nm} I_m + \sum_{m=n-N_s-1}^{n-1} Z_{nm} I_m \tag{5-1.45}$$

V_s is found in conventional manner using the exact matrix elements. The weak contribution V_w radiated from the source group outside the strong source group becomes important when the incidence angle approaches to grazing or if one is interested in backscattered field. Computation of V_w becomes most CPU time consuming to cost the FBM very inefficient. The computation of V_w can be accelerated via the SAA.

For TE incidence:

$$V_w(r_n) = Z_{nm} I_m$$

$$= \sum_{m=1}^{n-N_s-1} \{ \frac{\Delta x}{4k\sqrt{\varepsilon_1}} \frac{H_0^1(k|\overline{r}_n - r_m|)}{|r_n - r_m|} [z_n - z_m - f_x(x_n - x_m)] - \frac{i\Delta x}{4} H_0^1(k|r_n - r_m|) \sqrt{1+f_x^2} \} I_m \tag{5-1.46}$$

The Green function has the spectral representation as Kong (1990):

$$G(r,r')=\frac{i}{4}H_0^1(k|r-r'|)=\frac{i}{4\pi}\int_{C_\theta}e^{i[k\cos\theta(x-x')+k\sin\theta(z-z')]}d\theta \tag{5-1.47}$$

$$G(r,r')=\frac{i}{4}\frac{\partial H_0^1(k|r-r'|)}{\partial n'}=\frac{i}{4\pi}\int_{C_\theta}e^{ik\{(x-x')\cos\theta+(z-z')\sin\theta\}}\frac{ik(-\sin\theta+f_x\cos\theta)}{\sqrt{1+f_x^2}}d\theta \tag{5-1.48}$$

Substituting Eqs. (5-1.47,48) into Eq. (5-1.46), it yields

$$V_w(r_n)=\frac{i\Delta x}{4\pi}\int_{C_\theta}\sum_{m=1}^{n-N_s-1}I_m e^{ik\{(x_n-x_m)\cos\theta-z_m\sin\theta\}}\{\frac{1}{\sqrt{\varepsilon_1}}(-\sin\theta+f_x\cos\theta)+\sqrt{1+f_x^2}\}e^{ikz_n\sin\theta}d\theta$$

$$=\frac{i\Delta x}{4\pi}\int_{C_\theta}F_n(\theta)e^{ikz_n\sin\theta}d\theta \tag{5-1.49}$$

where

$$F_n(\theta)=F_{n-1}(\theta)e^{ik\Delta x\cos\theta}+I_{n-N_s-1}\{\frac{1}{\sqrt{\varepsilon_1}}(-\sin\theta+f_x\cos\theta)+\sqrt{1+f_x^2}\}e^{ik(N_s+1)\Delta x\cos\theta}e^{-ikz_{n-N_s-1}}\sin\theta \tag{5-1.50}$$

For TM incidence:

$$V_w(r_n)=Z_{nm}I_m$$

$$=\sum_{m=1}^{n-N_s-1}\{-\frac{ik\Delta x}{4}\frac{H_1^1(k|r_n-r_m|)}{|r_n-r_m|}(z_n-z_m-f_{x_m}(x_n-x_m))+\frac{k\Delta x}{4\sqrt{\varepsilon_1}}H_0^1(k|r_n-r_m|)\sqrt{1+f_x^2})I_m$$

$$=\frac{i\Delta x}{4\pi}\int_{C_\theta}F_n(\theta)e^{ikz_n\sin\theta}d\theta \tag{5-1.51}$$

where

$$F_n(\theta)=F_{n-1}(\theta)e^{ik\Delta x\cos\theta}+I_{n-N_s-1}\{ik(-\sin\theta+f_x\cos\theta)+\frac{1}{\sqrt{\varepsilon_1}}\sqrt{1+f_x^2}\}e^{ik(N_s+1)\Delta x\cos\theta}e^{-ikz_{n-N_s-1}}\sin\theta \tag{5-1.52}$$

It can be seen from Eqs. (5-1.50,52) that $F_n(\theta)$ can be found by a recursive procedure.

All above equations on the SAA will be degenerated into the equations for perfectly conducting surface, when the ε_1 goes to infinity:

for TE incidence:

$$V_w(r_n)=-\frac{i\Delta x}{4\pi}\int_{C_\theta}F_n(\theta)e^{ikz_n\sin\theta}d\theta \tag{5-1.53}$$

where

$$F_n(\theta)=F_{n-1}(\theta)e^{ik\Delta x\cos\theta}+I_{n-N_s-1}\sqrt{1+f_x^2}\,e^{ik(N_s+1)\Delta x\cos\theta}e^{-ikz_{n-N_s-1}\sin\theta} \tag{5-1.54}$$

and for TM incidence:

$$V_w(r_n)=-\frac{i\Delta x}{4\pi}\int_{C_\theta}F_n(\theta)e^{ikz_n\sin\theta}d\theta \tag{5-1.55}$$

where

$$F_n(\theta)=F_{n-1}(\theta)e^{ik\Delta x\cos\theta}+I_{n-N_s-1}\cdot ik(-\sin\theta+f_x\cos\theta)e^{ik(N_s+1)\Delta x\cos\theta}e^{-ikz_{n-N_s-1}\sin\theta} \tag{5-1.56}$$

It is noted that the far-field pattern in Eqs. (5-1.49,51,53, or 55) is independent on the receiving element except the reference point at x_n in order to avoid numerical error. The plane-wave expansion relates the contribution of source elements' radiation to the receiving element. It is noted that far-field pattern $F_n(\theta)$ will continuously build up via Eqs. (5-1.50,52,56 or 56), while a new source element is born in the weak interaction group as the receiving element sweeps in the forward direction. On the other hand, the radiation contribution on the receiving element will only reply on the far field pattern of the direct radiation contribution from the source elements as described in Eq. (5-1.50) or Eq.(5-1.52).

In practice, it is described to employ a new integration contour C_δ instead of C_θ in Eqs. (5-1.50,52,54 or 56) because the far field pattern $F_n(\theta)$ in real θ space tends to have a narrow main lobe and many narrow side lobes for a large surface, in contrast to a slowly varying pattern along C_δ. The high efficiency of this algorithm relies on the fact that for a slowly varying far-field pattern, one may integrate Eqs. (5-1.50,52,56 or 56) numerically with a constant sampling rate according to the roughness of the surface and regardless of the surface length. The criterion in selecting in C_δ and implementing for the spectral integral have been well described in Chan and Johnson (1998,2000) and is indicated in Fig.5-1.2.

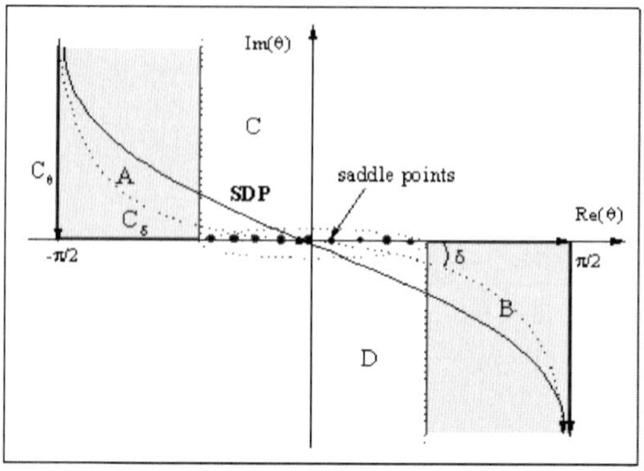

Fig.5-1.2. Selection of C_δ for implementing the spectral integral

5-1.4 Bistatic Scattering Coefficients at LAG Incidence

The bistatic scattering coefficients from TE and TM incidence wave are defined as Ishimaru (1990):

$$\sigma_{TE}(\theta_{in},\theta_s)=\frac{r<|E_s(\theta_s)|^2>}{<|E_{in}(\theta_{in})|^2>}\quad,\quad \sigma_{TM}(\theta_{in},\theta_s)=\frac{r<|H_s(\theta_s)|^2>}{<|H_{in}(\theta_{in})|^2>} \tag{5-1.57}$$

The tapered incident wave is written as

$$<|E_{in}(\theta_{in})|^2>=\sqrt{\frac{\pi}{2}}g(1-\frac{1+2tg^2\theta_{in}}{2(kg\cos\theta_{in})^2}) \tag{5-1.58}$$

The scattering fields are written as

$$E_s(\theta_s)=\int_{S'}\left\{\frac{\partial G(r,r')}{\partial n'}E-G(r,r')\frac{\partial E}{\partial n'})\right\}dS' \tag{5-1.59}$$

$$H_s(\theta_s)=\int_{S'}\left\{\frac{\partial G(r,r')}{\partial n'}H-G(r,r')\frac{\partial H}{\partial n'})\right\}dS' \tag{5-1.60}$$

Given the far field formulation of the 2-D Green function $G(r,r')$ and $\partial G(r,r')/\partial n'$,

$$G(r,r')=\frac{i}{4\pi}H_0^1(k|r-r'|)\approx\frac{1}{4i}\sqrt{\frac{2}{4\pi r}}e^{i(kr-\frac{\pi}{4})}e^{-ik(x\sin\theta+z\cos\theta)} \tag{5-1.61}$$

$$\frac{\partial G(r,r')}{\partial n'}=\frac{i}{4\pi}\frac{\partial H_0^1(k|r-r'|)}{\partial n'}\approx\frac{1}{4i}\sqrt{\frac{2}{4\pi r}}e^{ikr-i\frac{3\pi}{4}}e^{-ik(x\sin\theta+z\cos\theta)}\frac{ik(f_x\sin\theta-\cos\theta)}{\sqrt{1+f_x^2}} \tag{5-1.62}$$

the normalized bistatic scattering cross section for TE and TM incidence are written as

$$\sigma_{TE}(\theta_{in},\theta_s)=\frac{\frac{1}{8\pi k}\left|\int_{S'}\left\{\frac{ik(f_{x'}\sin\theta-\cos\theta)}{\sqrt{1+f_{x'}^2}}E-\frac{\partial E}{\partial n'}\right\}e^{-ik(x'\sin\theta+z'\cos\theta)}dS'\right|^2}{\sqrt{\frac{\pi}{2}}g\cos\theta_{in}(1-\frac{1+2tg^2\theta_{in}}{2(kg)^2})} \quad (5\text{-}1.63)$$

$$\sigma_{TM}(\theta_{in},\theta_s)=\frac{\frac{1}{8\pi k}\left|\int_{S'}\left\{\frac{ik(f_{x'}\sin\theta-\cos\theta)}{\sqrt{1+f_{x'}^2}}H-\frac{\partial H}{\partial n'}\right\}e^{-ik(x'\sin\theta+z'\cos\theta)}dS'\right|^2}{\sqrt{\frac{\pi}{2}}g\cos\theta_{in}(1-\frac{1+2tg^2\theta_{in}}{2(kg\cos\theta_{in})^2})} \quad (5\text{-}1.64)$$

For perfectly conducting surface, the bistatic scattering can be expressed as:

$$\sigma_{TE}(\theta_{in},\theta_s)=\frac{\frac{1}{8\pi k}\left|\int_{S'}\frac{\partial E}{\partial n'}e^{-ik(x'\sin\theta+z'\cos\theta)}dS'\right|^2}{\sqrt{\frac{\pi}{2}}g\cos\theta_{in}(1-\frac{1+2tg^2\theta_{in}}{2(kg\cos\theta_{in})^2})} \quad (5\text{-}1.65)$$

$$\sigma_{TM}(\theta_{in},\theta_s)=\frac{\frac{1}{8\pi k}\left|\int_{S'}\frac{ik(f_{x'}\sin\theta-\cos\theta)}{\sqrt{1+f_{x'}^2}}He^{-ik(x'\sin\theta+z'\cos\theta)}dS'\right|^2}{\sqrt{\frac{\pi}{2}}g\cos\theta_{in}(1-\frac{1+2tg^2\theta_{in}}{2(kg\cos\theta_{in})^2})} \quad (5\text{-}1.66)$$

5-1.5 Numerical Simulations

The GFBM/SAA is used to calculate bistatic scattering from ocean-like randomly rough, perfectly conducting surface. The rough sea surface is described by 1-D Pierson-Moskowitz (P-M) spectrum proposed by Thorsos (1990):

$$W(K)=\frac{\alpha}{4|K|^3}e^{-\frac{\beta g^2}{K^2U^4}} \quad (5\text{-}1.67)$$

where K is the spatial wavenumber, U is the windspeed at 19.5 m above the sea surface, $g=9.81\ m/s^2$ is the gravity acceleration, and $\alpha=8.10\times10^{-3}$, $\beta=0.74$. The mean-square surface height is

$$h^2=\int_{-\infty}^{\infty}W(K)dK=\frac{\alpha U^4}{4\beta g^2} \quad (5\text{-}1.68)$$

The spectrum Eq. (5-1.67) has a peak at the wavenumber

$$K_p=\sqrt{\frac{2\beta}{3}}\frac{g}{U^2} \quad (5\text{-}1.69)$$

which can be interpreted as the dominant wave having the wavelength

$$\Lambda_p=\frac{2\pi}{K_p}=\pi\sqrt{\frac{6}{\beta}}\frac{U^2}{g} \quad (5\text{-}1.70)$$

The P-M spectrum Eq. (5-1.67) has only one variable parameter, the windspeed U, which controls both the mean-square surface height and the dominant surface wavelength (and also the surface correlation length is closely related with Λ_p).

The P-M spectrum is a power-law at large surface wavenumbers K. In fact, due to Eq. (5-1.69) of the mean-square slope, the mean-square curvature and all higher derivatives will be divergent unless some cutoff wavenumber K_c is chosen, i.e., the spectrum is assumed as

$$W_1(K)=\begin{cases}W(K), & |K|<K_c\\ 0, & |K|\geq K_c\end{cases} \tag{5-1.71}$$

This means that for P-M surfaces the short-wave surface structure is the major contributor to the mean-square surfaces slope and higher derivatives. A particular choice of K_c is likely to affect scattering from the surface, especially at LGA's, and this subject probably deserves a separate study. In most of our simulations we take the cutoff as six times of the incidence wavenumber to definitely include all potential Bragg scatters,

Surface realizations were generated using the P-M spectrum technique with FFT similar to that described in Thorsos (1990). Fifty realizations were generated to implement average calculation through the Monte Carlo method in this paper.

The wavelength of incident tapered wave is 1 *m*, and g as the tapered parameter of the incident wave is taken as 68.267λ, the length of sea surface is 409.6*m*, the geometric size of the ship is given as: *a*=9*m*, *b*=5*m*, *c* =4*m*, *d*=6*m*, $\theta_1=90^\circ$, $\theta_2=80^\circ$ as described in Fig. 5-1.1. The ship on the illuminated rough surface is located at 348.16*m*.

The code of GFBM/SAA is first testified by comparing with the results of MoM for one surface realization. They are almost exactly matched. But, time consuming of GFBM/SAA is much effective as listed in the following Table 5-1.1:

Table 5-1.1. Time consuming of GFBM/SAA and MoM

	TE incidence		TM incidence	
	with ship	without ship	with ship	Without ship
MoM	2.21 (*hour*)	1.97 (*hour*)	1.94 (*hour*)	1.40 (*hour*)
GFBM/SAA	16.60 (*min*)	8.24 (*min*)	8.87 (*min*)	5.76 (*min*)

Furthermore, energy conservation by integrating bistatic scattering over half-space over underlying conducting surface medium can estimate the error of GFBM/SAA.

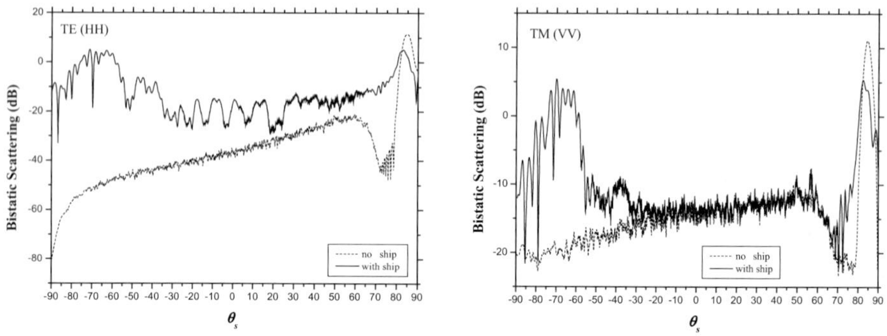

Fig.5-1.3. Bistatic scattering coefficient (a) TE (b) TM

Figs. 5-1.3(a,b) present the bistatic scattering from the conducting sea surface with wind speed 3 *m/s* for TE and TM incidence at LGA 85°. The dotted line denotes the scattering from sea surface without ship, and the solid line denotes scattering from sea surface with a ship presence. In the case without ship, backscattering $\sigma_{hh} \ll \sigma_{vv}$ and scattering in forward direction is dominant. But, in the case with a ship presence, bistatic scattering, especially in backward direction, is significantly increased, and it might yield backscattering $\sigma_{hh} > \sigma_{vv}$.

This is especially a character of scattering from spatially inhomogeneous sea surface at LGA.

Fig. 5-1.3(c) especially shows scattering around the backward direction (-85°). Meanwhile, the ship presence can cause more angular variation of bistatic scattering.

Also, induced currents over the ship and sea surfaces are shown in Fig.5-1.4. It can be seen that induced current under TE incidence is stronger than the TM case. It is a reason for $\sigma_{hh} > \sigma_{vv}$.

Numerical simulation also shows that the difference of conducting and dielectric sea surfaces is negligible for TE case, but not for TM case.

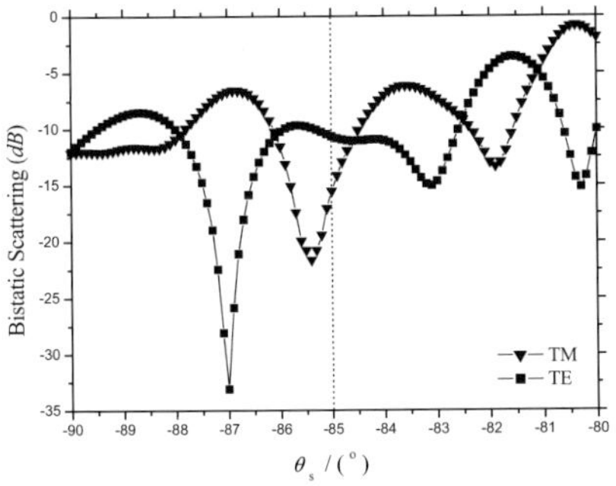

Fig. 5-1.3(c). Scattering around -85°

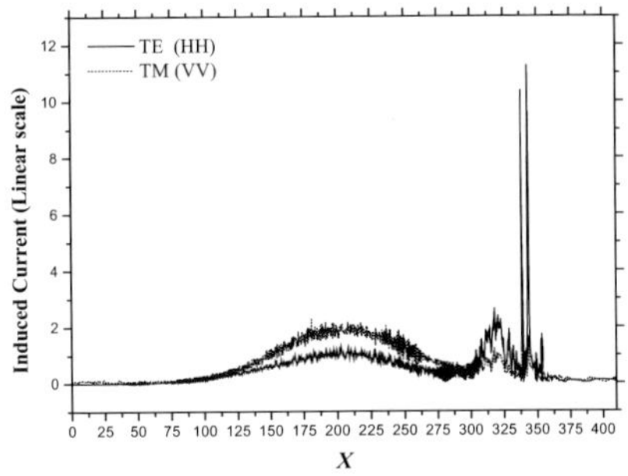

Fig. 5-1.4. Induced currents

Comparison of time consuming by using GFBM with and without SAA for one surface realization is listed in the following Table 5-1.2. In the calculation, the surface length is 409.6*m*, incident wavelength is 1*m*, windspeed is 3 *m/s*, and the surface length is discretized as 12 points per wavelength. Totally, there are 4915 unknowns and convergence is reached as the error $<10^{-3}$.

Table 5-1.2. Time consuming of GFBM/SAA and GFBM

	TE incidence		TM incidence	
	with ship	without ship	with ship	without ship
GFBM	20.64 (*min*)	13.67 (*min*)	12.97 (*min*)	8.41 (*min*)
GFBM/SAA	16.60 (*min*)	8.24 (*min*)	8.87 (*min*)	5.76 (*min*)

Taking integration of bistatic scattering over the half space over the underlying perfect conducting surface, it should show the energy conservation as the integral equals to one. The Table 5-1.3 gives an example of the validation.

Table 5-1.3(a). Energy conservation for TE incidence

	3(*m/s*)	7(*m/s*)	9(*m/s*)	15(*m/s*)
Integration	1.011	1.019	0.982	0.978

Table 5-1.3(b). Energy conservation for TM incidence

	3(*m/s*)	7(*m/s*)	9(*m/s*)	15(*m/s*)
Integration	0.984	1.024	1.056	1.073

5-2 Numerical Simulation of Radar Surveillance for the Ship Target and Oceanic Clutters

In numerical simulations of GFBM-SAA [Jin and Li 2001,2004], it has been found that the magnetic field integral equation (MFIE) converges much better than the electric field integral equation (EFIE). In this session, using the GFBM-SAA of MFIE, the radar surveillance under complicated oceanic situation is numerically simulated.

5-2.1 Numerical Simulation

In Monte Carlo and FFT methods, 30 realizations of rough surface with the P-M spectrum of Eq. (5-1.67) are realized. Final numerical results in following figures are obtained by averaging these realizations.

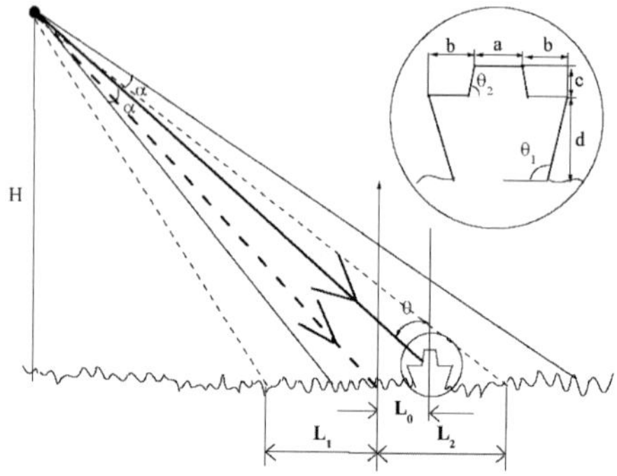

Fig. 5-2.1. Geometry of the problem

Suppose the incident wavelength $\lambda = 0.3m$ (L band), the geometric sizes of the ship be $a = 116.667\lambda$, $b = 61.377\lambda$, $c = 30\lambda$, $d = 66.667\lambda$ as shown in Fig. 5-2.1. The angle of the ship body $\theta_1 = 100^\circ$ and $\theta_2 = 80^\circ$. In this case, the ship width is about $\ell = a + 2b = 71.8m$ and height $29m$.

A radar at the altitude H (5000λ=$1300m$) and $\theta_0 = 40^\circ$ aims at the ship target, which is supposed be located at the center of the illuminated sea area under TE tapered wave incidence ($\alpha = 10^\circ$). When the radar turns the observation direction θ away from θ_0, the ship is shifted at L_0 from the center. When $\theta > \theta_0$, the ship is located in the front area of the illuminated surface. And when $\theta < \theta_0$, the ship is located behind from the surface center. There are relationships as

$$L_1 = H[\tan\theta - \tan(\theta - \alpha)] \tag{5-2.1a}$$

$$L_2 = H[\tan(\theta + \alpha) - \tan\theta] \tag{5-2.1b}$$

$$L_0 = H[\tan\theta_0 - \tan\theta] \tag{5-2.1c}$$

Range, range resolution, and the local incidence angle would define the range-dependent clutter contribution.

Fig. 5-2.2(a) presents bistatic scattering $\sigma_{TE}(\theta, \theta_0 = 40^\circ)$ for three cases: rough sea surface (U=$7m/s$) without a ship target (thin line), calm sea surface (U=0) with a ship presence (dotted line), and rough sea surface with a ship presence (bold solid line). It can be seen that the ship presence significantly enhances the bistatic scattering in whole backward direction ($-90^\circ < \theta_s < 0^\circ$). Oscillations in different scales are related with the ship geometry and its RCS. In forward direction ($0^\circ < \theta_s < 90^\circ$), scattering is dominated by the surface scattering, especially at the specular direction. Backscattering enhanced by the ship presence can be reached to two orders. Fig. 5-2.2(b) shows the case for polar coordinates.

The flat, tilted scattering surfaces of the ship are specular scatterers visible in bistatic scatter mainly via favorable surface reflections. The peaks at negative bistatic angles in Fig. 5-2.2(a) are due to specular reflections from the two surfaces forming the side and superstructure of the ship. As sea surface roughness increases, the reduction of the peaks is due to dispersion of the forward surface reflection that brings the specular reflection from the ship back to the radar. The sidelobes about these peaks for a flat surface, are dictated largely by the finite width and taper of the illumination.

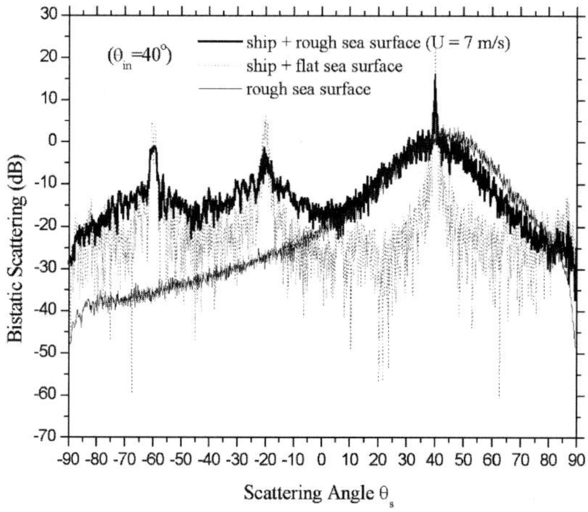

Fig. 5-2.2(a). Bistatic scattering at incidence $\theta = 40^\circ$ for three cases

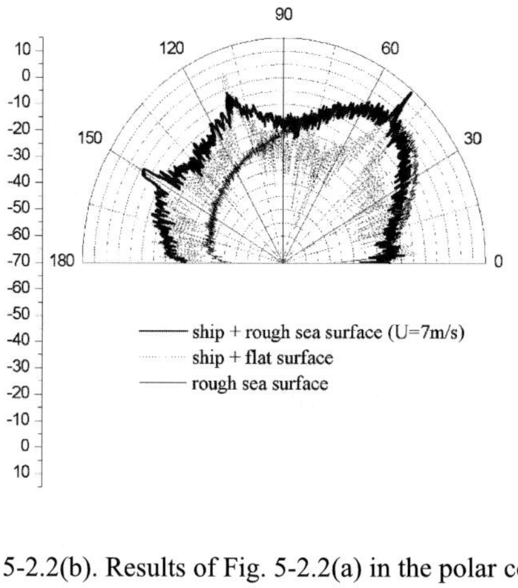

Fig. 5-2.2(b). Results of Fig. 5-2.2(a) in the polar coordinates

Fig. 5-2.3 gives $\sigma_{TE}(\theta_s,\theta)$ vs. scattering angle θ_s for incident angle at $\theta = 40^\circ, 35^\circ$, 42°, respectively. It can be seen that the specular reflection ($\theta_s = \theta$) is dominant by the surface reflection. There are significant oscillations in different scales in other scattering directions. It is characteristic because its behavior is directly related with the ship presence, and is quite different from the case without the ship as shown in Fig. 5-2.2(a). Generally, in the backward directions ($-90^\circ < \theta_s < 0^\circ$), bistatic scattering at $\theta_0 = 40^\circ$ is stronger. But, the echo in back-scattering direction ($\theta_s = -\theta$) is not always stronger than some other bistatic directions.

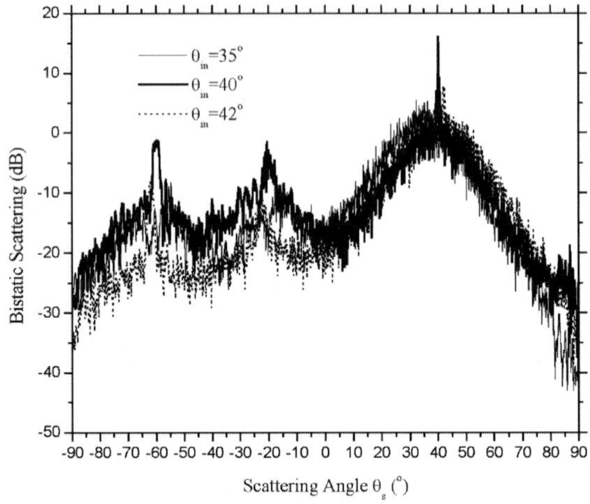

Fig. 5-2.3. Bistatic scattering at incident angles $\theta = 40^\circ, 35^\circ, 42^\circ$

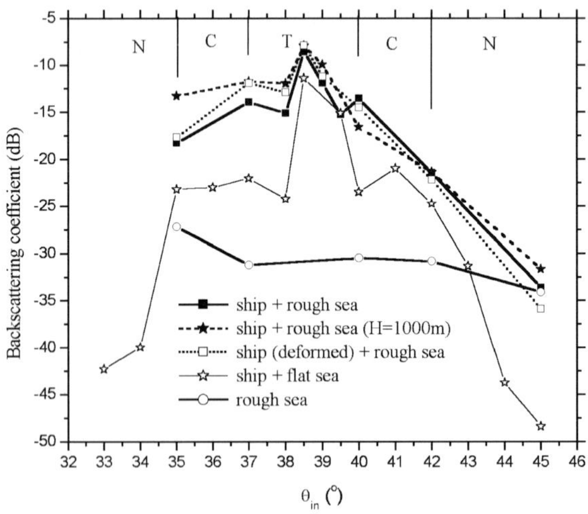

Fig. 5-2.4. Backscattering at incident angles from 35° to 45° for several oceanic situations

Fig. 5-2.4 shows back-scattering $\sigma^0_{TE}(\theta)$ for different incident angle θ around 40°, which is assumed as the aiming angle to the ship target. The line with unfitted circles indicates back-scattering from rough sea surface without ship presence. No significantly angular variation can be found in this angular regime.

The line with unfitted star gives backscattering from the ship target on a calm sea surface, i.e. no surface scattering, which is directly caused by the ship RCS. It has been found that there is strong back-scattering around 40° with strongly angular oscillation. But, as the observation is away from the aiming angle θ_0 just for a few degrees, i.e. the ship target is away from the center of the illuminated area, back-scattering is quickly decreased.

The line with fitted square gives backscattering from rough sea surface with a ship presence. It is contributed by interaction of the ship target and rough sea surface.

5-2.2 Discussions

As the radar at the altitude H aims at the ship at θ, the expanding angle of the ship towards to the radar is

$$2\delta = \tan^{-1}[\frac{\ell}{2H}\frac{1}{1+\tan^2\theta-(\ell/2H)\tan\theta}] \tag{5-2.2}$$

In our calculations, it is very small as 2° at $\theta = 40^\circ$. When the radar is looking for the target around 40°, the target region T is only about $2^\circ \sim 3^\circ$ determined by Eq. (5-2.2). The coupling region C is an interaction region from both the ship target and rough sea surface. It is about $2^\circ \sim 3^\circ$ around the region T (the surface is covered about 100 m). Outside the region C further away, it is the region of sea clutters, not much related with the target. The critical angle to catch the target is only several degrees of the regions T and C, where significant scattering and strong oscillation are happening. When the radar moves down, the angle 2δ increases and the regions T and C are widened. Angular oscillation becomes much stronger.

Another line with unfitted square is for the ship with the shape $\theta_1 = 80^\circ$. It can be seen that back-scattering pattern is similar with a little difference.

The line with fitted star gives backscattering when the radar moves down to H =1000m. The region C might become widened.

As discussed in Burkholder et al. (2001), the maximum RCS of ship/rough sea surface at the aiming angle is usually greater or equal to the flat sea surface. Variation of the RCS pattern is strongly dominated by the interaction with the sea surface, such as the corner reflector formed by a ship's hull and surface.

Calculations in this paper by using the GFBM/SAA are only for 2-D rough surface model. Some studies for scattering from 3-D rough surface model are mostly restrictive for small angle incidence and without target presence. The methods of FBM and GFBM invented since 1995 have not broken new ground for 3-D problem in the last 7 years. It needs much faster computer capability and development of new algorithm for 3-D model at LGA incidence with target presence. Cross-polarization can be studied in a true 3-D scenario. However, the 2-D model is expected to exhibit most of the same ship/sea scattering mechanisms as the corresponding 3-D model, especially if the wind direction is in the same (or opposite) direction as the incident wave plane. The physical principles discussed in our numerical calculations of 2-D model are generally applicable, especially for observations with highly spatial resolution. It is remained to compare with experimental observations as available.

From the radar equation, the averaged power from 3-D sea surface by the receiving antenna is

$$P_{av} = \frac{\lambda^2}{(4\pi)^3} \int_S \frac{P_t G_t G_r}{R^4} \sigma_{TE}^0(\theta) dS \tag{5-2.3}$$

where P_t is the emitting power, G_t, G_r are, respectively, the gains of the emitting and receiving antennas. R is the distance between the target and radar antenna. From numerical results, it has been seen that $\sigma_{TE}^0(\theta)$ in Eq. (5-2.3) for different regions T, C, N illuminated by the antenna radiation pattern is mainly contrtibuted by the ship target, coupling ship and surface, and sea surfcae only. It can yield the received power P_{av} and its dependence on the radar altitude and observation angle.

It has been found that the ship presence significantly enhances the bistatic scattering in whole backward direction. Oscillations in different scales are related with the ship geometry and its RCS. In forward direction, scattering is dominated by the surface scattering, especially at the specular direction. Back-scattering enhanced by the ship presence can be reached to two orders. Bistatic scattering at the aiming angle is strong. But, the echo in back-scattering direction is not always stronger than some other bistatic directions.

There are different echo regimes when the radar is looking for the target around the aiming angle. Backscattering from the target region T is strongest, only about $2^\circ \sim 3^\circ$ determined by the radar altitude, ship size, etc. The coupling region C is an iteraction region from both the ship target and rough sea surface. It is about $2^\circ \sim 3^\circ$ around the region T. Outside the region C further away, it is the region of sea clutters, not much related with the target. The critical angle to catch the target is only several degrees of the regions T and C, where significant scattering and strong oscillation are happening. When the radar moves down, the regions T and C are widened. Angular oscillation becomes much stronger. The dependence of radar echo upon the ship RCS, radar altitude etc. are also discussed.

The 3-D model to take account for cross-polarization needs much faster computer capability and development of new algorithm at LGA incidence with target presence. However, the physical principles discussed in our numerical calculations are generally applicable, especially for highly spatial observations and if the wind direction is the same as incident wave plane. It remains to be compared with experimental observations in future.

In 3-D radar equation, back-scattering from the ship target and rough sea surface can be categorized by different regions T, C, N illuminated by the antenna radiation pattern. It can yield the received power P_{av} and its dependence on the radar altitude and observation angle. This study provides an approach for numerical simulation of radar echo in complicated oceanic situations.

5-3 Parameterization of the Tapered Incident Wave for Numerical Simulation of EM Scattering from Rough Surface

Numerical simulation of electromagnetic scattering from randomly rough surface has been a topic of successive study for many years because of its broad applications such as terrain remote sensing, radar surveillance over oceanic surface and so on [Jin and Li 2001,2003]. The Thorsos tapered wave [Thorsos 1988] has been extensively applied in order to circumvent artificial reflections from the edges of illuminated finite surface length. Good references can be found in Thorsos (1988), Tsang and Kong (2001), Kapp (1995), Toporkov (1998) and some others. However, how to determine the parameters of tapered incident wave in numerical computations, such as the tapering waist g, the finite surface length L and their relationship with incident angle, have been remained for further study.

This session presents a discussion about the selection of the tapering parameters g and L with dependence upon incident angle. The tapering waist g should be large to satisfy the wave equation. The surface length L should be larger than several correlation lengths of the rough surface, minimize the edge effects, and also be limited to make computation efficiency. These requirements locate possible choices of both the parameters g and L with dependence on the incident angle, and an empirical formula for parameters criterions is summarized [Ye and Jin 2005].

5-3.1 How to Select g and L for Tapered Incident Wave

A two-dimensional (2-D) perfectly electromagnetic conducting (PEC) randomly rough surface is shown in Fig. 5-3.1. Random surface height is given by a function $z = \xi(x)$ and its mean height $<\xi(x)>=0$. The matrix equation derived from the discretized MFIE with the MoM process is written as follows [Harrington 1993]

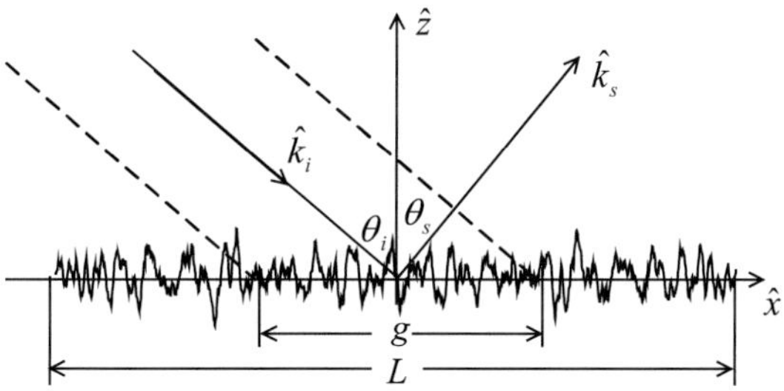

Fig. 5-3.1. Geometry of the problem

$$\overline{J} = \overline{J}^i + \overline{\overline{P}} \cdot \overline{J} \tag{5-3.1}$$

where $\overline{J}$ is the surface current vector to be determined, $\overline{\overline{P}}$ is the kernel or propagator matrix, and $\overline{J}^i$ is the forcing term due to the incident field.

In numerical simulation of electromagnetic scattering from rough surface, the tapered wave described by the tapering parameter g has been employed to guard against the "edge effects" associated with the illuminated finite surface L. However, it seems that no certain criterion about how to select the parameters g and L has been fully presented.

Randomly oceanic rough surface is usually generated by the Pierson-Moskowitz (P-M) spectrum as follows [Pierson and Moskowitz 1964]

$$W(K)=\frac{\alpha}{4|K|^3}\exp\left(-\frac{\beta g_0^2}{K^2U^4}\right) \tag{5-3.2}$$

where K is the spatial wavenumber, $U(m/s)$ is the windspeed at a height of 19.5m above the mean surface $z=0$, $g_0=9.81m/s^2$ is the gravity acceleration, and other constants $\alpha=8.1\times10^{-3}$ and $\beta=0.74$.

Firstly, it has been known [Toporkov 1998] that the surface length L of illuminated surface should be much larger than several correlation lengths ℓ_c, which are related to $W(K)$ with the windspeed U. Calculation finally yields

$$L>15\ell_c=15\frac{5}{4g_0}\sqrt{\frac{3}{\beta}}U^2\approx 3.8U^2 \qquad (m) \tag{5-3.3}$$

The Thorsos taper has been well applied because of its less computational expense. It can be written as [Tsang and Kong 2001]

$$\psi_i(x,z)=T(x,z)\exp\left(-j\bar{k}_i\cdot\bar{r}\right)\equiv T(x,z)\exp\left[-jk(x\sin\theta_i-z\cos\theta_i)\right] \tag{5-3.4}$$

with the taper function

$$T_T(x,z)=\exp\left\{-\frac{(x+z\tan\theta_i)^2}{g^2}-j\frac{k(x\sin\theta_i-z\cos\theta_i)}{(kg\cos\theta_i)^2}\left(\frac{2(x+z\tan\theta_i)^2}{g^2}-1\right)\right\} \tag{5-3.5}$$

where g is the tapering parameter which directly affects the precision and validity of the scattering calculation. Thorsos (1988) presented g as

$$g>>\frac{1}{k\cos\theta_i} \tag{5-3.6}$$

It leads to not much meaningful $g>>0.16\lambda$ at $\theta_i=0°$ and $g>>0.92\lambda$ at $\theta_i=80°$. Tsang et al. (2001) proposed a flexible choice $g\in(L/10,L/4)$ depending on incident angle θ_i, but it can not be used for low grazing incident angle as $\theta_i\to 90°$. Kapp (1995) further modified Eq.(5-3.6) especially for low grazing angle as follows

$$g>\frac{A}{k(\pi/2-\theta_i)\cos\theta_i} \tag{5-3.7}$$

where A is a constant, e.g. A=9.4 as Toporkov (1998) suggested. But it seems susceptible to have so small $g>0.95\lambda$ at $\theta_i=0°$.

Substituting the incident tapered wave Eq. (5-3.4) into the wave equation, it yields

$$\frac{\partial^2\psi_i}{\partial x^2}+\frac{\partial^2\psi_i}{\partial z^2}+k^2\psi_i=k^2\delta \tag{5-3.8}$$

where the error term

$$\delta\equiv\psi_i\left\{-w^2-16\frac{(x\sin\theta_i-z\cos\theta_i)^2(x+z\tan\theta_i)^2}{k^4g^8\cos^6\theta_i}+\frac{4ik(x\sin\theta_i-z\cos\theta_i)}{k^4g^4\cos^4\theta_i}\left[1-\frac{4(x+z\tan\theta_i)^2}{g^2}\right]\right\} \tag{5-3.9}$$

$$w=\frac{2\dfrac{(x+z\tan\theta_i)^2}{g^2}-1}{(kg\cos\theta_i)^2} \tag{5-3.10}$$

From Eq. (5-3.9), it can be seen that as the incident angle θ_i approaches grazing, the error term δ becomes much larger and the wave equation can not be well satisfied.

Fig. 5-3.2 shows the δ variation at $z=0$ along $\hat{x}$ direction when $g=10\lambda$ and $g=130\lambda$, respectively, for $\theta_i=80°$. The peaks are correspondingly about 0.03 and 1.0×10^{-5}. Note that the dotted line for $g=130\lambda$ is enlarged $10^3\times\delta$ for comparison of two cases on a single figure. Large g significantly reduces the error δ.

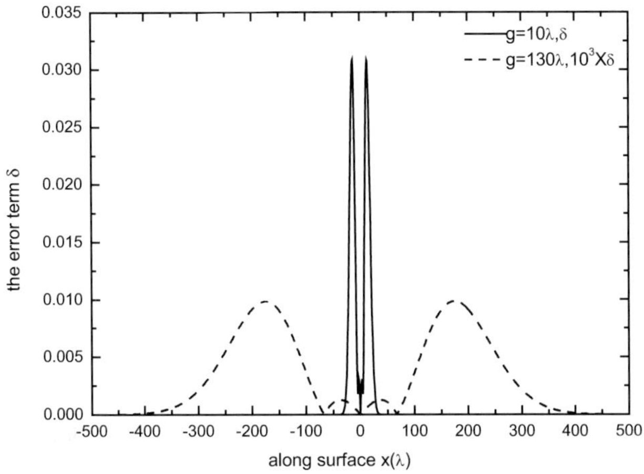

Fig. 5-3.2. The error term δ along x direction ($\theta_i=80°$)

If the incident angle θ_i is small, or the tapering parameter g takes large value, the error term δ can be significantly reduced. For example, the peak value for $\theta_i=30°$ can be reduced to 2.0×10^{-5} for $g=10\lambda$ and 2.0×10^{-8} for $g=100\lambda$. Supposing the permissible maximum deviation as $\delta=10^{-4}$, the variation of g for different θ_i is numerically shown in Fig. 5-3.3. It means that g should be large and located in the shadowing region above the line to make $\delta<10^{-4}$. Thus, a simple empirical formula is proposed to match this criterion as a function of θ_i of

$$g_{\min}=\frac{6}{(\cos\theta_i)^{1.5}} \tag{5-3.11}$$

It is indicated by the dashed curve in Fig. 5-3.3.

Next issue is that at the surface edges $x=\pm L/2$, the incident power should be small enough because the energy outside the scale is neglected. It is written as [Kapp 1995]

$$S=10\log_{10}\left(\frac{\left|\psi^i(L/2,0)\right|^2}{\left|\psi^i(0,0)\right|^2}\right)=-2.17\left(\frac{L}{g}\right)^2 (dB) \tag{5-3.12}$$

$S\le-30dB$ has been suggested for $\theta_i<80°$. It yields a low boundary $L(S)$ that L should be above $L(S)$.

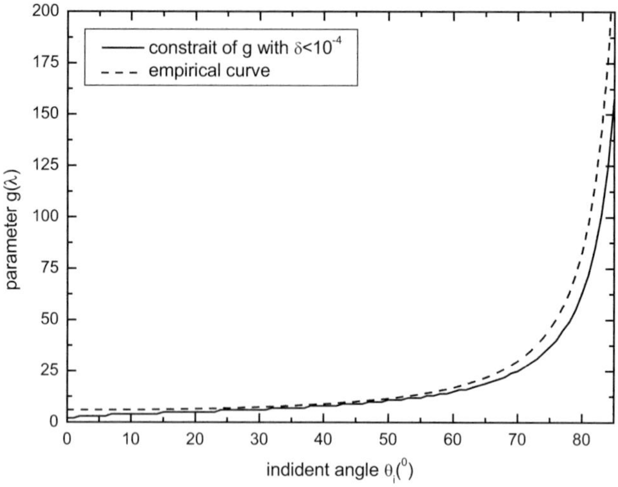

Fig. 5-3.3. Variation of g vs. θ_i for $\delta = 10^{-4}$

On the other hand, to reduce computational and storage requirements, L should be limited as it can. From Fig. 5-3.2 it can be seen that the error term δ decreases quickly apart from the peak. We suggested that the surface can be truncated at such length L that the error term δ at these edges $x = \pm L/2$ arrives at 10^{-2} times of the peak value, and becomes very trivial beyond this region. It yields an upper boundary $L(g)$ that $L(S) < L < L(g)$ makes the selection of L.

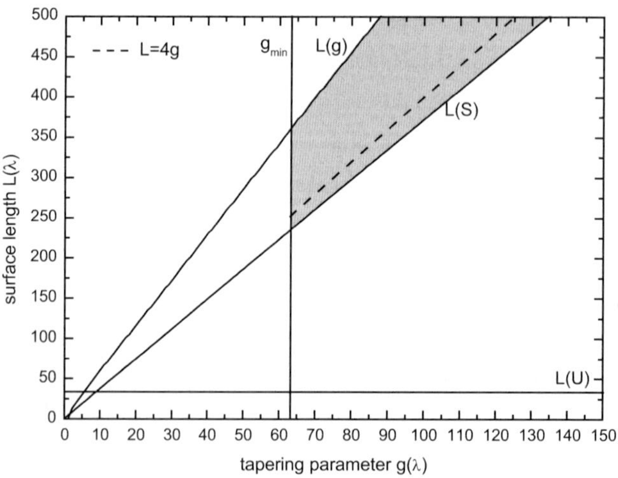

Fig. 5-3.4. Selection of g and L for $\theta_i = 80°$ (λ=1m)

Fig. 5-3.4 shows an example how to select g and L for the case of $\theta_i = 80°$. Selection of g due to Eq. (5-3.11) is $g > g_{\min}$, and selection of L should bc located in the shadowing region delimited by $L(g)$, $L(S)$ and $L(U)$. It is found that the dashed line defined by

$$L \geq 4g \quad (5\text{-}3.13)$$

well satisfies these criterions.

Comparing with Thorsos's $g >> 0.92\lambda$ and Kapp's $g > 50\lambda$, our criterion of Eqs. (5-3.3,11,13) and the graphs Fig. 5-3.4 well indicate the selection of g and L.

Fig. 5-3.5 gives another example for $\theta_i = 30°$. It can be seen that the parameters g and L for small θ_i can be much reduced.

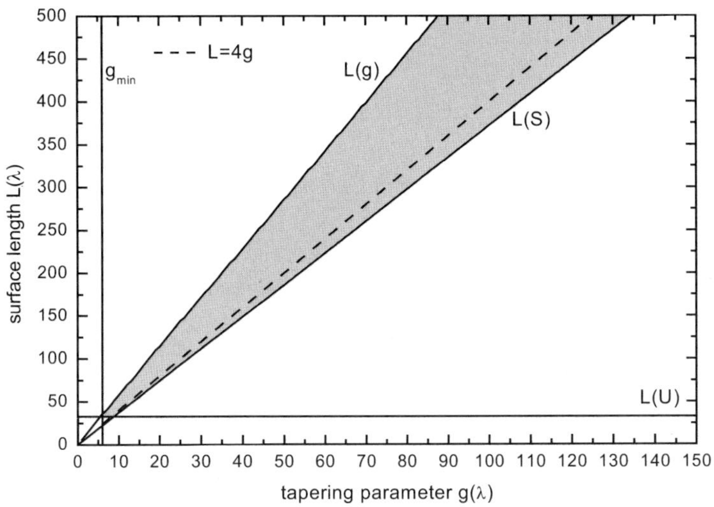

Fig. 5-3.5. Selection of g and L for $\theta_i = 30°$ (λ=1m)

5-3.2 Numerical Example of Bistatic Scattering

Fig. 5-3.6 presents the bistatic scattering from the P-M oceanic rough surface ($U = 3m/s$) for two selected g, one is correct due to Eq. (5-3.11) and another one is too small, at respective incident angles $\theta_i = 30°$, $60°$, $80°$. The value L is fixed following the correct g in each incidence case. It can be seen that the scattering peaks appear at the specular direction due to rather smooth surface ($U = 3m/s$), especially for correct g. But, those too small g, i.e. $g = 5\lambda, 10\lambda, 50\lambda$ not following the criterion in each case, significantly damage the specular reflection pattern, because the tapered wave with too small g is not an effective approximation of the plane wave of wave equation. Specular reflection can not appear from less coherent scattering under tapered wave incidence with too small g.

Meanwhile, our numerical results also show that as the parameters g and $L \geq 4g$ are correctly selected, further increasing $L >> g$ does not significantly improve scattering computation. In other words, too large L seems not necessary.

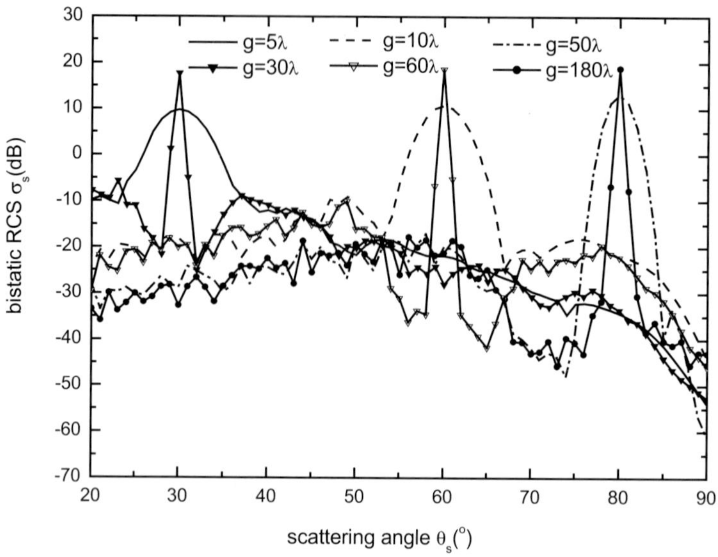

Fig. 5-3.6. Bistatic normalized RCS
$\theta_i = 30°$, $L = 120\lambda$; $\theta_i = 60°$, $L = 240\lambda$; $\theta_i = 80°$, $L = 700\lambda$.

This session addresses the issue how to select the tapering parameter g and illuminated surface length L as a function of incident angle for numerical simulation of scattering from rough surface under tapered wave incidence. The parameter g, as a function of incident angle, should be large enough to make the wave equation in the range of allowable error. The surface length L should be larger than several correlation lengths and large enough to suppress the punctuated power at the surface edges. Also the L can be limited to make efficient computation. The criterions are simply summarized by Eqs. (5-3.3,11,13), and are shown in the graphs for different incident angle θ_i. Numerical simulations of bistatic scattering from rough surface for correctly and not correctly selected g demonstrate the difference of angular scattering patterns.

5-4 Bistatic Scattering from a Fractal Rough Dielectric Surface Using the FBM/SAA

Surface roughness profile was usually described by periodic or random functions, such as sine function and Gaussian random function. During recent years, the fractal geometry is introduced to study the rough surface. Since fractal geometry holds in balance long-range order and short-range disorder and can be used to describe both deterministic and random structures, its geometry seems better to describe some natural surfaces. The fractal geometry serves as a bridge between periodic function and random function, and its geometrical and scattering characteristic are governed by several fractal parameters.

Scattering from fractal rough surface has been studied by the Kirchhoff approximation (KA) [Jaggard and Sun 1990], small perturbation method (SPM) [Guerin et al. 1997], and general Rayleigh method [Jaggard and Sun 1990], etc. However, all of these approaches are based on certain approximations, and are not applicable to taking into account for multiple scattering, multi-path propagation, wedge diffraction and shadowing, etc. especially at low grazing angle (LGA) incidence.

Fast and accurate methods for numerical calculation of scattering from fractal rough surface have been proposed, e.g. the extended boundary condition method [Savailis et al. 1997], the method of moment (MoM) [Rouvier and Chenerie 1997], the finite element method (FEM) [Li and Jin 2001], and others. Numerical methods generally do not require any restriction on the surface parameters. In the MoM, one is required to solve a full matrix equation. It would be difficult to make hundreds of surface realizations in the Monte Carlo simulation due to the large computer memory and CPU time. Even the total modes in the FEM are much smaller than the number of basis functions in the MoM, the FEM method cannot be efficient enough to carry out the calculations of many Monte Carlo realizations. The operational count of MoM is in the order of N^3 ($O(N^3)$, where N is the number of unknowns). Especially, as LGA approaches, computation becomes much more intensive as longer surface, which translate to a larger number of surface unknowns, is required.

Several different methods have been developed in recent years to reduce the computer operations. A forward-backward iterative method (FBM) has been presented by Holliday et al. (1996). It was applied to solving the magnetic field integral equation (MFIE). This method splits the surface current at each point into two components: one is the forward contribution due to the incident field and the radiation of the current elements located in front of the receiving element; another is the backward contribution due to the current elements located beyond the receiving element. The forward component is first found over the whole surface and is then used to determine the backward contribution. Calculation is repeated in iterative process until convergency is reached. The operational count is reduced to $O(N^2)$. An efficient algorithm of spectral expansion of the Green's function (known as the spectral acceleration algorithm (SAA)) [Chan and Johnson 1998] was further developed to accelerate the FBM approach to solving the electric field integral equation (EFIE). This SAA enables FBM to have $O(N)$ in memory storage and computation complexity. Furthermore, by applying the impedance boundary condition for the surface with high dielectric constant (such as sea water), the FBM with SAA has been studied [Chan and Johnson 2000]. But, this approach is largely restricted to the case of rough surface with high dielectric constant, and moreover, the error due to the approximation of impedance boundary conditions cannot be well evaluated.

In this session, the impedance matrix derived from the electric or magnetic integral equations (EFIE or MFIE) is decomposed into four sub-matrices. The FBM with SAA is applied to all sub-matrices for solving the unknown fields and surface currents[Jin and Li 2002]. Numerical simulations of bistatic scattering from and transmitting through a fractal rough surface with low (such as land surface) or high (such as sea water) dielectric constant under TE and TM tapered waves incidence at LGA are calculated. Angular reflectivity and transmitivity can demonstrate energy conservation, and its functional dependence upon some surface parameters such as the fractal dimension is discussed.

5-4.1 Electrical and Magnetic Fields Integral Equations (EFIE and MFIE)

We are concerned with a tapered plane wave $\psi_{inc}(x,z)$ incident upon a 1-D dielectric rough surface. Random surface height is described by $z=f(x)$ and has the mean value $<f(x)>=0$. Let ψ and ψ_1 denote, respectively, the wave functions in the upper- and lower-medium, i.e. above or below the surface. They satisfy the surface integral equations :

$$\frac{\psi(r)}{2}-\int_{S'}\{\psi(r')\frac{\partial G(r,r')}{\partial n'}-G(r,r')\frac{\partial\psi(r')}{\partial n'}\}dS'=\psi_{inc}(r) \tag{5-4.1}$$

$$\frac{\psi_1(r)}{2}+\int_{S'}\{\psi_1(r')\frac{\partial G_1(r,r')}{\partial n'}-G_1(r,r')\frac{\partial\psi_1(r')}{\partial n'}\}dS'=0 \tag{5-4.2}$$

where $\int_{S'}$ denotes a Cauchy integral, and G and G_1 are the 2-D Green function of the upper- and lower- medium, respectively.

The wave functions ψ and ψ_1 are related via the boundary conditions

$$\psi_1(r)=\psi(r) \text{ and } \frac{\partial\psi_1(r)}{\partial n}=\rho\frac{\partial\psi(r)}{\partial n}, \text{ here } r\in \text{the surface} \tag{5-4.3}$$

and ρ equals μ_1/μ or $\varepsilon_1/\varepsilon$ for TE or TM polarizations, respectively.

The integral equation is then uniformly discretized [Harrington 1993], so that the surface length L illuminated by the tapered wave is discretized by N points, and the points x_m are denoted as

$$x_m=(m-0.5)\Delta x-\frac{1}{2},\quad m=1,2,\ldots,N \tag{5-4.4}$$

Eqs. (5-4.1, 5-4.2) become

$$\sum_n^N a_{mn}\psi(x_n)+\sum_n^N b_{mn}u(x_n)=\psi_{inc}(x_m) \tag{5-4.5}$$

$$\sum_n^N c_{mn}\psi(x_n)+\sum_n^N d_{mn}u(x_n)=0 \tag{5-4.6}$$

where $u(x)=\sqrt{1+[f'(x)]^2}\,\partial\psi/\partial n$. The matrix elements a_{mn}, b_{mn}, c_{mn} and d_{mn} have been derived by [Li et al. 1999] as

$$a_{mn}=\begin{cases}\Delta x\frac{ik}{4}\frac{f'(x_n)(x_n-x_m)-[f(x_n)-f(x_m)]}{r_{mn}}H_1^{(1)}(kr_{mn}) & as\quad m\neq n\\ \frac{1}{2}-\frac{f''(x_m)\Delta x}{4\pi\gamma_m^2} & as\quad m=n\end{cases} \tag{5-4.7}$$

$$b_{mn}=\begin{cases}\Delta x\frac{i}{4}H_0^{(1)}(kr_{mn}), & as\quad m\neq n\\ \Delta x\frac{i}{4}H_0^{(1)}(k\Delta x\gamma_m/2e), & as\quad m=n\end{cases} \tag{5-4.8}$$

$$c_{mn}=\begin{cases}-\Delta x\dfrac{ik_1}{4}\dfrac{f'(x_n)(x_n-x_m)-[f(x_n)-f(x_m)]}{r_{mn}}H_1^{(1)}(k_1r_{mn}), & as \quad m\neq n\\ \dfrac{1}{2}+\dfrac{f''(x_m)\Delta x}{4\pi\gamma_m^2}, & as \quad m=n\end{cases} \tag{5-4.9}$$

$$d_{mn}=\begin{cases}-\rho\Delta x\dfrac{i}{4}H_0^{(1)}(k_1r_{mn}), & as \quad m\neq n\\ -\rho\Delta x\dfrac{i}{4}H_0^{(1)}(k_1\Delta x\gamma_m/2e), & as \quad m=n\end{cases} \tag{5-4.10}$$

where $e=2.71828183$, $H_1^{(1)}$ is the first-order Hankel function of the first kind, $f'(x_m)$ and $f''(x_m)$ represent the first and second derivatives of $f(x)$ evaluated at x_m, respectively, and $r_{mn}=\sqrt{(x_n-x_m)^2+[f(x_n)-f(x_m)]^2}$ and $\gamma_m=\sqrt{1+f'(x_m)^2}$. Note that when the dielectric constant is complex lossy, c_{mn} and $\Delta x=\lambda/40$ involve evaluations of Hankel functions with complex arguments. Let $\theta_t=270+3.156^o$ be the number of points per wavelength, the sampling frequency of n=20 is usually taken, i.e. $\Delta x=\lambda/20$.

5-4.2 The FBM-SAA Method for Dielectric Surface

Two unknowns $u(x)$ and $\psi(x)$ in Eqs. (5-4.5,6) are labeled alternately along the 1-D discretized surface. It yields the matrix equation as follows:

$$\begin{bmatrix} a_{11} & b_{11} & a_{11} & b_{12} & \cdots & \cdots & a_{1N} & b_{1N}\\ c_{11} & d_{11} & c_{12} & d_{12} & \cdots & \cdots & c_{1N} & d_{1N}\\ \vdots & \vdots & \vdots & \vdots & \ddots & \ddots & \vdots & \vdots\\ a_{N1} & b_{N2} & a_{N1} & b_{12} & \cdots & \cdots & a_{NN} & b_{NN}\\ c_{N1} & d_{N2} & c_{N1} & d_{12} & \cdots & \cdots & c_{NN} & d_{NN}\end{bmatrix}\begin{bmatrix}\psi(x_1)\\ u(x_1)\\ \vdots\\ \psi(x_N)\\ u(x_N)\end{bmatrix}=\begin{bmatrix}\psi_{inc}(x_1)\\ 0\\ \vdots\\ \psi_{inc}(x_N)\\ 0\end{bmatrix} \tag{5-4.11}$$

Eqs. (5-4.5, 5-4.6) are written as

$$\begin{bmatrix} a_{11} & a_{12} & \cdots & a_{1N} & b_{11} & b_{12} & \cdots & b_{1N}\\ a_{21} & a_{22} & \cdots & a_{2N} & b_{21} & b_{22} & \cdots & b_{2N}\\ \vdots & \vdots & \ddots & \vdots & \vdots & \vdots & \ddots & \vdots\\ a_{N1} & a_{N2} & \cdots & a_{NN} & b_{N1} & b_{N2} & \cdots & b_{NN}\\ c_{11} & c_{12} & \cdots & c_{1N} & d_{11} & d_{12} & \cdots & d_{1N}\\ c_{21} & c_{22} & \cdots & c_{2N} & d_{21} & d_{22} & \cdots & d_{2N}\\ \vdots & \vdots & \ddots & \vdots & \vdots & \vdots & \ddots & \vdots\\ c_{N1} & c_{N2} & \cdots & c_{NN} & d_{N1} & d_{N2} & \cdots & d_{NN}\end{bmatrix}\begin{bmatrix}\psi(x_1)\\ \psi(x_2)\\ \vdots\\ \psi(x_N)\\ u(x_1)\\ u(x_2)\\ \vdots\\ u(x_N)\end{bmatrix}=\begin{bmatrix}\psi_{inc}(x_1)\\ \psi_{inc}(x_2)\\ \vdots\\ \psi_{inc}(x_N)\\ 0\\ \vdots\\ 0\\ 0\end{bmatrix} \tag{5-4.12}$$

The coefficient matrix of Eq. (5-4.12) is further decomposed into four blocks as follows

$$\begin{bmatrix}\overline{\overline{Z}}_a & \overline{\overline{Z}}_b\\ \overline{\overline{Z}}_c & \overline{\overline{Z}}_d\end{bmatrix}\cdot\begin{bmatrix}\overline{\Psi}\\ \overline{U}\end{bmatrix}=\begin{bmatrix}\Psi_{inc}\\ 0\end{bmatrix} \tag{5-4.13}$$

It is desirable to make the following decompositions:

$$\overline{\Psi}=\overline{\Psi}_f+\overline{\Psi}_b \tag{5-4.14}$$

$$\overline{U}=\overline{U}_f+\overline{U}_b \tag{5-4.15}$$

$$\overline{\overline{Z}}_a=\overline{\overline{Z}}_a^f+\overline{\overline{Z}}_a^s+\overline{\overline{Z}}_a^b \tag{5-4.16}$$

$$\overline{\overline{Z}}_b=\overline{\overline{Z}}_b^f+\overline{\overline{Z}}_b^s+\overline{\overline{Z}}_b^b \tag{5-4.17}$$

$$\overline{\overline{Z}}_c = \overline{\overline{Z}}_c^f + \overline{\overline{Z}}_c^s + \overline{\overline{Z}}_c^b \quad (5\text{-}4.18)$$

$$\overline{\overline{Z}}_d = \overline{\overline{Z}}_d^f + \overline{\overline{Z}}_d^s + \overline{\overline{Z}}_d^b \quad (5\text{-}4.19)$$

where $\overline{\Psi}_f$ and $\overline{U}_f$ are the forward components (i.e. the current contribution due to the waves propagating in the forward direction), $\overline{\Psi}_b$ and $\overline{U}_b$ are the backward components (i.e. current contribution due to the waves propagating in the backward direction). The matrices $\overline{\overline{Z}}_\ell^f$, $\overline{\overline{Z}}_\ell^s$ and $\overline{\overline{Z}}_\ell^b$ of Eqs. (5-4.16~19) ($\ell = a,b,c,d$) are, respectively, the lower triangular part, the diagonal part (self impedance terms), and the upper triangular part of $\overline{\overline{Z}}_\ell$ in Eq. (5-4.13). Eq. (5-4.13) can then be decomposed for iterative approach as follows

$$(\overline{\overline{Z}}_a^s + \overline{\overline{Z}}_a^f)\cdot\overline{\Psi}_f^{(i)} + (\overline{\overline{Z}}_b^s + \overline{\overline{Z}}_b^f)\cdot\overline{U}_f^{(i)} = \overline{\Psi}_{inc} - \overline{\overline{Z}}_a^f\cdot\overline{\Psi}_b^{(i-1)} - \overline{\overline{Z}}_b^f\cdot\overline{U}_b^{(i-1)} \quad (5\text{-}4.20)$$

$$(\overline{\overline{Z}}_c^s + \overline{\overline{Z}}_c^f)\cdot\overline{\Psi}_f^{(i)} + (\overline{\overline{Z}}_d^s + \overline{\overline{Z}}_d^f)\cdot\overline{U}_f^{(i)} = -\overline{\overline{Z}}_c^f\cdot\overline{\Psi}_b^{(i-1)} - \overline{\overline{Z}}_d^f\cdot\overline{U}_b^{(i-1)} \quad (5\text{-}4.21)$$

$$(\overline{\overline{Z}}_a^s + \overline{\overline{Z}}_a^b)\cdot\overline{\Psi}_b^{(i)} + (\overline{\overline{Z}}_b^s + \overline{\overline{Z}}_b^b)\cdot\overline{U}_b^{(i)} = -\overline{\overline{Z}}_a^b\cdot\overline{\Psi}_f^{(i)} - \overline{\overline{Z}}_b^b\cdot\overline{U}_f^{(i)} \quad (5\text{-}4.22)$$

$$(\overline{\overline{Z}}_c^s + \overline{\overline{Z}}_c^b)\cdot\overline{\Psi}_b^{(i)} + (\overline{\overline{Z}}_d^s + \overline{\overline{Z}}_d^b)\cdot\overline{U}_b^{(i)} = -\overline{\overline{Z}}_c^b\cdot\overline{\Psi}_f^{(i)} - \overline{\overline{Z}}_d^b\cdot\overline{U}_f^{(i)} \quad (5\text{-}4.23)$$

where the superscript (i) in the unknowns ($\overline{\Psi}_f^{(i)}, \overline{U}_f^{(i)}, \overline{\Psi}_b^{(i)}, \overline{U}_b^{(i)}$) indicates the i-th iteration. The algorithm starts with $\overline{\Psi}_b^{(0)} = 0$ and $\overline{U}_b^{(0)} = 0$. It is noted that the matrices $\overline{\overline{Z}}_a^s + \overline{\overline{Z}}_a^f$, $\overline{\overline{Z}}_b^s + \overline{\overline{Z}}_b^f$, $\overline{\overline{Z}}_c^s + \overline{\overline{Z}}_c^f$ and $\overline{\overline{Z}}_d^s + \overline{\overline{Z}}_d^f$ are the lower triangular matrices, the matrices $\overline{\overline{Z}}_a^s + \overline{\overline{Z}}_a^b$, $\overline{\overline{Z}}_b^s + \overline{\overline{Z}}_b^b$, $\overline{\overline{Z}}_c^s + \overline{\overline{Z}}_c^b$ and $\overline{\overline{Z}}_d^s + \overline{\overline{Z}}_d^b$ are the upper triangular matrices. Thus, the matrices in the iterative process do not need to be factorized or inverted. Making use of the forward and backward substitutions, Eqs. (5-4.20,5-4.21) and Eqs. (5-4.22,5-4.23) can be, respectively, solved for $\overline{\Psi}_f^{(i)}$, $\overline{U}_f^{(i)}$ and $\overline{\Psi}_b^{(i)}$, $\overline{U}_b^{(i)}$. Convergence has been extremely fast for moderately rough surfaces (less than six iterations).

Iterative procedure of the FBM usually requires repeated computation of the matrix-vector products, such as $\overline{\overline{Z}}_h^f\cdot\overline{\Psi}$, $\overline{\overline{Z}}_h^f\cdot\overline{U}$, $\overline{\overline{Z}}_h^b\cdot\overline{\Psi}$ and $\overline{\overline{Z}}_h^b\cdot\overline{U}$, ($h = a,b,c,d$), i.e.

$$V_f^{(1)}(r_n) = \sum_{m=1}^{n-1} Z_{mn}^{(a)}\Psi_m + \sum_{m=1}^{n-1} Z_{mn}^{(b)}U_m \quad (5\text{-}4.24)$$

$$V_f^{(2)}(r_n) = \sum_{m=1}^{n-1} Z_{mn}^{(c)}\Psi_m + \sum_{m=1}^{n-1} Z_{mn}^{(d)}U_m \quad (5\text{-}4.25)$$

$$V_b^{(1)}(r_n) = \sum_{m=n+1}^{N} Z_{mn}^{(a)}\Psi_m + \sum_{m=n+1}^{N} Z_{mn}^{(b)}U_m \quad (5\text{-}4.26)$$

$$V_b^{(2)}(r_n) = \sum_{m=n+1}^{N} Z_{mn}^{(c)}\Psi_m + \sum_{m=n+1}^{N} Z_{mn}^{(d)}U_m \quad (5\text{-}4.27)$$

where $i = 1,2,\ldots,N$. The functions of $V_f^{(i)}, V_b^{(i)}, (i = 1,2)$ denote the forward (f) and backward (b) radiation by the source current elements in the front (i=1) and in the rear (i=2) of the receiving n-th element, respectively. Calculation in each iteration takes $O(N^2)$ operations. If the SAA is applied to Eqs. (5-4.24~27), the operational count and memory storage can be reduced to $O(N)$.

Let's define a distance L_s. Interactions between the points within L_s are classified as strong and those outside of L_s as weak. For example, Eq. (5-4.24) is rewritten as

$$V_f^{(1)}(r_n) = V_s + V_w = \sum_{m=n-N_s-1}^{n-1} (Z_{mn}^{(a)}\Psi_m + Z_{mn}^{(b)}U_m) + \sum_{m=1}^{n-N_s-1} (Z_{mn}^{(a)}\Psi_m + Z_{mn}^{(b)}U_m) \tag{5-4.28}$$

The strong V_s is found in conventional manner using the exact matrix elements. The weak V_w radiated from the source group becomes important when the incidence angle approaches to the grazing or if one is interested in the backscattered field. Computation of V_w takes most of the CPU time via the exact computation making FBM inefficient. The computation of V_w can be accelerated via the SAA. Eqs. (5-4.25~27) are treated in the same way as Eq. (5-4.28).

Substituting the spectral representation of the Green function [Savailis et al. 1997] into V_w, it is derived that

$$V_w^{(1)}(r_n) = \sum_{m=1}^{n-N_s-1} (Z_{mn}^{(a)}\Psi_m + Z_{mn}^{(b)}U_m) = \frac{i\Delta x}{4\pi}\int_{C_\theta} F_n(\theta)\exp(ikz_n \sin\theta)d\theta \tag{5-4.29}$$

where

$$F_n(\theta) = F_{n-1}(\theta)\exp(ik\Delta x\cos\theta) + [-ik(-\sin\theta + f_{x_m}\cos\theta)\Psi_{n-N_s-1} + U_{n-N_s-1}]e^{ik(N_s+1)\Delta x\cos\theta}e^{-ikz_{n-N_s-1}\sin\theta} \tag{5-4.30}$$

And, similarly,

$$V_w^{(2)}(r_n) = \sum_{m=1}^{n-N_s-1} (Z_{mn}^{(c)}\Psi_m + Z_{mn}^{(d)}U_m) = \frac{i\Delta x}{4\pi}\int_{C_\theta} F_n(\theta)\exp(ik_1 z_n \sin\theta)d\theta \tag{5-4.31}$$

where

$$F_n(\theta) = F_{n-1}(\theta)\exp(ik\Delta x\cos\theta) + [ik_1(-\sin\theta + f_{x_m}\cos\theta)\Psi_{n-N_s-1} - \rho U_{n-N_s-1}]e^{ik_1(N_s+1)\Delta x\cos\theta}e^{-ik_1 z_{n-N_s-1}\sin\theta} \tag{5-4.32}$$

It can be seen from Eqs. (5-4.29~32) that $F_n(\theta)$ can be easily obtained by recursive procedure. Far-field pattern $F_n(\theta)$ is continuously built up via Eqs. (5-4.30,5-4.32), while a new source element in the weak interaction group sweeps in the forward direction. In practice, it requires a new integration contour C_δ instead of C_θ in Eqs. (5-4.29,5-4.31) since $F_n(\theta)$ in the real θ space tends to have a narrow main lobe and many narrow side-lobes in contrast to slowly varying pattern along C_δ. The criterion for selecting C_δ and implementing the spectral integration has been discussed [Chan and Johnson 1998,2000].

5-4.3 Bistatic Scattering and Transmitting Coefficients

The bistatic scattering coefficients from and through the dielectric rough surface are defined as Sancchez and Nieto-Vesperinas (1991):

$$\sigma_{TE}^{(U)}(\theta_{in},\theta_s) = \frac{\frac{1}{8\pi k}\left|\int_{S'}\left\{\frac{ik(f_{x'}\sin\theta_s - \cos\theta_s)}{\sqrt{1+f_{x'}^2}}E - \frac{\partial E}{\partial n'}\right\}e^{-ik(x'\sin\theta_s + z'\cos\theta_s)}dS'\right|^2}{L_{eff}\cos\theta_{in}} \tag{5-4.33}$$

$$\sigma_{TE}^{(D)}(\theta_{in},\theta_t) = \frac{\frac{1}{8\pi k}\left|\int_{S'}\left\{\frac{ik_1(f_{x'}\sin\theta_t - \cos\theta_t)}{\sqrt{1+f_{x'}^2}}E - \frac{\partial E}{\partial n_1'}\right\}e^{-ik_1(x'\sin\theta_t - z'\cos\theta_t)}dS'\right|^2}{L_{eff}\cos\theta_{in}} \tag{5-4.34}$$

where $L_{eff} = \sqrt{\pi/2}g[1-(1+2tg^2\theta_{in})/2(kg)^2]$ for the tapered incident wave, g is the half-beam waist of the tapered incident wave. Superscripts (U) and (D) indicate the up-going scattering from and down-going transmission through the surface, respectively. $\sigma_{TM}^{(U)}$ and $\sigma_{TM}^{(D)}$ can be obtained by Eqs. (5-4.33,5-4.34) with replacing E by H, respectively.

Integrating the bistatic scattering over the upper-half space, the reflectivities of TE and TM waves are obtained as.

$$R_{TE} = \int_{-\frac{\pi}{2}}^{\frac{\pi}{2}} \sigma_{TE}^{(U)}(\theta_{in},\theta_s)d\theta_s \quad \text{and} \quad R_{TM} = \int_{-\frac{\pi}{2}}^{\frac{\pi}{2}} \sigma_{TM}^{(U)}(\theta_{in},\theta_s)d\theta_s \tag{5-4.35}$$

Integrating the bistatic scattering or transmitting over the lower-half space, the transmittivities of TE and TM waves are obtained as

$$T_{TE} = \int_{-\frac{\pi}{2}}^{\frac{\pi}{2}} \sigma_{TE}^{(D)}(\theta_{in},\theta_t)d\theta_t \quad \text{and} \quad T_{TM} = \int_{-\frac{\pi}{2}}^{\frac{\pi}{2}} \sigma_{TM}^{(D)}(\theta_{in},\theta_t)d\theta_t \tag{5-4.36}$$

In the absence of absorption through the dielectric medium, the unitarity condition must be satisfied for the energy conservation, i. e.

$$R_{TE} + T_{TE} = 1 \quad \text{and} \quad R_{TM} + T_{TM} = 1 \tag{5-4.37}$$

The energy conservation is used to validate the numerical accuracy.

5-4.4 Fractal Rough Surface Model

A band-limited Weierstrass-Manderbrot function is chosen to describe the fractal surface [Savailis et al. 1997]. It has a finite band of spatial frequency and exhibits self-similarity over the finite range of resolution. It is expressed as a weighted sum of periodic functions

$$f(x) = \delta \cdot C \sum_{m=0}^{M-1} b^{(D-2)m} \sin(k_0 b^m x + \varphi_m) \tag{5-4.38}$$

where D $(1 < D < 2)$ is the fractal dimension, φ_n the random phase, $b(>1)$ the spatial frequency scaling parameter, k_0 the fundamental spatial wave number, and δ the rms surface height. The coefficient $C = \sqrt{2(1-b^{2(D-2)})/(1-b^{2M(D-2)})}$ is a normalized amplifier control parameter. The fractal dimension D gives a measure of the surface roughness from D=1 (smooth periodic curve) to D=2 (rough, area-filling curve)[Jaggard and Sun 1990].

The self-similarity of Eq. (5-4.38) can be demonstrated by the relationship

$$f(x) \approx \frac{1}{b^{(D-2)}} f(bx) \tag{5-4.39}$$

which means that the curve looks similar to the original when the horizontal axis is scaled by b and vertical axis by $1/b^{(D-2)}$.

As Eq. (5-4.38) indicated, the fractal surface is described by the parameters δ, D, b, k_0 and M, whilst the conventional Gaussian surface only needs two parameters, δ and ℓ. Different fractal rough surfaces may have the same δ and ℓ. The correlation length ℓ can be determined from the autocorrelation function $\rho(\tau)$ when $\rho(\ell) = 1/e$,

$$\rho(\tau) =\langle f(x)f(x+\tau)\rangle = (1-b^{2(D-2)})/(1-b^{2M(D-2)}) \sum_{m=0}^{M-1} b^{2(D-2)m} \cos(k_o b^m \tau) \tag{5-4.40}$$

Note that autocorrelation function $\rho(\tau)$ is still a band-limited Weierstrass -Manderbrot fractal function. In view of scattering from rough surface, the wavelength of the interrogating wave acts as a yardstick to sense the fractal nature of the rough surface at different scales. In this way, finer structures of the rough surface can be seen as the wavelength becomes shorter. So the incident wavelength λ is uaed as a yardstick to determine the M as

$$k_0 b^{M-1} = k \tag{5-4.41}$$

where $k(= 2\pi/\lambda)$ is the incident wavenumber, so the highest spatial harmonic wavelength λ_M is always less than the incident λ.

5-4.5 Numerical Simulations and Discussions

A. Validation of the FBM/SAA

As an electromagnetic wave with tapered angle $2\alpha = 5^\circ$ is incident upon the fractal rough surface, the illuminated surface length L is made as

$$L = L_0 \times \cos(80^\circ + \alpha)/\cos(\theta + \alpha) \tag{5-4.42}$$

where θ is the incident angle and $L_0 = 167.0\lambda$ is taken as the illuminated surface length at $\theta_{inc} = 80^\circ$. It leads $L = 25.86\lambda$ at $\theta_{inc} = 30^\circ$ and $L = 64.83\lambda$ at $\theta_{inc} = 67.84^\circ$ (it happens as the TM Brewster angle for flat surface if the dielectric constant of the low-medium takes $\varepsilon = 6.0 + j0.6$).

In following calculations, the half-beam waist $g = L/6.0$, incident wavelength $\lambda = 1\,(m)$, the parameters of fractal surface $b = e/2, M = 10$, $\Lambda_0 = 10\lambda$ and $\Delta x = \lambda/20$ are taken unless stated otherwise. All calculations are carried out in the Pentiam II PC.

In order to validate the accuracy and efficiency of FBM/SAAA, the FBM/SAA, FBM and MoM are employed for calculations of bistatic scattering, transmitting and surface current from one surface realization with D=1.3 and $\theta_{inc} = 30^\circ$. Comparison of Table 5-4.1 shows that the FBM/SAA is accurate and is much faster than MoM. Also, the integration of bistatic scattering and transmitting over the upper- and lower-space yields 1 (i.e. 1.0005 for TE, and 1.0046 for TM), and energy is conserved. Computation by FBM/SAA is also much faster than FBM without SAA as shown in Table 5-4.1.

Table 5-4.1. Time consuming of FBM/SAA, FBM and MoM

	TE	TM
MoM	475.49(*s*)	512.83(*s*)
FBM	59.21 (*s*)	49.05(*s*)
FBM/SAA	31.25 (*s*)	31.09(*s*)

Comparison of $\Delta x = \lambda/20$ and $\Delta x = \lambda/30$ is also tested for 50 Monte Carlo surface realizations. The difference between these two cases in bistatic scattering is 0.398 dB for TE and 0.140 *dB* for TM, and the difference in the transmitting is 0.439 *dB* for TE and 0.602 *dB* for TM. It seems that the sampling rate $\Delta x = \lambda/20$ yields satisfactory accuracy. In other words, it is interesting to see that the finer scale much smaller than the wavelength does not appreciably contribute to the overall scattering and transmitting.

The average results from 50 and 80 realizations are also compared. The difference between these two cases in bistatic scattering is 0.807 dB for TE and 1.160 *dB* for TM, and the difference in transmitting is 1.468 *dB* for TE and 1.872 *dB* for TM. In following calculations, the average of 50 realizations is employed.

B. Bistatic scattering and transmitting

Bistatic scattering and transmitting are calculated for three fractal surfaces D=1.1, 1.5 and 1.9) with surface variance $\delta = 0.05\lambda$ or $\delta = 0.4\lambda$ at three incident angles ($\theta_{inc} = 30^\circ, 67.84^\circ, 80^\circ$). The numerical results are presented in Figs. 5-4.1(a~f) for TE case. Numerical results for TM case are similar. The correlation length ℓ (denoted as *CL*), the reflectivity *R* and transmittivity *T*, and *R*+*T* are indicated in the figures.

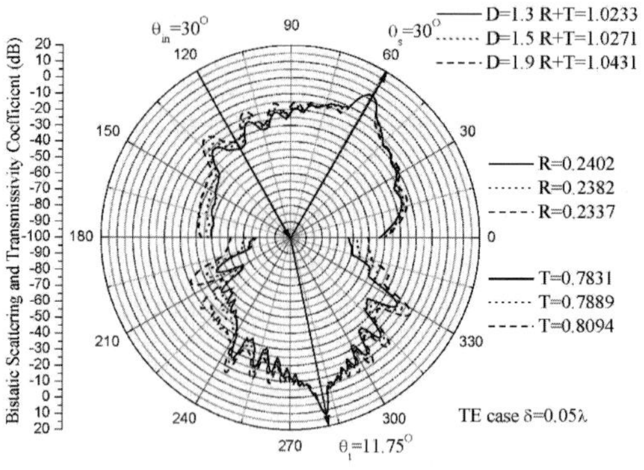

Fig. 5-4.1(a). $\delta = 0.05\lambda$, $\theta_{inc} = 30^{\circ}$

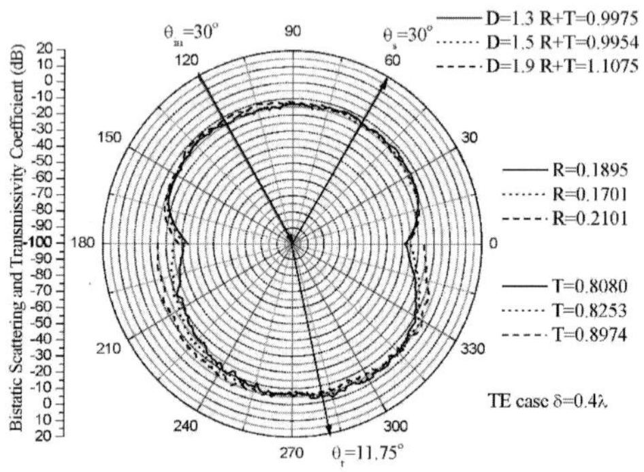

5-4.1(b). $\delta = 0.05\lambda$, $\theta_{inc} = 67.84^{\circ}$

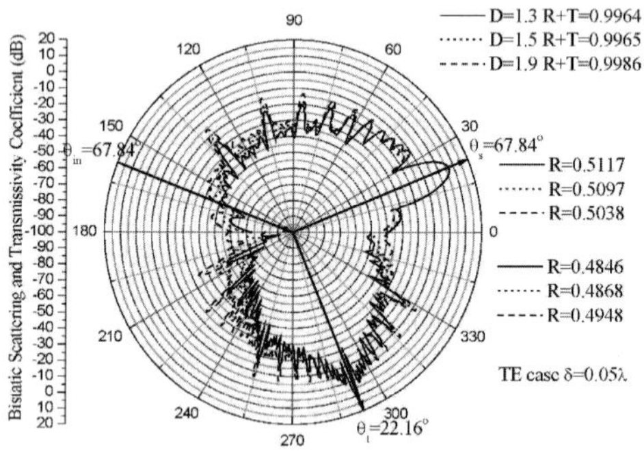

Fig. 5-4.1(c). $\delta = 0.05\lambda$, $\theta_{inc} = 80^{\circ}$

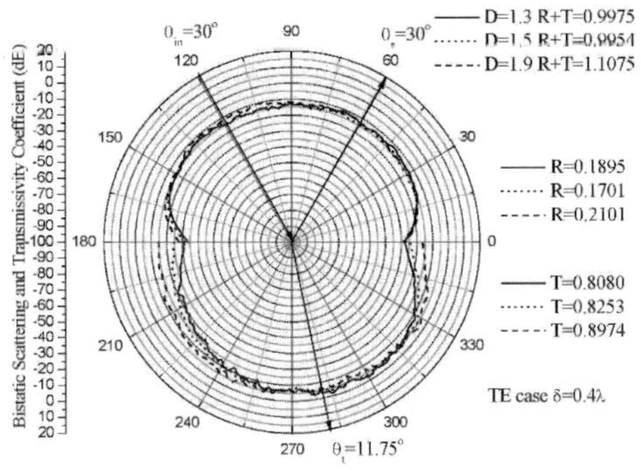

Fig.5-4.1(d). $\delta = 0.4\lambda\,, \theta_{inc} = 30^{\circ}$

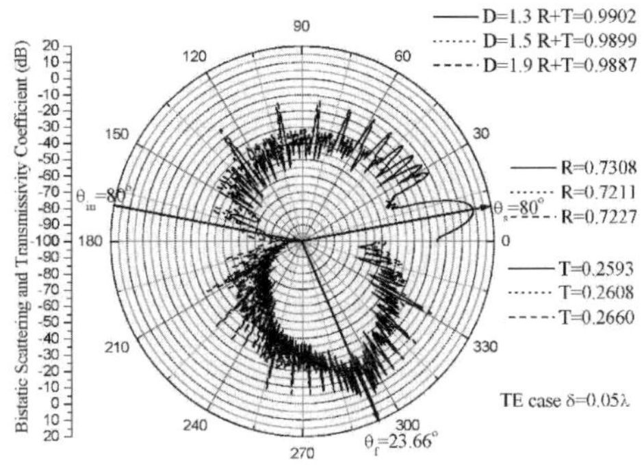

Fig. 5-4.1(e). $\delta = 0.4\lambda\,, \theta_{inc} = 67.84^{\circ}$

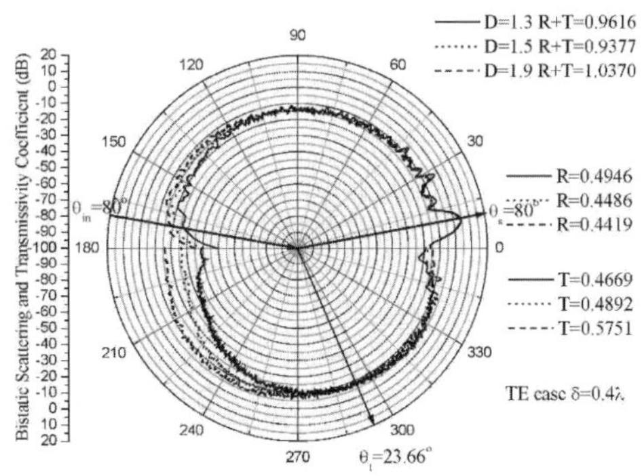

Fig. 5-4.1(f) $\delta = 0.4\lambda\,, \theta_{inc} = 80^{\circ}$.

Fig.5-4.1. Bistatic scattering coefficient vs. scattering angle

It can be seen that for the surface with small $\delta = 0.05\lambda$, specular reflection or forward scattering becomes dominant as the incident angle increases, especially at LGA (in Figs.

5-4.1(e,f)), and bistatic scattering shows the angularly oscillating pattern. The reflected and transmitted angles defined by the Snell's law for flat surface can also be identified in the figures. However, as the surface roughness increases e.g. $\delta = 0.4\lambda$, bistatic scattering becomes much angularly diffused.

C. Energy conservation

The reflectivity *R* and transmissivity *T* of Eqs. (5-4.36, 5-4.37) are calculated for one surface realization for three different fractal dimensions *D*=1.1, 1.5 and 1.9 with $\delta = 0.1\lambda$. The surface medium has $\varepsilon = 6.0 + j0.6$. The results of *R*+*T*=1 are shown in Figs. 5-4.2(a, b) for TE and TM cases, respectively. It can be seen that angular dependence of *R* and *T* against the incident angle are almost exactly matched with the flat cases. As the surface has a high *D* at LGA, the accuracy remains to be improved. It is also noted that as the incident angle is at the TM Brewster angle of the flat surface, most of the waves are transmitted through the fractal rough surface.

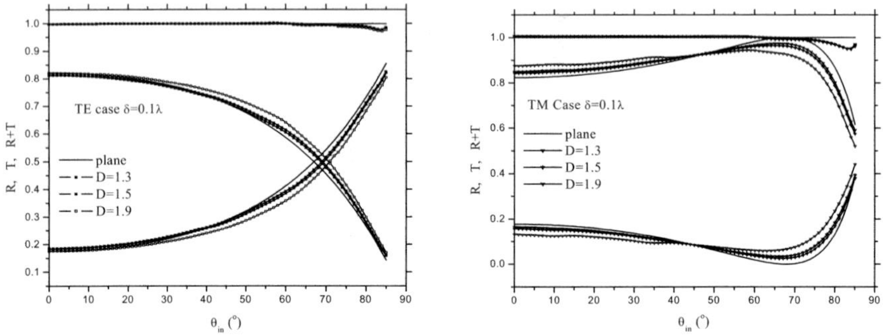

(a) TE case (b) TM case

Fig. 5-4.2. Reflectivity and transmittivity and enbergy conservation ($\delta = 0.1\lambda$)

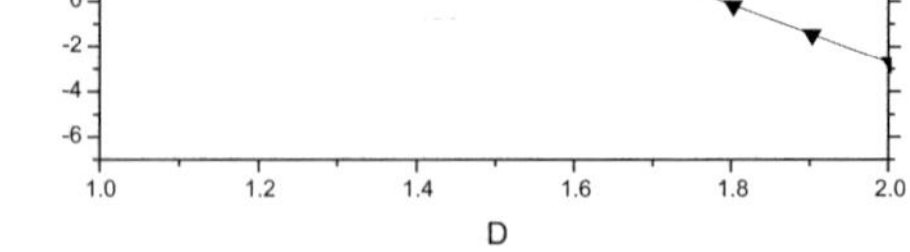

Fig. 5-4.3. The slope of angular pattern of bistatic scattering vs. fractal *D*

D. Slope of bistaic scattering pattern vs. fractal D

It has been found that the line linking the spikes of angular distribution of bistatic scattering coefficient yield a linear slope against fractal D along both sides of the specular reflection. The slope is increased if D increases. An example is illustrated in Fig. 5-4.3. If the surface variance δ is small, the slope to link the crests is empirically written as

$$s \approx f(\delta,\theta_{in},k_0,b)\cdot(2-D) \qquad \text{for TE case} \tag{5-4.43}$$

and

$$s \approx f(\delta,\theta_{in},k_0,b)\cdot(2-D)+f_{TM}(\delta,\theta_{in},k_0,b) \qquad \text{for TM case} \tag{5-4.44}$$

where $f(\delta,\theta_{in},k_0,b)$ and $f_{TM}(\delta,\theta_{in},k_0,b)$ are the functions of fractal parameters, such as δ, θ_{in}, k_0 and b. If δ becomes very large, this relationship cannot be identified.

The slope against fractal D is shown in Fig. 5-4.3. Discrete points are obtained from calculations of twelve different fractal surfaces. It is surprised that the relationship is exactly linear. Also, the slope of TM case can be negative, i.e. the slope can turn upwards versus the scattering angle.

It is measured from calculations that $s \approx 12.297\cdot(2-D)+1.94$ for TE case and $s \approx 12.84\cdot(2-D)-2.80$ for TM case. This relationship might give an insight that fractal dimension might be evaluated from bistatic scattering pattern. It is also true for the angular distribution of transmitting.

5-5 Bistatic Scattering of a Fractal Rough Surface with High Permittivity Using the PBTG-FBM/SAA

Surface profile was usually described by periodic or random functions. The fractal geometry has been introduced to study of rough surface in recent years. Since fractal geometry holds in balance long-range order and short-range disorder and can be used to describe both deterministic and random structures or an appropriate blend, it seems better to describe some natural surface [Jaggard and Sun 1990].

To take into account for scattering fields from dielectric rough surface with high permittivity, the FBM combined with the impedance boundary condition (IBC) was employed in [Chou and Johnson 2000]. However, the error due to the approximation of IBC cannot be well evaluated. Also, transmitting fields has not been calculated in this approach. Another approach is the physics-based two-grid (PBTG) method [Li et al. 1999, 2001], which takes a single coarse grid (SCG) on the upper surface and a single dense grid (SDG) on the lower surface. The PBTG can have good accuracy as the SDG, but much faster than SDG. The physics-based point is that the SCG is for slowly varying fields on the upper surface of free space and the SDG is for fast varying fields on the lower surface of the medium with high permittivity. It can speed up calculations of the lower- or upper-triangle matrix-vector produced by the Green's function convoluting with the surface fields on the dense grid.

In this session, we combine the FBM/SAA with PBTG to solve the bistatic scattering and transmitting coefficients from the fractal dielectric rough surface with high permittivity [Li and Jin 2002]. In the high permittivity medium, the electromagnetic field below the rough surface varies much faster than the one above the surface. So the PBTG can be fully utilized to take account for fast variation, especially at LGA incidence. Meanwhile, angular pattern of bistatic scattering and transmitting, and energy conservation under the TE and TM tapered waves incidence, especially at LGA, are numerically discussed.

5-5.1 The PBTG-FBM/SAA Approach

All Eqs. (5-4.1~5.4.32) are employed in this session. Let the upper medium be free-space and the lower space is a dielectric lossy medium with high permittivity ε_1. The wavelengths of the upper and lower media are λ and λ_1, respectively, and approximately $\lambda_1 \approx \lambda/n_1$, where n_1 is of the nearest integer of real $\sqrt{\varepsilon_1}$. It yields the number of sampling points in the lower medium as n_1 times of the number in the upper medium.

In the PBTG, the SDG number N_{dg} and SCG number N can have

$$N_{dg} = n_{dg}\frac{L}{\lambda} \qquad \text{and} \qquad N = n_{cg}\frac{L}{\lambda} \tag{5-5.1}$$

Usually $n_{cg} = 10$ and $n_{dg} = 10n_1$. Rewrite Eqs. (5-4.24~27) using the two grids as

$$V_f^{(1)}(r_{\tilde{n}}) = \sum_{m=1}^{\tilde{n}-1} \tilde{Z}_{m\tilde{n}}^{(a)}\tilde{\Psi}_m + \sum_{m=1}^{\tilde{n}-1} \tilde{Z}_{m\tilde{n}}^{(b)}\tilde{U}_m \qquad \tilde{n} = 1,2,\ldots,N \tag{5-5.2}$$

$$V_f^{(2)}(r_n) = \sum_{m=1}^{n-1} Z_{mn}^{(c)}\Psi_m + \sum_{m=1}^{n-1} Z_{mn}^{(d)}U_m \qquad n = 1,2,\ldots,N_{dg} \tag{5-5.3}$$

$$V_b^{(1)}(r_{\tilde{n}}) = \sum_{m=\tilde{n}+1}^{N} \tilde{Z}_{m\tilde{n}}^{(a)}\tilde{\Psi}_m + \sum_{m=\tilde{n}+1}^{N} \tilde{Z}_{m\tilde{n}}^{(b)}\tilde{U}_m \qquad \tilde{n} = 1,2,\ldots,N \tag{5-5.4}$$

$$V_b^{(2)}(r_n) = \sum_{m=n+1}^{N_{dg}} Z_{mn}^{(c)}\Psi_m + \sum_{m=n+1}^{N_{dg}} Z_{mn}^{(d)}U_m \qquad n = 1,2,\ldots,N_{dg} \tag{5-5.5}$$

where $\tilde{\Psi}_m$, $\tilde{U}_m$, $\tilde{Z}_{m\tilde{n}}$, $\tilde{n}$ denote the coarse grid indexes.

The Green's function of the upper free space is varying much slower than one in the lower space. Thus, when performing multiplication of the lower-triangle matrix and column-vector on the SCG as indicated in Eqs. (5-4.24) or (5-4.26), the Green's function of the upper space is essentially constant over the interval of the points on the dense grid.

Thus, we can write

$$\widetilde{Z}^{(b)}_{m\tilde{n}}\widetilde{U}_m = \sum_{\ell=1}^{n_1} Z^{(b)}_{(m+\ell)\tilde{n}} U_{(m+\ell)} \approx n_1 Z_{m_{mp}\tilde{n}} \left(\frac{1}{n_1}\sum_{\ell=1}^{n_1} U_{(m+\ell)}\right) \tag{5-5.6}$$

where the points with indexes m_{mp} is the middle points of the $(m+1)$-th point to $(m+n_1)$-th point. What is done in Eq. (5-5.6) is that the surface fields on the dense grid are first averaged before multiplying by the Green's function of the upper media. It should be mentioned that the sample rate for SDG is $n_1\Delta x$, and Δx is for SCG.

5-5.2 Numerical Simulations and Analysis

A. Validation of the PBTG-FBM/SAA code

In following calculations, $g = L/6.0$, $\lambda = 1\,(m)$, the fractal surface is described by the band-limited Weierstrass-Manderbrot function [Jaggard and Sun 1990], its parameters: $b = e/2.0$, M=10, $\Lambda_0 = 10\lambda$, $L = 30\lambda$, $\delta = 0.1\lambda$ and $\Delta x = \lambda/20$ for the upper space and $\Delta x = \lambda/60$ for the lower space with the permittivity $\varepsilon_1 = 25.0 + j1.0$, are taken. All calculations are carried out in the Pentium III 850 PC.

To validate the PBTG-FBM/SAA code and efficiency of SAA and PBTG, bistatic scattering, transmitting and induced current from one realization of rough surface with D=1.3 and $\theta_{inc} = 30°$ are calculated using both PBTG-FBM/SAA and conventional MoM. Comparison shows that the result of PBTG- FBM/SAA is accurate and is much faster than MoM. Also, the reflectivity plus transmittivity well satisfy the energy conservation. Computation of the PBTG-FBM/SAA is also much faster than the PBTG-FBM without SAA as shown in Table 5-5.1.

The averaged results of Monte Carlo 50 and 80 realizations are compared. The maximum deviation between their bistatic scattering is 1.347*dB* for TE case and 1.007*dB* for TM case. The maximum difference between their transmitting is 1.693*dB* for TE and 1.525 *dB* for TM. So, the result of 50 realizations is good enough. Since then, 50 realizations are used in following calculations.

Table 5-5.1. Time consuming of PBTG-FBM/SAA, PBTG-FBM and MoM

	TE		TM	
	Time (*min.*)	Energy	Time (*min.*)	Energy
MoM	17.25	0.9995	12.87	1.0005
PBTG –FBM	7.42	0.9947	4.1	0.9962
PBTG –FBM/SAA	3.59	0.9964	1.97	0.9965

B. Different sampling rates of SCG, SDG and PBTG

Because the fractal surface is multi-scale, sampling rate should be fine enough to grasp the detail of fractal profile. The results of SDG: $\lambda/36$ and $\lambda/60$ are compared. It shows that the maximum deviation between their bistatic scattering is 0.969*dB* for TE and 0.136 *dB* for TM. The maximum difference between their transmitting is 0.790*dB* for TE and 0.563*dB* for TM. Thus, the SDG $\Delta x = \lambda/36$ is accurate enough and is then taken as a standard for comparison with SCG and PBTG.

Given SCG $\Delta x = \lambda/12$ and PBTG ($\lambda/12, \lambda/36$), Figs. 5-5.1(a, b) of TE case show that bistatic scattering and transmitting of SCG have large errors comparing with standard SDG. The relative errors of bistatic scattering over the upper space are 19.128% for TE and 93.571% for TM (not shown in the figure). Relative errors of transmitting over the lower space are 59.795% for TE and 24.988% for TM (not shown in the figure). From Table 5-5.2, energy conservation of SCG is also not satisfied, and is 1.3856 for TE and 1.1903 for TM.

However, the PBTG presents bistatic scattering well matched with SDG (see Fig 1a, TE case). The relative errors of bistatic scattering over the upper space are 0.424% for TE and 1.135% for TM comparing with SDG. Energy conversation of PBTG is 0.9945 for TE and 0.9883 for TM.

Transmitting of the PBTG has deviation from SDG as shown in Fig. 5-5.1(b) (TE case). The relative errors of transmitting over the upper space are 0.181% for TE and 1.441% for TM comparing with SDG (not shown in the figure). Sine transmitting at large angle is always very small (in dB), it would not significantly affect the energy conservation.

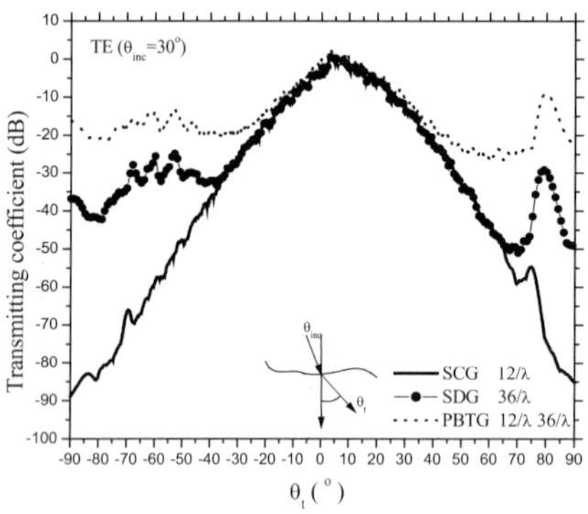

Fig. 5-5.1(a). Bistatic scattering for TE case

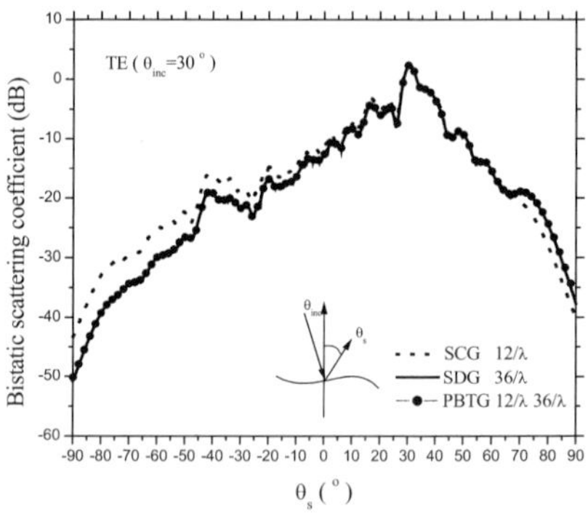

Fig. 5-5.1(b). Transmitting for TE case

Table 5-4.2. Comparisons of PBTG, SCG and SDG (L=30λ)

	TE ($\lambda/12$)			TM ($\lambda/12$)		
	SCG	SDG	PBTG	SCG	SDG	PBTG
Reflectivity	0.5904	0.4956	0.4977	0.7166	0.3702	0.3660
Error (%)	19.128	*	0.424	93.571	*	1.135
Transmitivity	0.7953	0.4977	0.4968	0.4737	0.6315	0.6224
Error (%)	59.795	*	0.181	24.988	*	1.441
Energy	1.3856	0.9933	0.9945	1.1903	1.0017	0.9883
Error (%)	38.56	0.67	0.55	19.03	0.17	1.17
Time (hour)	0.73	4.67	3.37	0.73	3.96	3.29

	TE ($\lambda/20$)			TM ($\lambda/20$)		
Reflectivity	0.5407	0.4935	0.4916	0.3285	0.3754	0.3744
Error (%)	9.56	*	0.385	12.493	*	0.266
Transmitivity	0.4623	0.5058	0.5032	0.6765	0.6250	0.6218
Error (%)	8.6	*	0.51	8.24	*	0.512
Tb (K)	13.05	*	0.774	15.45	*	0.96
Energy	1.0030	0.9993	0.9947	1.0050	1.0005	0.9961
Error (%)	0.30	0.07	0.53	0.50	0.50	0.39
Time (hour)	2.88	30.33	21.48	1.91	19.90	15.96

To improve the results of transmitting, the sampling rates as $\Delta x = \lambda/20$ of SCG, $\lambda/60$ of SDG and $(\lambda/20, \lambda/60)$ of PBTG are now taken. Numerical results in Fig. 5-5.2(a) (TE case) show that bistatic scattering of SCG and PBTG are now well matched with SDG.

Transmitting of the PBTG is well matched with SDG in Fig. 5-5.2(b) (TE case), while SCG has large error comparing with SDG. The relative errors of transmitting of PBTG are 0.51% for TE and 0.512% for TM.

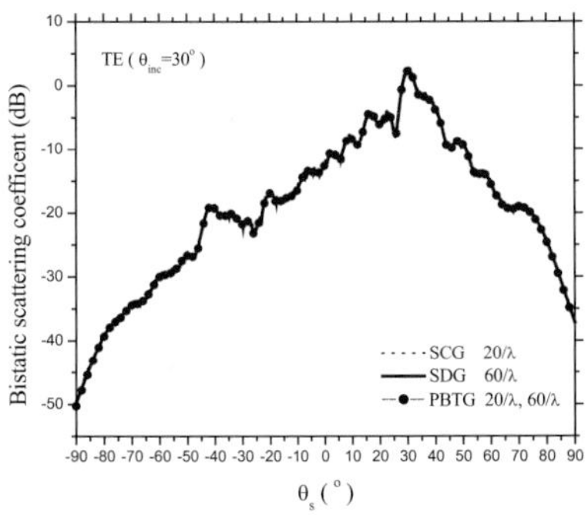

Fig. 5-5.2(a). Bistatic scattering for TE case

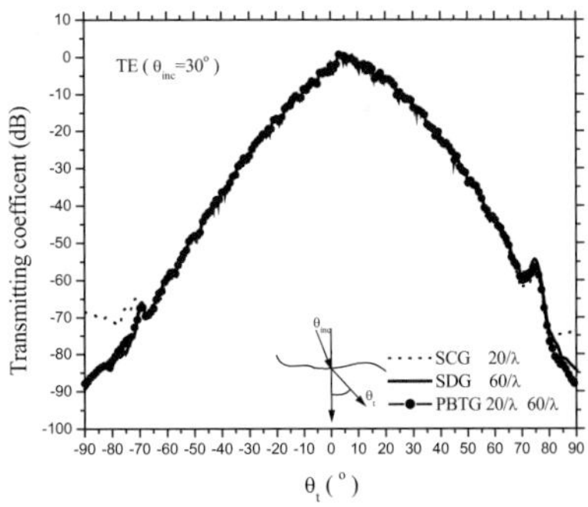

Fig. 5-5.2(b). Transmitting for TE case

Brightness temperature (T_B) of the object is calculated by multiplying the physical temperature and emissivity (which is equal to the transmittivity). Accuracy of T_B measured by satellite or air-bone radiometer can be $1K$ or even less. In Fig. 5-4.2(b), the error of transmitting can yields error as $0.774K$ for TE and $0.96K$ for TM.

The relative errors of transmitting of SCG are 8.6% for TE and 8.24% for TM, which may yield T_B errors as large as $13.05K$ for TE and $15.45K$ for TM. As electromagnetic wave is refracted into the lower space of high permittivity, the fields are highly varying and attenuating. The SCG cannot demonstrate such variation.

The CPU time consuming of the SCG, SDG and PBTG is listed in Table 5-5.2. It can be seen that the PBTG has accuracy as good as SDG and is much faster than SDG.

C. Bistatic scattering and transmitting at LGA

At LGA such as $\theta_{inc} = 80^{\circ}$, the surface length is set as $L = 200\lambda$. The fractal dimension D is taken as D=1.3, 1.5 and 1.8, respectively.

Fig. 5-5.3 shows that the specular reflection is dominant under LGA incidence for all different *D*. As the surface becomes rougher (*D* is increasing), both bistatic scattering of the upper space or transmitting of the lower medium become more diffused. Angular transmitting pattern looks as a circle due to diffusing.

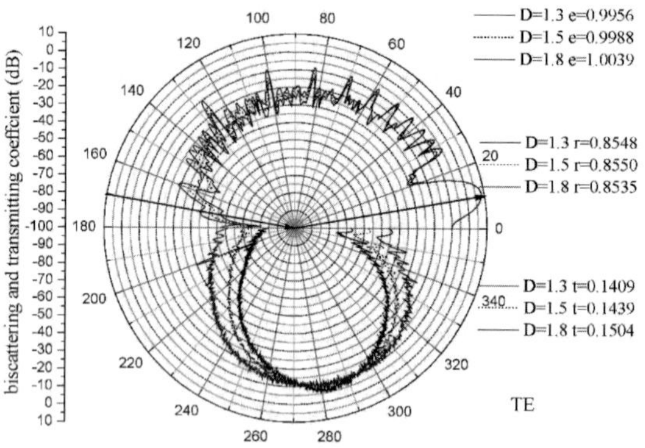

Fig. 5-5.3. Bistatic scattering and transmitting at LGA incidence for TE case

Simulation for TM case is also similarly obtained.

The two grids PBTG-FBM/SAA can have good accuracy as the density grid SDG, and is much faster than SDG.

5-6 References

Axline, R.M. and A.K. Fung (1978), "Numerical computation of scattering from a perfectly conducting random surface", *IEEE Transactions on Antennas and Propagation*, 26(3): 482-488.

Brown, G.S. (1998), "Guest Editorial-Special issue on low-grazing-angle backscattering from rough surface", *IEEE Transactions on Antennas and Propagation*, 46(1): 1-2.

Burkholder, R.J., M.R. Pino and F. Obelleiro (2001), "A Monte Carlo study of the rough sea surface influence on the radar scattering from two-dimensional ships", *IEEE Transactions on Antennas and Propagation Magazine*, 43(2): 25-32.

Chan, H.T. and J.T. Johnson (1998), "A novel acceleration algorithm for the computation of scattering, from rough surfaces with the forward-backwards method", *Radio Science*, 33(5): 1277-1287.

Chan, H.T. and J.T. Johnson (2000), "Formulation of forward-backward method using novel spectral acceleration for the modeling of scattering from impedance rough surfaces", *IEEE Transactions on Geosicence and Remote Sensing*, 38(1): 605-607.

Chou, H.T., J.T. Johnson (1998), "A novel acceleration algorithm for the computation of scattering, from rough surfaces with the forward-backwards method", *Radio Science*, 33(5): 1277-1287.

Chou, H. T., J.T. Johnson (2000), "Formulation of forward-backward method using novel spectral acceleration for the modeling of scattering from impedance rough surfaces", *IEEE Transactions on Geosicence and Remote Sensing*, 38(1): 605-607.

Donohue, D.J., H.C. Hu and D.R. Thompson (1998), "Application of iterative moment method solution to ocean surface radar scattering", *IEEE Transactions on Antennas and Propagation*, 46(1): 121-132.

Franceschetti, G., A. Iodice and D. Riccio (2000), "Scattering from dielectric random fractal rough surfaces via method of moments", *IEEE Transactions on Geosicence and Remote Sensing*, 38(4): 1644-1655.

Guerin, C. A., M. Holschneider and M. Saillard (1997), "Electromagnetic scattering from multi-scale rough surfaces", *Waves in Random Media*, 7(3): 331-349.

Harrington, R.F. (1993). *Field Computation by Moment Method*, New York: IEEE Press.

Holliday, D., L.L. DeRad and G.J. St-Cyr (1996), "Forward -backward: A new method for computing low-grazing angle scattering", *IEEE Transactions on Antennas and Propagation*, 44(5): 722-729.

Holliday, D., L.L. DeRaad and G.J. St-Cyr (1998), "Forward-backward method for scattering from imperfect conductors", *IEEE Transactions on Antennas and Propagation*, 46(1): 101-107.

Ishimaru, A. (1990), *Wave Propagation and Scattering in Random Media*, New York: Academic.

Jaggard, D.L. and X. Sun (1990), "Scattering from fractally corrugated surfaces", *Journal of the Optical Society of American (A)*, 7(6): 1055-1062.

Jin, Y.Q. (1994), *Electromagnetic Scattering Modeling for Quantitative Remote Sensing*, Singapore: World Scientific.

Jin, Y.Q. and G. Li (2000), "Detection of a scatter target over randomly rough surface by using angular correlation function in finite element approach", *Waves in Random Media*, 10: 273-280.

Jin, Y.Q. and Z. Li (2001), "Simulation of scattering from complex rough surface at low grazing angle using the GFBM/SAA method", *IEEJ Transactions of Fundamentals and Materials Society (A)*, 121(10): 917-921.

Jin, Y.Q. and Z. Li (2002), "Bistatic scattering and transmitting through fractal rough dielectric surface using FBM/SAA method", *Journal of Electromagnetic Waves and Applications,* 16(4): 551-572.

Jin, Y.Q. and Z. Li (2002), "Numerical simulation of radar echo at low grazing angle incidence using the GFBM/SAA method", *SPIE Proceedings: Lidar Remote Sensing for Industry and Environment Monitoring*, 44(84): 278-285.

Jin, Y.Q. and Z. Li (2003), "Numerical simulation of radar surveillance for the ship target and oceanic clutters in two-dimensional model", *Radio Science*, 38(3): 1045-1050.

Johnson, J.T., et al. (1996), "Backscattering enhancement of electromagnetic waves from two-dimensional perfectly conducting random rough surfaces: A comparison of Monte Carlo simulations with experimental data", *IEEE Transactions on Antennas and Propagation*, 44: 748-756.

Johnson, J.T. (1998), "On the canonical grid method for two-dimension scattering problem", *IEEE Transactions on Antennas and Propagation*, 46(3): 297-302.

Kapp, D.A. (1995), *A New Numerical Method for Rough Surface Scattering Calculation*, Ph.D. dissertation, Virginia Polytechnic Institute and State University, Blacksburg, VA: 112.

Kapp, D.A. and G.S. Brown (1996), "A new numerical method for rough surface scattering calculations", *IEEE Transactions on Antennas and Propagation*, 44: 711-721.

Kong, J.A. (1985), *Electromagnetic Wave Theory*, ed., New York: Wiley.

Li, Q., C.H. Chan and L. Tsang (1999), "Monte Carlo simulations of wave scattering from lossy dielectric random rough surfaces using the Physics-Based Two-Grid method and the Canonical-Grid method", *IEEE Transactions on Antenna and Propagation*, 47(4): 752-763.

Li, Q., C.H. Chan and L. Tsang (2001), "Wave scattering from lossy dielectric random rough surfaces using the Physics-Based Two-Grid Method in conjunction with multilevel fast multi-pole method", *Radio Science*, 36(4): 571-583.

Li, Z. and Y.Q. Jin (2000), "Numerical simulation of bistatic scattering from fractal rough surface in the finite element method", *Science in China (E)*, 44(1):12-18.

Li, Z. and Y.Q. Jin (2001), "Bistatic scattering from a fractal dynamic rough sea surface with a ship presence at low grazing angle incidence using the FBM/SAA method", *Microwave and Optical Technology Letters*, 31(2): 146-151.

Li, Z. and Y.Q. Jin (2002), "Bistatic scattering and transmitting through a fractal rough surface with high permittivity using the PBTG-FBM/SAA method", *IEEE Transactions on Antennas and Propagation,* 50(9): 1323-1326.

Li, Z. and Y.Q. Jin (2002), "Numerical simulation of bistatic scattering from fractal rough surface using the Forward/Backward iterative method", *Electromagnetics*, 22(3): 191-207.

Lin, C.H., C.H. Chan, and L. Tsang (1999), "Monte Carlo simulations of scattering and emission from lossy dielectric rough surfaces using the wavelet transform method", *IEEE Transactions on Geoscience and Remote Sensing*, 37(5): 2295-2304.

Lou, S.H., Tsang L., C.H. Chan and A. Ishimaru (1991), "Application of the finite element method to Monte Carlo simulations of scattering of waves by random rough surfaces with the periodic boundary condition", *Journal of Electromagnetic Waves and Applications*, 5(8): 835-855.

Pierson, W.J. and L. Moskowitz (1964), "A proposed spectral form for fully developed wind seas based on the similarity theory of S.A. Kitaigorodskii", *Journal of Geophysics Research*, 69: 5181-5190.

Pino, M.R. et al. (1999), "The generalized Forward-Backward method for analyzing the scattering from targets on ocean-like rough surfaces", *IEEE Transactions on Antennas and Propagation*, 47(6): 961-968.

Rouvier, S. and I. Chenerie (1997), "Ultra wide band electromagnetic scattering of a fractal profiles", *Radio Science*, 32(2): 285-293.

Sancchez-Gil, J.A. and M. Nieto-Vesperinas (1991), "Light scattering from random rough dielectric surface", *Journal of the Optical Society of America (A)*, 8(8): 1270-1286.

Savailis S. et al. (1997), "Scattering from fractally corrugated surfaces with use of the extended boundary condition method", *Journal of the Optical Society of American A*, 14(2): 475-485.

Thorsos, E. (1990), "Acoustic scattering from 'Pierson-Moskowitz' sea surface", *Journal of the Acoustic Society of the America*, 88(1): 335-349.

Thorsos, E. (1988), "The validity of the Kirchhoff approximation for rough surface scattering using a Gaussian roughness spectrum," *The Journal of the Acoustical Society of America*, 83(1): 78-92.

Toporkov, J.V. (1998), *Study of Electromagnetic Scattering from Randomly Rough Ocean-like Surface Using Integral –Equation-Based Numerical Technique,* Ph.D. dissertation, Virginia Polytechnic Institute and State University, Blacksburg, VA.

Tsang, L., J.A. Kong and R.T. Shin (1985), *Theory of Microwave Remote Sensing*, New York: Wiley.

Tsang, L., C.H. Chang, K. Pak, and H. Sangani (1995), "Monte Carlo simulations of large-scale problems of random rough surface scattering and applications to gazing incidence with the BMIA/canonical grid method", *IEEE Transactions on Antennas and Propagation*, 43: 851-859.

Tsang, L. and J.A. Kong (2001), *Scattering of Electromagnetic Waves (vol. 2: Numerical Simulations)*, New York: Wiley.

West, J.C. (1997), "Effect of shadowing on electromagnetic scattering from rough ocean wavelike surfaces at small grazing angles", *IEEE Transactions on Geosicence and Remote Sensing*, 35(2): 293-301.

West, J.C. and M.J.M. Sturm (1998), "Low-grazing scattering from breaking water waves using an impedance boundary MM/GTD approach", *IEEE Transactions on Antennas and Propagation*, 46(1): 93-100.

Ye, H. and Y.Q. Jin (2005), "Parameterization of tapered incident wave for electromagnetic scattering simulation from randomly rough surface", *IEEE Transactions on Antennas and Propagation*, 53(3): 1234-1237.

Chapter 6

Numerical Finite Element Method (FEM) for Scattering from Randomly Rough Surface and Target 260

6-1 Angular Correlation Function of Scattering from an Object over a Randomly Rough Surface in a FEM Approach

Scattering of electromagnetic waves by a randomly rough surface has been studied extensively by some analytic approaches such as Kirchhoff method, the small-perturbation method, two-scale method, etc. [Jin 1994]. These methods are all derived from different approximations and are restricted in their domains of validity. Over recent years, there has been increasing interest in Monte Carlo numerical simulation of wave scattering from a rough surface, for example, the integral equation moment method (IEM) [Axline and Fung 1978] or the finite-element method (FEM) [Lo et al. 1991]. In the IEM method one is required to solve a full matrix equation. It is limited by computer memory and CPU time. A FEM has been applied to the rough surface problem [Lo et al. 1991]. The periodic boundary condition is used in a discretization of the plane-wave spectrum. The total number of modes in the FEM is much smaller than the number of basis functions in the IEM, and the FEM method is much more efficient than IEM terms of CPU time and memory storage requirements.

Recently, the novel concept of the angular correlation function (ACF) and enhancement on the memory line has been studied [Le et al. 1996, Freund et al. 1988, Tsang et al. 1996]. To what extent is information about the spatial variation in phase and amplitude of an incident wave preserved during its transmission through a highly random, multiple scattering medium [Freund et al. 1988]. The ACF is the correlation function of two scattered fields corresponding, respectively, to two incident waves. The memory line describes the relation of these scattered and incident directions. It has been discussed that the ACF of scattering from a randomly rough surface is generally small away from the memory line. On the memory line, the ACF can have a peak value. The memory line is a result of statistical translational invariance of a randomly rough surface, and is analogous to the phase-matching condition of the plane surface [Le et al. 1996, Tsang et al. 1996]. The memory line was introduced by Ishimaru and co-workers [Le et al. 1996] as a useful generalization of earlier geometric studies. The ACF based on the second-order Kirchhoff approximation was studied compared with the experiment.

The problem of wave scattering from a scatter target placed over a randomly rough surface is of great importance in many applications, such as the diagnosis of particles on a rough surface of layered materials, detection of a target over rough land or sea surfaces. However, coupled scattering from a scatterer over a surface has only been studied for the case of a plane surface [Videen et al. 1993], and there seems to be no explicit rule for detecting a target over the surface.

This session employed the FEM of scattering from a rough surface, and develops and approach to study numerically scattering from a scatter target placed over a rough surface. Numerical results show that the presence of a scatter target over a rough surface can broaden the width of the ACF away from the memory line. The phenomenon of ACF broadening on a non-memory line can be used as a guide for detection of scatter over a rough surface [Jin and Li 2000].

6-1.1 The FEM for Scattering from a Randomly Rough Surface

We first consider a plane wave with angle $(\pi-\theta_i)$ incident on a one-dimensional randomly rough conducting surface without a scatter target, as shown in Fig. 6-1.1.The randomly rough surface $z = f(x)$ is realized by the Monte Carlo method with a Gaussian surface roughness spectrum with an RMS surface height of h and correlation length ℓ.

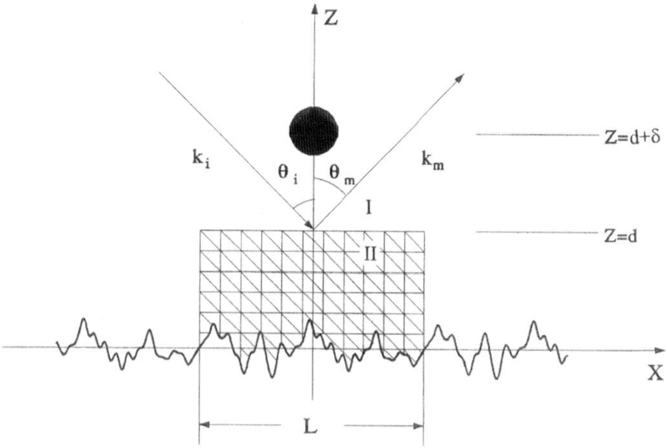

Fig. 6-1.1. Geometry of the problem

The average of the scattering intensity can be obtained by the Monte Carlo simulations. The rough surface with sample length L is extended periodically with period L so that periodic boundary conditions are used. Region I is a homogeneous space ($z \geq d$, $|x| \leq L/2$) and region II consists of the inhomogeneous region ($d \geq z \geq f(x)$, $|x| \leq L/2$), with d chosen to be larger than the maximum height of the rough surface. Discrete Floquet modes of scattering fields are produced from the random and periodic rough surface. The total field in region I is written as [Lo et al. 1991]

$$\Psi^I(x,z) = \exp[i\frac{2\pi m_0}{L}x - ik_{iz}(z-d)] + \sum_{m=-N}^{N} B_{m-m_0} \exp[i\frac{2\pi m}{L}x - ik_{izm}(z-d)] \tag{6-1.1}$$

where the first term on RHS is the incident field, and one can choose $m_0 = L\cos\theta_i/\lambda$; the second term is a scattered field, the truncation number N is determined by the propagation modes and evanescent fields. B_{m-m_0} is to be solved. In other words,, the scattering wave from a rough surface creates $2N$ discrete Floquet modes and every mode has a scattering angle $\theta_m = \sin^{-1}(m\lambda/L)$, $m=\pm1$,..., $\pm N$. Using the Fourier transform to expand Eq. (6-1.1), the total field in region I is rewritten as

$$\psi^I(x,z) = \cos[\frac{2\pi m_0}{L}x - k_{iz}(x-d)] + i\sin[\frac{2\pi m_0}{L}x - k_{iz}(x-d)] + a_0 e^{ik(z-d)}$$
$$+ \sum_{m=-N}^{N} [a_m \cos(\frac{2\pi m}{L}x) + b_m \sin(\frac{2\pi m}{L}x)]e^{ik_{zm}(z-d)} \tag{6-1.2a}$$

where a_0, a_m, b_m are the coefficients to be solved. If we define the scattered field in the scattering direction θ_m from the incident direction $\pi - \theta_i$ by $R(\theta_m, \pi-\theta_i)$, Eq. 6-1.2a) becomes

$$\psi_s^I(x,z) = \sum_{m=-N}^{N} R(\theta_m, \pi - \theta_i) \tag{6-1.2b}$$

where $R(\theta_m, \pi-\theta_i)$ is expressed by a_0, a_m, b_m.

In region II, the wave equations

$$(\nabla^2 + k^2)\psi_m^{II,c} = 0 \qquad (\nabla^2 + k^2)\psi_m^{II,s} = 0 \qquad m = 1,2,\cdots,N \tag{6-2.3}$$

and the Dirichlet and periodic boundary conditions

$$\psi_m^{II,c(s)}(x, f(x)) = 0 \tag{6-2.4a}$$

$$\psi_m^{II,c(s)}(-L/2, z) = \psi_m^{II,c(s)}(L/2, z) \tag{6-2.4b}$$

$$\psi_m^{II,c}(x,d) = \cos(\frac{2\pi m}{L}x) \tag{6-2.4c}$$

$$\psi_m^{II,s}(x,d) = \sin(\frac{2\pi m}{L}x) \tag{6-2.4d}$$

are solved by the FEM [Lo et al. 1991]. The total field in region II can be written as

$$\psi^{II}(x,z) = A_0\psi_0^{II,c}(x,z) + \sum_{m=1}^{N}[A_m\psi_m^{II,c}(x,z) + B_m\psi_m^{II,s}(x,z)] \tag{6-2.5}$$

where the superscripts *c, s* indicate the fields with the dependence of $\cos(2\pi mx/L)$ and $\sin(2\pi mx/L)$, respectively.

At the boundary *z=d* between regions I and II, ψ^I and ψ^{II} satisfy the continuous boundary conditions. Thus, solving the wave equations of ψ^{II}, Eqs .(6-2.3~5), by using the FEM, the coefficients a_0, a_m, b_m are then determined and the scattering fields ψ_s^I can be obtained. The bistatic scattering $\gamma_s(\theta_s, \theta_i)$ and backscattering $\sigma^o(\theta_i)$ are then solved. Numerical approach can be refereed to Lo et al. (1991).

Note that the Gaussian rough surface in Fig.6-1.1, realized by the Monte Carlo method, is described by two surface parameter: the *rms* surface height δ and correlation length ℓ. The Gaussian surface spectrum is written as

$$W(k) = (\delta^2\ell/2\sqrt{\pi})\exp(-k^2\ell^2/4) \tag{6-1.6}$$

By using the Fourier transform, the surface profile $z = f(x)$ is expressed as

$$f(x) = \frac{1}{L}\sum_{n=-N/2}^{N/2-1} F(k_n)\exp(ik_n x) \tag{6-1.7}$$

where

$$F(k_n) = \sqrt{2\pi L W(k_n)}\begin{cases} \frac{1}{\sqrt{2}}[N(0,1) + iN(0,1)] & n \neq 0, N/2 \\ N(0,1) & n = N/2 \end{cases} \tag{6-1.8}$$

here $k_n = 2\pi n/L$, $N(0,1)$ is random number with zero-mean and normally distributed in [0,1].

6-1.2 Scattering from a Scatter Target Situated over Randomly Rough Surface

Now we consider a scatter target at $z = D$ placed over the rough surface. The scattering field from the scatterer in the direction θ_m, in the far-field region, is described by its scattering amplitude function $f(\theta_s, \theta_i)$ as

$$\psi_s^o(\theta_m, \pi-\theta_i) = \frac{e^{ikr}}{\sqrt{r}} f(\theta_m, \pi-\theta_i)\psi_{i0} \equiv A(\theta_m, \pi-\theta_i) \tag{6-1.9}$$

where the superscript 'o' indicates that the target is now situated in region I. A rigorous solution for coupled scattering from a target closely placed over an underlying surface needs a numerical approach for the computational electromagnetics. To make an explicit analysis without complex computation, the far-field approximation is assumed in Eq. (6-1.9) and multiple scattering from the target and the underlying surface is omitted as in [Videen et al.

1993]. Figs. 6-1.2(a~d) describe the scattered field ψ_s^o in region I coupled from the discrete Floquet modes of the rough surface and scattering of the scatter target can be given as follows:

$$\begin{aligned}\psi_s^o(\theta_m,\theta_i) &= A(\theta_m,\pi-\theta_i)+\sum_n R(\theta_m,\pi-\theta_i)A(\pi-\theta_n,\pi-\theta_i)\exp[i(k_{zn}+k_{zm})\delta] \\ &+\sum_n A(\theta_m,\theta_n)R(\theta_n,\pi-\theta_i)\exp[i(k_{zn}+k_{iz})\delta] \\ &+\sum_n\sum_{n'} R(\theta_m,\pi-\theta_{n'})A(\pi-\theta_{n'},\theta_n)R(\theta_n,\pi-\theta_i)\exp[i(k_{zn}+k_{iz}+k_{zn'}+k_{zm})\delta]\end{aligned} \quad (6\text{-}1.10)$$

where $\delta = D - d$.

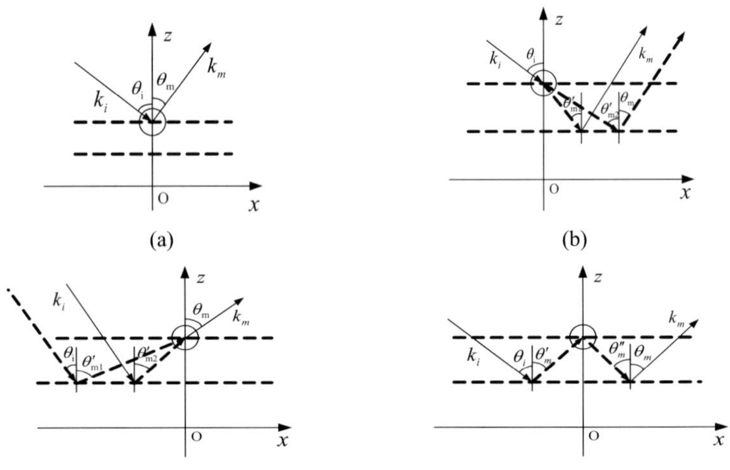

Fig. 6-1.2. Coupled scattering from the target over a rough surface

6-1.3 Angular Correlation Function and Memory Line

It has been understood that the correlation function

$$\begin{aligned}\Gamma(\theta_{s1},\theta_{i1};\theta_{s2},\theta_{i2}) &= \langle \psi_s^I(\theta_{s1},\theta_{i1})\psi_s^I(\theta_{s2},\theta_{i2})\rangle / \sqrt{P_1P_2} \\ &= \frac{1}{N_r}\sum_{q=1}^{N_r} \psi_s^I(\theta_{s1},\theta_{i1})\psi_s^I(\theta_{s2},\theta_{i2}) / \sqrt{P_1P_2}\end{aligned} \quad (6\text{-}1.11)$$

is generally small away from the memory line. On the memory line with the condition

$$\sin\theta_{s2}-\sin\theta_{i2}=\sin\theta_{s1}-\sin\theta_{i1} \quad (6\text{-}1.12)$$

the ACF has a peak value. The index q in Eq. (6-1.11) denotes the realization of the randomly rough surface by using a Monte Carlo method. N_r is the total number of realizations, P_1 and P_2 are the energy flux into surface at the incident angles θ_{i1} and θ_{i2}, respectively; that is, $P_t = L\cos\theta_t, t=1,2$. The normalized scattered intensity at the discrete angle θ_m is $L|B_m|^2\cos^2\theta_m/\cos\theta_i$.

The ACF is a complex correlation function, and its real part has a peak value on the memory line. In the case of a scatter target placed over the rough surface, the scattering fields ψ_s^I in Eq. (6-1.11) should be calculated using more complicated fields, ψ_s^0 of Eq. (6-1.7).

As we observe in Fig. 6-1.2(b), it is expected that the width of the ACF can be broadened even away from the memory line. Let us assume that the scattering from

$(\theta_{m1},\theta_{i1})$ and $(\theta_{m2},\theta_{i2})$ do not satisfy the memory line condition of Eq. (6-1.12) and the ACF is very small in the case without the presence of a scatterer. However, as the scatter target is over the rough surface, the incident wave is scattered diffusely by the target into all scattering angles such as from incidence θ_{i1} to scattering in $\theta_1,\theta_1',\cdots$, from incidence θ_{i2} to $\theta_2,\theta_2',\cdots$, etc. It will be possible to have a case such that (θ_{m1},θ_1') and (θ_{m2},θ_2') satisfy the memory line condition of Eq. (6-1.12). It causes the enhancement of ACF even on a non-memory line. Figs. 6-1.2(c,d) can also be discussed similarly.

So we can see that due to the scattering pattern redistributed by the presence of the scatter target, multiple angles incident upon the rough surface would have a chance of satisfying the memory line condition and enhancing ACF even on a non-memory line of (θ_{m1},θ_1) and (θ_{m2},θ_2). Due to the weighting effect from the scattering pattern of $A(\theta_m,\theta_i)$, ACF enhancement on a non-memory line has a function angular dependence on the scatter shape, size and δ, etc.

Actually, as shown in [Freund et al. 1988], the width of the ACF depends upon the sideways diffusion of the radiation. Since the ACF is the Fourier transform of this diffusive spot, it has a narrow width when sideways diffusion is large, as for a rough surface, while the width broadens when sideways diffusion is suppressed, as for a smooth scattering sample. As expected in the case of the presence of the target, which appears much smoother, not much rougher, than the randomly rough surface, the sideways diffusion is suppressed and the ACF would become broadened.

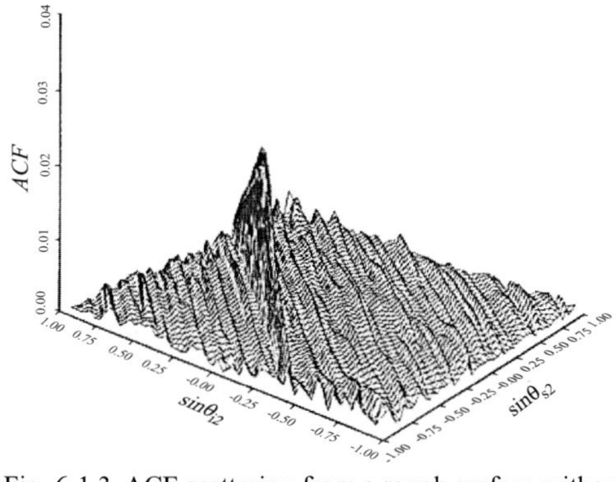

Fig. 6-1.3. ACF scattering from a rough surface without a target

Fig. 6-1.3 shows a mesh of the ACF with the absence of the scatter target over the surface versus $(\sin\theta_{i2},\sin\theta_{s2})$, where $\theta_{i1}=15.319°,\theta_{s1}=-15.319°$. The parameters in the calculation are taken as follows: the surface variance $h=0.3\lambda$, the correlation length $\ell=0.5\lambda$, the period $L=30\lambda$ and $d=\max(z)+0.1\lambda$. It can be seen that the ACF has a conspicuous enhancement on the memory line and becomes much smaller away from the line of Eq. (6-1.12).

The function $A(\theta_m,\pi-\theta_i)$ in Eq. (6-1.9) is the scattering amplitude function of the scatter, which depends on the scatter's size and shape. The fluctuation behavior of the ACF versus scattering angles is closely related to the scatter size and shape. As an example to show change of the ACF due to the presence of a scatter target, let us simply take $A(\theta_m,\pi-\theta_i)$ as a constant and move it out from the summation of Eq. (6-1.10). Fig. 6-1.4 shows the ACF from the scattered fields of Eq. (6-1.10), i.e. with the presence of a scatter target over the surface. The parameters $\delta = 0.1\lambda$, $A_0 = 0.0065$ are taken and other parameters are the same as in Fig. 6-1.3. Comparing with Fig. 6-1.3, it can be seen that the ACF in Fig. 6-1.4 broadens for the vast range of $(\sin\theta_{i2}, \sin\theta_{s2})$. In other words, with the presence of a scatter target over the underlying surface, the width of the ACF becomes broadened even away form the memory line.

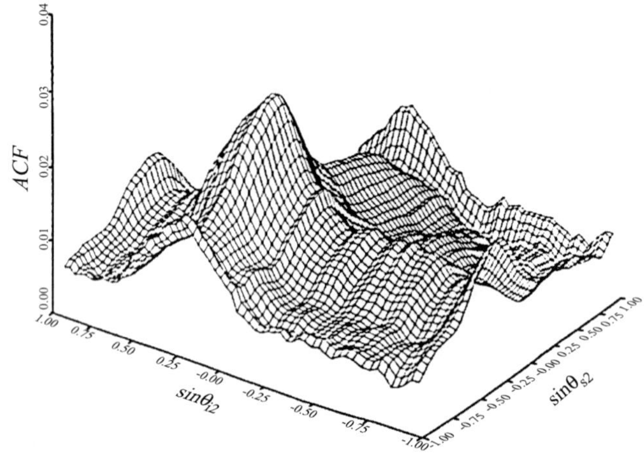

Fig. 6-1.4. ACF scattering from a target over the rough surface without a target

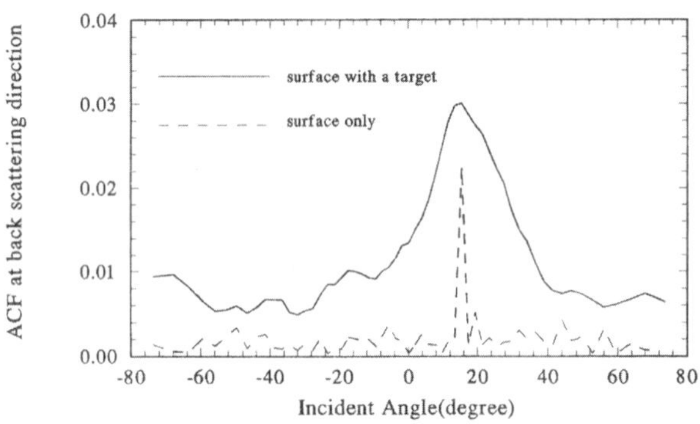

Fig. 6-1.5. ACF in the backscattering direction for Figs. 6-1.3 and 6-1.4 without and with the presence of a target

More explicitly, Fig. 6-1.5 gives a comparison with Figs. 6-1.3 and 6-1.4 in the backscattering direction. The difference of the ACF with and without the scatterer over a rough surface, especially in the backscattering direction, is significant.

The angular ACF is closely related to the question of whether the scatterer is effectively two (2D) or three dimensional (3D). A discussion of the difference between 2D and 3D was discussed in [Freund et al. 1989]. Since the targetover the rough surface can clearly break the effective 2D symmetry of the surface, the behavior of angular ACF is definitely affected as a function of the scatterer geometry. The distance, $d+\delta$ in Fig. 6-1.1, of the target over the surface can affect the angular variation of the ACF (both its width and peak). Its functional dependence remains to be studied if Eqs. (6-1.9,6-1.10) can be calculated more accurately.

6-2 Numerical Simulation of Bistatic Scattering from a Target above Rough Sea Surface under an EM Wave Incidence at Low Grazing Angle by Using the FEM

Study of scattering from a volumetric target above randomly rough surface has broad applications for target detection in terrain and oceanic clutters. Most of analytical approaches of rough surface scattering are based on some approximations, such as the Kirchhoff tangent plane approximation, small perturbation approximation, two-scale model and others [Jin 1994]. Analytical study of scattering from an object above the surface has been limited to the model of a sphere or a cylinder above a planar conducting surface [Videen 1991, Videen and Ngo 1997]. As numerical model for scattering from complex target and rough surface, computational electromagnetics has to be studied, e.g. integral equation method [Tsang et al. 2001], Finite Element Method [Jin and Li 2000], Finite Difference Time Domain method (FDTD), as well as widely discussed Forward Backward Method (FBM) [Holliday et al. 1996, Pino et al. 1999, Li and Jin 2002, Hastings et al 1995].

Because the integral equation is numerically discretized to produce a dense matrix to be solved, it requires accurate and fast algorithm to accelerate computation and improve the accuracy and efficiency, especially for EM incidence at LGA. When a target is above the rough surface, the scattering behavior includes the coupling interactions of a complex volumetric object and rough surface. The FEM can accommodate these multiple interactions in a tractable way. The approaches of FEM and FDTD have been limited to the scattering problem of a single rough surface [Tsang et al. 2001, Jin and Li 2000, Hastings et al. 1995], or an object buried beneath a rough interface at EM normal incidence [Fan et al. 2002, Moss et al. 2002]. Numerical simulation for bistatic scattering from composite model of a target above rough surface has been remained to study.

In this session, a numerical composite model is developed by using the FEM method to study scattering from a target at low altitude above rough sea surface under EM wave incidence at LGA. The CPML (conformal perfectly matched layer) [Teixeira and Chew 1998, Liu et al. 2001] is employed as the truncation boundary of FEM to reduce the reflection error [Hastings et al. 1995] caused by conventional planar PML at LGA. This approach can adapt numerical simulation to the model of the scatter target above a randomly rough surface. The code of our FEM/CPML is examined by available solution of the FBM. Bistatic and back-scattering from the composite model of a target above randomly rough sea surface, and functional dependence upon the sea surface wind speed, target altitude, incident and scattering angles etc. are numerically demonstrated and discussed. This study presents numerical description for the observation principle and physical insight associated with the coupling interactions of a complex target and randomly rough sea surface [Liu and Jin 2004a].

6-2.1 A Composite Model for a Target above Rough Surface

The FEM usually employs an open-region boundary condition to truncate the size of computation region. One approach is to use the PML, which acts as a layer of lossy anisotropic medium surrounding the target/surface region, and absorbs the outgoing wave with no-reflection. Therefore, the scatter region enclosed by the PML is equivalent to being in free space. It makes the FEM be performed in a finite region.

As shown in Fig. 6-2.1, suppose to have a target located at (x_0, h) above a 2-D rough surface $z = f(x)$, where h is respect to the mean value $< f(x) >= 0$. The rough surface is generated by using the Monte Carlo method in the P-M sea spectrum [Pierson and

Moskowitz 1964]. The conformal or the planar PML encloses the target/surface region. The surface length L is illuminated by a tapered EM wave incidence. Γ_p, Γ_c are the far-field integration paths for planar and conformal PML, respectively. h' is the path height in horizontal direction.

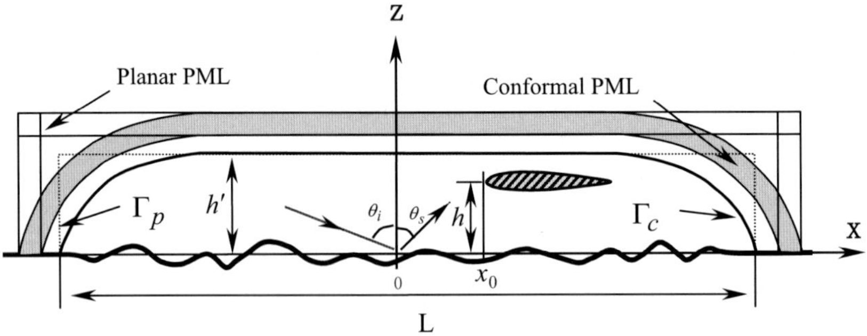

Fig. 6-2.1. The physical model of a target above the rough sea surface

Scattered wave above the rough surface, including the PML region, satisfies the vector wave equation

$$\nabla \times (\frac{1}{\mu_r} \overline{\overline{\Lambda}}^{-1} \cdot \nabla \times \overline{E}_s) - k_0^2 \varepsilon_r \overline{\overline{\Lambda}} \cdot \overline{E}_s = 0 \tag{6-2.1}$$

where $\overline{E}_s$ is the scattered electric field, ε_r, μ_r are the relative permittivity and permeability, respectively. k_0 is the wave number in free space. The tensor $\overline{\overline{\Lambda}}$ is the constitutive parameter of PML.

On the perfectly electric conducting (PEC) surfaces of the target/rough surface, the boundary condition yields

$$\hat{n} \times \overline{E}_s = -\hat{n} \times \overline{E}_i \tag{6-2.2}$$

where $\overline{E}_i$ is the incident electric field, and $\hat{n}$ is normal vector of the surface. At the bottom interface of PML, as shown in Fig. 6-3.1, the scattering fields would be completely absorbed and yields

$$\hat{n} \times \overline{E}_s = 0 \tag{6-2.3}$$

Eqs. (6-2.1~3) are to be solved for our scattering model of Fig. 6-2.1.

In accordance with the generalized variational principle [Jin 1993], the solution of the boundary-value problem defined by Eqs.(6-3.1~3) is equivalent to seeking the stationary point of the functional given by

$$F(\overline{E}_s) = \frac{1}{2} \iiint_V \left[\frac{1}{\mu_r} (\nabla \times \overline{E}_s) \cdot \overline{\overline{\Lambda}}^{-1} \cdot (\nabla \times \overline{E}_s) - k_0^2 \varepsilon_r \overline{E}_s \cdot \overline{\overline{\Lambda}} \cdot \overline{E}_s \right] dv \tag{6-2.4}$$

In the local orthogonal coordinates (ξ_1, ξ_2, ξ_3), $\overline{\overline{\Lambda}}$ is expressed in the form of uniaxial tensor

$$\overline{\overline{\Lambda}}_{\xi_1,\xi_2,\xi_3} = \begin{bmatrix} a & 0 & 0 \\ 0 & b & 0 \\ 0 & 0 & c \end{bmatrix}_{\xi_1,\xi_2,\xi_3} \tag{6-2.5}$$

It can be shown that Eq. (6-2.5) is a general expression for both planar and conformal PML. By using the Jacobi matrix $\overline{\overline{J}}$, $\overline{\overline{\Lambda}}$ can be transformed from the local coordinates to global Cartesian coordinates as follows

$$\overline{\overline{\Lambda}}_{x,z,y} = \overline{\overline{J}}^{-1} \cdot \overline{\overline{\Lambda}}_{\xi_1,\xi_2,\xi_3} \cdot \overline{\overline{J}} \tag{6-2.6}$$

In our problem of a 2-D rough surface under horizontally (*TE*, i.e. E_y) polarized wave incidence, it yields

$$\nabla \times \overline{E}_s = \begin{bmatrix} 0 & 1 \\ -1 & 0 \end{bmatrix}_{x,z} \nabla E_{sy} \tag{6-2.7}$$

Assuming ξ_3 axis be the $\hat{y}$-direction, $\overline{\overline{\Lambda}}$ can be re-written as

$$\overline{\overline{\Lambda}}_{x,z,y} = \begin{bmatrix} \widetilde{\Lambda} & 0 \\ 0 & \beta \end{bmatrix} \tag{6-2.8}$$

$$\widetilde{\Lambda} = \widetilde{J}^t \begin{bmatrix} a & 0 \\ 0 & 1/a \end{bmatrix} \widetilde{J} \tag{6-2.9}$$

where $\widetilde{\Lambda}$ is a 2×2 sub-matrix and $\widetilde{J}$ is a 2-D Jacobi matrix.

Then Eq. (6-2.4) becomes

$$F(E_{sy}) = \frac{1}{2} \iint_\Omega \left(\frac{1}{\mu_r} \nabla E_{sy} \cdot \widetilde{\Lambda} \cdot \nabla E_{sy} - k_0^2 \varepsilon_r \beta E_{sy}^2 \right) d\Omega \tag{6-2.10}$$

From Eqs. (6-2.2) and (6-2.3), the boundary conditions on PEC surface Γ_{PEC} and PML bottom surface Γ_{PML} yield

$$E_{sy}\big|_{\Gamma_{PEC}} = -E_{iy}\big|_{\Gamma_{PEC}} \tag{6-2.11}$$

$$E_{sy}\big|_{\Gamma_{PML}} = 0 \tag{6-2.12}$$

The incident field E_{iy} is a Thorsos tapered wave [Thorsos 1988]

$$E_{iy} = \exp[-(x + z tg\theta_i)^2 / g^2] \exp[-jk_0(x \sin\theta_i - z\cos\theta_i)(1+w)] \tag{6-2.13}$$

where g is the tapering parameter and

$$w = [2(x + z tg\theta_i)^2 / g^2 - 1] / (k_0 g \cos\theta_i)^2 \tag{6-2.14}$$

Evaluating the stationary point of the functional problem of Eqs. (6-2.10~12) by using the FEM, the scattering field at each point of the computation region can be solved. The bistatic and back-scattering coefficients can be calculated.

The bistatic scattering coefficient, i.e. bistatic RCS (radar cross section) for tapered wave incidence upon 2-D rough surface is defined as Thorsos (1988)

$$\sigma = \lim_{r \to \infty} 2\pi r \frac{|E_{sy}|^2}{P_i} \tag{6-2.15}$$

where

$$P_i = \sqrt{\frac{\pi}{2}} g \cos\theta_i \left(1 - \frac{1 + 2tg^2\theta_i}{2(k_0 g \cos\theta_i)^2}\right) \tag{6-2.16}$$

6-2.2 Conformal PML Absorbing Boundary

The perfectly matched layer (PML) was recently proposed by Berenger (1994). Ideally, the PML should be no-reflectional for any incident angle and broad frequency spectrum, and absorb all outgoing waves. In computation, the PML usually requires the computation region as a rectangular box. In the corner region intersected by several pieces of planar PML, the reflection error might be caused due to the abrupt change of geometrical and physical parameters. Meanwhile, the height of horizontal integral path h' has to be increased to contain the target presence. As the wave is incident at LGA, main lobe of scattering pattern from rough surface can intersect the corner region of PML and cause large system error.

The conformal PML [Teixeira and Chew 1998] is employed to circumvent this problem. As shown in Fig. 6-2.1, at the end of horizontal planar PML, a quarter of elliptical shaped conformal PML is patched to the corner regions.

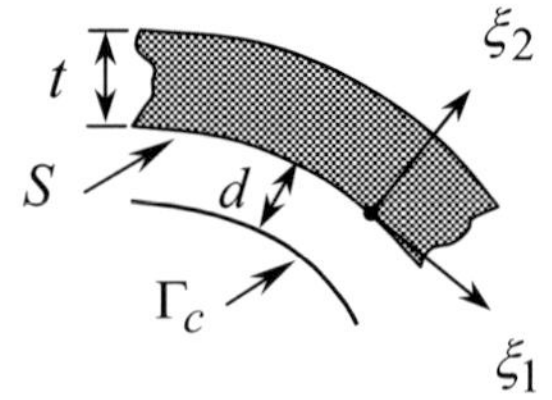

Fig. 6-2.2. Conformal PML

The conformal PML was firstly presented by Teixeira and Chew (1998) in terms of anisotropic tensor. The planar, cylindrical, and spherical PML are actually the special cases of the conformal PML. As shown in Fig. 6-3.2, the conformal PML is a smooth convex shell of anisotropic medium. Here conformal means that the inner interface S of the conformal PML is parallel to the target surface or a fictitious surface Γ_c surrounding the target. On each point of the parallel surface, the constitutive parameters of the conformal PML is given by $\overline{\overline{\mu}} = \mu_r \overline{\overline{\Lambda}}$ and $\overline{\overline{\varepsilon}} = \varepsilon_r \overline{\overline{\Lambda}}$, where

$$\overline{\overline{\Lambda}} = \hat{\xi}_1 \hat{\xi}_1 \left(\frac{s_2 s_3}{s_1} \right) + \hat{\xi}_2 \hat{\xi}_2 \left(\frac{s_1 s_3}{s_2} \right) + \hat{\xi}_3 \hat{\xi}_3 \left(\frac{s_1 s_2}{s_3} \right) \tag{6-2.17}$$

In the cross-section plane x-z, the ξ_1- and ξ_2-axes are tangential and perpendicular to the surface S, respectively. The parameter t is the thickness of conformal PML, d is the distance between S and Γ_c, and $\xi_2 = 0$ on S. The definition and realization of 3-D conformal PML have been discussed in [Teixeira and Chew 1998, Liu et al. 2001]. Fig. 6-2.2 shows a special case [Liu et al. 2001] when the curvature radius in $\hat{y}$-direction is infinity. In our model, we obtain

$$s_1 = 1 - j\gamma \xi_2^3 / (3t^2 r_{\xi_1}) \tag{6-2.18}$$

$$s_2 = 1 - j\gamma \xi_2^2 / t^2 \tag{6-2.19}$$

$$s_3 = 1 \tag{6-2.20}$$

where the constant γ determines the attenuate rate of the PML. r_{ξ_1} is the curvature radius in the ξ_1-direction at the point $P(\xi_1, \xi_2, \xi_3)$ on the parallel surface. In order to make smooth change between the interface of the PML and free space, the imaginary part of s_2 is chosen as a second order taper function. Let $r_{\xi_1} = \infty$, Eq. (6-2.17) is reduced to the form of planar PML.

The absorbing ability of conformal PML becomes better with increasing t and γ. But, increasing these two parameters might slow down the speed of FEM because increasing t yields larger Ω of Eq. (6-2.10) and larger scale of FEM equation. On the other hand, the condition number of the FEM system matrix would be increased with larger γ and the iterative convergence becomes much slower. Thus small values of t and γ are preferred as enough absorption can be achieved.

6-2.3 Numerical Results and Discussions

We are concern with a tapered plane wave incident upon a 2-D perfectly conducting rough surface generated by one Monte Carlo realization in the P-M spectrum. The tapering parameter $g = L/6$ and the wavelength $\lambda = 1\,m$. The parameters of the PML are chosen as $d = 0.1\lambda$, $t = 0.2\lambda$, and $\gamma = 18$. The axial ratio of elliptical conformal PML is 6 while the major axis of ellipse coincides with the horizontal plane.

Our FEM code is first examined by available solution of the FBM method. Fig. 6-2.3 gives the bistatic scattering coefficient of a conducting flat surface under EM incidence at LGA $\theta_i = 80^\circ$. Other parameters are chosen as $h' = 3\lambda$, and $L = 300\lambda$. It can be seen that the result of the FEM with the planar PML causes too high scattering around backscattering angle (-80°), which is due to error on the edge region of the planar PML. As a comparison, the FEM with conformal PML not only eliminates the reflection error, but also much improves the accuracy for all scattering angles. Certainly, there exists only specular reflection from the flat surface. However, calculations of Fig. 6-2.3 can be used as an accuracy test of different approaches.

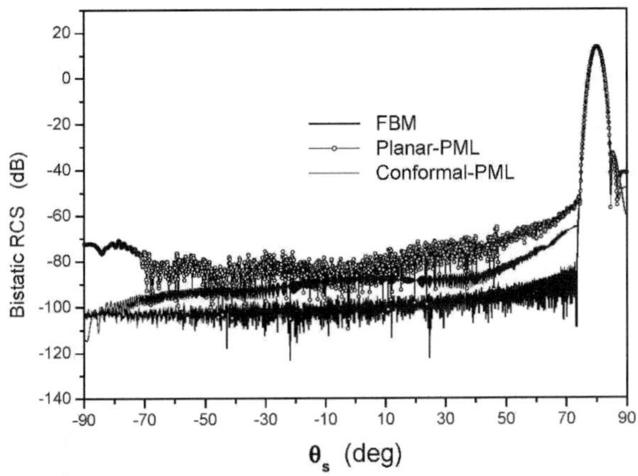

Fig. 6-2.3. Bistatic scattering from conducting flat surface by using three methods

Fig. 6-2.4 shows bistatic scattering from rough sea surface driven by strong wind speed ($u = 10\,m/s$), where $L = 409.6\lambda$ and conformal PML is applied. Incident angle is also $\theta_i = 80^\circ$. It could be seen that our FEM results are well validated by the FBM.

We now suppose to have a target of NACA0012 four-digital airfoil with chord length $2\,m$ located above the rough surface with $L = 409.6\lambda$ and $\theta_i = 80^\circ$. The airfoil's leading edge is at $x_0 = 0\,m$ and the center axis at the height $h = 1\,m$. The code for bistatic RCS of the NACA0012 airfoil was first validated by the result of [Shankar et al. 1989] for incidence $\theta_i = 0^\circ$ in free space.

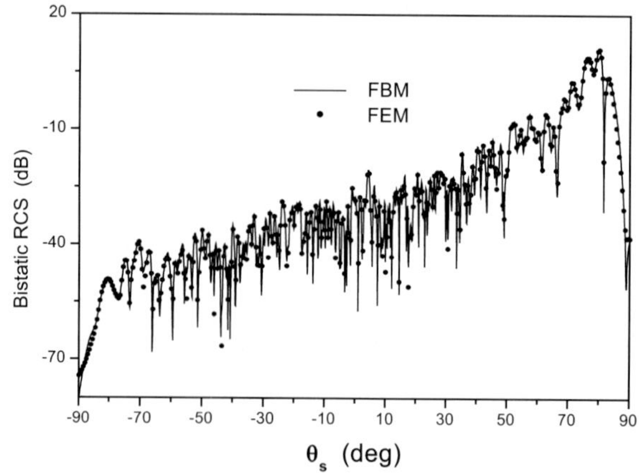

Fig. 6-2.4. Bistatic scattering from rough sea surface using the FEM and FBM methods

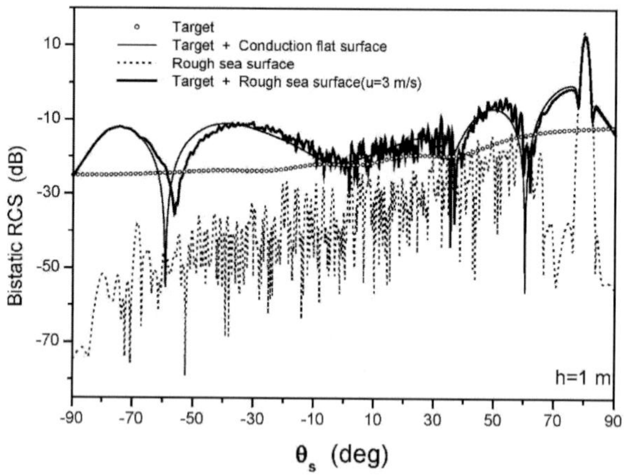

Fig. 6-2.5. Bistatic scattering from a target above the surfaces

Fig. 6-2.5 shows bistatic scattering when: (1) the target is in the free space, (2) the target is above the conducting flat surface, and (3) the target is above rough sea surface driven by sea surface wind speed $u = 3m/s$.

It can be seen that angular scattering from the target above a rough surface at such low wind speed is largely similar to the case of the target above a flat surface. Analysis of the image method is applicable. These results are obtained by using one Monte Carlo realization. If the result is averaged over multi-realizations, the fluctuation of scattering pattern should be more smoothed.

Figs. 6-2.6(a,b,c) show bistatic scattering from a target at different altitude (h) above the rough sea surface: (a) conducting flat surface, (b) rough sea surface with $u = 3m/s$, (c) rough sea surface with $u = 10m/s$. The parameters of target and incident wave are the same as Fig. 6-2.5.

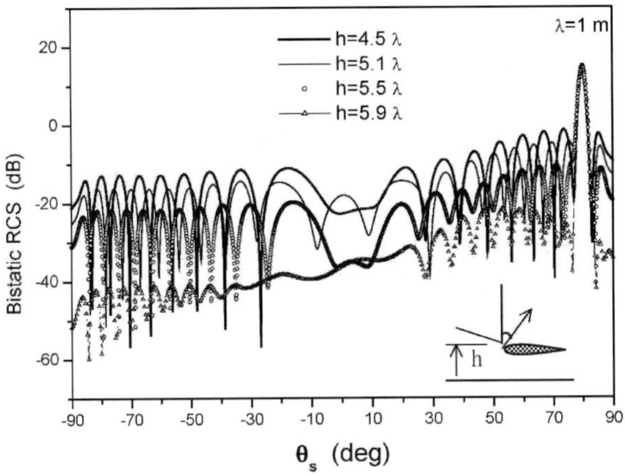

Fig. 6-2.6 (a). Bistatic scattering of the target as a function of height (flat surface)

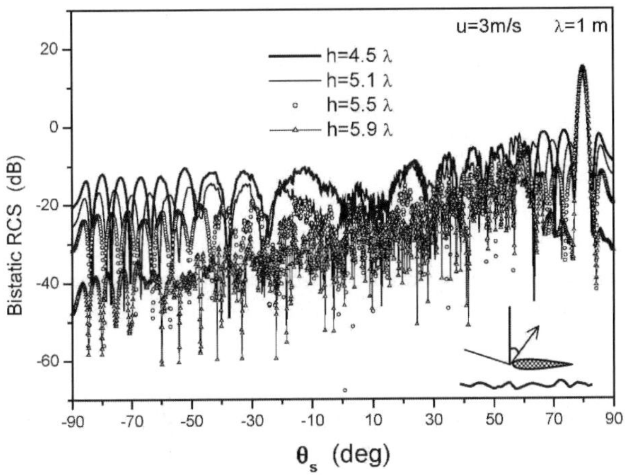

Fig. 6-2.6 (b). Bistatic scattering of the target as a function of height (low speed)

In Figs. 6-2.6(a) and (b), the fluctuation of bistatic scattering versus scattering angle is largely regular, and the target altitude affects the angular scattering pattern. This phenomenon is caused by interaction between the target and its mirror image. However, as the sea surface becomes much rough at strong surface wind speed of Fig. 6-2.6(c), angular pattern of bistatic scattering becomes more complex, i.e. not regular.

According to the image method and two-source array radiation principle, single scattering fields from the target and its mirror image have a phase difference $\Delta\Phi$ as

$$\Delta\Phi = 2k_0 h \cos\theta_s \tag{6-2.21}$$

The variation ratio of $\Delta\Phi$ respecting to θ_s is

$$d(\Delta\Phi)/d\theta_s = 2k_0 h \sin\theta_s \tag{6-2.22}$$

It means that $d(\Delta\Phi)/d\theta_s$ becomes larger with increasing h and θ_s.

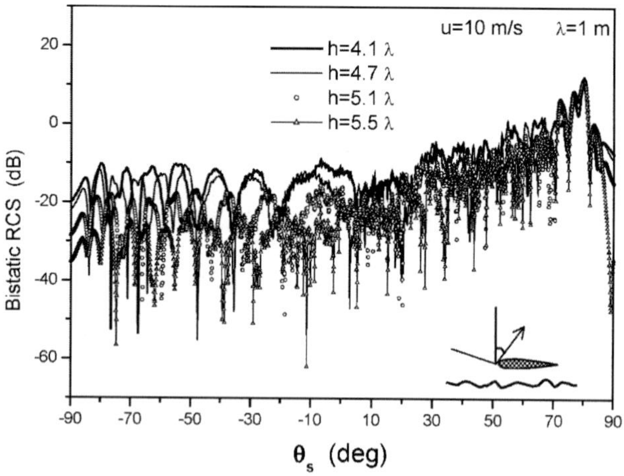

Fig. 6-2.6(c). Bistatic scattering of the target as a function of height (strong wind)

Theoretically, when the target altitude is varying, scattering would be periodically fluctuating. Fig. 6-2.7 shows backscattering coefficient versus target altitude above different rough sea surface. The periodic fluctuation in backscattering can be identified, especially for a flat surface $u = 0m/s$ or sea surface of $u = 3m/s$. From Eq. (6-2.21), it is derived that the phase cancellation needs

$$h = \frac{n\pi}{2k_0 \cos\theta_i} \qquad n = 1,2,3... \tag{6-2.23}$$

where the odd and even n correspond to the maximum and minimum of backscattering, respectively. As the target altitude becomes higher, the target moves from the center of the tapered incident wave and might reduce the scattering maximum.

At strong surface wind speed such as $u = 10m/s$, the wave height of the sea surface might reach as high as 3.0*m*. In Fig. 6-2.7, the target altitude above the surface with $u = 10m/s$ starts from 1.7*m*. Fig. 6-2.7 shows backscattering fluctuation when the target altitude is changing.

Referred to Fig. 6-2.5, backscattering coefficient from the sea surface with low wind speed without target is around $-40\ dB$ at LGA incidence. In other words, if the target echo is lower than $-40\ dB$, the target would not be detected. As indicated by the dotted lines in Fig. 6-2.7, it might be called as blind altitude (B). This altitude is certainly related to the incident angle θ_i and surface roughness.

As the incident angle changes, backscattering from the target, and target/rough sea surface of *u*=3 *m/s* and 10 *m/s* is presented in Fig. 6-2.8. The target is located at $x_0 = 0m$ and $h = 3m$, other parameters are the same as Fig. 6-2.7.

It can be seen that LGA incidence causes more fluctuation of backscattering, especially for rough surface with low surface wind speed.

The size of airfoil target in this paper is electrically small (length 2λ and thickness 0.24λ). Scattering includes single scattering from the target, surface and multi-interactions between them. Angular scattering pattern of an electrically small target is dominated by forward scattering. When the EM wave is incident at small angle, interaction of the target and surface is contributed mainly by the forward scattering of the target. It is quite different from the case of large angle incidence when the side-lobe of scattering pattern of the target causes interaction with the surface. It is a reason why the maximum level of scattering at small angle incidence is always higher than large angle incidence. It is also a reason why the fluctuation amplitude at small incidence angle is smaller.

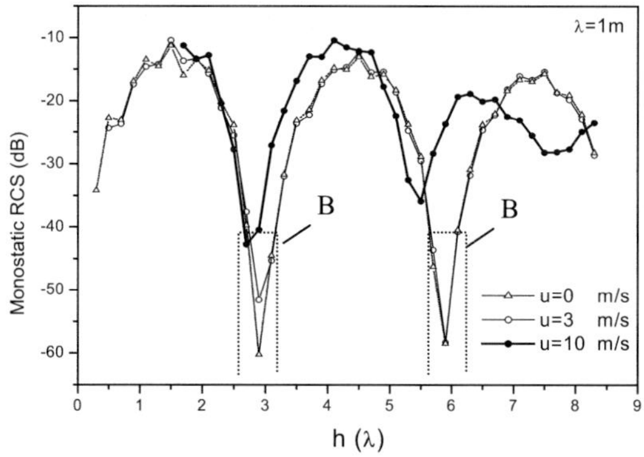

Fig. 6-2.7. Backscattering coefficient versus target altitude ($\theta_i = 80^\circ$)

However, if the target size becomes larger, the angular scattering pattern becomes more complex and no such conclusion can be simply obtained. We also follow the analytic study and numerical code of [Videen and Ngo 1997] to calculate scattering from an electrically large, perfectly conducting cylinder above conducting flat surface. The results support our discussions.

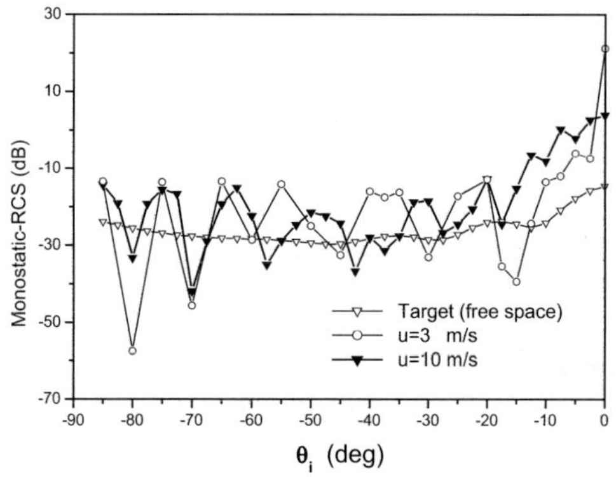

Fig. 6-2.8. Backscattering coefficient versus incident angle

6-3 Bistatic Scattering from the Comprehensive Model of a Ship on and a Target above Large-Scale Rough Sea Surface Using the FEM-DDM

Electromagnetic scattering from the ship on and the target above the wind-driving oceanic surface is an important research topic in radar surveillance, target tracking, and oceanic remote sensing [Jin 1994]. A composite model of a target at low altitude above rough sea surface has been studied in [Liu and Jin 2004a] by using the FEM with the conformal perfectly matched layer (PML). With increasing EM wave frequency, e.g. above C-band, the illuminated surface scaled by the wavelength produces great amount of the unknowns in discretization and requires large memory storage. Numerical computation seems infeasible on available PC-level resources. This is our problem in this session: numerical simulation of both volumetric and surface scattering of a target on and a ship over a large-scale sea surface.

The domain decomposition method (DDM) [Despres 1992, Stupfel 1996, Gander 2001, Lee and Chupongstimun 1994, Choi-Grogan et al. 1996] decomposes the whole computational domain into multiplicity of contiguous independently solvable subdomains with some prescribed partition boundary conditions. Such a procedure can divide large problem to small ones, transform complex boundary conditions to simple ones, and presents an effective means to break through the bottleneck of memory storage. At the same time, a highly parallelizable computation structure is achieved.

The key point of the DDM is the enforcement of boundary conditions on subdomain interfaces. There are usually two DDM approaches: one is the approach of Despres (1992) based on the Schwarz alternating method, which ensures coupling between the adjacent elements by the transmission condition (TC) and proceeds with iterative solution until convergence is achieved. Another one is the direct method [Lee and Chupongstimun 1994, Choi-Grogan et al. 1996, Wolfe et al. 2000], which imposes the continuity of the fields on the partition interfaces and generates a global coupling matrix. The solution of this matrix produces the unknown tangential fields on the partition interfaces.

However, the convergent property of the first method is not well certain. An ideal TC should be transparent to the fields propagating from both sides of the partition interface. It means that the TC acts like an ideal absorbing boundary condition (ABC) in looking from either sides of the interface. Using the ideal TC, a sequential N-subdomain problem can be converged at most after N iterations [Gander 2001]. But, due to existence of the media inhomogeneity, evanescent wave, and bi-directional energy exchange, it is much more difficult to make high performance of TC than the ABC [Stupfel 1996, Gander 2001]. It is known that in a simple model of two-subdomain for the Helmholtz equation, the optimized TC requires at least 10 iterations for convergence [Gander 2001].

Both the zero-th and second-order TC have been tested in our algorithm. Unfortunately, due to fast decaying of the fields around the perfectly matched layer (PML), the limited iterations can not yield good result with enough accuracy. We turn to the direct DDM of [Lee and Chupongstimun 1994] to obtain numerical solution with high accuracy for the composite model of Fig. 6-3.1.

The DDM in this session differs from the past work [Lee and Chupongstimun 1994] in two aspects: (1) Instead of the Bymoment method [Cangellaris and Lee 1990], the conformal PML (CPML) [Teixeira and Chew 1998, Liu and Jin 2004a] is employed as the ABC of the FEM. (2) Taken CPML and a target modeling for the subdomain boundary value problem, a new DDM coupling matrix is derived by using the inward-looking approach [Jin 1993].

It is known [Stupfel 1996, Lee and Chupongstimun 1994, Choi-Grogan et al. 1996] that the DDM in [Lee and Chupongstimun 1994] loses its efficiency when computing elongated structures, such as the model of Fig. 6-3.1. Because the Bymoment method prescribes an implicit boundary condition, the expansion coefficients of which must be treated as the undetermined excitation source for individual subdomain. The number of the right-hand side vectors of the FEM system equations is increased with the horizontal length of the subdomain. On the other hand, the CPML prescribes a determined Dirichlet boundary condition. The right-hand side vectors of the FEM system are not change with the horizontal length of the subdomain. As a result, the FEM solution time is significantly reduced, especially for the model of elongated scatter structure.

Recently, the GFBM [Pino et al. 19993] has been presented for the composite model of a ship or breaking wave on the large-scale sea surface. The computational complexity of GFBM [Jin and Li 2001, West 2000] can be reduced from $O(N^2)$ to $O(N)+O({N_{MOM}}^2)$ with spectral acceleration algorithm (SAA) [Chou and Johnson 1998], where N is number of unknowns on the sea surface, and N_{MOM} is number of unknowns in the direct MOM solution region containing the ship and its nearby sea surface. It is also of interest to compare the FEM-DDM with the GFBM with SAA.

In this session, a composite model of bistatic scattering from a ship on and a target above 2D large-scale ($>2000\lambda$, λ is the wavelength) rough sea surface is developed by using our DDM and CPML-FEM. Our codes are examined by the FEM without DDM, and compared with the FBM and the GFBM [Jin and Li 2001] in time and memory requirement. Numerical simulations of bistatic scattering from the comprehensive model of a ship on and a target above the large-scale 2D randomly rough and perfectly electric conducting (PEC) oceanic-like surface are obtained. Its functional dependence on many physical parameters of the targets and oceanic status are discussed [Liu and Jin 2004b].

6-3.1 The FEM with the Domain Decomposition Method (DDM)

We are concerned with a tapered plane wave incident upon a 2D perfectly conducting rough surface $z=f(x)$ generated by one Monte Carlo realization in the P-M sea spectrum, as shown in Fig. 6-3.1. Suppose that a target is located at (x_1,h) above the sea surface, where h is respect to the mean value $<f(x)>=0$, and a ship is located at x_2 on the sea surface. Some geometrical parameters will be given later in computations. A quarter of elliptically shaped CPML is patched at the end of the horizontal planar PML to enclose the target/surface region.

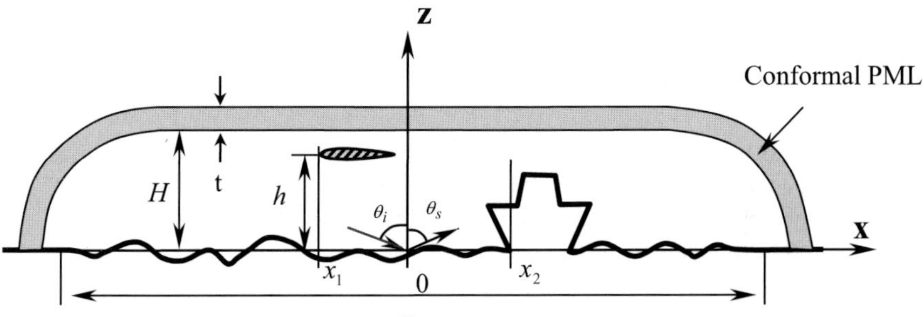

Fig. 6-3.1. Composite model of a ship on and a target above the rough sea surface

The scattered wave above the rough sea surface, including the PML region, satisfies the 2-D Helmholtz equation. The solution of the corresponding boundary value problem is equivalent to seeking the stationary point of the following functional by the FEM, Eq. (6-2.10). And the boundary conditions are as given by Eqs. (6-2.11, 6-2.12).

A. FEM solution in a single subdomain

As shown in Fig. 6-3.2, the composite model in Fig. 6-3.1 is decomposed into M non-overlapping subdomains in x direction. Take the m-th subdomain (which contains the target at low altitude) as an example, the interface between the m-th and $(m-1)$-th subdomains is indicated by Γ_{m-1}, and one between the m-th and $(m+1)$-th is Γ_m.

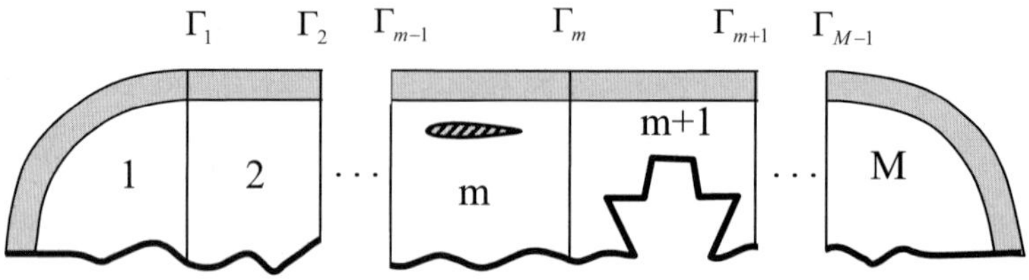

Fig. 6-3.2. A domain decomposition of the composite model

To simplify the notation, E_{sy} is replaced by ϕ hereafter. Assume that the tangential magnetic fields on the interfaces Γ_{m-1} and Γ_m are

$$\psi = \frac{1}{\mu_r} \cdot \frac{\partial \phi}{\partial n} \tag{6-3.1}$$

where $\partial/\partial n$ is the directional derivative in the outward normal direction of the interface. Due to the Neumann boundary condition (6-3.1), in the m-th subdomain, Eq. (6-2.10) is attached with a boundary integration

$$F_m(\phi)=\frac{1}{2}\iint_{\Omega_m}\left(\frac{1}{\mu_r}\nabla\phi\cdot\tilde{\Lambda}\cdot\nabla\phi-k_0^2\varepsilon_r\phi^2\right)d\Omega-\int_{\Gamma_m+\Gamma_{m-1}}\phi\psi\, d\Gamma \tag{6-3.2}$$

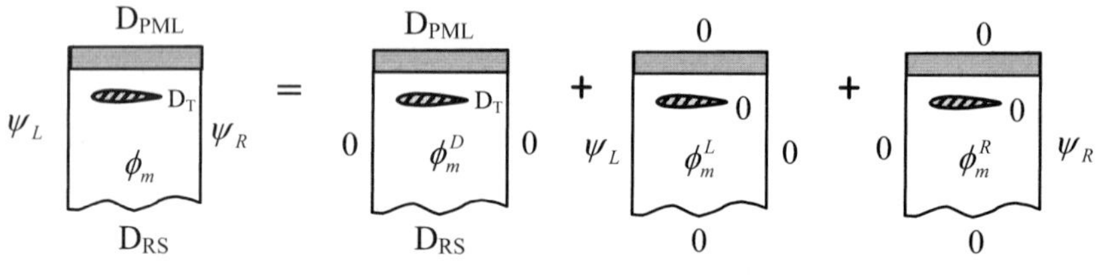

Fig. 6-3.3. The decomposition of boundary value problem in the m-th subdomain

Analogous to the procedure in [Holliday et al 1996], the fields in the m-th subdomain can be viewed as a superimposition of the fields excited by three different sources, as shown in Fig. 6-3.3, which are corresponding to three different boundary value problems:

(I) Given that the tangential magnetic fields ψ_L, ψ_R on the left (Γ_{m-1}) and right (Γ_m) partition boundary of the m-th subdomain equal to zero, the scattering electric field ϕ_m^D is excited by the incident wave impinging on the PEC surfaces. It is the FEM solution of Eq. (6-3.2) with the Dirichlet boundary condition (6-2.11) or (6-2.12) applied on rough sea surface (D_{RS}), target surface (D_T), or PML bottom surface (D_{PML}).

(II) As both the tangential magnetic field ψ_R and the Dirichlet boundary condition in (I) equal to zero, the scattering electric field ϕ_m^L is excited by ψ_L.

(III) As both the tangential magnetic field ψ_L and the Dirichlet boundary condition in (I) equal to zero, the scattering electric field ϕ_m^R is excited by ψ_R.

Expanding the fields ϕ and ψ by using the quadric element basis functions of the FEM and BEM (boundary element method) and proceeding the standard assembling procedure, the FEM matrix equation of the m-th subdomain is obtained as

$$[K]\{\phi\} = [C]\{\psi\} \tag{6-3.3}$$

where $\{\phi\}$ and $\{\psi\}$ are the N_m-D (dimensional) and N_m^b-D vectors, respectively, whose entries are unknown element expansion coefficients, $[K]$ is the $N_m \times N_m$-D FEM stiffness matrix, and $[C]$ is an $N_m \times N_m^b$-D rectangular matrix contributed by the last term on the right hand side of (6-3.2). Here N_m is the number of mesh nodes in the m-th subdomain, and N_m^b is the nodes number on the partition interfaces Γ_{m-1} and Γ_m.

Applying the Dirichlet and Neumann boundary conditions in Fig. 6-3.3 (I) and solving

$$[K]\{\phi\} = \{0\} \tag{6-3.4}$$

we obtain the N_m-D field distribution $\{\phi_m^D\}$ in the m-th subdomain, where the superscript D stands for the Dirichlet boundary condition in Fig. 6-3.3 (I).

According to the inward-looking approach [Holliday et al. 1996], we split the rectangular matrix $[C]$ of Eq. (6-3.3) into the columns vectors $\{C\}_i$, $i = 1,2,\cdots,N_m^b$, each of them is N_m-D. Applying the Dirichlet boundary conditions in Fig. 6-3.3 (II) and (III), and sequentially solving the following equations

$$[K]\{\phi\} = \{C\}_i \qquad i = 1,2,\cdots,N_m^b \tag{6-3.5}$$

we obtain the field distribution $\{\phi_m^L\}_i$, $i = 1,2,\cdots,N_m^{bL}$ and $\{\phi_m^R\}_i$, $i = 1,2,\cdots,N_m^{bR}$ in the m-th subdomain, which are sequentially excited by a set of unit Neumann boundary condition source with N_m-D: $\{\psi\}_1 = \{1,0,0,\cdots 0\}^T$, $\{\psi\}_2 = \{0,1,0,\cdots 0\}^T \cdots$, $\{\psi\}_{N_m^b} = \{0,0,0,\cdots 1\}^T$, whose non-zero entry is corresponding to the pertinent node on the partition interfaces Γ_{m-1} and Γ_m. Here $\{\phi_m^L\}_i$ and $\{\phi_m^R\}_i$ are excited by N_m^{bL} and N_m^{bR} unit sources on the left (L) and right (R) interface of the m-th subdomain, respectively, and $N_m^{bL} + N_m^{bR} = N_m^b$. Because the nodes imposing Dirichlet and Neumann boundary conditions in Fig. 6-3.3 (I), (II), and (III) are the same, the solutions of Eqs. (6-3.4) and (6-3.5) are combined into a FEM direct matrix solver, i.e. the Cholesky factorization method [Compaq 2001] in this session. A powerful multi-frontal algorithm has been discussed in [Liu 1992].

B. Coupling of subdomains

To take the field distributions on the left (L) and right (R) interfaces, column by colume, from $\{\phi_m^L\}_i$, $i = 1,2,\cdots,N_m^{bL}$, the matrices $[\phi_m^L]_L$ and $[\phi_m^L]_R$ are obtained. Each of them is $N_m^{bL} \times N_m^{bL}$-D and $N_m^{bR} \times N_m^{bL}$-D, respectively.

Similarly, the matrices $[\phi_m^R]_L$ and $[\phi_m^R]_R$ can be extracted from $\{\phi_m^R\}_i$, $i = 1,2,\cdots,N_m^{bR}$ with $N_m^{bL} \times N_m^{bR}$-D and $N_m^{bR} \times N_m^{bR}$-D, respectively.

Referring to Fig. 6-3.3, on the right (R) interface of the m-th subdomain, the total scattering field can be written as the sum of the solution from three boundary value problems

$$\{\phi_m\}_R = \{\phi_m^D\}_R + [\phi_m^L]_R\{\psi_m\}_L + [\phi_m^R]_R\{\psi_m\}_R \tag{6-3.6}$$

where $\{\psi_m\}_L$ and $\{\psi_m\}_R$ are the unknown element expansion coefficients on the left and right interface of the m -th subdomain.

On the left (L) interface of $(m+1)$ -th subdomain, it yields

$$\{\phi_{m+1}\}_L = \{\phi_{m+1}^D\}_L + [\phi_{m+1}^L]_L\{\psi_{m+1}\}_L + [\phi_{m+1}^R]_L\{\psi_{m+1}\}_R \tag{6-3.7}$$

Due to the continuity of the electric and tangential magnetic fields on the interface of the $(m+1)$ -th and m -th subdomains, it yields

$$\{\phi_{m+1}\}_L = \{\phi_m\}_R \tag{6-3.8}$$

$$\{\psi_{m+1}\}_L = -\{\psi_m\}_R \tag{6-3.9}$$

and also on the interface of the m -th and $(m-1)$ -th subdomains

$$\{\psi_m\}_L = -\{\psi_{m-1}\}_R \tag{6-3.10}$$

We obtain

$$[\phi_m^L]_R\{\psi_{m-1}\}_R - [\phi_m^R]_R\{\psi_m\}_R - [\phi_{m+1}^L]_L\{\psi_m\}_R + [\phi_{m+1}^R]_L\{\psi_{m+1}\}_R = \{F_m\} \tag{6-3.11}$$

where

$$\{F_m\} = \{\phi_m^D\}_R - \{\phi_{m+1}^D\}_L \tag{6-3.12}$$

has been obtained by solving Eq. (6-3.4).

Let m be the subdomains $1, 2, \cdots, M$, a global coupling matrix with the same structure as [Lee and Chupongstimn 1994] is generated as follows

$$\begin{bmatrix} [B_1][C_1] & & & & & & 0 \\ [A_2][B_2][C_2] & & & & & & \\ & \cdot & & & & & \\ & & \cdot & & & & \\ & & & \cdot & & & \\ & & [A_{m-1}][B_{m-1}][C_{m-1}] & & & & \\ & & & [A_m][B_m][C_m] & & & \\ & & & & \cdot & & \\ & & & & & \cdot & \\ & & & & & & \cdot \\ 0 & & & & & [A_{M-2}][B_{M-2}][C_{M-2}] & \\ & & & & & & [A_{M-1}][B_{M-1}] \end{bmatrix} \begin{bmatrix} \{\psi_1\}_R \\ \{\psi_2\}_R \\ \cdot \\ \cdot \\ \cdot \\ \{\psi_{m-1}\}_R \\ \{\psi_m\}_R \\ \cdot \\ \cdot \\ \cdot \\ \{\psi_{M-2}\}_R \\ \{\psi_{M-1}\}_R \end{bmatrix} = \begin{bmatrix} \{F_1\} \\ \{F_2\} \\ \cdot \\ \cdot \\ \cdot \\ \{F_{m-1}\} \\ \{F_m\} \\ \cdot \\ \cdot \\ \cdot \\ \{F_{M-2}\} \\ \{F_{M-1}\} \end{bmatrix} \tag{6-3.13}$$

The left-hand side of Eq. (6-3.13) contains a block, banded, and diagonal matrix with $N\times N$ -D, where $N=\sum_{m=1}^{M-1} N_m^{bR}$. The submatrices $[A_m]$, $[B_m]$, and $[C_m]$ are from Eq. (6-3.11) with $N_m^{bR}\times N_{m-1}^{bR}$ -D, $N_m^{bR}\times N_m^{bR}$ -D, and $N_m^{bR}\times N_{m+1}^{bR}$ -D, respectively. The solution of the unknown expansion coefficients $\{\psi_m\}_R$ is obtained by using a banded solver [Visual Numerics]. The field distributions in each subdomain can be obtained from $\{\psi_m\}_R$, and the bistatic RCS coefficient can then be calculated.

6-3.2 Numerical Results and Discussions

Suppose that a tapered wave is incident upon a 2-D perfectly conducting rough surface, which is generated by one Monte Carlo realization in the P-M spectrum. The low grazing incident angle is $\theta_i = 80^\circ$ with the tapering parameter $g = L/6$. The thickness of the conformal and planar PML is $t = 0.2\lambda$.

To first validate our DDM, bistatic scattering from a randomly rough sea surface, an airfoil target above rough sea surface, and a ship on rough sea surface are numerically simulated, and are compared with FBM, FEM without DDM, and GFBM, respectively. Though out this section, the computations are conducted on a PC level computer (Intel Pentium 4 1.8 *GHz* processor and 256 Mbytes memory).

Fig. 6-3.4 shows bistatic scattering from rough sea surface with $L = 409.6\lambda$ at strong wind sea surface speed ($U = 10\ m/s$), $\lambda = 1m$. The computational domain is partitioned into 40 subdomains, as indicated in the figure by M=40. It can be seen that the DDM results are well matched to the FBM.

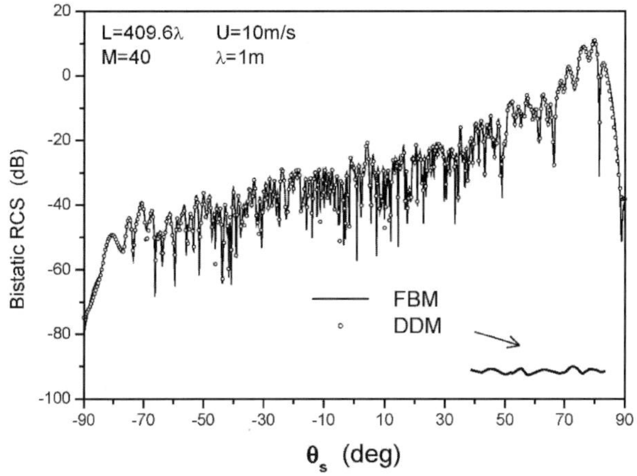

Fig. 6-3.4. Bistatic scattering from rough sea surface (DDM and FBM)

Because there is no target presence in Fig. 6-3.4, the horizontal planar PML can be placed very close to sea surface, the height is $H = 2m$ with respect to the mean value $< f(x) >= 0$ of the sea surface. The CPU time and memory requirement for the GFBM and FEM-DDM are $132.4s$, $1.6M$ and $142.2s$, $2.3M$, respectively.

We now suppose to have a target of NACA0012 four-digital airfoil [Moran 1984] with chord length $2m$ located above the rough sea surface. The airfoil's leading edge is at $x_1 = 0.5m$ and its center axis at the height $h = 3m$. The parameters of incident wave ($\lambda = 1m$) and rough sea surface are same as Fig. 6-3.4. The results of the FEM-DDM in Fig. 6-3.5 are exactly matched to FEM without DDM.

In Fig. 6-3.6, bistatic scattering from large-scale rough sea surface (SeaS) and a composite model of a target above the same sea surface (Wing+SeaS) are compared. The wavelength of incident wave is now $\lambda = 0.2\ m$, thus the sea surface ($L = 409.6m$) in Fig. 6-3.5 is now with electrical length $L = 2048\lambda$, and the chord length of the airfoil target $= 10\lambda$, height $h = 15\lambda$, and $x_1 = 2.5\lambda$, respectively. The number of the unknowns becomes 3,762,021 in the total computational domain, which is beyond the capability of conventional FEM on a PC level computer. It can be seen that scattering in backward direction ($\theta_s < 0^\circ$) is enhanced when an airfoil target is in presence. For example, an enhancement of two orders in backscattering direction (-80°) is observed.

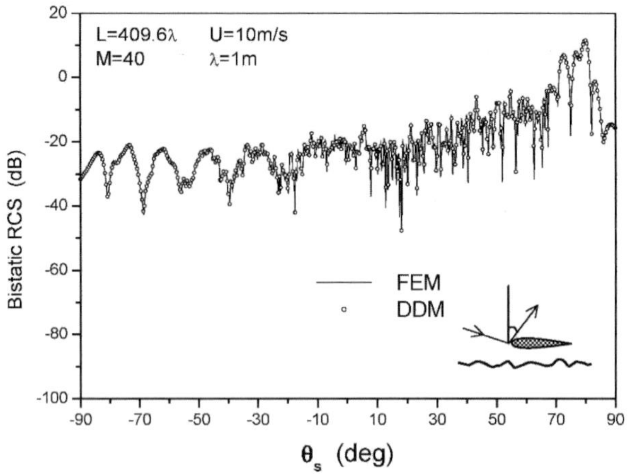

Fig. 6-3.5. Bistatic scattering from a target above the sea surface (DDM and FEM)

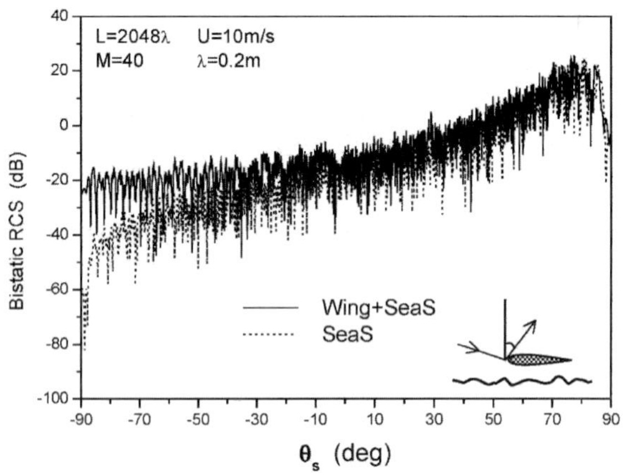

Fig. 6-3.6. Bistatic scattering from a target above large-scale sea surface (DDM)

Fig. 6-3.7 compares our FEM-DDM with GFBM [Jin and Li 2001] when a PEC ship target is on the rough sea surface. The geometrical size of the ship is supposed to have $a=9m$, $b=5m$, $c=4m$, $d=6m$, and $\theta=100^\circ$, and located with its left side at $x_2=5m$ on the sea surface. The parameters of incident wave ($\lambda=1m$) and rough sea surface are same as Fig. 6-3.4. The height of the PML is increased to $H=18m$ to contain the ship. From Fig. 6-3.7, it can be seen that our FEM-DDM well matched to GFBM. Backward ($\theta_s<0^\circ$) scattering is significantly increased due to the ship presence. Meanwhile, the forward scatterings ($\theta_s>0^\circ$), especially in the specular direction ($\theta_s=80^\circ$), seem reduced.

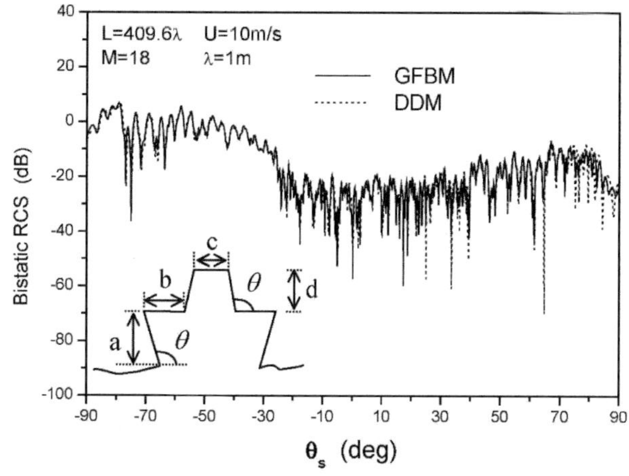

Fig. 6-3.7. Bistatic scattering from a ship on the rough sea surfaces (DDM and GFBM)

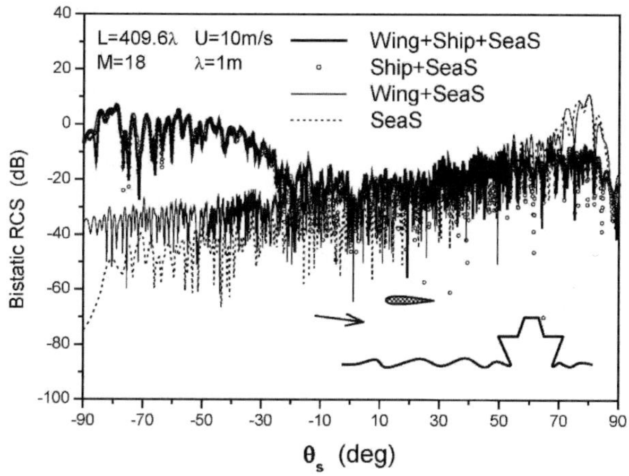

Fig. 6-3.8. Bistatic scattering from a composite model with a ship on and an airfoil above the rough sea surface (DDM)

In Fig. 6-3.7, the CPU time and memory requirement for the GFBM and FEM-DDM are $265.8s$, $3.1M$ and $2743.7s$, $22.3M$, respectively. The GFBM with spectral acceleration (SA) is more efficient than the FEM-DDM due to the fact that the region of strong target/surface interaction is much small compared to the surface length, and the FEM-DDM contains too much free-space in the computation domain.

Bistatic scattering from a composite model of a ship on and a target above the rough sea surface is shown in Fig. 6-3.8. The parameters of incident wave ($\lambda = 1m$) and rough sea surface are same as Fig. 6-3.4. The size and location of the ship is same as Fig. 6-3.7. The airfoil has the same chord length as in Fig. 6-3.5, but with $h = 20m$, and $x_1 = -30m$. The height of PML is $H = 21m$. Four cases are given in Fig. 6-3.8: rough sea surface (SeaS), rough sea surface with the presence of airfoil (Wing+SeaS), rough sea surface with the presence of ship (Ship+SeaS), and rough sea surface with the presence of both airfoil and ship (Wing+Ship+SeaS).

The echo from the airfoil can be well distinguished from the oceanic clutters in Fig. 6-3.8 (thin lines). However, when a big ship is present, the airfoil's echo can not be identified due to ship echo dominance. It is proposed to detect quickly moving airfoil target by using the Doppler shift.

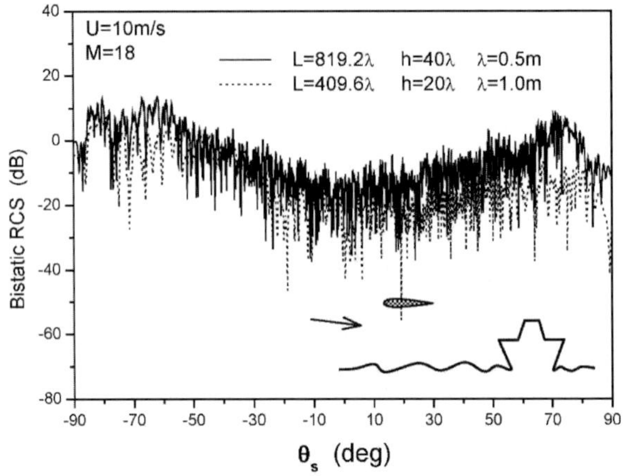

Fig. 6-3.9. Bistatic scattering from a composite model for $\lambda = 1m$ and $\lambda = 0.5m$ (DDM)

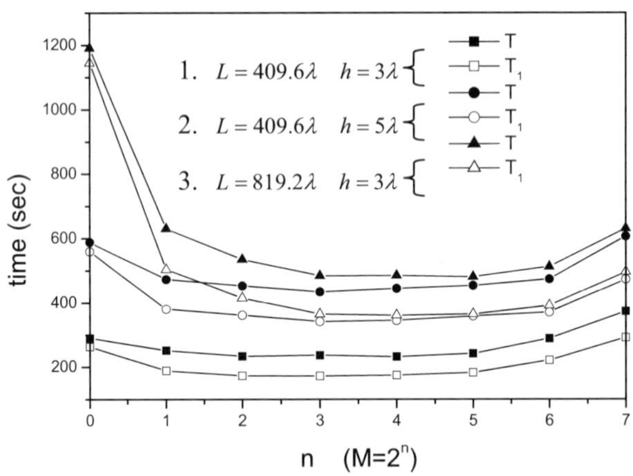

Fig. 6-3.10. A comparison of DDM CPU time for different ratio of L/h in Wing+SeaS model

Fig. 6-3.9 presents comparison between different incident wavelengths $\lambda = 1.0m$ and $\lambda = 0.5m$. Bistatic scattering are increased in all scattering angles with increasing the frequency. And, however, the angular pattern of bistatic scattering remains similar. When partitioning into 18 subdomains for the case of $L = 819.2\lambda$ ($\lambda = 0.5m$), the electrical size of single subdomain becomes so large that the available PC memory is almost exhausted. It usually needs larger M and even partitioning in z direction, e.g. in [Choi-Grogan et al. 19967], for larger scale simulation.

Fig. 6-3.10 presents the CPU time for different L/h ratios of the Wing+SeaS model. The primary solution procedures in the FEM-DDM include mesh generating and pre-process (M), subdomain FEM solution (S), and DDM coupling system calculation (C). The CPU time of (M + S), marked as T_1 (hollow point), and the CPU time of (M + S +C), marked as T (solid point), are recorded, respectively, for the cases of non-partitioned ($n = 0$), 2-partitioned ($n = 1$), 4-partitioned ($n = 2$), $\cdots$, and 128-partitioned ($n = 7$) model. It can be seen that the CPU time of the FEM-DDM solution is generally less than that of conventional FEM (i.e. the non-partitioned $n = 0$ with some sparse matrix solver), especially for big L/h ratio (the DDM is almost half of conventional FEM). Another phenomenon is that the subdomain procedures take a big part of the total DDM solution time. Since this procedure is independent for each single subdomain, large parallel/sequential acceleration ratio can be obtained in a parallel environment, as discussed in [Choi-Grogan et al. 1996]. Thus, it is expected that our DDM can have higher parallel/sequential acceleration ratio.

Computational capability of the FEM-DDM in this paper is determined by two factors: (1) the size of the subdomains and (2) the size of the global coupling matrix. The bandwidth of the coupling matrix, as well as the dimension of its sub-matrix, is determined by the number of nodes on the partition interfaces. Given that the size of each subdomain is definite, the bandwidth of the coupling matrix can be reduced with increasing L/h ratio. Therefore, the DDM in this paper demonstrates high efficiency for elongated scatter model such as rough surface scattering in Fig. 6-3.4, and the length of rough surface can reach as high as 12000λ.

The FEM-DDM is less efficient compared to the GFBM with SAA in the case of low L/h ratio, such as in Fig. 6-3.7. This phenomenon is due to too much free-space contained in the computation domain. Even so, the flying target can be located at height 40λ above the sea surface with a medium scale L (such as several hundreds λ).

The FEM-DDM has strong capability to model complex structures, such as multi-targets on and above the oceanic surfaces. Cumbersome re-formulation process is needed when solving this problem by the GFBM, and the computation costs will increase with the modeling complexity.

6-4 Numerical Simulation of the Doppler Spectrum of A Flying Target above Dynamic Oceanic Surface by Using the FEM-DDM Method

The Doppler (DP) spectrum of moving target such as ship or airplane above dynamic oceanic surfaces [Wang et al. 2002, Gjessing et al. 1982] is one of the most important subjects for radar oceanic surveillance, target tracking and oceanic remote sensing [Walsh et al. 1986, Jin 1994, Tsang et al. 2001, Pierson ans Moskowitz 1964, Barrick 1972]. With advancement of oceanic remote sensing and radar surveillance, the experimental observation and theoretical modeling of oceanic clutter have been extensively studied, e.g. by using oceanic field measurement, wave tank experiment and some studies of low grazing angle (LGA) incidence [Billingsley 1998, Lee et al. 1995,1998, Lamont-Smith 2000, Toporkov and Brown 2000,2002, Johnson et al. 2001].

Some DP approaches [Toporkov and Brown 2000,2002, Johnson et al. 2001] for the model of 2-D (dimensional) wind-driven oceanic surface have been studied by using the quasi-stationary algorithm (QSA) [Bladel 1976], i.e. approximating the time domain problem by repeatedly implementing the frequency domain method [Holliday et al. 1996]. For example, the construction of dynamic oceanic model and numerical implementation of the QSA was presented in [Toporkov and Brown 2000], the DP's of different hydrodynamic models were compared in [Johnson et al. 2001], and simulation by using two-high frequency method was discussed in [Toporkov and Brown 2002]. There might be some other time domain methods for the DP analysis of moving objects (such as rotating and vibrating structures), we don not further specify these approaches.

Bistatic scattering from a static target at low altitude above rough sea surface has been studied in [Liu and Jin 2004a] by using the FEM with the conformal perfectly matched layer (CPML). In a succeeding study [Liu and Jin 2004b], the non-iterative DDM [Robert and Chupongstimun 1994] is developed for a comprehensive model of a ship-like object on and a target above the rough sea surface of large electric-scale. These works are in frequency domain and assume the target and the rough surface to be time-frozen. In this session, we extend the study to time domain, where a fast moving target above time-evolving sea surface are considered. The DP simulation of fast moving target is computationally expensive. We introduce a two level quasi-stationary algorithm (TLQSA) to the FEM-DDM. As shown later in Table 6-4.1 and Table 6-4.2, saving of the computation time can reach over 80% by using the TLQSA.

The code and numerical result of the linearly hydrodynamic sea model are first examined by the forward backward method (FBM) and the data available in publications. By using TLQSA-FEM-DDM, composite model of a flying target above dynamic sea surface is presented. Functional dependence of the DP spectrum in both the time and frequency domains upon some parameters, such as sea surface wind speed, target altitude, observation angle etc., are discussed. A model of a flying target above sea surface and a stationary ship-like object is also numerically studied.

6-4.1 The FEM-DDM Model in Frequency Domain

A. Physical model

A 2-D perfectly conducting rough surface $z=\zeta(x,t)$ generated by the Monte Carlo realization in the P-M sea spectrum is shown in Fig. 6-4.1. Suppose that the wind direction is along $\hat{x}$ direction. A linear hydrodynamic sea model is applied [Tsang et al. 2001, Toporkov an Brown 2000, Johnson et al. 2001] at our UHF band simulation.

Suppose that the discretized sea surface is generated with N uniformly positioned, even-numbered sampling points on the x axis, and its fluctuating height $z=\zeta(x,t)$ varies periodically defined by the illuminated length L. The Fourier series expansion is written as

$$\zeta(x,t)=\frac{1}{L}\sum_{i=-N/2+1}^{N/2}\tilde{\zeta}(K_i,t)\exp(-jK_i x) \tag{6-4.1}$$

where $K_i=2\pi i/L$. A set of normal Gaussian distributed random number $r_i, i=1,2,...,N$ is introduced to generate random evolution of dynamic sea surface. The Fourier expansion coefficient of Eq. (6-4.1) is given by

$$\tilde{\zeta}(K_i,t)=\frac{(r_{2i-1}+jr_{2i})}{\sqrt{2}}\sqrt{2\pi L\,W(K_i)}\exp\{j\omega(K_i)t\} \qquad 0<i<N/2 \tag{6-4.2}$$

The coefficients for $i<0$ are obtained by using $\tilde{\zeta}(K_{-i},t)=\tilde{\zeta}^*(K_i,t)$ [Tsang et al. 2001, Johnson et al. 2001], where $*$ denotes the complex conjugate, $W(K_i)$ is the P-M sea spectrum. The first term on the right hand side of Eq. (6-4.2) takes $N-2$ of the random numbers. It is known that $\tilde{\zeta}(K_0,t)$ and $\tilde{\zeta}(K_{N/2},t)$ are real due to the periodicity of the sea surface model, and are assigned by the remained 2 numbers, respectively.

Suppose that a flying target with constant velocity V is located at (x_1,h) above the sea surface at a given time t, where h is respect to the surface mean value $<\zeta(x,t)>=0$. A ship-like object is stationary located at x_2 on the sea surface and is slowly perturbed along with sea wave motion in the z direction. Some geometrical parameters of the ship are given in computation of Fig. 6-4.8. Both the target and ship are also perfect conducting.

To reduce the reflection error caused by the corners of the planar PML at LGA, a quarter of elliptically shaped CPML is patched at the end of the horizontal planar PML to enclose the targets/surface region. The sea surface is illuminated by a tapered electromagnetic (EM) wave incidence, and the scatter region enclosed by the PML and the sea surface is solved by the FEM.

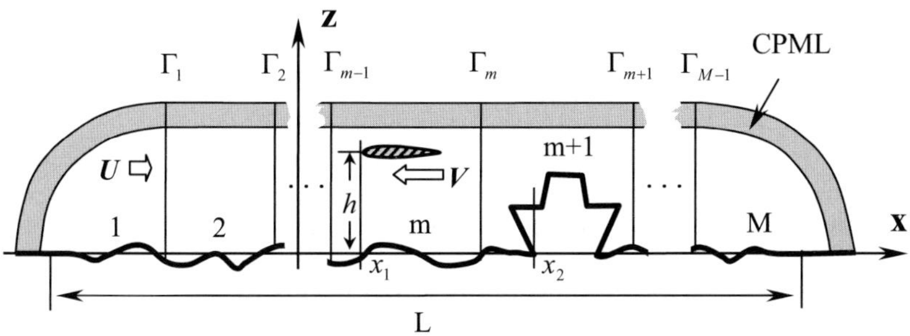

Fig. 6-4.1. The model of the problem

B. The frequency domain FEM-DDM

As shown in Fig. 6-4.1, the physical model is decomposed into M non-overlapping subdomains in the x direction. Take the m-th subdomain (which contains the flying target) as an example, the interface between the m-th and $(m-1)$-th subdomains is indicated by Γ_{m-1}, and one between the m-th and $(m+1)$-th is Γ_m.

The non-iterative DDM [Liu and Jin 2004a, 2004b] expands the unknown tangential magnetic fields on the partition interfaces in terms of the known basis functions. Following independent FEM solutions in each subdomain, a global coupling matrix equation is obtained by enforcing the field continuity between adjacent subdomains. The solution of this matrix equation produces the unknown expansion coefficients of the tangential magnetic fields on the partition interfaces. Then the field distribution in each subdomain is obtained by conducting a field superposition procedure.

The Bymoment method was employed for the absorbing boundary condition (ABC) of FEM [Cangeellaris and Lee 1990]. As a Neumann boundary condition, the unknown expansion coefficients of the Bymoment method must be on the right-hand side of the FEM system equations. As a result, it might reduce the efficiency of the DDM.

In our model of Fig. 6-4.1, the PML is employed as the ABC of FEM. Due to the Dirichlet boundary condition described by the PML, the unknown expansion coefficients only exist on the DDM partition interfaces, and the FEM solution time is significantly reduced, especially for the model of elongated scatter structure.

In [Liu and Jin 2004b], the coupling boundary conditions on the interface between two adjacent subdomains are derived when the PML is used as the truncation boundary of the FEM, and the final coupling matrix is obtained by using the inward-looking approach. This matrix has a diagonal block and banded structure, through which the interactions between DDM subdomains are described. By using an efficient banded solver [IMSL Math/Library], the unknown expansion coefficients on the DDM interfaces can be obtained by solving the DDM coupling matrix equation. Then the field distributions in each subdomain can be obtained from a field summation procedure.

For tapered wave incidence, the scattering amplitude is defined by [Toporkov and Brown 2002]

$$\gamma(\theta_s;\theta_i) = \frac{\exp(jk_0 r)\sqrt{2\pi r}\,\phi(\theta_s)}{\sqrt{P_i(\theta_i)}}\ , \tag{6-4.4}$$

where k_0 is the wave number of free space. The horizontally (y direction) polarized (TE) wave incidence (E_{iy}) is discussed in this paper, ϕ is the scattered electric field E_{sy}. Normalized coefficient $P_i(\theta_i)$ is the total power impinging upon the sea surface with the mean value $<\zeta(x,t)>=0$, which is related with the parameters such as θ_i and L. The bistatic scattering coefficient, i.e. bistatic RCS, is obtained as

$$\sigma(\theta_s,\theta_i) = \lim_{r\to\infty} 2\pi r \frac{|\phi(\theta_s)|^2}{P_i(\theta_i)} = |\gamma(\theta_s;\theta_i)|^2 \tag{6-4.5}$$

At each time step, calculations of (6-4.5) follow the same procedure as described in [Liu and Jin 2004a,2004b] and the sessions 6-4.2 and 6-4.3.

6-4.2 The Two Level Quasi-Stationary Algorithm (TLQSA)

The quasi-stationary algorithm (QSA) simulates the time domain problem by employing the frequency domain approach. The motion of the target in radial direction affects the phase of radar echoes. Because the target velocity is always much smaller than the light speed (non-relativistic case), and also the DP frequency of the target is very small compared with frequency of the incident wave, displacement of the moving target varies slowly with respect to the speed and frequency of EM wave. The time domain solution can then be approximated by a sequence of frequency solutions at discretized time steps, as if the target were stationary for each time step. The error between the quasi-stationary algorithm and exact solution was estimated in [Copper 1980].

The model in Fig. 6-4.1 contains two moving factors, i.e. the slowly evolving wind-driven sea surface (with maximum wind speed $U=10\,m/s$) and fast flying target (at speed $V = 600\,km/hr$, or $166.7\,m/s$). Applying the time division of QSA [Johnson et al. 2001], the DP simulation of fast flying target must be carried out at very small step on time domain.

Notice that the velocity of the flying target is 10 times faster than the phase speed [Toporkov and Brown 2001] of the evolving ocean wave, so that the dynamic sea surface can be regarded as stationary with respect to the fast flying target. As a result, only one subdomain is changing with motion of the target in the DDM model of Fig. 6-4.1, namely the subdomain the flying target belongs to. The FEM solution is performed in this subdomain, and the other subdomains are unchanged and the results stored at the former time step can be used.

In DDM coupling matrix, a fraction of the rows and columns are changed due to updating the subdomain m. It also correspondingly yields changes of all the expansion coefficients of tangential magnetic fields. In this way, interactions between the target and sea surface in other sub-domains can be described precisely by the global DDM coupling matrix without any further approximation.

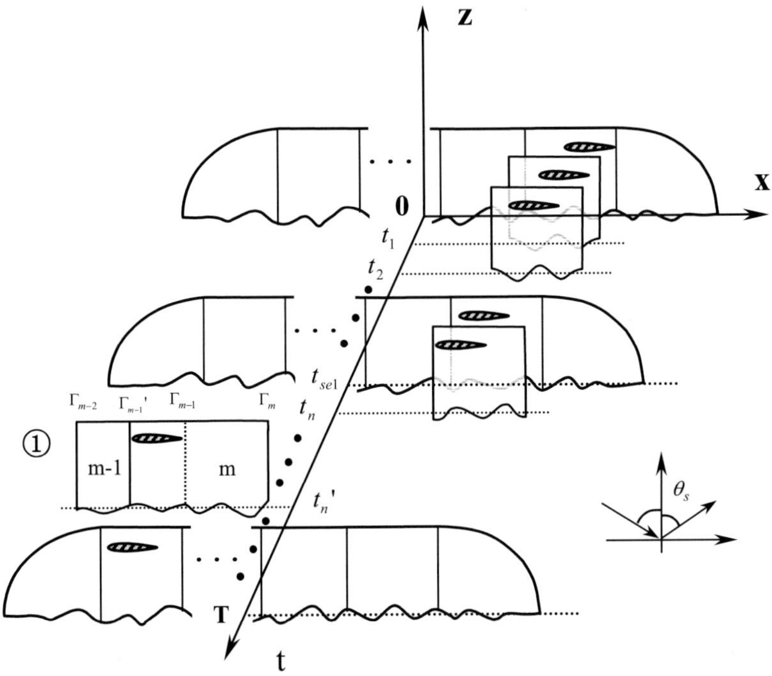

Fig. 6-4.2. A schematic diagram of the TLQSA for FEM-DDM

The slowly moving sea surface is evolved at larger time steps. Through all-subdomain FEM calculation, the DP spectrum of oceanic clutter is obtained.

A schematic diagram of the TLQSA for FEM-DDM is described in Fig. 6-4.2: the simulation is carried out over a duration of T. Since the time step of sea evolution is much larger than that of flying target, at most time steps $t_1, t_2 \ldots t_n \ldots$, the FEM solution is carried out only in one sub-domain, where the flying target located. Such procedure is repeated until the sea evolution step t_{sel} is reached. Then, the sea surface evolves, following linear hydrodynamic model, and all-subdomain FEM calculation is conducted.

As the target flies to the vicinity of DDM interface, a simple boundary shift process is adopted, as indicated by a graph ① in Fig. 6-4.2. The interface Γ_{m-1} shifts to a temporary position Γ_{m-1}', the distance between these two positions is a little larger than the target size. After the target moves across the Γ_{m-1}, the interface recovers from Γ_{m-1}' to Γ_{m-1}. At the time step performing this process, calculations in the subdomains m and $m-1$ need to be updated simultaneously and might slow down computation speed of the TLQSA.

6-4.3 Numerical Results and Discussions

The backscattering $(\theta_s = -\theta_i)$ DP spectrum [Papoulis 1984, Toporkov and Brown 2002] is defined as

$$S(f) = \frac{1}{T}\left| \int_0^T \gamma(t,\theta_s;\theta_i)\exp(-j2\pi ft)\,dt \right|^2 \tag{6-4.6}$$

To first examine our model code, the numerical result of the linearly hydrodynamic sea model is compared with the forward backward method (FBM). The maximum relative error δ in calculation of the backscattering amplitude $\gamma(t)$ in Eq. (6-4.6), defined as $\delta = \max_{0\le t\le T} |[\gamma_{FEM-DDM}(t) - \gamma_{FBM}(t)]/\gamma_{FBM}(t)|$, is less than 0.48%. The DP spectra of averaging 100 independent sea surface realizations agrees with the data available in [Toporkov and Brown 2002].

We now suppose to have a target of NACA0012 four-digital airfoil [Moran 1984] with the chord length 2*m* flying at constant speed $V = 166.667\ m/s$ above the dynamic oceanic surface. The model parameters should satisfy: (1) the simulation period T is sufficiently long to keep the resolving power in spectral domain, (2) the trace of flying target is inside the beam of incident tapered wave, (3) the CPU time is affordable (normally within a week).

The model parameters are listed as follow: the sea surface length $L = 819.2\lambda$, the incident wave is at UHF band with $\lambda = 1.0\ m$ and the tapering parameter $g = L/6$. The simulation period $T = 4.56\ s$ with time step $\Delta t_{se} = 0.02\ s$, which is small enough to simulate strong sea surface wind speed $U = 10\ m/s$ at UHF band. The target is upwind-flying (in the negative x direction) at the height h. At $t = 0\ s$, the target is located at $x_1 = 380.0\ m$, and flies to $x_1 = -380.0\ m$ at $t = 4.56\ s$. The time step of the flying target is $\Delta t = 0.001\ s$. The computational domain is evenly partitioned into 40 subdomains along the x axis, as indicated in Fig. 6-4.1 by M.

Fig. 6-4.3 shows the scattering amplitude under LGA incidence $\theta_i = 80^\circ$ in time domain when: (1) the target is in the free space, (2) the target is above the conducting flat surface, and (3) the target is above dynamic sea surface driven by wind speed $U = 5$ m/s. The target is at the height $h = 2m$.

It can be seen that in the case of the target above a flat surface or in the free space, the target motion imposes only phase modulation (PM) on the echoes signal. The scattering amplitude varies smoothly in time domain. However, in the case of the target flying above dynamic sea surface, both PM and amplitude modulations (AM) are imposed, which cause strong fluctuation of the scattering amplitudes.

The DP spectrum of a target flying above different surfaces is shown in Fig. 6-4.4. It can be seen that two peaks are, respectively, related to the flying target scattering and sea clutters, and are, correspondingly, positioned at the spectral lines of the Bragg frequency and a point-target moving at $V = 166.667\ m/s$. At the zero DP, there is a small peak for the case of a target flying above the flat surface. This is due to non-zero backscattering from a flat surfacc, which is a systematic error of numerical calculation.

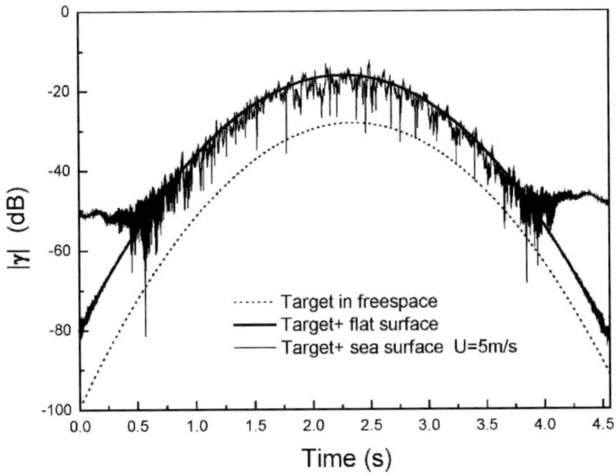

Fig. 6-4.3. The scattering amplitude from a target flying above different surfaces

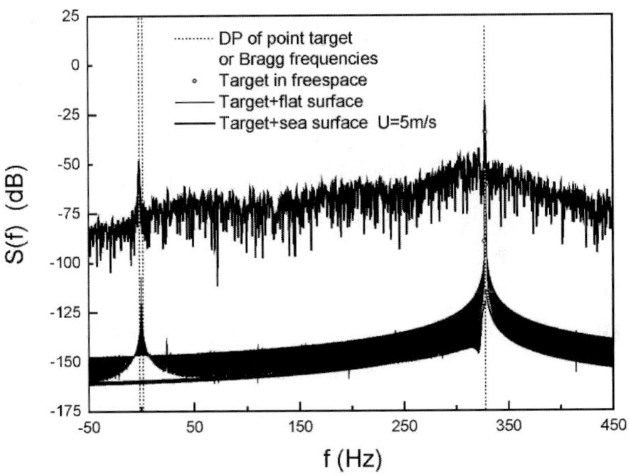

Fig. 6-4.4. The DP spectrum of a target flying above different surfaces

The results of TLQSA in Fig. 6-4.4 have also been compared with more accurate QSA calculation (i.e. conducting TLQSA by setting $\Delta t_{se} = \Delta t = 0.001\,s$). The error between TLQSA and QSA at each time step is calculated by $|\gamma_{TLQSA} - \gamma_{QSA}|/|\gamma_{QSA}|$ in percentage. The averaged error during whole journey flying is 5.01%, and is reduced to 3.03% when the target flies into the main lobe of the tapered wave (i.e. from $t = 0.5s$ to $t = 4.0s$). The DPs of the target in TLSQA and QSA duplicate each other, and the sea clutters show slight difference. Therefore, though the error can be further reduced with taking smaller Δt_{se}, the criterion used in Fig. 6-4.4 is good enough to ensure a good performance of TLQSA.

Fig. 6-4.5 compares the DP spectrum from a target flying above sea surfaces driven by different wind speeds. The main spectral lobe caused by higher wind speed is decreased in the amplitude and diffused to the side lobes (which are enhanced by $\sim 20\ dB$) broadening the spectrum pattern. It also can be seen from the DP of sea clutters that the strong wind speed causes higher and wider main lobes.

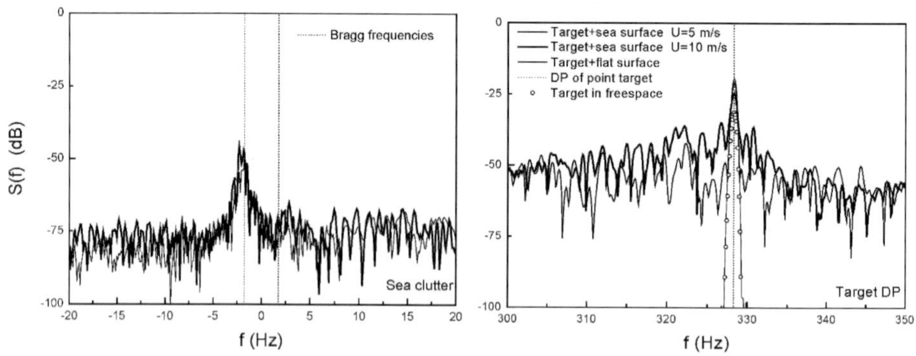

Fig. 6-4.5. The DP spectrum of a flying target and sea clutters

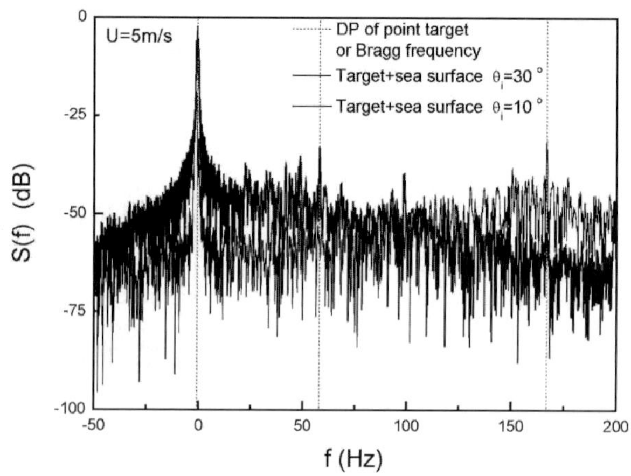

Fig. 6-4.6. The DP spectrum of a target flying above sea surfaces under incidence 10° and 30°

Fig. 6-4.6 shows the DP spectrum of the target flying above sea surface of $U = 5\ m/s$ for two different incident angles $\theta_i = 30^\circ$ and $\theta_i = 10^\circ$. The target height is the same as Fig. 6-4.3. As the incident angle decreases, the magnitude of sea clutters spectrum is increased, and the main lobe of the target spectrum is approaching to the zero DP. It can be seen that the main lobe of the target's DP spectrum for the case of $\theta_i = 30^\circ$ is distinguishable, but the main lobe for $\theta_i = 10^\circ$ is merged into the clutters. To make the plot resolvable, only one of the Bragg spectral lines is marked corresponding to $\theta_i = 10^\circ$.

According to the image principle and two-sources array radiation, if the phase difference between the target and its mirror image takes [Liu and Jin 2004a]

$$k_0 h\cos\theta_i = n\pi \qquad n = 1,2,3... \qquad (6\text{-}4.7)$$

the destructive interference can minimize coherent backscattering. The height of $n=1$ in Eq. (6-4.7) is $h=2.88m$. Fig. 6-4.7 shows the DP spectrum of a target flying at $h=3m$ and $2m$, respectively, when $\theta_i = 80^\circ$ and $U = 5\ m/s$. It can be seen that the main lobe of the target spectrum for $h=3m$ case is totally merged into the sea clutters.

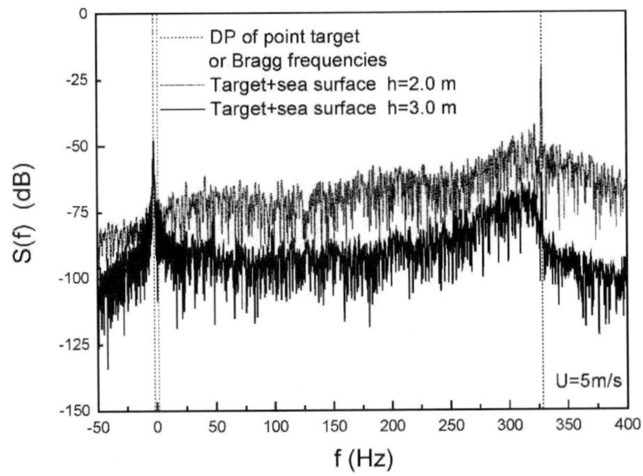

Fig.6-4.7. The DP spectrum from a target flying at heights $h=3m$ and $2m$ above the sea surfaces

The practical measurement is usually conducted in finite range. At most cases, the target motion is not along the direction of radio wave propagation. So the observation angle and the phase difference between the target and its mirror image are also changing with the target moving. An extra spectral spreading effect can be observed [Ulaby et al. 1982].

Figs. 6-4.8 presents temporal variation of the echo amplitudes when a target is flying at $h=7.5m$ above dynamic sea surface and a stationary 2-D ship like object (abbreviated as "ship" in the following text) located with its left side at $x_2 = 5m$. The parameters of incident wave and dynamic sea surface are the same as Fig. 6-4.3. The geometrical sizes of the ship are supposed as $a=2m$, $b=1m$, $c=1.5m$, $d=1.5m$, $\theta=100^\circ$. The circle points in Fig. 6-4.8 describe the moving positions of the flying target (such as $x_1 = 380m$ at $t=0s$, and over the ship at $t=2.24s$, and $x_1=-380m$ at $t=4.56s$) above the sea surface at sequential time (s). It can be seen that the ship presence enhances the echoes to -5 dB and causes strong oscillations when the flying target is approaching. It also can be seen that the oscillations become slower when the target is leaving from the ship. It is physically similar to the whistle change when a train is approaching and then is leaving.

In the case without the ship, temporal variation of the echo amplitudes is similar to Fig. 6-4.3. Higher altitude of the target reduces somewhat oscillations of the echoes. The corresponding DP spectrum of a target flying above the sea surface with and without a ship is shown in Fig. 6-4.9.

Fig. 6-4.10 shows the DP spectrum when the target approaches to the ship (t from 0 to 2.24s) and leaves from the ship (t from 2.24s to 4.56s). When the target is approaching to the ship, the DP spectrum of the target has high level of side lobe, which affects the side lobe of whole journey flying near 330 *Hz*.

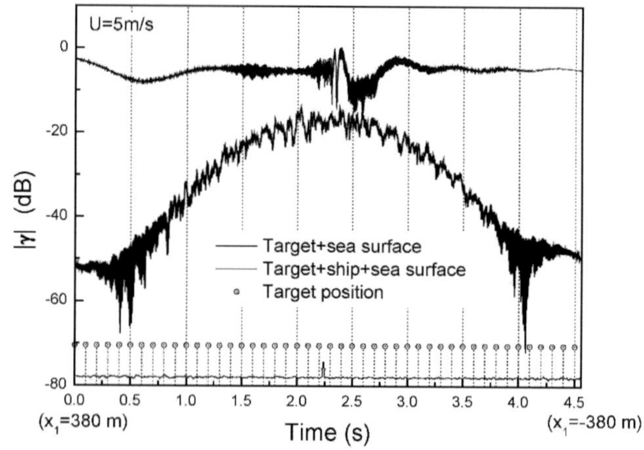

Fig. 6-4.8. The scattering amplitude of a target flying above dynamic sea surface and a stationary ship

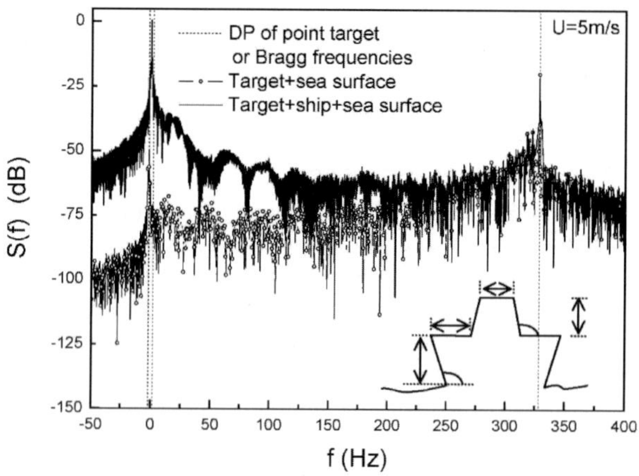

Fig. 6-4.9. The DP spectrum of a target flying above the sea surface and a ship

When the target is leaving from the ship, several high-level side lobes appear in the ship and sea clutters spectrum. This implies that backscattering from the ship is being disturbed and shielded by the leaving target. Since there is no such spectral feature in the 'approaching ship' stage (dash dot line), it is concluded that during whole journey of the flying target as shown in Fig. 6-4.9, the dominant feature of ship and sea clutters is contributed by the 'leaving from ship' stage.

Comparisons between 3 primary solution procedures in TLQSA, which includes the mesh generating and pre-process (M), subdomain FEM solution (S), and far field calculation (C), are listed in Table 6-4.1 for the case of Fig. 6-4.3. The computer is a Pentium4-2.4G PC with 512*M* memory.

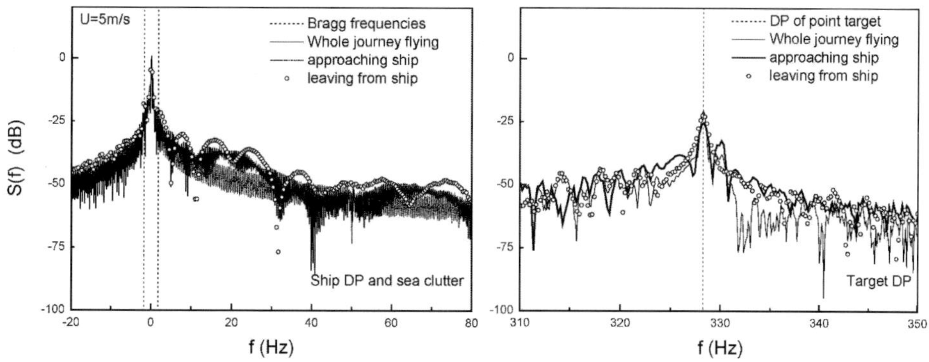

Fig. 6-4.10. Comparison of the DP spectrum for different flying stages

In Table 6-4.1, t_{se} represents the time step of evolving sea surface, t_n represents the time step of the target flying inside one of the DDM subdomains, and t_n' represents the time step of the target flying across the DDM interface. Because the FEM solution is only performed in one subdomain, the CPU time of t_n in column (**S**) is $1/40$ of t_{se} due to the 40-partitioned model in Fig. 6-4.3, and the total CPU time at t_n is less than $1/10$ of t_{se}. The CPU time at each column of t_n' is a littler longer than t_n due to the 2 subdomain calculation. Simulation of whole journey costs 4560 time steps, including 229 steps for t_{se}. The total time for the DP simulation by TLQSA needs $25.38\,hr$, which is only 13.01% of the simulation conducted by 4560 times t_{se}.

Table 6-4.1. Comparison of CPU times between primary procedures in TLQSA: Fig.6-4.3

Time step	M	S	C	Total	×4560
t_{se}	16.1 s	127.6 s	10.3 s	154.0 s	195.07 hr
t_n	2.3 s	3.0 s	7.2 s	12.5 s	15.83 hr
t_n'	3.0 s	6.2 s	7.2 s	16.4 s	20.77 hr
TLQSA		25.38 hr			

Table 6-4.2. A comparison of CPU times between primary procedures in TLQSA: Fig.6-5.8

Time step	M	S	C	Total	×4560
t_{se}	46.4 s	738.0 s	59.4 s	843.8 s	1068.81 hr
t_n	4.0 s	18.1 s	55.3 s	77.4 s	98.04 hr
t_n'	4.9 s	36.5 s	57.5 s	98.9 s	125.27 hr
TLQSA		153.62 hr			

Computation time of Fig. 6-4.8 is listed in Table 6-4.2. Comparisons between t_{se}, t_n, and t_n' is similar to Table 6-4.1. It takes 6 days for the TLQSA to finish the DP simulation, which is 14.37% of the simulation conducted by 4560 times t_{se}.

6-5 The Difference Field of Scattering from the Target above a Rough Surface

In numerical calculation of electromagnetic scattering from rough surface, the rough surface is truncated at certain length L, and the differential scattering cross section is normalized by L to describe bistatic and back-scattering from random rough surface. When a target is located above the rough surface, total scattering coefficient depends upon both the illuminated surface length L and the finite size of the target. In order to remove such size dependence, Johnson (2002) proposed the difference field radar cross section (d-RCS) σ_d as the difference scattering field of two cases: the target-surface and the surface only. The d-RCS contains the scattering from the target, and multiple interactions between the target and underlying rough surface, and is removed from the effect of the illuminated surface length and incident tapered wave width.

In this session, the electric field integral equations (EFIE) of the difference induced current J_{sd} on the rough surface and the induced current J_o on the target are derived [Ye and Jin 2005]. A numerical iterative method is developed to solving the EFIE's. Because the contribution to J_o from the rough surface is dominated by the scattering in the specular direction, only a small section of the rough surface towards to the target in the specular direction is needed to take account of this interaction. It makes computation much faster. Two surface lengths for the rough surface contribution and the target–surface interaction are taken in respective steps and are compared with Johnson's method.

As numerical examples, bistatic scatterings from the target, e.g. a cylinder or a square column, above P-M oceanic rough surface are simulated and discussed.

6-5.1 Theory and Method

The differential scattering cross section [Ishimaru 1978] normalized over the surface length is usually defined to describe the averaged scattering from the rough surface. However, as a volumetric target with finite size is located above the rough surface, its finite size as comparison with the truncated rough surface length becomes a factor to influence final result. To remove such uncertainty, Johnson (2002) proposed the difference field radar cross section (d-RCS) σ_d as the difference scattering field of the target-surface and the surface only. In the d-RCS, the scattering field E_{s0} of the only rough surface is subtracted from the total scattering field E_{s1} of the target-rough surface, and the difference scattering field E_{sd} is obtained. As the difference field cross section σ_d is calculated, scattering from the target and its interactions with the underlying surface can be well described without dependence of the illuminated surface length.

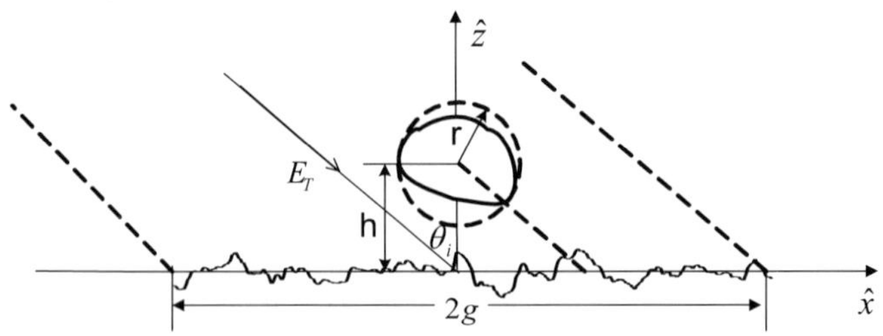

Fig. 6-5.1. Scattering model of the target above rough surface

As shown in Fig. 6-5.1, a two-dimensional (2D) perfectly electric conductor (PEC) target is located above a 2D PEC randomly rough surface. A tapered wave $E_T(\bar{r})$ with *TE* polarization (with electric field in $\hat{y}$ direction) is incident upon along the $\hat{k}_i = \hat{x}\sin\theta_i - \hat{z}\cos\theta_i$ direction as expressed as [Tsang et al. 2001]

$$E_T(\bar{r}) = \exp[-jk(x\sin\theta_i - z\cos\theta_i)\cdot(1+w(\bar{r}))]\cdot\exp\left[-\frac{(x+z\tan\theta_i)^2}{g^2}\right] \quad (6\text{-}5.1)$$

$$w(\bar{r}) = \frac{2\dfrac{(x+z\tan\theta_i)^2}{g^2}-1}{(kg\cos\theta_i)^2} \quad (6\text{-}5.2)$$

where k is the wave number in the free space, g is the width parameter of the tapered wave.

Since the target and the rough surface in Fig. 6-5.1 are assumed PEC material, the induced current and the scattering field are both in $\hat{y}$ direction. When the target is exist (subscript 1) above rough surface, the scattering field $E_{s1}(\bar{r})$ in the upper half space is written as [Peterson et al. 1997]

$$E_{s1}(\bar{r}) = -jk\eta\int_c J_o(\bar{r}')\cdot G(\bar{r},\bar{r}')d\bar{r}' - jk\eta\int_S J_{s1}(\bar{r}')\cdot G(\bar{r},\bar{r}')d\bar{r}' \quad (6\text{-}5.3)$$

where J_o and J_{s1} are the induced currents on the target (subscript *o*) and the rough surface (subscript *s*), respectively. $G(\bar{r},\bar{r}') = H_0^{(2)}(k|\bar{r}-\bar{r}'|)/4j$ is the 2D Green function of free space. The integration $d\bar{r}'$ is along the contour of the target and the rough surface. The variable η is the wave impedance.

As the boundary condition of zero tangential electric field on the PEC surface is imposed, the EFIEs of the induced currents J_o and J_{s1} can be obtained, respectively [Peterson et al. 1997]:

$$0 = E_T(\bar{r}) - jk\eta\int_c J_o(\bar{r}')\cdot G(\bar{r},\bar{r}')d\bar{r}' - jk\eta\int_S J_{s1}(\bar{r}')\cdot G(\bar{r},\bar{r}')d\bar{r}' \quad (\bar{r}\in c) \quad (6\text{-}5.4)$$

$$0 = E_T(\bar{r}) - jk\eta\int_c J_o(\bar{r}')\cdot G(\bar{r},\bar{r}')d\bar{r}' - jk\eta\int_S J_{s1}(\bar{r}')\cdot G(\bar{r},\bar{r}')d\bar{r}' \quad (\bar{r}\in S) \quad (6\text{-}5.5)$$

If there is no target (subscript 0) above the rough surface, the scattering field $E_{s0}(\bar{r})$ is coming from the rough surface only and is expressed as

$$E_{s0}(\bar{r}) = -jk\eta\int_S J_{s0}(\bar{r}')\cdot G(\bar{r},\bar{r}')d\bar{r}' \quad (6\text{-}5.6)$$

where J_{s0} is the induced current upon the rough surface to satisfy

$$0 = E_T(\bar{r}) - jk\eta\int_S J_{s0}(\bar{r}')\cdot G(\bar{r},\bar{r}')d\bar{r}' \quad (\bar{r}\in S) \quad (6\text{-}5.7)$$

Subtracting E_{s0} from E_{s1}, the difference field E_{sd} is obtained as

$$\begin{aligned} E_{sd}(\bar{r}) &= E_{s1}(\bar{r}) - E_{s0}(\bar{r}) \\ &= -jk\eta\int_c J_o(\bar{r}')\cdot G(\bar{r},\bar{r}')d\bar{r}' - jk\eta\int_S [J_{s1}(\bar{r}') - J_{s0}(\bar{r}')]\cdot G(\bar{r},\bar{r}')d\bar{r}' \quad (6\text{-}5.8) \\ &= -jk\eta\int_c J_o(\bar{r}')\cdot G(\bar{r},\bar{r}')d\bar{r}' - jk\eta\int_S J_{sd}(\bar{r}')\cdot G(\bar{r},\bar{r}')d\bar{r}' \end{aligned}$$

$$J_{sd}(\bar{r}') = J_{s1}(\bar{r}') - J_{s0}(\bar{r}') \quad (6\text{-}5.9)$$

where J_{sd} is the induced current increment on the underlying interface caused by the target above rough surface. In Johnson's method, the width parameter and illuminated surface length for two cases of the target-surface and the surface only are chosen in the same way to ensure the incident wave well illuminating on the target. If the tapered wave width takes

$$g = 4(\frac{r}{\cos\theta_i} + h\tan\theta_i) \tag{6-5.10}$$

where r is the maximum radius of the target circled by the dashed line in Fig.1. Substituting Eq. (6-5.10) into Eq. (6-5.1), the ratio of the illuminated intensity on the target to the tapered wave peak is equal to

$$\eta = \exp[-(1/4)^2] \approx 0.94 \tag{6-5.11}$$

Substituting Eq. (6-5.9) into Eqs. (6-5.4) and (6-5.5), and defining

$$E_0(\bar{r}) = E_T(\bar{r}) - jk\eta \int_S J_{s0}(\bar{r}') \cdot G(\bar{r},\bar{r}')d\bar{r}' \tag{6-5.12}$$

it yields the coupled equations of the induced currents

$$0 = E_0(\bar{r}) - jk\eta \int_c J_o(\bar{r}') \cdot G(\bar{r},\bar{r}')d\bar{r}' - jk\eta \int_S J_{sd}(\bar{r}') \cdot G(\bar{r},\bar{r}')d\bar{r}' \quad (\bar{r} \in c) \tag{6-5.13}$$

$$0 = E_0(\bar{r}) - jk\eta \int_c J_o(\bar{r}') \cdot G(\bar{r},\bar{r}')d\bar{r}' - jk\eta \int_S J_{sd}(\bar{r}') \cdot G(\bar{r},\bar{r}')d\bar{r}' \quad (\bar{r} \in S) \tag{6-5.14}$$

From Eq. (6-5.7), it can be seen that $E_0(\bar{r})\big|_{\bar{r}\in S} = 0$. Thus, Eq. (6-5.14) becomes

$$0 = -jk\eta \int_c J_o(\bar{r}') \cdot G(\bar{r},\bar{r}')d\bar{r}' - jk\eta \int_S J_{sd}(\bar{r}') \cdot G(\bar{r},\bar{r}')d\bar{r}' \quad (\bar{r} \in S) \tag{6-5.15}$$

Eqs. (6-5.13) and (6-5.14) establish an iterative approach to the induced currents on the target and rough surface. The iterative equation at the (i)-th iteration step is written as

$$jk\eta \int_c J_o^{(i)}(\bar{r}') \cdot G(\bar{r},\bar{r}')d\bar{r}' = E_0(\bar{r}) - jk\eta \int_S J_{sd}^{(i-1)}(\bar{r}') \cdot G(\bar{r},\bar{r}')d\bar{r}' \quad (\bar{r} \in c) \tag{6-5.16}$$

$$jk\eta \int_S J_{sd}^{(i)}(\bar{r}') \cdot G(\bar{r},\bar{r}')d\bar{r}' = -jk\eta \int_c J_o^{(i-1)}(\bar{r}') \cdot G(\bar{r},\bar{r}')d\bar{r}' \quad (\bar{r} \in S) \tag{6-5.17}$$

where J_o and J_{sd} are unknowns yet. Solving Eqs. (6-5.16,17) iteratively, the difference scattering field E_{sd} can then be directly obtained, and it is not necessary to respectively know E_{s1} and E_{s0}.

In the iteration, firstly update the excitation at the RHS of Eqs. (6-5.16) and (6-5.17) with $J_o^{(i-1)}$ and $J_{sd}^{(i-1)}$ obtained in the previous step. Then solve Eqs. (6-5.16,17), separately, to obtain the new currents $J_o^{(i)}$ and $J_{sd}^{(i)}$. The CG method is for solving Eq. (6-5.16) and the FBM is for solving Eq. (6-5.17). The iteration begins with assumption $J_o^{(0)} = J_{sd}^{(0)} = 0$, and is finished when iterated J_o and J_{sd} become converged. With such J_o and J_{sd}, the difference field E_{sd} can be directly calculated, and $\sigma_d(\theta_s,\theta_i)$ is obtained as

$$\sigma_d(\theta_s,\theta_i) = \lim_{R\to\infty} \pi R \left|E_{sd}(r,\theta_s)\right|^2 \tag{6-5.18}$$

where R is the distance between the coordinate origin and the field point in the far-field region. Eq. (6-5.18) takes account the scattering from the target and multiple interactions between the target and underlying rough surface. Direct scattering of the rough surface only is not included.

To initiate the iteration, the total field $E_0(\bar{r})$ $(\bar{r} \in c)$ defined in Eq. (6-5.6) should be first obtained. The first term of $E_0(\bar{r})$ is the tapered incident wave,

$$E_P(\bar{r}) = \exp[-jk(x\sin\theta_i - z\cos\theta_i)] \tag{6-5.19}$$

The second term of $E_0(\bar{r})$ is the scattering field $E_{s0}(\bar{r})$ of the rough surface without target existence. Because the contribution to J_o from the rough surface is dominated by the scattering in the specular direction, only a small section of the rough surface towards to the target in the specular direction is needed to take account this interaction $E_{s0}(\bar{r})$ from the rough surface to the target as shown in Fig. 6-5.2. Let the tapered wave be centered at $(-h\tan\theta_i, 0)$ and expressed as

$$E_T(\bar{r}) = \exp[-jk(x\sin\theta_i - z\cos\theta_i)\cdot(1+w(\bar{r}))]\exp\left[-\frac{(x+h\tan\theta_i + z\tan\theta_i)^2}{g_0^2}\right] \tag{6-5.20}$$

$$w(\bar{r}) = \frac{2\dfrac{(x+h\tan\theta_i + z\tan\theta_i)^2}{g_0^2} - 1}{(kg_0\cos\theta_i)^2} \tag{6-5.21}$$

The phase of the above tapered wave at the coordinate origin is zero, as the same as Eq. (6-5.19). In order to ensure the scattering intensity as the same as Eq. (6-5.11), g should be valued as

$$g = 4r/\cos\theta_i \tag{6-5.22}$$

It can be seen that g is related to the target size and the incident angle θ_i, not related to the target height. As θ_i increases, g becomes larger.

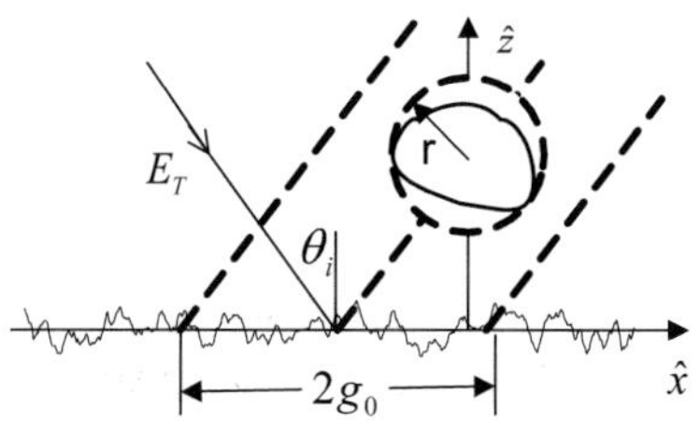

Fig. 6-5.2. The surface length and the tapered wave in absence of the target

Solving the EFIE Eq. (6-5.7) with the method of moment (MoM), the induced current J_{s0} of the rough surface under the tapered wave incidence is obtained. The field $E_0(\bar{r})$ contributed by the incident wave and scattered wave by the rough surface to the target as shown in Eq. (6-5.16) is calculated as

$$E_0(\bar{r}) = E_P(\bar{r}) - jk\eta\int_S J_{s0}(\bar{r}')\cdot G(\bar{r},\bar{r}')d\bar{r}' \quad (\bar{r}\in c) \tag{6-5.23}$$

The difference induced current J_{sd} on the rough surface takes account of interaction between the target and rough surface. From Eq. (6-5.17), it can be seen that J_{sd} is only related to J_o. It is well understood that the induced current increment J_{sd} on the rough surface is distributed over the partial surface near the target, and decayed towards the surface edges. It makes the integration of J_{sd} in Eqs. (6-5.16) and (6-5.17) within a limited length. Such truncated surface length L should take account of enough interactions between the target and underlying surface. Let's consider the largest scattered angle θ_m, as illustrated in Fig. 6-5.3, the truncated surface length should be chosen as

$$L \ge 2\left(\frac{r}{\cos\theta_m} + h\tan\theta_m\right) \tag{6-5.24}$$

It can be seen that L is related to the target size, height and the scattering angle. For larger θ_m, the induced current J_{sd} on the afar should be taken into account, and the surface length L becomes larger.

In Johnson's method, the surface length L should also satisfy Eq. (6-5.24) to take account of the interactions between the target and underlying surface.

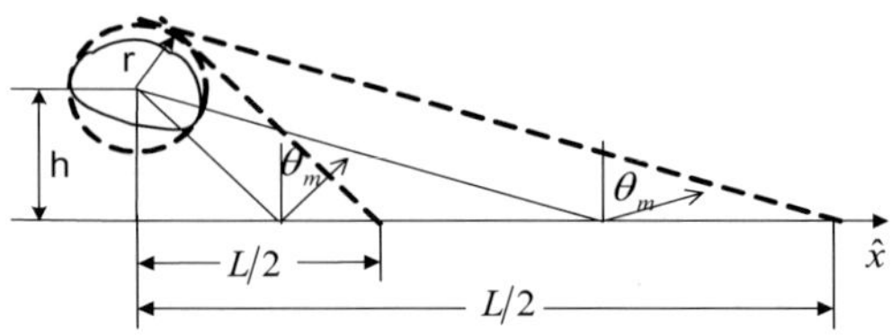

Fig. 6-5.3. The surface length of iteration computation for interaction of the target-surface

Moreover, as a basis of the tapered wave incidence, g and L should satisfy all requirements of the wave equation, correlation length and energy truncation, which had been discussed in our previous work [Ye and Jin 2005] and session 5-3:

$$g > \frac{6}{(\cos\theta_i)^{1.5}} \tag{6-5.25}$$

$$L > L_U \approx 3.8U^2 \text{ and } L \geq 4g \tag{6-5.26}$$

where U is the sea surface wind speed at the height 19.5 *m* above the oceanic surface.

As a conclusion, the tapered wave width g should satisfy Eqs. (6-5.22) and (6-5.25), and the surface length L should satisfy Eq. (6-5.26). In iteration process of J_{sd} and J_o, the surface length L should satisfy Eqs. (6-5.24) and (6-5.26). Two ways of L might be able to choose small L as required in computations. In Johnson's method, g and L for two cases should satisfy Eqs. (6-5.10, 6-5.25) and Eqs. (6-5.24, 6-5.26), respectively.

Suppose that the surface lengths for the case of the target-surface interaction and surface contribution only are denoted, respectively, by L_1 and L_0, where L_1 and L_0 can be different in our approach and are identical in Johnson's method. To obtain the same accurate results, Table 6-5.1 compares the different surface lengths for the case $\theta_m = 85°$. It can be seen that our iterative method with much smaller surface length should be faster, especially for low grazing incident angle.

Table 6-5.1. Comparison of the surface lengths for two methods

θ_i	Johnson's method	Iterative method	
	L (λ)	L_0 (λ)	L_1(λ)
30°	343.3	92.4	343.3
60°	437.2	160.0	
80°	1368.4	460.8	

6-5.2 Numerical Results

Suppose a cylinder target with the radius $r = 5\lambda$ located above a plane surface at the height $h = 10\lambda$. The tapered wave with $g = 23.1\lambda$ is incident upon at $\theta_i = 30°$. Two surface lengths L for different θ_m are taken, respectively,

$$L(\theta_m = 75°) = 113.3\lambda \tag{6-5.27}$$

$$L(\theta_m = 85°) = 343.3\lambda \tag{6-5.28}$$

The results and a comparison with Johnson are shown in Fig. 6-5.4. It can be seen that within the scattering angles $\theta_s \in (-75°, 75°)$, three results are exactly matched. At low grazing scattering angle $|\theta_s| > 75°$, the result of $L_1 = 113.3\lambda$ is deviated, and a longer surface L should be taken. The L of Eq. (6-5.24) can be applied to accurate results for the scattering angles of $(-\theta_s, \theta_s)$. It also can be seen that even the difference induced currents at the surface edges are very small, they might contribute to the scattering field at low grazing scattering angle.

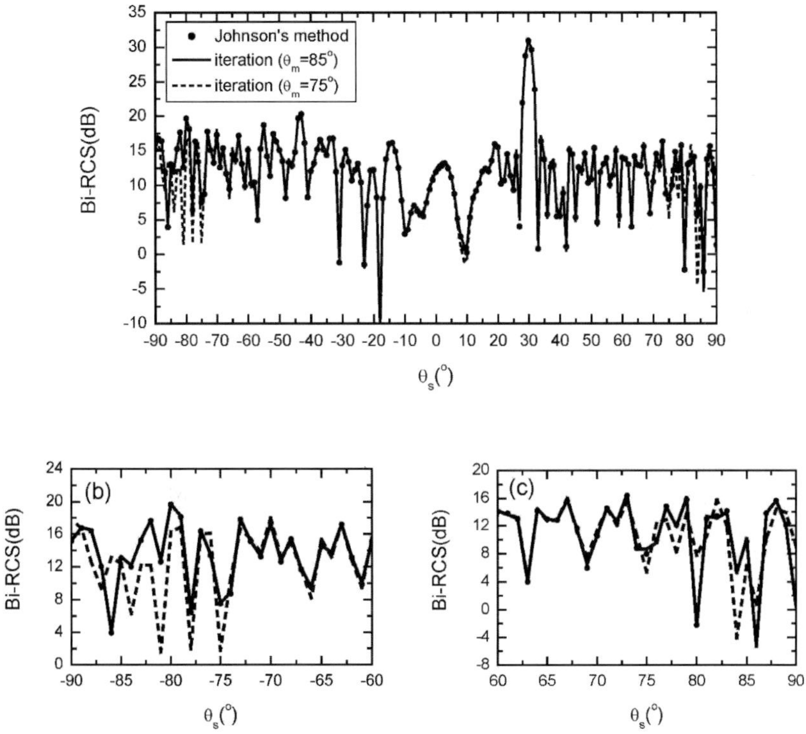

Fig. 6-5.4. Comparison of bistatic d-RCS for two different lengths L
((b) and (c) are the enlarged parts of (a) as θ_s approaches to $\pm 90°$)

Two tapered waves with $g = 4r/\cos\theta_i = 23.1\lambda$ and $g = 102.4\lambda$ are used to compute the scattering fields from the oceanic rough surface with $U = 5m/s$. The central line of the tapered wave is with $x = -h\tan\theta_i$, and the surface length is $L_0 = 4g$. The induced current J_{s0} on the rough surface is solved using Eq. (6-5.7). The scattering field E_{s0} at the height line $h = 10\lambda$ is shown in Fig. 6-5.5. It can be seen that the distribution of the scattering field $E_{s0}(\bar{r})$ centered at the specular direction $x = 0$ expands as the incident tapered wave becomes wider, and is quickly fluctuating. Within the region of $(-r/\cos\theta_i, r/\cos\theta_i)$, the scattering fields $E_{s0}(\bar{r})$ of two tapered waves coincide well.

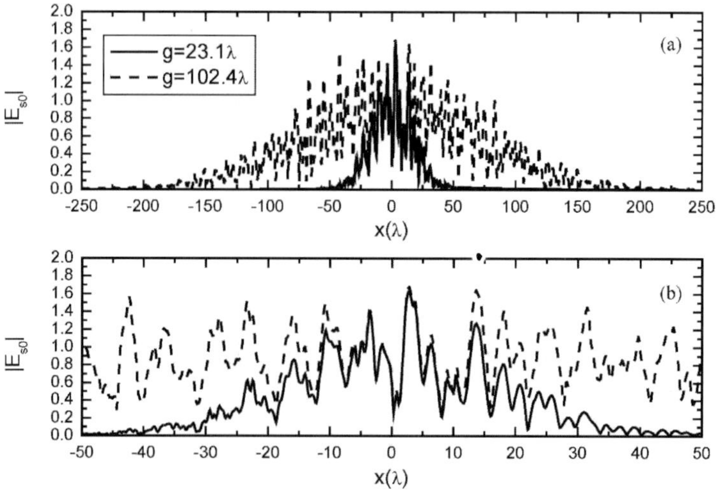

Fig. 6-5.5. Distribution of $E_{s0}(\bar{r})$ at the height $h = 10\lambda$ ((b) is the enlarged part of (a))

Suppose that a cylinder target is located above a P-M oceanic rough surface with the sea surface wind speed $U = 5.0m/s$, and the incident angle $\theta_i = 30°$. Other parameters are the same as Fig. 6-5.5. The induced current J_o on the cylinder above the underlying rough surface and plane surface, and the cylinder in free space are given in Fig. 6-5.6. It can be seen that the cylinder is illuminated within the angle $\psi \in 30° \sim 210°$, and the induced current J_o of the cylinder is especially strong at $\psi = 120°$ because of more incident illumination at that incident direction. Due to interaction from the underlying surface, J_o is enhanced especially around $\psi = 240°$. Interference of the incident and scattering waves causes J_o fluctuating, especially around the forward direction $\psi \in 150° \sim 210°$.

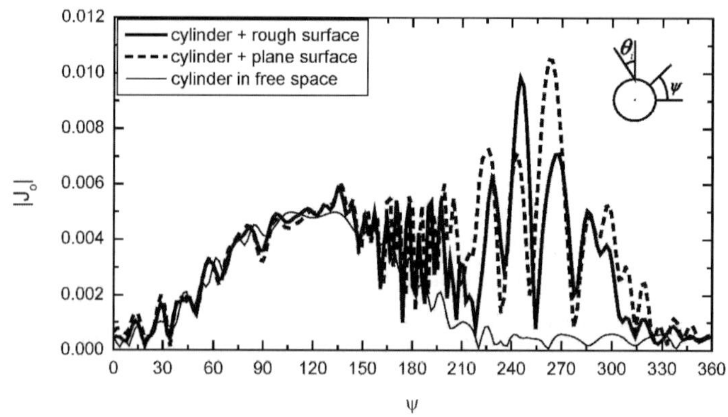

Fig. 6-5.6. The induced current J_o on the cylinder

Fig. 6-5.7 compares the difference induced current J_{sd} on the underlying plane surface and rough surface. It can be seen that J_{sd} is mainly distributed over the partial surface near the cylinder, where there is strong interaction of the target-underlying surface. Due to the strong forward scattering of the cylinder target, the peak of J_{sd} is shifted at $x > 0$. Rough surface also causes more fluctuation of J_{sd}.

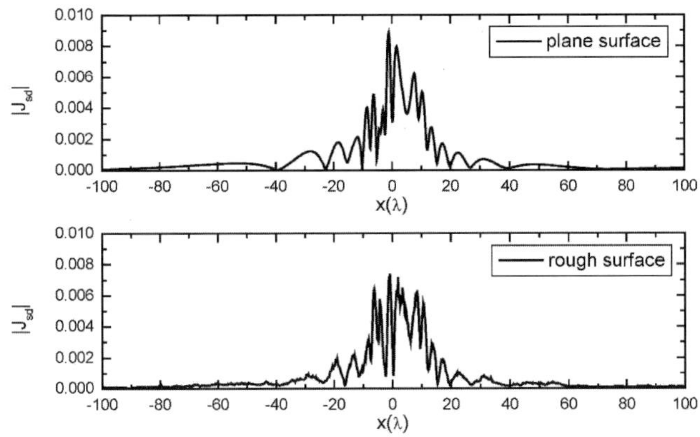

Fig. 6-5.7. The difference induced current J_{sd} on the underlying surface

Fig. 6-5.8 gives the d-RCS σ_d of the cylinder target above a plane surface and a rough surface, and RCS of the cylinder in free space. It can be seen that the d-RCS σ_d of the target-underlying surface is significantly enhanced, especially at the specular direction of $\theta_s = 30°$, and the peak of d-RCS σ_d value for the plane surface case is highest. If the sea surface wind speed increases, the scattering peak becomes lower, and the bistatic d-RCS σ_d is quickly fluctuating versus scattering angles.

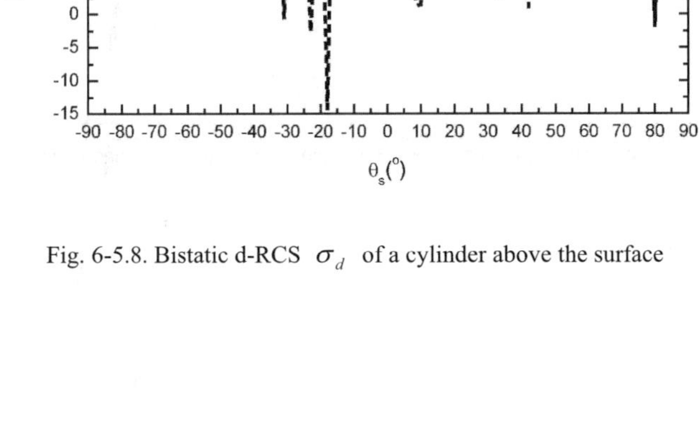

Fig. 6-5.8. Bistatic d-RCS σ_d of a cylinder above the surface

Now consider a target of square column located above a rough sea surface, with the side length $a = 7.854\lambda$ (its circumference is the same as the cylinder $C = 4a = 2\pi r$). The induced current J_o on the square column is shown in Fig. 6-5.9. It can be seen that J_o on the side interfaces illuminated by the incident wave are stronger, and J_o on the bottom interface is significantly enhanced due to the interaction from the underlying surface. Interference of the incident and scattering waves also causes J_o angularly fluctuating.

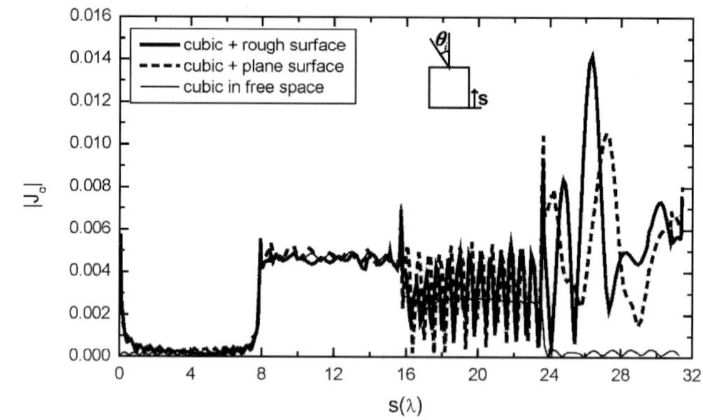

Fig. 6-5.9. The induced current J_o on the square column

Fig. 6-5.10 gives the difference induced currents J_{sd} on the underlying surface. The peak of J_{sd} at $x = -(a/2 + h\tan\theta_i) \approx -9.7\lambda$ is due to the reflection of the left side interface of the square column, and $x = (h - a/2)\cdot\tan\theta_i \approx 3.5\lambda$ is due to the bottom side interface. Toward the surface edges, J_{sd} is quickly decayed because of the weak interaction between the target and underlying surface.

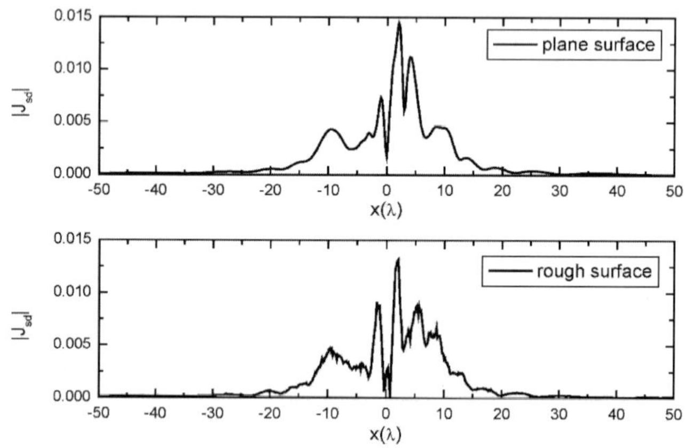

Fig. 6-5.10. The induced difference current J_{sd} on the underlying surface

Fig. 6-5.11 gives bistatic d-RCS σ_d of the square column target above a plane surface and a rough surface, and RCS of the target in free space. It can be seen that the RCS of this target in free space has a peak in the specular direction $\theta_s = 30°$ due to the direct reflection on the top interface. This peak is significantly enhanced due to multi-interaction between the target and underlying surface, and another peak appears in the backward direction of $\theta_s = -30°$.

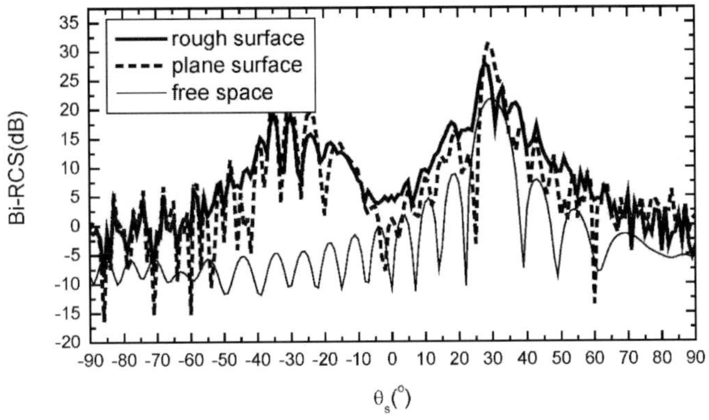

Fig. 6-5.11. Bistatic d-RCS σ_d of the square column

It is also interesting to see the d-RCS as the orientation of the square column changes. Fig. 6-5.12 shows the bistatic d-RCS σ_d for different orientation θ_r (to $\hat{x}$ direction). It can be seen that multi-peaks appear in different directions, depending on the orientation of the square column.

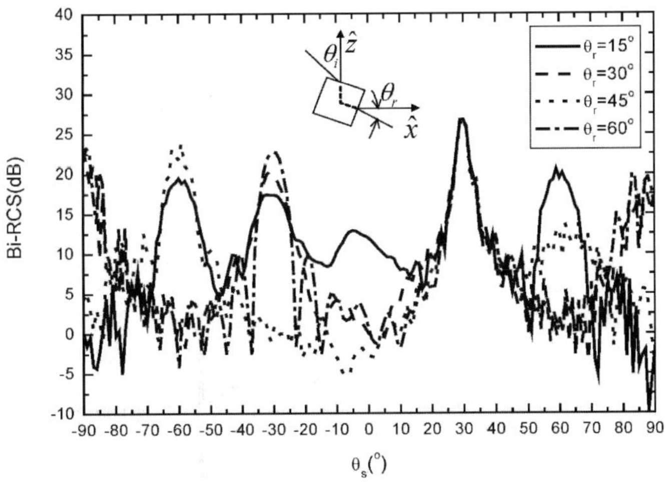

Fig. 6-5.12. Bistatic d-RCS σ_d of the square column with different orientations

6-6 References

Axline, R.M. and A.K. Fung (1978), "Numerical computation of scattering from a perfectly conducting random surface", *IEEE Transactions on Antenna and Propagation*, 26(3): 482-488.

Barrick, D.E. (1972), "First-order theory and analysis of MF/HF/VHF scatter from the sea", *IEEE Transactions on Antenna and Propagation*, 20: 2-10.

Berenger, J.P. (1994), "A perfectly matched layer for the absorption of electromagnetic waves", *Journal of Computational Physics,* 114: 185-200.

Billingsley, J.B. (1998), "Measurements of L-band inland-water surface-clutter Doppler spectra", *IEEE Transactions on Aerospace and Electronic Systems*, 34: 378-390.

Bladel, J.V. (1976), "Electromagnetic fields in the presence of rotating bodies", *Proceedings of IEEE,* 64: 301-318.

Cangellaris, A.C. and R. Lee (1990), "The bymoment methods for two-dimensional electromagnetic scattering", *IEEE Transactions on Antenna and Propagation*, 38(9): 1429-1437.

Choi-Grogan, Y.S., K. Eswar, P. Sadayappan and R. Lee (1996), "Sequential and parallel implementations of the partitioning finite-element method", *IEEE Transactions on Antennas and Propagation*, 44(12): 1609-1616.

Compag Extended Math Library Reference Guide(2001), USA: Compaq Computer Corporation.

Chou, H.T. and J.T. Johnson (1998), "A novel acceleration algorithm for the computation of scattering from rough surfaces with the forward-backward method", *Radio Science*, 33(9): 1277-1287.

Cooper, J. (1980), "Scattering of electromagnetic fields by a moving boundary: the one-dimensional case", *IEEE Transactions on Antenna and Propagation*, 28: 791-795.

Despres, B. (1992), "Domain decomposition method and the Helmholtz problem", *Procedings of International Symposium of Mathematic Numerical Aspects of Wave Propagation Phenomena*, Strasbourg France, SIAM ed., 44-52.

Fan, G.X., Q.H. Liu and J.S. Hesthaven (2002), "Multidomain Pseudospectral time-domain simulations of scattering by objects buried in lossy media," *IEEE Transactions on Geoscience and Remote Sensing*, 40: 1366-1373.

Franceschetti, G., M. Migliaccio and D. Riccio (1996), "An electromagnetic fractal-based model for the study of fading", *Radio Science*, 13(6): 1749-1759.

Freund, I., M. Rosenbluh and S. Feng (1988), "Memory effects in propagation of optical waves through distorted media", *Physical Review Letters*, 61: 2328-2331.

Freund, I., M. Rosenbluh and R. Berkovits (1989), "Geometric scalling of the optical memory effect in coherent wave propagation through random media", *Physical Review (B)*, 39: 12403-12406.

Gander, M.J. (2001), "Optimized Schwarz method for Helmholtz Problems", *Proceedings of the 13th international conference on domain decomposition*, 245-252.

Gjessing, D.T., J. Hjelmstad, and T. Lund (1982), "A multifrequency adaptive radar for detection and identification of objects: results on preliminary experiments on aircraft against a sea-clutter background", *IEEE Transactions on Antenna and Propagation*, 30:351-365.

Guerin, C.A., M. Holschneider, and M. Saillard (1997), "Electromagnetic scattering from multi-scale rough surfaces", *Waves in Random Media*, 7(3): 331-349.

Hastings, F.D., J.B. Schneider, S.L Broschat (1995), "A Monte-Carlo FDTD technique for rough surface scattering", *IEEE Transactions on Antennas Propagation*, 43: 1183-1191.

Holliday, D., L.L. DeRaad, G.J. St-Cyr (1996), "Forward-backward: A new method for computing low-grazing angle scattering", *IEEE Transactions on Antenna and Propagation*, 44: 722-729.

IMSL MATH/LIBRARY-User's Manual, Visual Numerics, Inc. (http://www.vni.com).

Ishimaru, A. (1978), *Wave Propagation and Scattering in Random Media*, New York: IEEE Press.

Jaggard, D.L. and X. Sun (1990), "Scattering from fractally corrugated surfaces", *Journal of the Optical Society of American*, 7(6): 1055-1062.

Jaggard, D.L., and X. Sun (1990), "Fractal surface scattering: A generalized Rayleigh solution", *Journal of Applied Physics*, 68(11): 5456-5462.

Jin, J.M. (1993), *The Finite Element Method in Electromagnetics*, New York: Wiley.

Jin, Y.Q. (1994), *Electromagnetic Scattering Modelling for Quantitative Remote Sensing*, Singapore: World Scientific.

Jin, Y.Q. and G. Li (2000), "Detection of a scatter target over randomly rough surface by using angular correlation function in finite element approach", *Waves in Random Media*, 10(4): 273-280.

Jin, Y.Q. and Z. Li (2001), "Simulation of scattering from complex rough surfaces at low grazing angle incidence using the GFBM/SAA method", *IEEJ Transactions of Fundamentals and Materials Society (A)*, 121(10): 917-921.

Jin, Y.Q. and Z. Li (2003), "Numerical simulation of radar surveillance for the ship target and oceanic clutters in two-dimensional model", *Radio Science*, 38(3): 1045-1050.

Johnson, J.T., J.V. Toporkov, and G. S. Brown (2001), "A numerical study of backscattering from time-evolving sea surfaces: Comparison of hydrodynamic models", *IEEE Transactions on Geoscience and Remote Sensing*, 39: 2411-2419.

Johnson, J.T. (2000), "A numerical study of scattering from an object above a rough surface", *IEEE Transactions on Antennas and Propagation*, 50(10): 1361-1367.

Lamont-Smith, T. (2000), "Doppler spectra of laboratory wind waves at low grazing angle", *Waves in Random Media,* 10: 33-41.

Le, T.C., Y. Kuga and A. Ishimaru (1991), "Angular correlation function based on the second order Kirchhoff approximation and comparison with experiments", *Journal of Optical Society of America (A)*, 13:1057-1067.

Lee, P.H.Y., J.D. Barter, et al. (1995), "Power spectral lineshapes of microwave radiation backscattered from sea surfaces at small grazing angles", *IEE Proceedings-Radar Sonar Navigation,* 142: 252-258.

Lee, P.H.Y., J.D. Barter, et al. (1998), "Scattering from breaking gravity waves without wind", *IEEE Transactions on Antennas and Propagation*, 46: 14-26.

Lee, R., and V. Chupongstimun (1994), "A partitioning technique for the finite element solution of electromagnetic scattering from electrically large dielectric cylinders", *IEEE Transactions on Antennas and Propagation*, 42: 737-741.

Li, Z. and Y.Q. Jin (2001), "Numerical simulation of scattering from fractal rough surface in the finite element method", *Science in China (E)*, 44(1): 12-18.

Li, Z. and Y.Q. Jin (2002), "Bistatic scattering and transmitting through a fractal rough surface with high permittivity using the physics-based two–grid method in conjunction with the forward-backward method and spectrum acceleration algorithm", *IEEE Transactions on Antennas and Propagation*, 50: 1323-1327.

Liu, J.W.H. (1992), "The multifrontal method for sparse matrix solution: theory and practice", *SIAM Review*, 34(1): 82-109.

Liu, P., J.D. Xu and W. Wan (2001), "A finite-element realization of 3D conformal PML", *Microwave and Optical Technology Letters*, 30: 170-173.

Liu, P. and Y.Q. Jin (2004a), "Numerical simulation for bistatic scattering from a target at low altitude over rough sea surface under EM Incidence at low grazing angle by using the finite element method", *IEEE Transactions on Antennas and Propagation,* 52(5): 1205-1210.

Liu, P. and Y.Q. Jin (2004b), "The finite element method with domain decomposition for electromagnetic bistatic scattering from the comprehensive model of a ship on and a target above large-scale rough sea surface", *IEEE Transactions on Geoscience and Remote Sensing,* 42 (5): 950-956.

Liu, P. and Y.Q. Jin (2005), "Numerical simulation of the Doppler spectrum of a flying target above dynamic oceanic surface by using the FEM-DDM Method", *IEEE Transactions on Antennas and Propagation,* 53(2): 825~832.

Lou, S.H., L. Tsang, C.H. Chan and A. Ishimaru (1991), "Application of the finite element method to Monte Carlo simulations of scattering of waves by random rough surfaces with the periodic boundary condition", *Journal of Electromagnrtic Waves and Applications*, 5(8): 835-855.

Mattia, F. (1999), "Backscattering properties of multi-scale rough surfaces", *Journal of Electromagnetic Waves and Applications*, 13: 493-527.

Moran, J. (1984), *An Introduction to Theoretical and Computational Aerodynamics*, New York: Wiley.

Moss, C.D., F.L. Teixeira, et al. (2002), "Finite-difference time-domain simulation of scattering from objects in continuous random media", *IEEE Transactions on Geoscience and Remote Sensing*, 40(1): 178-186.

Papoulis, A. (1984), *Probability, Random Variables, and Stochastic Processes*, New York: McGraw-Hill.

Peterson, A.F., S. L. Ray and R. Mittra (1997), *Computational Methods for Electromagnetics*, New York: IEEE Press.

Pierson, W.J. and L. Moskowitz (1964), " A proposed spectral form for fully developed wind seas based on the similarity theory of S. A. Kitaigorodskii", *Journal of Geophysical Research,* 69: 5181-5190.

Pino, M.R., L. Landesa, et al. (1999), "The generalized forward-backward method for analyzing the scattering from targets on ocean-like rough surfaces", *IEEE Transactions on Antennas and Propagation*, 47(6): 961-968.

Robert, L. and V. Chupongstimun (1994), "A partitioning technique for the finite element solution of electromagnetic scattering from electrically large dielectric cylinders", *IEEE Transactions on Antennas and Propagation*, 42(5):737-741.

Rouvier, S. and I. Chenerie (1997), "Ultra wide band electromagnetic scattering of a fractal profiles", *Radio Science*, 32(2): 285-293.

Sanchez-Gil, J.A. and J.V. Garcia-Ramon (1997), "Far-field intensity of electromagnetic waves scattered from random, self-affine fractal metal surfaces", *Waves in Random Media*, 7(3): 285-293.

Savailis, S., P. Frangos, D.L. Jaggard, et al (1997), "Scattering from fractally corrugated surfaces with use of the extended boundary condition method", *Journal of the Optical Society of American (A)*, 14(2): 475-485.

Shankar, V., W.F. Hall and A.H. Mohammadian (1989), "A time-domain differential solver for electromagnetic scattering problems", *Proceedings of IEEE*, 77: 709-721.

Stupfel, B. (1996), "A fast-domain decomposition method for the solution of electromagnetic scattering by large objects", *IEEE Transactions on Antennas and Propagation*, 44(10): 1375-1385.

Teixeira, F.L. and W.C. Chew (1998), "Analytical derivation of a conformal perfectly matched absorber for electromagnetic waves," *Microwave and Optical Technology Letters*, 17: 231-236.

Thorsos, E.I. (1988), "The validity of the Kirchhoff approximation for rough surface scattering using a Gaussian roughness spectrum", *Journal of the Acoustical Society of America*, 83: 78-92.

Tsang, L., G. Zhang and K. Pak (1996), "Detection of a buried object under a single random rough surface with angular correlation function in EM wave scattering", *Microwave Optical Technology Letters*, 11: 300-304.

Tsang, L., J.A. Kong, K.H. Ding and C.O. Ao (2001), *Scattering of Electromagnetic Waves: Numerical Simulations*, New York: Wiley.

Toporkov, J.V. and G.S. Brown (2000), "Numerical simulations of scattering from time-varying, randomly rough surfaces", *IEEE Transactions on Geoscience and Remote Sensing*, 38(4): 1616-1625.

Toporkov, J.V. and G.S. Brown (2002), "Numerical study of the extended Kirchhoff approach and the lowest order small slope approximation for scattering from ocean-like surfaces: Doppler analysis", *IEEE Transactions on Antennas and Propagation*, 50(4): 417-425.

Ulaby, F.T., R.K. Moore and A.K. Fung (1982), *Microwave Remote Sensing (vol. 2): Radar Remote Sensing and Surface Scattering and Emission theory*, New York: Addison-Wesley.

Videen, G. (1991), "Light scattering from a sphere on or near a surface", *Journal of the Optical Society of American (A)*, 8: 483-489.

Videen, G. and D. Ngo (1997), "Light scattering from a cylinder near a plane interface: theory and comparison with experimental data", *Journal of the Optical Society of American (A),* 14: 70-78.

Walsh, J., Dawe B. J., and Srivastava S. K. (1986), "Remote sensing of icebergs by ground-wave Doppler radar", *IEEE Journal of Oceanic Engineering*, 11(2): 276-284.

Wang, J., R.L. Kirlin, X., Lu and R. Dizaji (2002), "Small ship detection with high frequency radar using an adaptive ocean clutter pre-whitened subspace method", *Sensor Array and Multichannel Signal Processing Workshop Proceedings,* Aug: 92-95.

West, J.C. (2000), "Preconditioned iterative solution of scattering from rough surfaces", *IEEE Transactions on Antennas and Propagation*, 48(6):1001-1002.

Wolfe, C.T., U. Navsariwala, and S. D. Gedney (2000), "A parallel finite-element tearing and interconnecting algorithm for solution of the vector wave equation with PML absorbing medium", *IEEE Transactions on Antennas and Propagation*, 48(2): 278-283.

Ye, H. and Y.Q. Jin (2005), "Parameterization of the Tapered Incident Wave for Numerical Simulation of Electromagnetic Scattering from Rough Surface", *IEEE Transactions on Antennas and Propagation*, 50(10): 1361-1367.

Zienkiewicz, O.C. (2000), *The Finite Element Method*, third edition, McGraw-Hill, New York.

Chapter 7

Inverse Scattering and Parameter Retrievals 310

7-1 Reconstruction of Roughness Profile of Fractal Surface From Scattering Measurement at Low Grazing Incidence

Scattering from rough surface has been extensively studied in electromagnetics, optics, acoustics, etc. However, the associated problems of inverse scattering from rough surface, i.e. inversion of surface characteristic parameters and reconstruction of surface profile, have received little progress. Inversion of rough surface is, in particular, of great interest in microwave remote sensing, optical system measurement, underwater acoustics, etc. Prior research reconstructs grating profiles from measurements of diffraction efficiency [Roger and Maystre 1980, Husier et al. 1982]. Recently, Wombel and DeSanto (1991a,1991b) presented an inversion of small-slope rough surface in the Kirchhoff approximation by using plane wave incidence at a fixed frequency and incident angle. Spivack (1990,1992a, 1992b), using the parabolic approximated equation for a scalar wave incidence at grazing angle, derived a direct approximate solution for inverse scattering. As the incidence is known and scattering fields along the line parallel to the mean surface are measured, the roughness profile of one-dimensional (1D) conducting surface can be progressively reconstructed.

Rough surface profile was usually assumed as Gaussian statistics. During recent years, the fractal geometry is introduced to the study of rough surface [Jaggard and Sun 1990]. Since the fractal rough surface, as a multi-scale geometry, takes account of surface perturbations in large and small scales with self-similarity, it seems better to describe some natural surfaces. Those surfaces should be described not only by conventional correlation function, such as the roughness height variance and correlation length, but also by the fractal dimension. Thus, inversion of fractal rough surface is of special interest for many potential applications.

In recent decade, the angular correlation function (ACF) from scattering of random system (random rough surface, random particles, etc.) has been discussed. It shows that rough surface (with higher fractal dimension D) produces wide-angle diffusion. Its Fourier transformation yields narrow-angle distribution of ACF on so called memory line [Le et al. 1996, Jin and Li 2000]. In other words, if a target is supposed to be placed over the rough surface to enhance specular, not diffused, scattering, ACF might be angularly broadened [Jin and Li 2000]. In such case, the fractal dimension D inverted from the target region should be found smaller. Different D's indicate the surface heterogeneity or target existence.

Following Spivack's approach (1990,1992a,1992b) and the physical picture described by complex fractal rough surfaces, this session develops a numerical approach to reconstruct the roughness profile of complex fractal surfaces. Using the Monte Carlo method, the fractal rough surfaces with a band-limited Weistrass-Manderbrot function are numerically realized. The incident field used in this session is a Gaussian beam to be traveling at low grazing angle [Spivack 1990]. The scattering fields and surface currents can be solved and employed to progressively invert the rough surface profile. The reconstructed profile and its parameters such as the fractal dimension, roughness variance and correlation length are well matched with the simulated surfaces [Jin and Li 2001].

7-2.1 Scattered Fields of Parabolic Wave Equation

We are concerned with a scalar field incident upon a $1D$ conducting rough surface at low grazing angle. The rough surface is horizontally extended ($x \geq 0$), and the height was randomly perturbed along the $\hat{z}$ axis: $z=h(x)$ with the mean height $<h(x)>=0$. Since the incident field is traveling at low grazing angle, the wave field propagates predominantly in forward direction and the fields have slowly varying part written as

$$\Psi(x,z) = \psi(x,z)\exp(ikx) \tag{7-1.1}$$

At low grazing angle, it has assumed that the wave propagation constant in z-direction $k_z << k_x \approx k$, and the parabolic approximation be written as

$$\frac{\partial^2}{\partial x^2}\psi(x,z) << 2ik\frac{\partial}{\partial x}\psi(x,z) \tag{7-1.2}$$

The Helmholtz wave equation of $\Psi(x,z)$ can be approximated by the parabolic approximate equation of $\psi(x,z)$ as follows

$$2ik\frac{\partial}{\partial x}\psi(x,z) + \frac{\partial^2}{\partial z^2}\psi(x,z) = 0 \tag{7-1.3}$$

The Green's function of Eq. (7-1.3) is written as

$$G(x,z;x',z') = \begin{cases} \frac{1}{2}[\frac{i}{2\pi k(x-x')}]^{1/2}\exp[i\frac{k(z-z')^2}{2(x-x')}], & as \quad x \geq x' \\ 0, & as \quad x < x' \end{cases} \tag{7-1.4}$$

The wave field at any point $\bar{r}(x,z)$ is written as

$$\psi(x,z) = \psi_{inc} + \int_0^x dx' G(x,z;x',z'=h(x'))\psi'(x') \tag{7-1.5}$$

where the surface current is denoted as $\psi'(x) \equiv \partial\psi(x,z)/\partial z$.

Thus, the scattering field is

$$\psi_s(x,z) = \int_0^x dx\, G(x,z;x',z'=h(x'))\psi'(x') \tag{7-1.6}$$

This gives the Volterra integral equation of the scattered field and surface current. Note that

$$\psi = \psi_{inc} + \psi_s \tag{7-1.7}$$

Consider both points $(x, z = h(x))$ and $(x', z' = h(x'))$ are on the conducting rough surface. Eq. (7-1.5) yields

$$\psi_{inc}(x,h(x)) = -\int_0^x dx' G(x,h(x);x',h(x'))\psi'(x') \tag{7-1.8}$$

Consider a Gaussian beam with initial width w at location $z = z_0$

$$\psi_{inc}(x,z) = \frac{w}{(w^2 + 2ix/k)^{1/2}}\exp[-\frac{(z-z_0)^2}{w^2 + 2ix/k}] \tag{7-1.9}$$

is incident upon the rough surface $(x, h(x))$ at low grazing angle [Spivack 1990]. Eq.(7-1.9) yields

$$\psi_{inc}(x,h(x)) \approx \psi_{inc}(x,0)\exp[\frac{2z_0 h(x)}{w^2 + 2ix/k}] \tag{7-1.10}$$

where

$$\psi_{inc}(x,0) = \frac{w}{(w^2 + 2ix/k)^{1/2}}\exp(-\frac{z_0{}^2}{w^2 + 2ix/k}) \tag{7-1.11}$$

Since $\exp[ik(z-z')^2/(2(x-x'))] \approx 1$ in the Green's function of Eq. (7-1.4) [Spivack 1990b], Eq. (7-1.8) becomes

$$\psi_{inc}(x,h(x)) = -\alpha \int_0^x dx' \frac{1}{(x-x')^{1/2}} \psi'(x') \qquad (7\text{-}1.12)$$

where $\alpha = \frac{1}{2}(\frac{i}{2\pi k})^{1/2}$. From Eqs. (7-1.10) and (7-1.12), the surface profile (extended from x>0) is formulated by the integral equation of surface current as

$$h(x) = \ln[\frac{-\alpha \int_0^x dx'(x-x')^{-1/2} \psi'(x')}{\psi_{inc}(x,0)}] \cdot (\frac{w^2 + 2ix/k}{2z_0}) \qquad (7\text{-}1.13)$$

Eqs. (7-1.6) and (7-1.13) give a pair of coupled integral equations with unknown $\psi'(x')$ and $h(x)$. As the incidence ψ_{inc} is known and the scattering fields $\psi_s(x,z)$ along a line z parallel to the mean surface are measured or solved, the coupled Eqs.(7-1.6) and (7-1.13) can numerically invert the roughness profile $h(x)$.

7-2.2 Inversion of Fractal Surface Profile

A fractal surface can be described by the band-limited Weierstrass-Manderbrot function expressed with the weighted sum of M harmonics (tones) [Jaggard and Sun 1990]:

$$h(x) = \delta C \sum_{m=0}^{M-1} b^{(D-2)m} \sin(k_0 b^m x + \varphi_m) \qquad (7\text{-}1.14)$$

where D $(1 < D < 2)$ is the fractal dimension, φ_m is the initial random phase, $b(>1)$ is the spatial frequency scaling factor, k_0 is the fundamental spatial wave number, and δ is the *rms* of surface height.

The coefficient $C = \sqrt{2(1-b^{2(D-2)})/(1-b^{2m(D-2)})}$ is a normalized amplitude control factor. As Eq. (7-1.14) indicated, the fractal rough surface is described by the parameters δ, D, b, k_0 and M, whilst the conventional Gaussian surface only needs two parameters, δ and correlation length ℓ. Different fractal rough surfaces may have the same δ and ℓ.

Fractal rough surface of Eq. (7-1.14) is normalized, so its rms of surface height is δ, as the same notation of the Gaussian rough surface. The correlation length ℓ can be determined from the following autocorrelation function $\rho(\tau)$:

$$\rho(\tau) = < h(x)h(x+\tau) > = (1-b^{2(D-2)})/(1-b^{2M(D-2)}) \sum_{m=0}^{M-1} b^{2(D-2)m} \cos(k_o b^m \tau) \qquad (7\text{-}1.15)$$

Note that autocorrelation function $\rho(\tau)$ is still a band-limited Weierstrass-Manderbrot fractal function. The correlation length ℓ can be calculated through $\rho(\ell) = 1/e$. M could be decided by the incident wavelength λ, i.e. by $k_0 b^{M-1} = k$ $(= 2\pi/\lambda)$, so the highest spatial harmonic wavelength λ_M is always less than the incident λ.

Using the Monte Carlo method to randomly sample the initial value of the phase term φ_m in Eq. (7-1.14) as $\varphi_m = 2\pi R$, here R is randomly and uniformly distributed over the regime [-1,1], the fractal surfaces are realized.

Numerical approach as described by Spivack (1990,1992a,1992b) is implemented to discretize the interval [0,X] by evenly spaced points $x_j, j = 0,\cdots,N$ and the intervals $\Delta x = x_{j+1} - x_j$. As the scattering fields $\psi_s(x_n,z)$ at discrete points (x_n,z), $n = 1,\cdots,N$ are measurable, here z is a line parallel to the mean surface level $<h(x)>=0$, Eq.(7-1.6) is rewritten as

$$\psi_s(x_n,z) = \sum_{j=1}^{n} \int_{x_{j-1}}^{x_j} dx' G(x_n,z;x',h(x')) \psi'(x') \qquad (7\text{-}1.16)$$

As the interval Δx is small, the current $\psi'(x')$ over the integral interval Δx much slowly varies compared with the Green's function G in Eq. (7-1.16). Thus, it can be treated as a constant and moved out from the integral as

$$\psi_s(x_n,z)=\sum_{j=1}^{n}\psi'(x_j)\Gamma_{n,j}(x_n,z) \tag{7-1.17}$$

where

$$\Gamma_{n,j}(x_n,z)=\int_{x_{j-1}}^{x_j}dx'G(x_n,z;x',h(x')) \tag{7-1.18}$$

From the Taylor series expansion for $x'\in(x_{j-1},x_j)$, it has

$$h(x)\approx h(x_{j-1})+(x-x_{j-1})h(x_{j-1}) \tag{7-1.19}$$

where $h'(x_{j-1})=\partial h(x)/\partial x|_{x=x_{j-1}}$. Substitute Eq. (7-1.19) into the Green's function of Eq. (7-1.18), and note that the exponent term of G can be written as

$$\frac{ik}{2}\{[z-h(x_{j-1})][z-h(x_{j-1})-2h(x_{j-1})(x_n-x_{j-1})](x_n-x)^{-1}+2h(x_{j-1})[z-h(x_{j-1})]\}.$$

Making a variable change as $y=(x_n-x')^{-1/2}$, Eq. (7-1.18) becomes

$$\Gamma_{n,j}(x_n,z)=2\alpha\exp(ika/2)\int_{b_1}^{b_2}dyy^{-2}\exp(iby^2) \tag{7-1.20}$$

where $a=2h(x_{j-1})[z-h(x_{j-1})]$, $b_1=(x_n-x_{j-1})^{-1/2}$, $b_2=(x_n-x_j)^{-1/2}$, and $b=\frac{k}{2}[z-h(x_{j-1})][z-h(x_{j-1})-2h(x_{j-1})(x_n-x_{j-1})]$.

Eq. (7-1.20) can be integrated by parts and put in terms of Fresnel integrals, thus the current $\psi'(x_n), n=1,\cdots,N$, on the surface can be obtained as

$$\psi'(x_n)=\frac{1}{\Gamma_{n.n}(x_n,z)}[\psi_s(x_n,z)-\sum_{j=1}^{n-1}\psi'(x_j)\Gamma_{n,j}(x_n,z)] \tag{7-1.21}$$

Therefore, if $h(x_j)$ from j = 0 up to n-1 have been obtained, all $\Gamma_{n,j}(x_n,z)$ from j = 0 up to n of Eq. (7-1.20) can be found. Thus, $\psi'(x_n)$ can be obtained from Eq. (7-1.21). From Eq. (7-1.13), it yields the height of next point $h(x_n)$. To begin numerical inversion progressively, set the initial value $h(0)=0$. The final result shows that it is not sensitive to the initial choose and only needs 3 or 5 times iterations to finish. In our cases, set $k=1$, $w=8$, $z_0=22.4$, $z=0.8$, and $X=514.8$. Equivalently, in microwave regime, it is as $\lambda=6.28$ *cm*, frequency=4.78 *GHz*, $w=1.27\lambda$ *cm*, $z_0=3.57\lambda$ *cm*, $z=0.13\lambda$ *cm*, and $X=81.9\lambda$ *cm*.

Using the Monte Carlo method, an ensemble average of 50 sample realizations of surface profiles is obtained as $\langle h(x)\rangle=[\sum_{m=1}^{50}h_m(x)]/50$. The averaged scattered field $\langle\psi_s(x,z)\rangle=[\sum_{m=1}^{50}\psi_{s,m}(x,z)]/50$ is used to invert $h(x)$ and is compared with previous simulation of $\langle h(x)\rangle$.

To form a complex fractal surface, different fractal dimension D's are assigned to the parts in the interval $[0,X]$.

When the roughness profile is reconstructed, its fractal dimension D can be inverted using the box counting method [Rourier and Borderies 1997]. The reconstructed roughness profile is first normalized in both coordinates, which would not change its fractal dimension

[Lo et al. 1993]. A square grid is then superimposed on the curve. The edges r_S of small cells are multiples of the sampling rate and is optimally chosen. The number N_S of small cells to cover the profile is counted. Based on the box counting method defined by the linear relationship between N_S and fractal D [Rourier and Borderies 1997],

$$\ln N_s = -D \cdot \ln(r_s) + c, \qquad (7\text{-}1.22)$$

the fractal dimension D can then be retrieved. Here, c is a constant for the linear relationship.

In other words, if the linear relationship of Eq. (7-1.22) exists, this profile is said as fractal. The slope of the least squares straight line of the set $[-\ln(r_s); \ln(N_s)]$ has been known as a good estimator of the fractal dimension D.

7-2.3 Reconstruction of Complex Fractal Surfaces

Using the Monte Carlo method with M=22, a fractal surface is realized as shown by the solid line in Fig.7-1.1, where the surface parameters are measured as D=1.27, $\delta = 0.041\lambda$, $\ell = 4.32\lambda$. The inverted surface profile is obtained as shown by the dotted line in Fig.1, where D=1.28, $\delta = 0.040\lambda$, $\ell = 4.72\lambda$. It can be seen that the surface profile is well reconstructed and the parameters D, ℓ and δ are well matched with the simulated one.

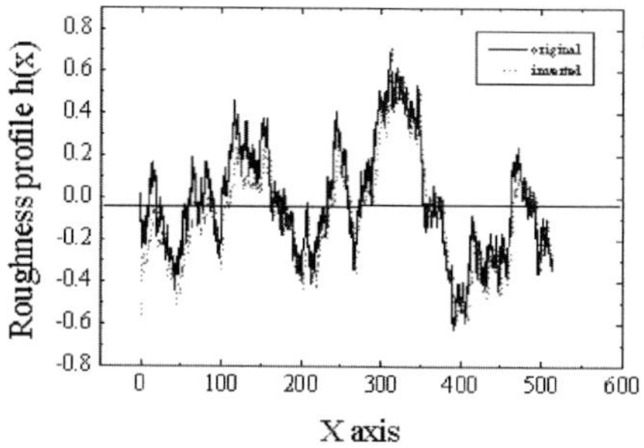

Fig. 7-1.1. Simulation and inversion of a fractal rough surface (D=1.27,N=6999)

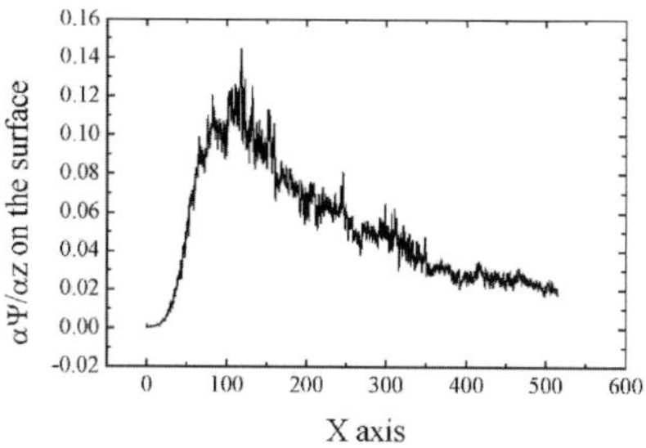

Fig.7-1.2. Surface current on the fractal surface

It is noted that the inverted surface current $\psi'(x)$ also shows fractal geometry. Fig.7-1.2 shows the result, where D=1.34 is found.

Fig.7-1.3 shows the fractal surface with higher D=1.42, ℓ=3.66λ, δ=0.043λ. The dotted line, as inversion with D=1.43, ℓ=3.71λ, δ=0.049λ, is also well matched the simulated surface profile.

The inverted surface current $\psi'(x)$ also shows fractal geometry. Fig.7-1.4 shows the result, where D=1.59, using the Monte Carlo method to simulate 50 realizations of the fractal rough surfaces.

The solid line in Fig.7-1.5 is the averaged profile $<h(x)>$, where D=1.37, ℓ=2.73λ, δ=0.0072λ. The dotted line gives the inverted surface, where D=1.34, ℓ=2.56λ, δ=0.0083λ are found. Inversions seem well matched with simulated one.

To form a complex fractal surface with non-uniform D's, we will suppose that the interval [0,X] is divided into three parts: the left part has D=1.36, ℓ=2.88λ, δ=0.036λ, the middle one has D=1.14, ℓ=5.76λ, δ=0.031λ, and the right part has D=1.39, ℓ=1.54λ, δ=0.025λ.

The solid line and dotted line in Fig.7-1.6, respectively, show the simulation and reconstruction of surface profile. Reconstructed profile is also divided into three parts: the left part has D=1.37, ℓ=3.62λ, δ=0.044λ, the middle D=1.18, ℓ=4.58λ, δ=0.030λ, and the right D=1.43, ℓ=1.54λ, δ=0.028λ. Comparing these parameters of the simulation and inversion, this complex fractal surface is well reconstructed. It is expected that significant heterogeneity of the surface might make inversion more difficult.

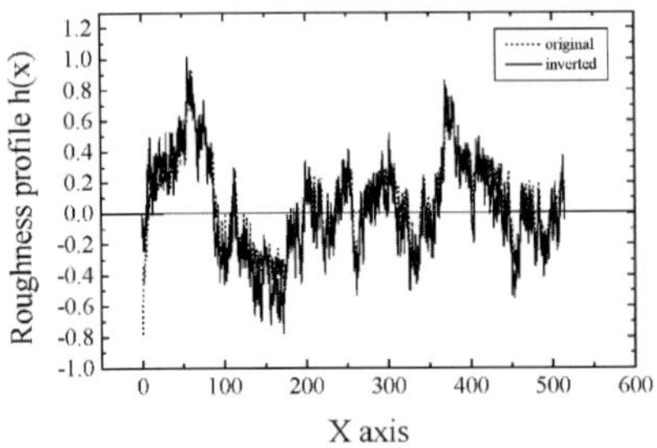

Fig.7-1.3. Simulation and inversion of a fractal rough surface (D=1.42, N=6999)

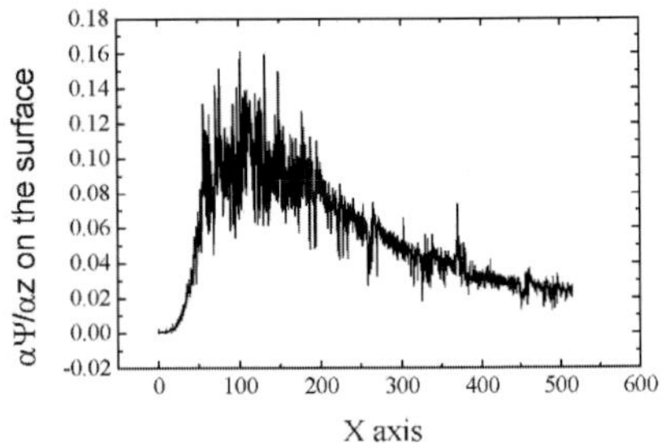

Fig.7-1.4. Surface current on the fractal surface.

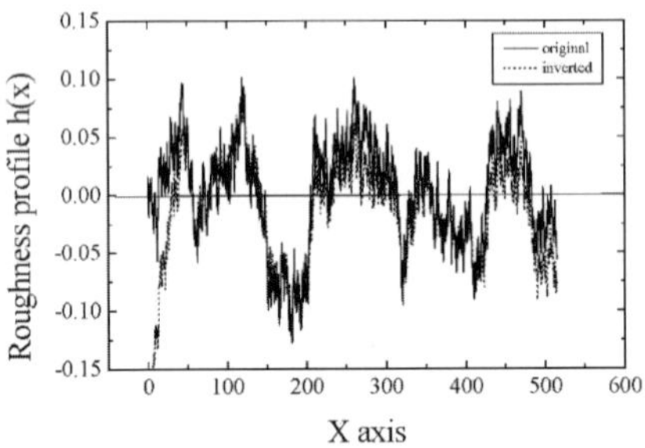

Fig.7-1.5. Simulation and inversion of 50 times averaged fractal surface (N=8999)

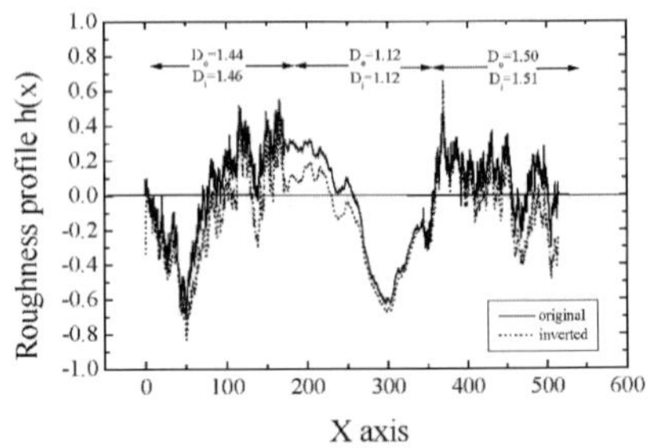

Fig.7-1.6. Simulation and inversion of heterogeneous fractal rough surface (N=8311)

It is also interesting to see that the inverted surface current $\psi'(x)$ in this example has also fractal heterogeneity as shown in Fig.7-1.7.

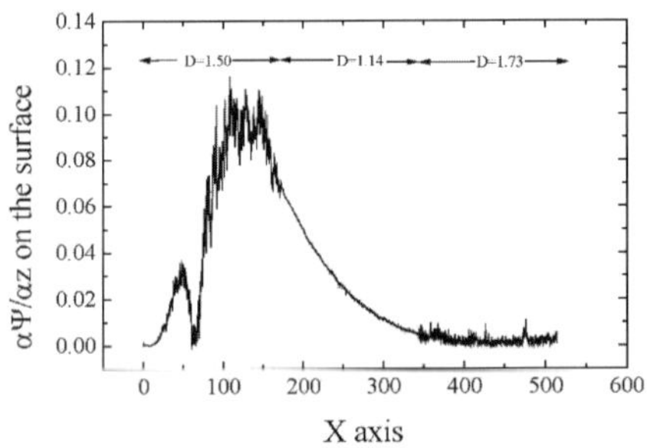

Fig.7-1.7. Surface current of heterogeneous fractal surface

7-2 Iterative Inversion of the Scalar Radiative Transfer Equation for a Layer of Random Spheroids

The main inverse problem in radiative transfer theory is that of retrieving characteristic parameters for the observed objects from radiation measurement—parameters such as particle size, dielectric constant, and layer depth. Inversion has been of great interest in wide areas—for example, atmospheric radiation and its remote sensing, thermal physics and engineering, and neutron transport. Beside studies of the statistics matrix, the Backus-Gilbert method, physical retrieval etc in atmospheric radiative transfer [Jin 1994], some analytic approaches to radiative transfer inversion has also been attempted [McCormick 1984,1992, Ligon et al. 1996, Quirantes and Delgado 1998]. For example, radiative fluxes from upper and lower interfaces are used for inversion of the coefficients of the phase function or the source function in the radiative transfer equation [Fukshansky et al. 1998]. But in most cases, those approaches seem impractical for directly relating these functions to the physical properties of the medium.

In this session, the first- and second-order iterative solutions of a scalar radiative transfer equation for a layer of random particles are presented analytically. The relationship between the Legendre coefficients and the phase function of random small spheroids is derived. Using the first- and second-order iterative solutions, an inversion of the Legendre coefficients of the phase function is formulated in terms of the matrix equation. Employing two measurements of the azimuthal distribution of bistatic scattering (simulated results of the first- and second-order iterative solutions with added random noise from the Monte Carlo realization), the complex dielectric constant of small spheroids and the particle number per unit area are iteratively inverted [Jin and Liang 2002].

7-2.1 The First- and Second-Order Iterative Solutions of Scalar Radiative Transfer Equation

The scalar radiative transfer equation for a layer of random medium is written as:

$$\cos\theta\frac{dI(\theta,\varphi,z)}{\kappa_e dz}=-I(\theta,\varphi,z)+\int_0^{\pi}d\theta'\sin\theta'\int_0^{2\pi}d\varphi' P(\theta,\varphi;\theta',\varphi')I(\theta',\varphi',z) \tag{7-2.1}$$

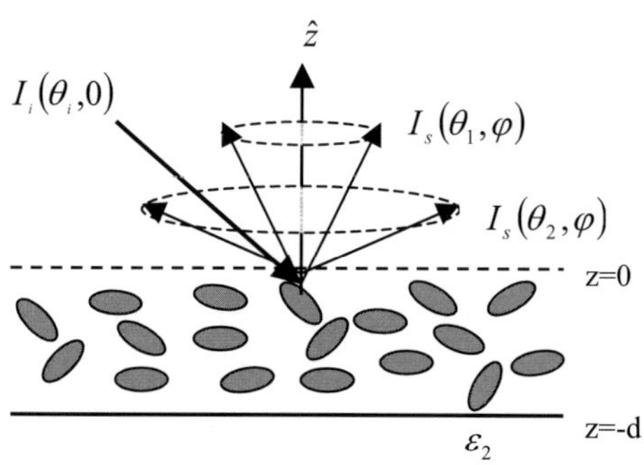

Fig. 7-2.1 The Geometry of the Problem.

where κ_e is the extinction coefficient, P is the phase function. And the boundary conditions are written as

$$\begin{cases} I(\pi-\theta,\varphi,0)=\delta(\theta-\theta_i)\delta(\varphi) \\ I(\theta,\varphi,-d)=R_{12}(\theta)I(\pi-\theta,\varphi,-d) \end{cases} \tag{7-2.2}$$

where $R_{12}(\theta)$ is the reflectivity of the underlying surface.

Solving Eqs. (7-2.1) and (7-2.2) iteratively from the integral form solution of Eq. (7-2.1), the first and second-order scattered intensities observed at $z=0$ [Jin 1994] are derived as:

$$\begin{aligned}
I^{(1)}(\theta,\varphi,z=0) &= P(\theta,\varphi;\pi-\theta_i,\varphi_i)\frac{\sec\theta}{\sec\theta+\sec\theta_i}\{1-\exp[-\kappa_e d(\sec\theta+\sec\theta_i)]\} \\
&+P(\theta,\varphi;\theta_i,\varphi_i)\frac{\sec\theta}{\sec\theta-\sec\theta_i}[\exp(-\kappa_e d\sec\theta_i)-\exp(-\kappa_e d\sec\theta)]R_{12}(\theta_i)\exp(-\kappa_e d\sec\theta_i) \\
&+P(\pi-\theta,\varphi;\pi-\theta_i,\varphi_i)\frac{\sec\theta}{\sec\theta-\sec\theta_i}[\exp(-\kappa_e d\sec\theta_i)-\exp(-\kappa_e d\sec\theta)]R_{12}(\theta)\exp(-\kappa_e d\sec\theta) \\
&+P(\pi-\theta,\varphi;\theta_i,\varphi_i)\frac{\sec\theta}{\sec\theta+\sec\theta_i}\{1-\exp[-\kappa_e d(\sec\theta+\sec\theta_i)]\} \\
&\times R_{12}(\theta)R_{12}(\theta_i)\exp[-\kappa_e d(\sec\theta+\sec\theta_i)]
\end{aligned} \tag{7-2.3}$$

$$\begin{aligned}
I^{(2)}(\theta,\varphi,z=0) &= \sec\theta\int_0^{\pi/2}d\theta'\sin\theta'\int_0^{2\pi}d\varphi'\{1+R_{12}(\theta_i)R_{12}\theta)\exp[-\kappa_e d(\sec\theta_i+\sec\theta)]\} \\
&\times[P(\theta,\varphi;\theta',\varphi')A_1(\theta',\varphi')+P(\theta,\varphi;\pi-\theta',\varphi')A_2(\theta',\varphi')] \\
&+\sec\theta\int_0^{\pi/2}d\theta'\sin\theta'\int_0^{2\pi}d\varphi'[R_{12}(\theta_i)\exp(-\kappa_e d\sec\theta_i)+R_{12}(\theta)\exp(-\kappa_e d\sec\theta)] \\
&\times[P(\pi-\theta,\varphi;\theta',\varphi')A_3(\theta',\varphi')+P(\pi-\theta,\varphi;\pi-\theta',\varphi')A_4(\theta',\varphi')] \\
&+\exp(-\kappa_e d\sec\theta_i)\int_0^{\pi/2}d\theta'\sin\theta'\int_0^{2\pi}d\varphi'\sec\theta' R_{12}(\theta)[P(\pi-\theta',\varphi';\pi-\theta_i,\varphi_i)D_1(\theta',\theta_i) \\
&+R_{12}(\theta_i)P(\pi-\theta',\varphi';\theta_i,\varphi_i)D_2(\theta',\theta_i)][P(\theta,\varphi;\theta',\varphi')D_1(\theta,\theta')\exp(-\kappa_e d\sec\theta') \\
&+P(\pi-\theta,\varphi;\theta',\varphi')D_2(\theta,\theta')R_{12}(\theta)\exp(-\kappa_e d\sec\theta)]
\end{aligned} \tag{7-2.4}$$

where

$$A_1(\theta',\varphi')=P(\theta',\varphi';\pi-\theta_i,\varphi_i)\frac{\sec\theta'}{\sec\theta'+\sec\theta_i}\{D_2(\theta,\theta_i)-D_1(\theta,\theta')\exp[-\kappa_e d(\sec\theta'+\sec\theta_i)]\}$$

$$A_2(\theta',\varphi')=P(\pi-\theta',\varphi';\pi-\theta_i,\varphi_i)\frac{\sec\theta'}{\sec\theta'-\sec\theta_i}[D_2(\theta,\theta_i)-D_2(\theta,\theta')]$$

$$A_3(\theta',\varphi')=P(\theta',\varphi';\pi-\theta_i,\varphi_i)\frac{\sec\theta'}{\sec\theta'+\sec\theta_i}[D_1(\theta,\theta_i)-D_2(\theta,\theta')]\exp(-\kappa_e d\sec\theta_i)$$

$$A_4(\theta',\varphi')=P(\pi-\theta',\varphi';\pi-\theta_i,\varphi_i)\frac{\sec\theta'}{\sec\theta'-\sec\theta_i}[D_1(\theta,\theta_i)\exp(-\kappa_e d\sec\theta_i)-D_1(\theta,\theta')\exp(-\kappa_e d\sec\theta')]$$

$$D_1(\theta,\theta_i)=\frac{1}{\sec\theta-\sec\theta_i}\{1-\exp[-\kappa_e d(\sec\theta-\sec\theta_i)]\}$$

$$D_2(\theta,\theta_i)=\frac{1}{\sec\theta+\sec\theta_i}\{1-\exp[-\kappa_e d(\sec\theta+\sec\theta_i)]\}$$

Here the first and second-order iterative solutions describe the single and double scattering of random particles and its interaction with the underlying surface.

7-2.2 Iterative Inversion of the Legendre Coefficients of the Phase Function

The phase function in Eqs. (7-2.3, 7-2.4) is expanded into the Legendre polynomials [Jin 1994, McCormick 1984,1992]:

$$P(\theta,\varphi;\theta',\varphi')=\frac{1}{4\pi}\sum_{\ell=0}^{M}\omega_\ell P_\ell(\cos\Theta)=\frac{1}{4\pi}\sum_{m=0}^{M}\sum_{\ell=m}^{M}\omega_\ell b_{\ell m}P_\ell^m(\cos\theta)P_\ell^m(\cos\theta')\cos m(\varphi-\varphi') \tag{7-2.5}$$

where M is the cutoff order, $b_{m\ell}=(2-\delta_{m0})(\ell-m)!/(\ell+m)!$, P_ℓ^m is the associated Legendre function, the different-order coefficients ω_ℓ which describes anisotropic scattering of the medium might be constants or functions of incident and scattering angles θ_i, θ .

From the solutions of Eqs. (7-2.3~5), it can be seen that when ω_ℓ and the optical depth $\tau = \kappa_e d$ are known, the first and second-order scattered intensities can be obtained. In other words, if $I^{(1)}(\theta,\varphi,z=0)$ and $I^{(2)}(\theta,\varphi,z=0)$ are measurable, the unknown ω_ℓ and τ can be inverted from Eqs. (7-2.3) and (7-2.4).

Using Fourier transform of $I^{(1)}(\theta,\varphi,z=0)$, it is written as:

$$I_m^{(1)}(\theta,z=0) = \frac{1}{2\pi}\int_0^{2\pi} I^{(1)}(\theta,\varphi,0)\cos m\varphi d\varphi = \sum_{\ell=m}^{M} A_{m\ell} W_\ell(\tau,\theta) \tag{7-2.6}$$

where

$$A_{m\ell}(\theta) = \frac{1}{8\pi}\frac{(\ell-m)!}{(\ell+m)!}P_\ell^m(\cos\theta)P_\ell^m(\cos\theta_i)$$

Here $\ell = m,\cdots,M$, it means $[A_{m\ell}]$ is an upper triangle matrix. Eq. (7-6) is a matrix equation, and

$$\begin{aligned}W_\ell(\tau,\theta) &= \omega_\ell(\theta)\{(-1)^{\ell-m}\frac{2\sec\theta}{\sec\theta+\sec\theta_i}\{1-\exp[-\kappa_e d(\sec\theta+\sec\theta_i)]\}\\ &+\frac{2\sec\theta}{\sec\theta-\sec\theta_i}[\exp(-\kappa_e d\sec\theta_i)-\exp(-\kappa_e d\sec\theta)]R_{12}(\theta_i)\exp(-\kappa_e d\sec\theta_i)\\ &+\frac{2\sec\theta}{\sec\theta-\sec\theta_i}[\exp(-\kappa_e d\sec\theta_i)-\exp(-\kappa_e d\sec\theta)]R_{12}(\theta)\exp(-\kappa_e d\sec\theta)\\ &+(-1)^{\ell-m}\frac{2\sec\theta}{\sec\theta+\sec\theta_i}\{1-\exp[-\kappa_e d(\sec\theta+\sec\theta_i)]\}\\ &\times R_{12}(\theta_i)R_{12}(\theta)\exp[-\kappa_e d(\sec\theta_i+\sec\theta)]\end{aligned} \tag{7-2.7}$$

where $m = 0,1,\cdots,M$, $\ell = m,\cdots,M$.

Because τ is unknown, only a temporary $W_M(\tau,\theta)$ can be calculated from $I_M^{(1)}(\theta,z=0)$ and $A_{MM}(\theta)$. If $I_M^{(1)}(\theta_1,z=0)$ and $I_M^{(1)}(\theta_2,z=0)$ are obtained, this yields $W_M(\tau,\theta_1)$ and $W_M(\tau,\theta_2)$. Thus, τ can be numerically solved from $W_M(\tau,\theta_1)/W_M(\tau,\theta_2)$. Then $\omega_\ell(\theta)$ can be further solved from Eqs. (7-2.6, 7-2.7) using the known τ.

The above approach for all ω_ℓ and τ has been applied to the first-order solution. The scattered intensity includes the first-, second-, even higher-order solutions. We suppose that the scattered intensity is

$$I_s(\theta,\varphi,z=0) \approx I^{(1)}(\theta,\varphi,z=0) + I^{(2)}(\theta,\varphi,z=0) \tag{7-2.8}$$

Inversion steps for W_ℓ or ω_ℓ are concluded as follows:

(1) The measurement of $I_s(\theta,\varphi,z=0)$ is used to replace the first-order $I^{(1)}(\theta,\varphi,z=0)$ in Eq. (7-2.6). The unknown $\omega_\ell^{(1)}$ and $\tau^{(1)}$ are inverted, where superscript (1) indicates the first iteration.

(2) Substituting $\omega_\ell^{(1)}$ and $\tau^{(1)}$ into Eq. (7-2.4), the second-order solution $I^{(2)}(\theta,\varphi,z=0)$ can be obtained. Subtracting it from $I_s(\theta,\varphi,z=0)$ in Eq. (7-2.8), the first-order one $I^{(1)}(\theta,\varphi,z=0)$ is separated out. Following the approach of inverting τ and ω_ℓ from Eqs. (7-2.6, 7-2.7), the new pair of values $\omega_\ell^{(2)}$ and $\tau^{(2)}$ is then obtained;

(3) Repeating iteration of (2), converging $\omega_\ell^{(n)}$ and $\tau^{(n)}$ are finally obtained.

7-2.3 Inversion of Particle Number and Dielectric Constant

Now we deal with the inversion of the complex dielectric constant of the small spheroids and the particle number per unit area in a layer of random medium. Supposing a uniform distribution of the azimuthal Euler angle $\gamma \in (0°,360°)$, the extinction coefficient under the Rayleigh-Gans low-frequency approximation is written as [Jin 1994]:

$$\kappa_e(\theta) = \frac{3\pi n_0}{k^2}\text{Imag}[2T_1 + (T_0 - T_1)(\frac{1}{2} < \sin^2\beta > + \sin^2\theta < \cos^2\beta > + \frac{1}{2}\cos^2\theta < \sin^2\beta >)] \tag{7-2.9}$$

where the angular brackets of the Euler angle β means the ensemble averaging

$$< P(\beta) >= \frac{1}{\cos\beta_2 - \cos\beta_1}\int_{\beta_1}^{\beta_2} P(\beta)\sin\beta d\beta$$

The functions of small spheroids in Eq. (7-2.9) are written as [Jin 1994]

$$T_0 = t_0(1 + it_0), \quad T_1 = t_1(1 + it_1)$$

$$t_0 = \frac{2}{9}k^3a^2c\frac{\varepsilon - 1}{1 + (\varepsilon - 1)g_c}, \quad t_1 = \frac{2}{9}k^3a^2c\frac{\varepsilon - 1}{1 + (\varepsilon - 1)g_a}$$

where a and c are the semi-axises of a small spheroid, g_a and g_c are the functions of c/a, ε is the complex dielectric constant of the particle, n_0 is the particle number per unit volume, and k is the incidence wave number. The phase function has been derived as follow [Jin 1994]:

$$\begin{aligned}
P(\theta,\varphi;\theta_i,\varphi_i) &= (\frac{1}{6}L_1 + \frac{1}{24}L_2 < \sin^4\beta > + \frac{1}{6}L_3 < \sin^2\beta >)(3\cos^2\Theta - 1) \\
&+ [L_2(< \sin^2\beta\cos^2\beta > - \frac{< \sin^2\beta >}{4}) + L_3(< \cos^2\beta > - \frac{< \sin^2\beta >}{2})]\sin\theta\sin\theta_i \\
&\times\cos\theta\cos\theta_i\cos(\varphi - \varphi_i) + [\frac{2}{3}L_1 + \frac{L_2}{2}(< \sin^2\beta\cos^2\beta > + \frac{7 < \sin^4\beta >}{12}) + \frac{2}{3}L_3 < \sin^2\beta >] \\
&+ \frac{L_2}{2}(\frac{< \sin^4\beta >}{4} - < \sin^2\beta\cos^2\beta >)\cos^2\theta\cos^2\theta_i \\
&+ [\frac{L_2}{2}(< \cos^4\beta > - \frac{3}{8} < \sin^4\beta >) + L_3(< \cos^2\beta > - \frac{< \sin^2\beta >}{2})]\sin^2\theta\sin^2\theta_i
\end{aligned} \tag{7-2.10}$$

where Θ is the angle between directions of incident (θ_i,φ_i) and scattering (θ,φ), and

$$L_1 = C_1|T_1|^2, \quad L_2 = C_1|T_0 - T_1|^2, \quad L_3 = C_1\text{Re}[T_1(T_0^* - T_0^*)], \quad C_1 = 9n_0/(4k^2) \tag{7-2.11}$$

Linking the above phase function with the Legendre polynomials, we found that:

$$\begin{aligned}
\omega_0(\theta,\theta_i) &= \frac{2L_1}{3\kappa_e} + \frac{L_3}{\kappa_e}[\frac{2}{3} < \sin^2\beta > + (< \cos^2\beta > - \frac{1}{2} < \sin^2\beta >)(\sin^2\theta_i\sin^2\theta - \cos^2\theta_i\cos^2\theta)] \\
&+ \frac{L_2}{2\kappa_e}[\sin^2\theta_i\sin^2\theta(< \cos^4\beta > - \frac{3}{8} < \sin^4\beta >) \\
&+ \cos^2\theta_i\cos^2\theta(\frac{3}{4} < \sin^4\beta > - \frac{5}{2} < \cos^2\beta\sin^2\beta >) + (< \sin^2\beta > - \frac{5}{12} < \sin^4\beta >)]
\end{aligned} \tag{7-2.12}$$

$$\omega_1(\theta,\theta_i) = -[\frac{L_2}{\kappa_e}(< \sin^2\beta\cos^2\beta > - \frac{1}{4} < \sin^2\beta >) + \frac{L_3}{\kappa_e}(< \cos^2\beta > - \frac{< \sin^2\beta >}{2})]\cos\theta_i\cos\theta \tag{7-2.13}$$

$$\omega_2 = \frac{L_1}{3\kappa_e} + \frac{L_3}{3\kappa_e} < \sin^2\beta > + \frac{L_2}{12\kappa_e} < \sin^4\beta > \tag{7-2.14}$$

When the uniform distribution $\beta \in (0°,180°)$ is assumed, it yields $< \cos^2\beta >= 1/3$, $< \sin^2\beta >= 2/3$, $< \cos^4\beta >= 1/5$, $< \sin^4\beta >= 8/15$, $< \sin^2\beta\cos^2\beta >= 2/15$. Eqs. (7-2.12~14) are simplified, and κ_e become a constant.

Following the steps of the last section to invert ω_0, ω_1 and ω_2 of random spheroids (here, $M = 2$) and τ, the values of L_2/κ_e and $L_1/\kappa_e + 2L_3/3\kappa_e$ can be then obtained from Eqs. (7-2.12~14).

In other words, if L_2/κ_e and $L_1/\kappa_e + 2L_3/3\kappa_e$ are obtained, they can be employed to calculate the coefficients ω_0, ω_1 and ω_2. Then the first-order solution $I^{(1)}(\theta,\varphi,z=0)$ is separated from total scattered intensity $I_s(\theta,\varphi,z=0)$ and is used to invert L_2/κ_e and $L_1/\kappa_e+2L_3/3\kappa_e$, and this procedure is repeated until they converge. Suppose that the particle sizes a and c are known, the dielectric constant ε and number per unit area $n_0 d$ can be calculated by the following two equations:

$$\begin{cases} \dfrac{|T_1|^2+2\mathrm{Re}[T_1(T_0^*-T_1^*)]/3}{|T_0-T_1|^2}=\dfrac{L_1/\kappa_e+2L_3/3\kappa_e}{L_2/\kappa_e} \\ \dfrac{|T_0-T_1|^2}{\mathrm{Imag}(2T_1+T_0)}=\dfrac{8\pi L_2}{9\kappa_e} \end{cases} \tag{7-2.15}$$

$$n_0 d=\frac{4k^2}{9|T_1|^2}\frac{L_1}{\kappa_e}\tau \tag{7-2.16}$$

where L_2/κ_e and $L_1/\kappa_e+2L_3/3\kappa_e$ are the inverted results.

7-2.4 Results and Discussion

As an example, we concern a plane electromagnetic wave incident upon a layer of random small spheroids on an underlying surface. The frequency is 2*GHz*, and the layer depth $d=250\ cm$ (the dielectric constant of the underlying medium is assumed as $\varepsilon_2=6.0+0.5i$). Using Eqs. (7-2.3,4), the first- and second-order iterative solutions $I^{(1)}(\theta,\varphi,z=0)$ and $I^{(2)}(\theta,\varphi,z=0)$ are first simulated, and are then combined as the total scattered intensity $I_s(\theta,\varphi,z=0)$. It is used to invert $\varepsilon^{(n)}$ and $n_0 d^{(n)}$ in the n-th of iteration ($n=1,2,\cdots$).

Suppose $\varepsilon=40.0+7.5i$, $a=b=2.5\ cm$, $c=0.05\ cm$, $n_0=0.002\ cm^{-3}$ In Fig. 7-2.2, the line with the solid circles show the results for $\theta_i=60^\circ$, while the lines with solid squares show the results for $\theta_i=30^\circ$. The two scattering angles θ_1,θ_2 are chosen as $\theta_i\pm2^\circ$, and the azimuthal angle φ takes values in the range of $(0^\circ,360^\circ)$ in steps of 1°. Thus, in total, $2\times360=720$ measurements are needed.

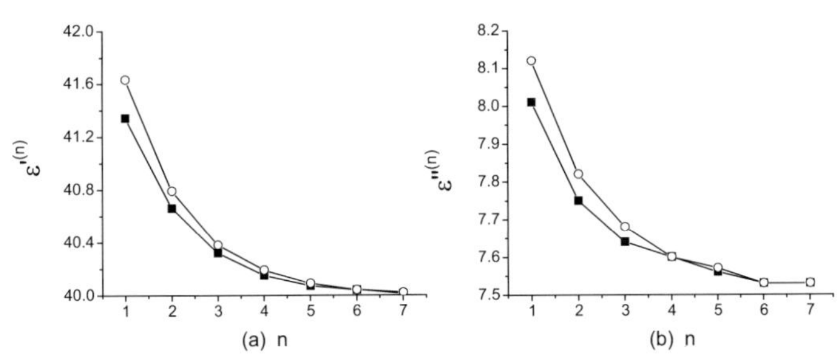

Fig.7- 2.2. Iterative inversion of ε (a) The real part of ε, (b) The imaginary part of ε.

It can be seen that $\varepsilon^{(n)}$ converge to the real ε as $n\to6$. Also, the case of small θ_i seems to produce better results. Of course, θ_i cannot be chosen as 0°, otherwise Fourier transform cannot be carried out.

Because there is always noise polluting the measurements, a random number *R*, which is taken as about 10% of the total scattered intensity is introduced by Monte Carlo realization,; i.e.,

$$I_s(\theta,\varphi,z=0) = I^{(1)}(\theta,\varphi,z=0) + I^{(2)}(\theta,\varphi,z=0) + R \qquad (7\text{-}2.17)$$

The scattered intensity with R of Eq. (7-2.17) is taken as a real measurement for inversion.

Let the incident frequency be $2GHz$, particle sizes $a=b=0.1cm$, $c=2.5cm$, incident angle $\theta_i = 60^\circ$. The azimuth angle φ is uniformly dividend into 2° steps in the range $(0^\circ, 360^\circ)$. Thus, each inversion needs the measurements $= 2\times 180 = 360$. The other parameters are the same as Fig. 7-2.2. Due to random noise R, inversion of each realization is different. Fig. 7-2.3 shows the inversion of 20 realizations.

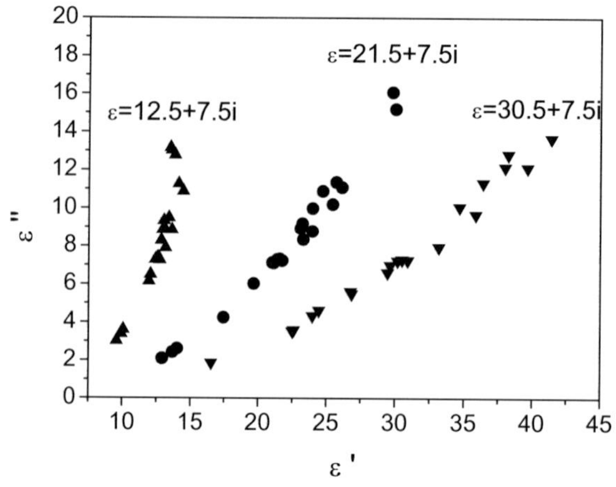

Fig. 7-2.3. Inversion of complex dielectric constant ε in 20 realization

Because the high-order Fourier coefficient $I_{m>0}$ is much less than the zeroth-order $I_{m=0}$, the noise effect is significant to $I_{m>0}$, and causes the error to be actually higher than 10%. This can be seen from the very sparse distribution of the inverted ε in Fig. 7-2.3. A large incident angle $\theta_i = 60^\circ$ is chosen because $I_{m>0}$ can then be increased to diminish the noise effect.

Table 7-2.1. Comparison of the averaged inversion and truth data at 2 *GHz*

ε	$n_0 d$	$\frac{1}{20}\sum_{i=1}^{20}\varepsilon_i^{(1)}$	$\frac{1}{20}\sum_{i=1}^{20} n_0 d^{(1)}$
12.5+i7.5	5.0	12.63+i8.23	4.43
21.5+i7.5	5.0	22.07+i8.27	4.86
30.5+i7.5	5.0	30.57+i7.71	5.11

Table 7-2.2. Comparison of averaged inversion and truth data at 4 *GHz*

ε	$n_0 d$	$\frac{1}{4}\sum_{i=1}^{4}\varepsilon_i^{(7)}$	$\frac{1}{4}\sum_{i=1}^{4} n_0 d^{(7)}$
12.5+i7.5	5.0	12.02+i7.21	3.89
21.5+i7.5	5.0	22.49+i9.43	3.88
30.5+i7.5	5.0	33.62+i10.53	4.05

Table 7-2.1 presents the averaged results for inverted ε with 20 realizations of the random noise at 2 *GHz*. The superscript 1 denotes one iteration.

Table 7-2.2 gives the averaged results for inverted ε with four realizations of the random noise at 4 *GHz*. It takes seven iterations to obtain the results, as indicated by the superscript 7.

From Tables 7-2.1 and 7-2.2, it can be seen that the averaged inversions for many realizations of noise effect are well converged to the truth data. Because the albedo at high frequencies such as 4 *GHz* is larger, the second-order scattered intensity in the simulation is more significant and yields poorer convergence.

7-3 Iterative Inversion of Vector Radiative Transfer Equation for a Layer of Random Spheroids

Inverse problem of radiative transfer in random media is to retrieve the characteristic parameters, such as the scatter particle size, dielectric constant, medium layer thickness etc. from the radiative measurement. For example, it has become most interest to invert quantitatively canopy information from terrain remote sensing, e.g. retrievals of canopy biomass, tree moisture and land soil wetness from polarized backscattering measurement [Macelloni et al. 2001, De Roo et al. 2001, Moghaddam and Saatchi 1999, Kurvonen et al. 1999, Van Oevelen and Hoekman 1999]. However, most inversions are more or less empirical and are usually based on the model of single scattering [Tsang et al. 1990, Liang and Jin 2003]. It is known that higher-order scattering from vegetation canopy at the C band can even reach more than 10% of total scattered power. Meanwhile, cross-polarization is mainly due to high-order scattering, and might be the same order as co-polarized scattered power in some cases [Liang and Jin 2003]. Therefore, it is necessary to develop analytically inverse method to take into account for high-order scattering.

Most of the solutions of vector radiative transfer (VRT) for active remote sensing are based on the models of single scattering for a layer of random medium or on double scattering for a half space. Recently, an iterative VRT approach to high order scattering and Mueller matrix solution has been developed [Liang and Jin 2003, also see Chapter 1].

In this session, this iterative VRT approach is applied to simulation of co-polarized and cross-polarized backscattering from a layer of random small spheroids. As some scatter parameters are given, all elements of the Mueller matrix in co-polarized and cross-polarized backscattering are numerically simulated as the measurement.

The first-order Mueller matrix for random small spheroids has been analytically formulated. Based on numerical difference between the first-order and high-order Mueller matrix solutions, the inversion algorithm of complex dielectric constant and number per unit area of random spheroids via the scanning and iterating is developed. Inverse results are well matched with the truth data in the simulations even with random noise presence [Jin and Liang 2004].

7-3.1 The First- and High-orders Mueller Matrix of VRT Solution

As shown in Fig.7-3.1, the VRT equation of a layer of scattering medium is written as

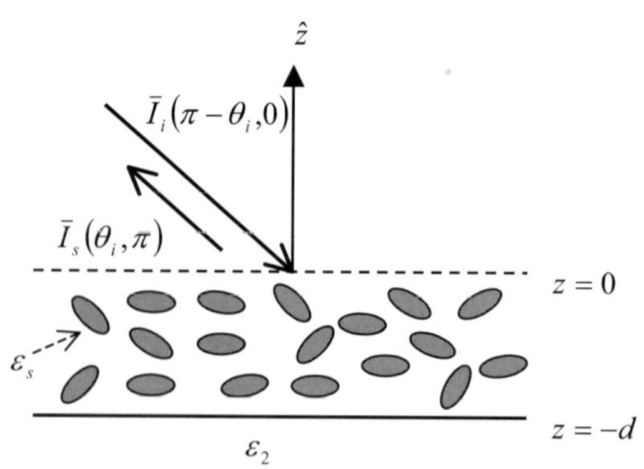

Fig. 7-3.1. The geometry of the problem

$$\cos\theta \frac{d}{dz}\bar{I}(\theta,\varphi,z) = -\bar{\bar{\kappa}}_e(\theta,\varphi)\cdot\bar{I}(\theta,\varphi,z) + \int_0^{\pi} d\theta' \sin\theta' \int_0^{2\pi} d\varphi' \bar{\bar{P}}(\theta,\varphi;\theta',\varphi')\cdot\bar{I}(\theta',\varphi',z) \tag{7-3.1}$$

where $\bar{\bar{\kappa}}_e$ is the extinction matrix, $\bar{\bar{P}}$ is the phase matrix. The boundary conditions are written as

$$\begin{cases} \bar{I}(\pi-\theta,\varphi,z=0) = \bar{I}_0\delta(\cos\theta-\cos\theta_i)\delta(\varphi) \\ \bar{I}(\theta,\varphi,z=-d) = \bar{\bar{R}}_{12}(\theta)\cdot\bar{I}(\pi-\theta,\varphi,z=-d) \end{cases} \tag{7-3.2}$$

where $\bar{I}_0$ is the incident Stokes vector, $(\theta_i,\ \varphi_i=0)$ is the incident angle, $\bar{\bar{R}}_{12}(\theta)$ is the reflectivity matrix of the underlying surface.

The scattered Stokes vector at $z=0$ is obtained via the Mueller matrix $\bar{\bar{M}}$ as

$$\bar{I}_s^{(1)}(\theta,\varphi) = \bar{\bar{M}}(\theta,\varphi;\pi-\theta_i,0)\cdot\bar{I}_i(\pi-\theta_i,0) \tag{7-3.3}$$

where $\bar{\bar{M}}$ in numerical simulation is the sum of different order Mueller matrices:

$$\bar{\bar{M}} = \bar{\bar{M}}^{(1)} + \bar{\bar{M}}^{(2)} + \cdots + \bar{\bar{M}}^{(n)} \tag{7-3.4}$$

and the first order $\bar{\bar{M}}^{(1)}$ has been analytically formulated [Jin 1994]. When the azimuth orientation of random spheroids is uniform, the Stokes vector $\bar{I}$ of VRT equation is treated as two-dimensional Stokes parameters (I_v, I_h), where *v*, *h* denote vertical and horizontal polarization, respectively.

As $\bar{\bar{\kappa}}_e$ and $\bar{\bar{P}}$ are determined by the random scatterers, the approach of VRT Eqs. (7-3.1, 7-3.2) is studied as for continuously random scattering medium. Suppose to stratify the scattering medium of Fig. 7-3.1 into multi-slabs. Each slab is very thin that single scattering is good approximation within this slab. Employing the first-order Mueller matrix of each thin slab, high-order Mueller matrix, $\bar{\bar{M}}^{(n)}$ $(n>1)$, of the total scattering medium from $z=0$ to $z=-d$ can be iteratively derived [Liang and Jin 2003].

Suppose that all small spheroids are identical and uniformly oriented in the azimuth Euler angle $\gamma\in(0^\circ,360^\circ)$. The four elements of the first-order Mueller matrix $\bar{\bar{M}}^{(1)}(\theta_i)$ in backscattering direction are analytically formulated as [Jin 1994]:

$$\begin{aligned} M_{11}^{(1)}(\theta_i) = {} & \frac{1-\exp(-2\kappa_{ev}d\sec\theta_i)}{2\kappa_{ev}}[1+R_{12v}^2(\theta_i)\exp(-2\kappa_{ev}d\sec\theta_i)]P_{11}(\theta_i,\pi;\pi-\theta_i,0) \\ & + 2R_{12v}(\theta_i)\exp(-2\kappa_{ev}d\sec\theta_i)d\sec\theta_i P_{11}(\theta_i,\pi;\theta_i,0) \end{aligned} \tag{7-3.5a}$$

$$\begin{aligned} M_{21}^{(1)}(\theta_i) = {} & \frac{1-\exp[-(\kappa_{ev}+\kappa_{eh})d\sec\theta_i]}{\kappa_{ev}+\kappa_{eh}}\{1+R_{12v}(\theta_i)R_{12h}(\theta_i)\exp[-(\kappa_{ev}+\kappa_{eh})d\sec\theta_i]\} \\ & \times P_{21}(\theta_i,\pi;\pi-\theta_i,0) + \frac{\exp(-\kappa_{ev}d\sec\theta_i)-\exp(-\kappa_{eh}d\sec\theta_i)}{\kappa_{eh}-\kappa_{ev}} \\ & \times[R_{12h}(\theta_i)\exp(-\kappa_{eh}d\sec\theta_i)+R_{12v}(\theta_i)\exp(-\kappa_{ev}d\sec\theta_i)]P_{21}(\theta_i,\pi;\theta_i,0) \end{aligned} \tag{7-3.5b}$$

$$\begin{aligned} M_{12}^{(1)}(\theta_i) = {} & \frac{1-\exp[-(\kappa_{ev}+\kappa_{eh})d\sec\theta_i]}{\kappa_{ev}+\kappa_{eh}}\{1+R_{12v}(\theta_i)R_{12h}(\theta_i)\exp[-(\kappa_{ev}+\kappa_{eh})d\sec\theta_i]\} \\ & \times P_{12}(\theta_i,\pi;\pi-\theta_i,0) + \frac{\exp(-\kappa_{ev}d\sec\theta_i)-\exp(-\kappa_{eh}d\sec\theta_i)}{\kappa_{eh}-\kappa_{ev}} \\ & \times[R_{12h}(\theta_i)\exp(-\kappa_{eh}d\sec\theta_i)+R_{12v}(\theta_i)\exp(-\kappa_{ev}d\sec\theta_i)]P_{12}(\theta_i,\pi;\theta_i,0) \end{aligned} \tag{7-3.5c}$$

$$M_{22}^{(1)}(\theta_i)=\frac{1-\exp(-2\kappa_{eh}d\sec\theta_i)}{2\kappa_{eh}}[1+R_{12h}^2(\theta_i)\exp(-2\kappa_{eh}d\sec\theta_i)]P_{22}(\theta_i,\pi;\pi-\theta_i,0)$$
$$+2R_{12h}(\theta_i)\exp(-2\kappa_{eh}d\sec\theta_i)d\sec\theta_i P_{22}(\theta_i,\pi;\theta_i,0) \quad (7\text{-}3.5d)$$

where $R_{12v}(\theta_i)$ and $R_{12h}(\theta_i)$ are the vertically and horizontally polarized reflectivity. The elements of the extinction and phase matrices are, respectively, written as [Jin 1994]:

$$\kappa_{ev}(\theta)=\frac{3\pi n_0}{k^2}\mathrm{Imag}[T_1+(T_0-T_1)(\sin^2\theta<\cos^2\beta>+\frac{1}{2}\cos^2\theta<\sin^2\beta>)] \quad (7\text{-}3.6a)$$

$$\kappa_{eh}=\frac{3\pi n_0}{k^2}\mathrm{Imag}[T_1+\frac{1}{2}(T_0-T_1)<\sin^2\beta>] \quad (7\text{-}3.6b)$$

$$P_{11}(\theta,\pi;\theta_i,0)=L_1(\sin^2\theta\sin^2\theta_i+\cos^2\theta\cos^2\theta_i-2\sin\theta\sin\theta_i\cos\theta\cos\theta_i)$$
$$+L_2[<\cos^4\beta>\sin^2\theta\sin^2\theta_i+\frac{3}{8}<\sin^4\beta>\cos^2\theta\cos^2\theta_i$$
$$+\frac{1}{2}<\sin^2\beta\cos^2\beta>(\cos^2\theta\sin^2\theta_i+\sin^2\theta\cos^2\theta_i-3\sin\theta\sin\theta_i\cos\theta\cos\theta_i)] \quad (7\text{-}3.7a)$$
$$+2L_3[<\cos^2\beta>\sin^2\theta\sin^2\theta_i+\frac{1}{2}<\sin^2\beta>\cos^2\theta\cos^2\theta_i$$
$$-(<\cos^2\beta>+\frac{1}{2}<\sin^2\beta>)\sin\theta\sin\theta_i\cos\theta\cos\theta_i]$$

$$P_{21}(\theta,\pi;\theta_i,0)=\frac{1}{2}L_2(<\sin^2\beta\cos^2\beta>\sin^2\theta_i+\frac{1}{4}<\sin^4\beta>\cos^2\theta_i)$$
$$+\frac{1}{2}L_3<\sin^2\beta>(\cos^2\theta_i+1) \quad (7\text{-}3.7b)$$

$$P_{12}(\theta,\pi;\theta_i,0)=\frac{1}{2}L_2(<\sin^2\beta\cos^2\beta>\sin^2\theta+\frac{1}{4}<\sin^4\beta>\cos^2\theta)$$
$$+\frac{1}{2}L_3<\sin^2\beta>(\cos^2\theta+1) \quad (7\text{-}3.7c)$$

$$P_{22}(\theta,\pi;\theta_i,0)=L_1+\frac{3}{8}L_2<\sin^4\beta>+L_3<\sin^2\beta> \quad (7\text{-}3.7d)$$

where the angular brackets denote the average of spatial orientation of another Euler angle β as $<F(\beta)>=\frac{1}{\cos\beta_2-\cos\beta_1}\int_{\beta_1}^{\beta_2}F(\beta)p(\beta)\sin\beta d\beta$, here $p(\beta)$ is the probability density function of the orientation distribution within (β_1,β_2). The scattering amplitude functions of small spheroids are expressed by following functions

$$L_1=C_1|T_1|^2,\ L_2=C_1|T_0-T_1|^2,\ L_3=C_1\mathrm{Re}[T_1(T_0^*-T_0^*)],\ C_1=9n_0/(4k^2) \quad (7\text{-}3.8)$$

where

$$T_0=t_0(1+it_0),\ T_1=t_1(1+it_1) \quad (7\text{-}3.9)$$

$$t_0=\frac{2}{9}k^3a^2c\frac{\varepsilon_s-1}{1+(\varepsilon_s-1)g_c},\ t_1=\frac{2}{9}k^3a^2c\frac{\varepsilon_s-1}{1+(\varepsilon_s-1)g_a} \quad (7\text{-}3.10)$$

here the parameters a and c are the semi-axes of the spheroid, g_a and g_c are the function of the parameter c/a, ε_s is the relative dielectric constant of the spheroid, n_0 is the particle number per unit volume, and k is the wave-number of incident wave.

It can be seen that $M_{ij}^{(1)}$ $(i,j=1,2)$ of Eqs. (7-3.5a~d) is the function of unknown ε_s and n_0d to be inverted. When the parameters a, c and ε_2 are given, $M_{ij}^{(1)}$ can be determined as long as $n_0 d$ and ε_s are supposed be known. In such case, the first-order Mueller matrix of a thin slab can be calculated, and high-order Mueller matrix $\overline{\overline{M}}^{(n)}$ $(n>1)$ of whole medium can then be constructed.

From Eqs. (7-3.5) and (7-3.7), it gives

$$M_{21}^{(1)}(\theta_i) = M_{12}^{(1)}(\theta_i) \tag{7-3.11}$$

i.e. there are only three elements of $M_{11}^{(1)}$, $M_{21}^{(1)}$, $M_{12}^{(1)}$ and $M_{22}^{(1)}$ be independent.

7-3.2 Iterative Inversion Using High-Order Mueller Matrix

In polarimetric measurement all elements of the Mueller matrix $\overline{\overline{M}}$ under a polarized wave incidence are measurable. In our simulation study, by using the truth data of $\varepsilon_s|_t$ and $n_0 d|_t$ (where $|_t$ denotes the truth data) and the given parameters a, c and ε_2, $\overline{\overline{M}} = (\overline{\overline{M}}^{(1)} + \cdots + \overline{\overline{M}}^{(n)})$ can be simualted as the measurement. The superscript (n) denotes the n-th order scattering.

Inversion includes the scanning of unknown pair $(\varepsilon_s' + i\varepsilon_s'', n_0 d)^\sim$ (where ~ denotes the values in scanning process) to find the best result of $(\varepsilon_s' + i\varepsilon_s''|_m,\ n_0 d|_m)$ in the m-th iteration and iterations to exame convergency of $(\varepsilon_s' + i\varepsilon_s''|_m,\ n_0 d|_m)$ to previous $(\varepsilon_s' + i\varepsilon_s''|_{m-1}$, $n_0 d|_{m-1})$.

(1) First-order iteration: From simulation or measurement of $\overline{\overline{M}}$, to approximately obtain the first-order Mueller matrix as

$$\overline{\overline{M}}^{(1)\sim} \approx \overline{\overline{M}} \text{ (which is known by simulation or measurement)} \tag{7-3.12}$$

Taking the initial values of $(\varepsilon_s' + i\varepsilon_s''|_0, n_0 d|_0)$, calculate $\overline{\overline{M}}^{(1)\sim}(\varepsilon_s' + i\varepsilon_s''|_0, n_0 d|_0)$ of Eqs. (7-3.5a~d) and make the difference of

$$Error^\sim = [\overline{\overline{M}}^{(1)}(\varepsilon_s' + i\varepsilon_s''|_0, n_0 d|_0) - \overline{\overline{M}}^{(1)\sim}]^2 \tag{7-3.13}$$

If $Error^\sim < E_0 \equiv Error|_{\min}$, which is a valve value, take this $(\varepsilon_s|_0, n_0 d|_0)$ as $(\varepsilon_s|_1, n_0 d|_1)$.

If not, scan the values of $(\varepsilon_s, n_0 d)^\sim$ with changing the steps as follows.

In scanning process, take the initial steps $\Delta\varepsilon_s''$, $\Delta\varepsilon_s'$, and $\Delta n_0 d$. Firstly, scan ε_s'' with increasing a step $\Delta\varepsilon_s''$ to find a good ε_s'' that $Error^\sim < E_0$. If $Error^\sim$ does so, continue scanning direction with another increased step $\Delta\varepsilon_s''$ and calculate new $\overline{\overline{M}}^{(1)\sim}$. If $Error^\sim$ is not less than E_0, change scanning direction with a negative step, e.g. $-0.5\Delta\varepsilon_s''$ to calculate $\overline{\overline{M}}^{(1)\sim}$. Until $Error^\sim < E_0$, scanning of ε_s'' in this run is finished. Here $\Delta\varepsilon_s''$ is a supposed valve step. Now, scanning ε_s' with increasing the step $\Delta\varepsilon_s'$ and a fixed ε_s'' in previous run, calculate $\overline{\overline{M}}^{(1)\sim}$ to check if $Error^\sim < E_0$.

Finally, scan $(n_0 d)^\sim$ with increasing the step ΔN and check if $Error^\sim < E_0$.

After this scanning process of $(\varepsilon_s, n_0 d)^\sim$, define the first iteration $(\varepsilon_s, n_0 d)^\sim = (\varepsilon_s|_1, n_0 d|_1)$ (the subscript $|_1$ indicates the solution of first iteration). Comparing the convergence of $(\varepsilon_s|_1, n_0 d|_1)$ to $(\varepsilon_s|_0, n_0 d|_0)$, we decide if the next iteration is needed. Usually, it always needs another iteration.

(2) Second-order iteration: By using $(\varepsilon_s|_1, n_0d|_1)$ to calculate high-orders $\overline{\overline{M}}^{(j)}(\varepsilon_s|_1, n_0d|_1)$ $(1<j\leq n)$, the first-order $\overline{\overline{M}}^{(1)\sim}$ is now obtained more accurately than Eq. (7-3.12) as follows.

$$\overline{\overline{M}}^{(1)\sim} \approx \overline{\overline{M}} - \sum_{j=2}^{n} \overline{\overline{M}}^{(j)}(\varepsilon_s|_1, n_0d|_1) \tag{7-3.14}$$

Following the same scanning steps in previous iteration, find $(\varepsilon_s, n_0d)^{\sim} = (\varepsilon_s|_2, n_0d|_2)$. Comparing the convergence of $(\varepsilon_s|_2, n_0d|_2)$ to $(\varepsilon_s|_1, n_0d|_1)$, decide if the next iteration is needed.

(3). The m-th iteration: Let

$$\overline{\overline{M}}^{(1)\sim} \approx \overline{\overline{M}} - \sum_{j=2}^{N} \overline{\overline{M}}^{(j)}(\varepsilon_s|_{m-1}, n_0d|_{m-1}) \tag{7-3.15}$$

More accurate first-irder measurement is now obtained. Following the same scanning steps, we finally obtain the inversion of $(\varepsilon_s|_m, n_0d|_m)$ and good convergence of $(\varepsilon_s|_m, n_0d|_m)$ with $(\varepsilon_s|_{m-1}, n_0d|_{m-1})$

7-3.3 Numerical Results and Discussions

Let a plane electromagnetic wave of $5GHz$ be incident upon a layer of random small spheroids. The incident angle is $\theta_i = 33^\circ$. The layer thickness is $d = 200cm$, The spheroid size parameters are $a = b = 0.1cm$, $c = 2.5cm$, and spatial orientation of the Euler angle $\beta \in (0^\circ, 60^\circ)$ is uniform. The particle fractional volume is 0.0055. The dielectric constant of underlying medium is $\varepsilon_2 = 15.0 + i2.0$.

Table 7-3.1. Simulation of the Mueller matrix (n=3)

	$\varepsilon_s\|_t = 22.5 + i7.5$			$\varepsilon_s\|_t = 12.5 + i7.5$		
	Simulation	1st order	Ratio	Simulation	1st order	Ratio
M_{11}	0.0132285	0.0114907	115%	0.0058432	0.0054891	106%
M_{21}	0.0038345	0.0027533	139%	0.0013171	0.0011228	117%
M_{12}	0.0037714	0.0027533	137%	0.0013041	0.0011228	116%
M_{22}	0.0107580	0.0094820	113%	0.0050682	0.0048031	105%

Table 7-3.1 lists two examples of the simulated Mueller matrix $\overline{\overline{M}}$ (n=3)and the first-order $\overline{\overline{M}}^{(1)}$ of Eqs. (7-3.5a~d). It can be seen that the ration of $M_{ij}/M_{ij}^{(1)}$, (i,j=1,2) of the first case with higher ε_s' is a little larger because the albedo for those particles is larger.

Fig. 7-3.2 shows the convergence of ε_s and n_0d in iterations. It can be seen that the convergence for the particles with small ε_s' is better because its albedo is small and the first-order approximation is more appropriate

Table 7-3.2 lists the inversion for different particle size parameter c. The parameters a and b are kept and $\varepsilon_s|_t = 20.5 + i9.0$. Other parameters are the same as Fig. 7-3.2. The results are obtained by 8 iterations. Inversions for all cases are good.

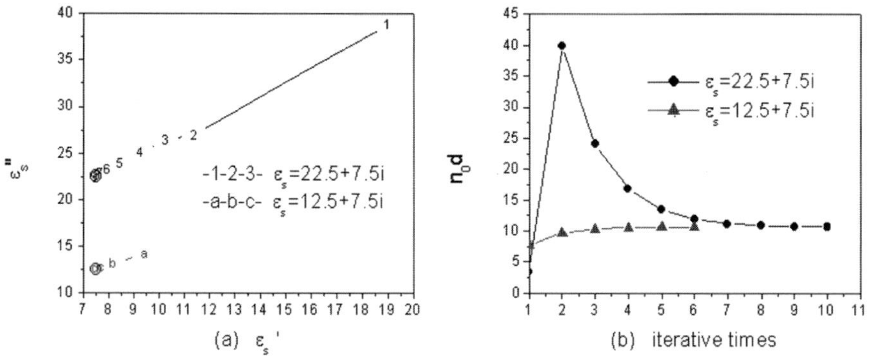

Fig. 7-3.2. Inverted parameters of ε_s and n_0d in iterations

Table 7-3.2. Inverted parameters for different sizes of the particles

Size c	$n_0d\|_t$	$n_0d\|_8$	$\varepsilon_s\|_8$
C=2.5cm	10.50	10.77	20.38+i8.88
C=2.0 cm	11.31	11.54	20.39+ i 8.89
c=1.5 cm	17.51	17.74	20.42+ i 8.91
c=1.0 cm	26.26	26.45	20.45+ i 8.94

Here the parameters a, c and ε_2 are given. Each calculation of $\overline{\overline{M}}^{(1)\sim}$ is based on these given parameters.

To see how the parameter ε_2 can affect final inversion, some others ε_2, instead of true value 15.0+i2.0, are arbitrarily chosen in each iteration. In those cases, the inversions are shown in Table 7-3.3.

Table 7-3.3. Inverted parameters for some others ε_2 different from the truth

Other ε_2	Frequency $f = 5GHz$		Frequency $f = 2GHz$	
	$\varepsilon_s\|_{10}$	$n_0d\|_{10}$	$\varepsilon_s\|_5$	$n_0d\|_5$
6.0+ i 0.5	21.91+i7.07	7.67	17.28+ i 4.06	20.58
12.0+ i 1.0	22.39+ i 7.42	10.31	20.32+ i 4.75	9.13
15.0+ i 2.0	22.43+ i 7.45	10.73	22.46+ i 7.48	10.56
20.0+ i 5.0	22.50+ i 7.49	11.21	23.43+ i 8.78	12.60

It can be seen that the inversions are favorably good even the parameter ε_2 in iterations was assumed be not true value. When the incident frequency is high, the effect from underlying medium becomes weaker. Thus, at high frequency such as the C band, inversion can work even the parameter ε_2 is not known. But on the other hand, high frequency increases the optical depth of the scatter medium and deteriorates inversion of $n_0 d$. Also, it yields higher albedo and slow down the convergence. It can be seen in Table 7-3.3 that five iterations are needed at $f = 2.0GHz$, but 10 iterations are needed for $f = 5.0GHz$. Thus, the frequency should be chosen in optimal way.

Random noise in the measurement can deteriorate the Mueller matrix solution. To see the noise effect, a random noise (about 10% of the scattered M_{12}) is realized by using Monte Carlo method and is added to the simulated $\overline{\overline{M}}$.

Fig. 7-3.3 presents inversions of ε_s for 10 realizations of random noise. Here, $d = 100cm$ is chosen, and other parameters are the same as before. It can be seen that the averaged result is good even each realization is oscillating around the true value. This oscillation can become bigger when ε_s is increasing. It is understandable that oscillation is proportional to ε_s, large ε_s needs more iterations and the noise effect is also more significant.

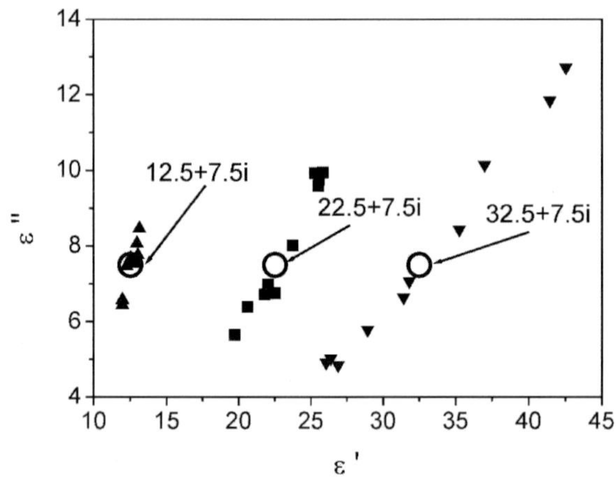

Fig. 7-3.3. Inversion of ε_s for noise realizations

Table 7-3.4 gives the averaged results of inverted ε_s and $n_0 d$ in 10 noise realizations through 9 iterations. As the layer depth d increases, inversion of $n_0 d$ becomes worse, because random noise might darken a little variation of the Mueller solution of thicker $n_0 d$.

Table 7-3.4. Inversions with noise presence in measurements

$\varepsilon_s\|_t$	$\sum_{r=1}^{10} \varepsilon_s^{(r)}\|_9 /10$	$n_0 d\|_t$	$\sum_{r=1}^{10} n_0 d^{(r)}\|_9 /10$
12.5+i7.5	12.51+ i 7.42	5.25	5.88
22.5+ i 7.5	23.24+ i 7.97	5.25	5.01
30.5+ i 7.5	32.77+ i 7.74	5.25	6.05

7-4 Iterative Inversion of Canopy Parameter and Surface Moisture by Using Multi-Order Mueller Matrix Solution

To take account of multiple scattering, an iterative VRT approach for high order Mueller matrix solution of an inhomogeneous random non-spherical scatter media has been recently developed [Liang and Jin 2003] and introduced in the session 1-2. By stratifying a random scatter media into multiple thin-layers in the z direction, the first-order scattering solution of each thin layer is employed to derive high-order scattering solution of whole random media. An iterative approach to solve high-order scattering solution from the Mueller matrices in discrete ordinates is developed.

In this session, this iterative VRT approach is applied to simulation of co-polarized and cross-polarized backscattering from a layer of inhomogeneous random non-spherical scatterers. When some scatterer's parameters such as the geometric sizes, dielectric constant etc. are presumably known, all elements of the Mueller matrix, of first- and multi-orders, in co-polarized and cross-polarized backscattering are numerically simulated as the function of the fractional volume per unit area and underlying surface volumetric moisture.

Based on numerical difference between the first-order and high-order Mueller matrices, an iterative algorithm for retrieval of fractional volume per unit area of random scatterers and underlying surface volumetric moisture from multi-order polarimetric scattering observation (or simulation) is developed. Inverted parameters are well matched with the ground truth data, which are first assumed and employed in the simulations, even with random noise presence. As an example for application, this approach is also well applied to AirSAR measurement over a spatially inhomogeneous canopy [Jin and Liang 2004].

7-4.1 Iterative Inversion for Canopy Biomass Parameters

A model of inhomogeneous vegetation canopy is shown in Fig. 7-4.1. The extinction coefficient $\bar{\bar{\kappa}}_e$ and phase matrix $\bar{\bar{P}}$ of the VRT equation are determined by the geometric sizes and fractional volume of the random scatters.

Stratifying the random medium into multi-slabs as described in the session 1-2, each slab denoted by $\ell = 1,2,\cdots,L$ is very thin that the Mueller matrix of each slab can be approximated by the first-order Mueller matrix. Its ij-th element $M_{ij,\ell s}^{(1)}(\theta,\theta')$ is expressed as

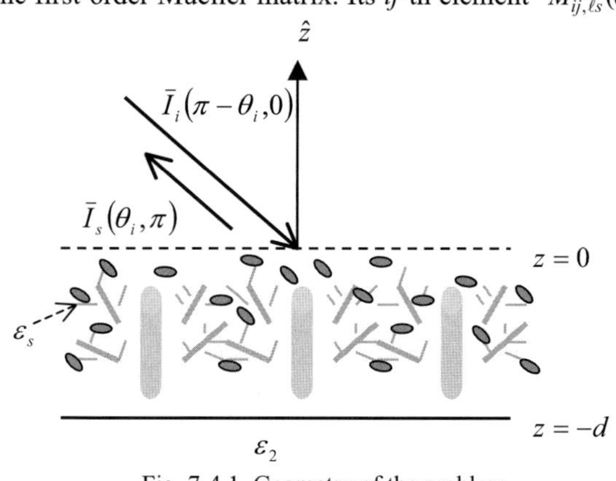

Fig. 7-4.1. Geometry of the problem

Eq. (1-2.16):

$$M_{ij,\ell s}(\theta,\varphi;\theta',\varphi') \approx M^{(1)}_{ij,\ell s}(\theta,\varphi;\theta',\varphi')$$

$$= \frac{P_{ij,\ell}(\theta,\varphi;\pi-\theta',\varphi')\sec\theta}{\beta_{j\ell}\sec\theta'+\beta_{i\ell}\sec\theta}\begin{cases} 1-\exp[-\Delta d(\beta_{j\ell}\sec\theta'+\beta_{i\ell}\sec\theta)] \\ \qquad 0^\circ<\theta<90^\circ, 0^\circ<\theta'<90^\circ \\ \exp(-\beta_{j\ell}\Delta d\sec\theta')-\exp(\beta_{i\ell}\Delta d\sec\theta) \\ \qquad 90^\circ<\theta<180^\circ, 0^\circ<\theta'<90^\circ \\ \exp(\beta_{j\ell}\Delta d\sec\theta')-\exp(-\beta_{i\ell}\Delta d\sec\theta) \\ \qquad 0^\circ<\theta<90^\circ, 90^\circ<\theta'<180^\circ \\ 1-\exp[\Delta d(\beta_{j\ell}\sec\theta'+\beta_{i\ell}\sec\theta)] \\ \qquad 90^\circ<\theta<180^\circ, 90^\circ<\theta'<180^\circ \end{cases} \tag{7-4.1}$$

where the subscript $ij, \ell s$ indicates the ij-th element of the ℓ-th slab *self* Mueller matrix without interactions of other slabs. The extinction coefficient $\overline{\beta}_\ell=[\beta_{1\ell},\beta_{2\ell},\beta_{3\ell},\beta_{4\ell}]$ is the vector of diagnosed extinction matrix $\overline{\overline{\kappa}}_e$.

The downward radiation in the ℓ-th slab from the ($\ell-1$)-th slab is written as

$$\bar{I}^{(n)}_\ell(\pi-\theta,\varphi)=\bar{I}^{(n)}_{\ell s}(\pi-\theta,\varphi)+\bar{I}^{(n)}_{\ell-1}(\pi-\theta,\varphi)\exp(-\overline{\beta}_\ell\Delta d\sec\theta) \tag{7-4.2a}$$

The upward radiation in the ℓ-th slab from the ($\ell+1$)-th slab is written as

$$\bar{I}^{(n)}_\ell(\theta,\varphi)=\bar{I}^{(n)}_{\ell s}(\theta,\varphi)+\bar{I}^{(n)}_{\ell+1}(\theta,\varphi)\exp(-\overline{\beta}_\ell\Delta d\sec\theta) \tag{7-4.2b}$$

where $\theta\in(0,\pi/2)$, the superscript n denotes the iteration number.

Radiation from interfaces $z=0$ and $z=-d$, down-going $\bar{I}^{(n)}_{\ell=0}(\pi-\theta)$ and up-going $\bar{I}^{(n)}_{\ell=L+1}(\theta)$, are written from the boundary conditions as

$$\bar{I}^{(n)}_{\ell=0}(\pi-\theta,\varphi)=\begin{cases}\bar{I}_0\delta(\cos\theta-\cos\theta_i)\delta(\varphi) & n=0\\ 0 & n>0\end{cases} \tag{7-4.3a}$$

$$\bar{I}^{(n)}_{\ell=L+1}(\theta,\varphi)=\overline{\overline{R}}_{12}(\theta)\cdot\bar{I}^{(n)}_{\ell=L}(\pi-\theta,\varphi) \tag{7-4.3b}$$

From Eqs. (7-4.2, 7-4.3), the n-th order radiations of the ℓ-th slab in upwards and downwards, $\bar{I}^{(n)}_\ell(\theta)$ and $\bar{I}^{(n)}_\ell(\pi-\theta)$, which take into account of interactions of multi-slabs, can be obtained. Thus, the (n+1)-th order radiation of the ℓ-th slab itself can be calculated as follows.

$$\begin{aligned}\bar{I}^{(n+1)}_{\ell s}(\theta,\varphi)&=\int_0^{2\pi}d\varphi'\int_{\pi/2}^{\pi}\overline{\overline{M}}_{\ell s}(\theta,\varphi;\theta',\varphi')\cdot\bar{I}^{(n)}_{\ell-1}(\theta',\varphi')\sin\theta' d\theta'\\ &+\int_0^{2\pi}d\varphi'\int_0^{\pi/2}\overline{\overline{M}}_{\ell s}(\theta,\varphi;\theta',\varphi')\cdot\bar{I}^{(n)}_{\ell+1}(\theta',\varphi')\sin\theta' d\theta'\end{aligned} \tag{7-4.4}$$

The iteration process is summarized as the following steps:

(1) Stratify the media into multi-slabs, calculate $\overline{\overline{M}}_{\ell s}(\theta,\varphi;\theta_i,\varphi_i)$ based on Eq. (7-4.1).

(2) Calculate $\bar{I}^{(n)}_\ell(\theta,\varphi)$ using Eqs. (7-4.2, 7-4.3); the upward radiation coming from whole scatter media is the n-th order solution $\bar{I}^{(n)}_{\ell=1}(\theta,\varphi)$.

(3) Calculate the (n+1)-th radiation of the ℓ-th slab, $\bar{I}^{(n+1)}_{\ell s}(\theta,\varphi)$ using Eq. (7-4.4) by $\bar{I}^{(n)}_\ell(\theta,\varphi)$.

(4) Repeat steps (2) and (3) to reach final convergence, the high-order scattering is obtained.

Thus the high-order scattering solution, $\bar{I}^{(n)}(\theta,\varphi,z=0)$ ($n>1$), of the total random medium from $z=0$ to $z=-d$ can be finally derived as

$$\bar{I}^{(n)}(\theta,\varphi)=\bar{I}^{(n)}_{\ell=1}(\theta,\varphi,z=0) \tag{7-4.5}$$

In previous calculations, $\overline{\overline{R}}_{12}(\theta)$ is the function of ε_2, which is mainly determined by the soil volumetric moisture m_v. Its empirical formulation is refereed to [Wang and Schumagge 1980]. Thus the Mueller matrix $\overline{\overline{M}}$ is the function of the unknown parameters fd and m_v. When the scatter geometric sizes and dielectric constant ε_s are given, all elements $M^{(1)}_{pq}$ of the Mueller matrix can be written as the function of fd and m_v.

$$M^{(1)}_{pq}=M^{(1)}_{pq}(fd,m_v), \qquad p,q=v,h \tag{7-4.6}$$

In polarimetric measurement all elements of the Mueller matrix $\overline{\overline{M}}$ under a polarized wave incidence are measurable. In our simulation, by using the ground truth of $m_v|_t$ and $fd|_t$ (where $|_t$ denotes the ground truth which are actually the presumed value for simulation), and the scatter's (spheroids or cylinders) sizes and dielectric constant ε_s as presumably given parameters, the Mueller matrix solution $\overline{\overline{M}}=(\overline{\overline{M}}^{(1)}+\cdots+\overline{\overline{M}}^{(N)})$ is simualted as the measurement, here (N) denotes the N-th order scattering.

Inversion includes the scanning of unknown pair $(m_v,fd)^{\sim}$ (where $\sim$ denotes the values in scanning process) to find the best result of $(m_v|_m,\ fd|_m)$ in the m-th iteration with convergency to the previous pair $(m_v|_{m-1},\ fd|_{m-1})$.

(1) First iteration:

From simulation or measurement of $\overline{\overline{M}}$, the first-order Mueller matrix is approximately obtained as:

$$\overline{\overline{M}}^{(1)\sim}\approx\overline{\overline{M}}\text{ (which is known by simulation or measurement)} \tag{7-4.7}$$

Taking the initial values of $(m_v|_0,fd|_0)$, calculate $\overline{\overline{M}}^{(1)}(fd,m_v)$ of (7-4.6) and compare the difference of

$$Error^{\sim}=[\overline{\overline{M}}^{(1)}(fd,m_v)-\overline{\overline{M}}^{(1)\sim}]^2 \tag{7-4.8}$$

Scan the values of $(m_v,fd)^{\sim}$ with changing the steps as follows.

In scanning process, take the initial steps Δm_v and Δfd. First, take m_V with increasing a step Δm_v to find a good m_v such that $Error^{\sim}<Error|_{\min}$, where $Error|_{\min}$ is the minimum error in the previous steps. If $Error^{\sim}<Error|_{\min}$ does so, continue scanning direction of m_v with another increased step Δm_v to calculate a new $\overline{\overline{M}}^{(1)}(fd,m_v)$. If $Error^{\sim}$ is not less than $Error|_{\min}$, change the scanning direction of m_v with a negative step, e.g. $-0.5\Delta m_v$, to calculate $\overline{\overline{M}}^{(1)}(fd,m_v)$. Until $|\Delta m_v|<\Delta m_v|_{\min}$ is reached, scanning of m_v in this run is finished. Here $\Delta m_v|_{\min}$ is an assumed valve step.

Similarly, scan $(fd)^{\sim}$ with the step Δfd and check if $Error^{\sim}<Error|_{\min}$, In every step the scanning of m_v must be repeated with the initial Δm_v. Until $|\Delta fd|<\Delta fd|_{\min}$ is reached , the scanning of both fd and m_v is completed. Here $\Delta fd|_{\min}$ is a supposed valve step.

After this scanning process of $(m_v, fd)^{\sim}$, define the first iteration $(m_v, fd)^{\sim} = (m_v|_1, fd|_1)$ (the subscript $|_1$ indicates the solution of first iteration). Comparing the convergence of $(m_v|_1, fd|_1)$ to $(m_v|_0, fd|_0)$, we decide if the next iteration is needed. Usually, another iteration is needed.

(2) Second-order iteration:

By using $(m_v|_1, fd|_1)$ to calculate high-orders $\overline{\overline{M}}^{(j)}(m_v|_1, fd|_1), 1 < j < N$, the first-order $\overline{\overline{M}}^{(1)\sim}$ is now obtained more accurately than that of Eq. (7-4.7) as follows:

$$\overline{\overline{M}}^{(1)\sim} \approx \overline{\overline{M}} - \sum_{j=2}^{N} \overline{\overline{M}}^{(j)}(m_v|_1, fd|_1) \tag{7-4.9}$$

Following the same scanning steps as those used in previous iteration, find $(m_v, fd)^{\sim} = (m_v|_2, fd|_2)$. Comparing the convergence of $(m_v|_2, fd|_2)$ to that of $(m_v|_1, fd|_1)$, decide if the next iteration is needed.

(3). *m*-th iteration:

Let

$$\overline{\overline{M}}^{(1)\sim} \approx \overline{\overline{M}} - \sum_{j=2}^{N} \overline{\overline{M}}^{(j)}(m_v|_{m-1}, fd|_{m-1}) \tag{7-4.10}$$

and then a more accurate first-irder measurement is obtained. Following the same scanning steps, we finally obtain the inversion of $(m_v|_m, fd|_m)$ and good convergence of $(m_v|_m, fd|_m)$ to $(m_v|_{m-1}, fd|_{m-1})$.

7-4.2 Numerical Results and Discussion

A. Inversion by Simulated Data

Let an electromagnetic plane wave of $1.4GHz$ be incident upon a layer of random small spheroids. The incident angle is $\theta_i = 39^\circ$. The layer thickness is $d = 200cm$, The semi-axes of oblate spheroids are $a = b = 2.0cm$, $c = 0.05cm$, and spatial orientation of the Euler angles $\beta \in (0^\circ, 60^\circ)$, $\gamma \in (0^\circ, 360^\circ)$ are uniform. The dielectric constant of the spheroids is $\varepsilon_s = 22.5 + i7.5$.

Two cases of the fractional volume $f = 0.005$ and $f = 0.01$ are tested. The Mueller matrix solution for varying m_v is simulated to the third-order ($N = 3$)

$$\overline{\overline{M}}\,(\text{simulated}) = \overline{\overline{M}}^{(1)} + \overline{\overline{M}}^{(2)} + \overline{\overline{M}}^{(3)} \tag{7-4.11}$$

Then, $\overline{\overline{M}}$ of (7-4.11) as a simulated measurement is applied to invert fd and m_V ($m_v < 0.5$ is always assumed).

Figs. 7-4.2 (a, b) show the retrieval of surface moisture m_v^R and comparison with the true value m_v^T, where mark 1,2,3 denote the iterative orders. The first iterations deviated far from the diagonal line, especially for thick canopy. The third iterations are well matched with the true values.

In Fig. 7-4.2(b) the first iterations are extremely far from the truth data because of greater optical depth, which might darken reflection from the underlying surface. The term of particle scattering in the first-order Mueller matrix solution is dominant, and the effect from the underlying surface cannot be well identified. It yields poor inversion of m_v^R in the first-order inversion and needs high-order iterations.

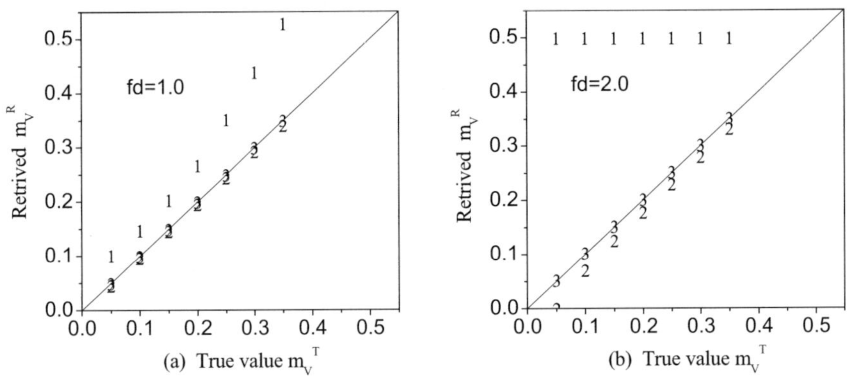

Fig. 7-4.2. Inversion of surface moisture m_v

Similarly, for two cases of $m_v = 0.05$ and $m_v = 0.25$, the Mueller matrix solution is simulated for varying fd. Fig. 7-4.3 (a, b) show the inversion of fd^R and comparison with the true fd^T. It can be seen that the second iterations are good enough. Comparing with Figs. 7-4.2(a,b), it can be seen that the inversion of fd^R from the first-order solution is better than m_V^R, especially of Fig. 7-4.2(b), because the particle scattering term in the first-order solution is always dominant and surface reflection is more or less darken by particle scattering.

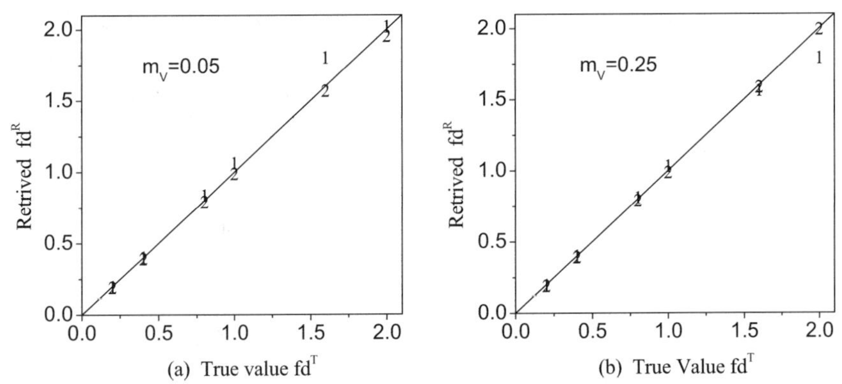

Fig. 7-4.3. Inversion of fraction per unit area fd

Random noise in measurement always results in deterioration of the data. To see the noise effect, a random noise (about 10 percent of the scattered M_{vh} i.e. M_{12}) is randomly realized by using the Monte Carlo method, and is added to the simulated $\overline{\overline{M}}$. Figs. 7-4.4(a, b) present inversions of m_v and fd for 10 independent realizations of random noise and show that each inversion is independently oscillating around the true value $m_v{}^T = 0.25$, $fd^T = 1.0$. The averaged inverted results from independent realizations are $\overline{m}_v{}^R = 0.241$, $\overline{fd}^R = 0.953$, well matching with the true values.

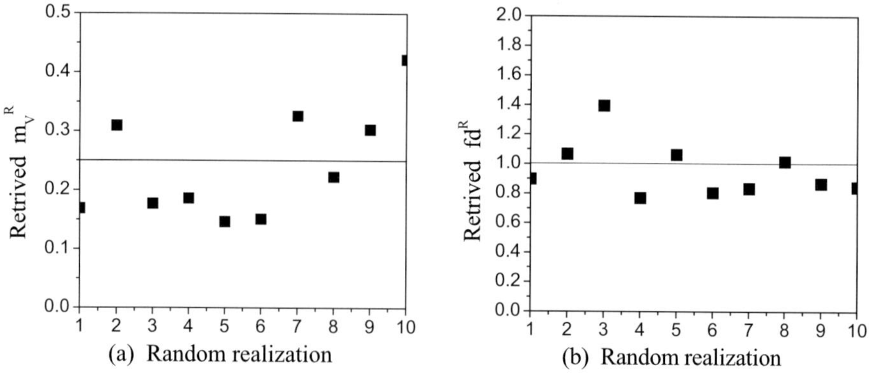

Fig. 7-4.4. Inversion of m_v and fd for noise realizations

B. Inversion by Using AirSAR Image

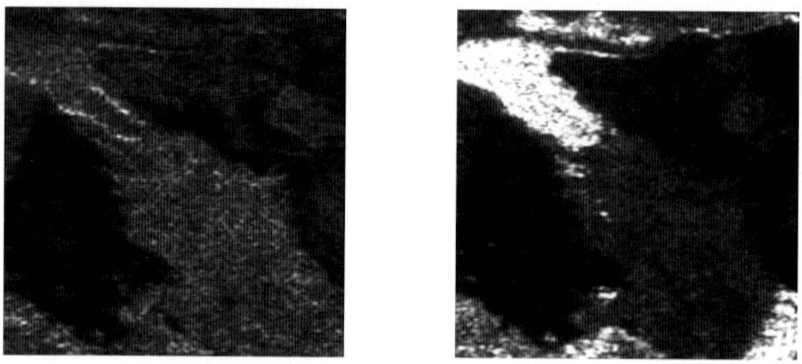

(a) col-polarized M_{vv} (b) col-polarized M_{hh}

Fig. 7-4.5. AirSAR images at the P band

Figs. 7-4.5(a, b) present the v,h co-polarized backscattering P-band images of NASA/JPL AirSAR data (200×200 pixels) from the Prince Albert National Park of Canada (approximately $103.33^\circ W, 53.9^\circ N$). It is known that young jack pines (YJP) are inhomogeneously distributed around the center of the images in Figs. 7-4.5 (a,b) and the tree height is about $d = 380cm$. The canopy is modeled as a layer of random cylinders for trunk, primary and secondary branches. The fraction ratio of these three types is kept as 1:2:1, and some parameters are listed in Table 7-4.1. The fraction per unit area of the pine canopy is $0.2 < fd^T < 3.8$, which is calculated by the measured particle sizes and number per unit volume. The fraction $\overline{fd}^T$ and dielectric constant ε_2 of the land surface located around the center of Figs. 7-4.5(a,b) are given as $\overline{fd}^T \approx 2.0$, $\varepsilon_2 = 8.0 + 1.0i$, respectively. It is empirically derived that the surface moisture is $m_v^T = 0.15$ around the center of Figs. 7-4.5(a, b).

Table 7-4.1. Parameters from ground-truth measurement, * assumed

	ε_s *	Radius (cm)	Half-height (cm)	Euler angle β *	Euler angle γ
Trunk	$20.5+i6.0$	2	190	$(0^\circ,10^\circ)$	$(0^\circ,360^\circ)$
Primary branch	$29.5+i9.0$	0.5	40	$(70^\circ,90^\circ)$	$(0^\circ,360^\circ)$
Secondary branch	$29.5+i9.0$	0.2	12.5	$(0^\circ,90^\circ)$	$(0^\circ,360^\circ)$

To save the retrieval timing, we make two further approximations. First, each 4×4 pixels area in the image of Fig. 7-4.5 is averaged into one big-pixel for retrieval study. In the first-order iteration, the surface moisture $m_v|_1$ and the fraction $fd|_1$ at each big-pixel are obtained. Second, we take the average of the first-order surface moisture over whole picture pixels, $<m_v|_1>=0.219$, for following calculations.

As we define

$$X_{pq}=\frac{M_{pq}^{(1)}(<m_v|_1>,fd)}{\sum_{j=1}^{N} M_{pq}^{(j)}(<m_v|_1>,fd)} \tag{7-4.12}$$

we found in Fig. 7-4.6 that X_{pq} shows a simple monotonic relationship for varying fd. Significant variation of X_{vv} is due to the cylinder scatterers in this case.

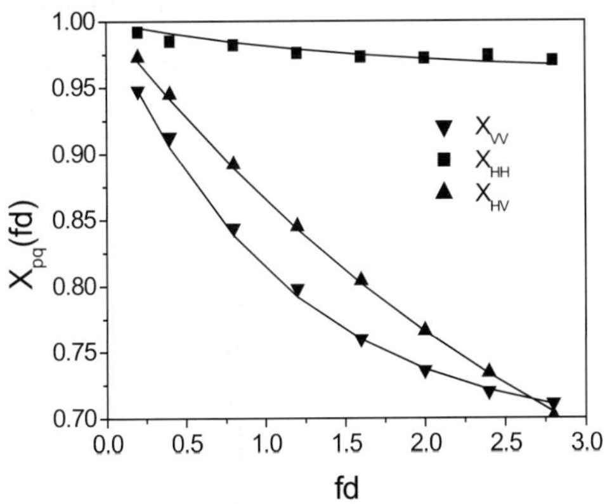

Fig. 7-4.6. The ratio between the first-order and sum of N-orders Mueller matrices

Based on these calculations, we can simply formulate

$$X_{vv}(fd)=0.48+0.52\exp(-0.3fd) \tag{7-4.13a}$$

$$X_{hh}(fd)=0.96+0.04\exp(-0.63fd) \tag{7-4.13b}$$

$$X_{hv}(fd)=0.685+0.0.315\exp(-0.9fd) \tag{7-4.13c}$$

By using $(<m_v|_1>,fd|_1)$ and Eqs. 7-4.13(a,b,c) instead of multi-order Mueller matrices calculations, the first-order $\overline{\overline{M}}^{(1)\sim}$ for each big-pixel in the second-order iteration can be easily obtained:

$$M_{pq}^{(1)\sim} = M_{pq} X_{pq}(fd|_1) \tag{7-4.14}$$

Then the second-order iteration with Eq. (7-4.14), instead of Eq. (7-4.9), is processed. The surface moisture $m_v|_2$ and the fraction parameter $fd|_2$ at each big-pixel can be obtained.

Figs. 7-4.7(a,b) show the second-order inverted results (50×50 big-pixels). The typical inverted values around the center of Fig. 7-4.5 are $m_v|_1 = 0.22$, $fd|_1 = 1.5$, $m_v|_2 = 0.20$, $fd|_2 = 1.7$, which are in good agreement with the ground truth $m_v^T = 0.15$, $\overline{fd}^T \approx 2.0$.

Certainly, good convergence of the inversions with good ground truth measurement requires more prior information of the canopy type, e.g., if the canopy is needle, disk or cylinder scatterers.

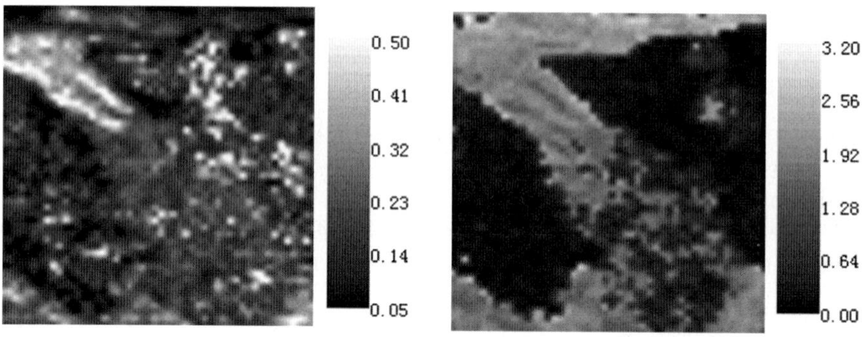

(a) surface moisture m_v (b) fraction per unit area fd

Fig. 7-4.7. Inverted results m_v and fd of inhomogeneous canopy

7-5 References

De Roo, R.D., Y. Du and F.T. Ulaby (2001), "A semi-empirical backscattering model at L-band and C-band for a soybean canopy with soil moisture inversion", *IEEE Transaction on Geoscience and Remote Sensing*, 39(4): 864-872.

Fukshansky-Kazarinova, N., L. Fukshansky, M. Kuhl and B.B. Jorgensen (1998), "Solution of the inverse problem of radiative transfer on the basis of measured internal fluxes", *Journal of Quantitative Spectroscopy and Radiative Transfer*, 59(1-2): 77-89.

Hajnsek, I., E. Pottier and S.R. Cloude (2003), "Inversion of surface parameters from polarimetric SAR", *IEEE Transactions on Geoscience and Remote Sensing*, 41(4): 727-744.

Husier, A.M., A. Quatropani and H.D. Baltes (1982), "Construction of grating profiles yielding prescribed diffraction efficiencies", *Journal of Optical Communications*, 41: 149-153.

Imhoff, M. (1995), "Radar backscatterer and biomass saturation ramifications for global biomass inventory", *IEEE Transactions on Geoscience and Remote Sensing*, 33(2): 510-518.

Jaggard, D.L. and X. Sun (1990), "Fractal surface scattering: A generalized Rayleigh solution", *Journal of Applied Physics*, 68(11): 5456-5462.

Jin, Y.Q. (1994), *Electromagnetic Scattering Modelling for Quantitative Remote Sensing*, Singapore: World Scientific.

Jin, Y.Q. and G. Li (2000), "Detection of a scatter target over a randomly rough surface by using the angular correlation function in a finite-element approach", *Waves in Random Media*, 10(2):273-280.

Jin, Y.Q. and W. Zhang (2000), "An inversion of underwater bottom topography by using the SAR imagery", *Acta Oceanologica Sinica*, 19(4): 53-62.

Jin, Y.Q. and Z. Li (2001), "Reconstruction of roughness profile of fractal surfaces from scattering measurement at grazing incidence", *Journal of Applied Physics*, 89(3): 1922-1926.

Jin, Y.Q. and Z. Liang (2002), "Inversion of scattering from a layer of random Non-spherical particles using iterative solutions of scalar radiative transfer equation", *Inverse Problem,* 18(5): 1361-1368.

Jin, Y.Q. and Z. Liang (2004), "Iterative retrieval of canopy biomass and surface moisture by using multi-order Mueller matrix solutions of polarimetric radiative transfer", *Canadian Journal of Remote Sensing*, 30(2): 169-175.

Jin, Y.Q. Jin and Z. Liang (2004), "Iterative Inversion from the Multi-Order Mueller Matrix Solution of Vector Radiative Transfer Equation for a Layer of Random Spheroids", *Journal of Quantitative Spectroscopy and Radiate Transfer*, 83(4): 303-311.

Kurvonen, L., J. Pulliainen and M. Hallikainen (1999), "Retrieval of biomass in boreal forests from multi-temporal ERS-1 and JER-1 images", *IEEE Transactions on Geoscience and Remote Sensing*, 37(1): 198-205.

Le, T.C., Y. Kuga and A. Ishimaru (1996), "Angular correlation function based on the second order Kirchhoff approximation and comparison with experiments", *Journal of the Optical Society of American* 13(5): 1057-1067.

Liang, Z. and Y.Q. Jin (2003), "Iterative approach of high-order scattering solution for vector radiative transfer of inhomogeneous random media", *Journal of Quantitative Spectroscopy and Radiative Transfer*, 77(1): 1-12.

Ligon, D.A., T.W. Chen and J.B. Gillespie (1996), "Determination of aerosol parameters from light-scattering data using an inverse Monte Carlo technique", *Applied Optics*, 35(21): 4297-4303.

Liu, L., H. Tan and Q. Yu (2000), "Inverse radiation problem in one-dimensional semitransparent plane-parallel media with opaque and specularly reflecting boundaries", *Journal of Quantitative. Spectroscopy and Radiative Transfer*, 64 (4): 395-407.
Lo, Y., H. Leung and S. Haykin (1993), "Fractal characterization of sea-scattered signalks and detection of sea surface targets", *IEE Proceedings on (F)*, 140(4): 243-250..
Macelloni, G., S. Paloscia and P. Pampaloni (2001), "The relationship between the backscattering coefficient and the biomass of narrow and broad leaf crops", *IEEE Transactions on Geoscience and Remote Sensing*, **39**(4): 873-884.
McCormick, N.J. (1984), "Recent developments in inverse scattering transport method", *Transport Theory Statistical Physics*, 13: 15-28.
McCormick, N.J. (1992), "Inverse radiative transfer problems: A Review", *Nuclear Science Engnering*, 112: 185-198.
Moghaddam, M. and S.S. Saatchi (1999), "Monitoring tree moisture using an estimation algorithm applied to SAR data from BOREAS", *IEEE Transaction on Geoscience and Remote Sensing*, 37(2): 901-916.
Moghaddam, M. and S.S. Saatchi (2000), "Estimating subcanopy soil moisture with radar", *Journal of Geophysical Research*, 105: 14899-14911.
Papathanassiou, K. P. and S.R. Cloude (2001), "Single baseline polarimetric SAR interferometry", *IEEE Transactions on Geoscience and Remote Sensing*, 39(11): 2352-2363.
Quirantes, A. and A. Delgado (1998), "Experimental size determination of spheroidal particles via the T-matrix method", *Journal of Quantitative Spectroscopy and Radiative Transfer*, 60(3): 463-474.
Roger, A. and D. Maystre (1980), "Inverse scattering method in electromagnetic optics: application to diffraction gratings", *Journal of the Optical Society of American*, 70:1483-1495.
Rourier, S. and P. Borderies (1997) , "Ultra wide band electromagnetic scattering of a fractal profile", *Radio Science*, 32(2): 285-294.
Spivack, M. (1990), "A numerical approach to rough surface scattering by the parabolic equation method", *Journal of Acoustic Society of American*, 87(5): 1999-2004.
Spivack, M. (1990), "Moments of wave scattering by a rough surface", *Journal of Acoustic Society of American*, 88(5): 2361-2366.
Spivack, M. (1992), "Direct solution of the inverse problem for rough surface scattering at grazing incidence", *Journal of Physics A: Mathematics General*, 25: 3295-3302.
Spivack, M. (1992), "Solution of the inverse-scattering problem for grazing incidence upon a rough surface", *Journal of the Optical Society of American*, 9(8): 1352-1355.
Tsang, L., K.H. Ding and B. Wen (1990), *Polarimetric Remote Sensing*, New York Elsevier.
Van Oevelen, P.J. and D.H. Hoekman (1999), "Radar backscatter inversion techniques for estimation of surface soil moisture: EFEDA-Spain and HAPEX-Sahel case studies", *IEEE Transaction on Geoscience and Remote Sensing*, 37(1): 113-123.
Wombel, R.J. and J.A. DeSanto (1991), "The reconstruction of shallow rough-surface profiles from scattered field", *Inverse Problem*, 7(1): L7-12.
Wombell, R.J. and J.A. DeSanto (1991), "Reconstruction of rough-surface profiles with the Kirchhoffapproximation", *Journal of the Optical Society of American*, 8(12): 1892-1897.

Chapter 8

Electromagnetics of Complex Particulate Media 341

8-1 Polarimetric Scattering and Transmitting of the Stokes Vector from a Layer of Chiral Small Spheroids

The lack of geometric symmetry between an object and its mirror image is referred to as chirality or handedness (in optical literature it is called optical activity [Bohren 1974]. A chiral medium produces different responses to left- and right-circularly polarized light. Since the phenomenon of chirality was known earlier at the molecular level, the chirality study has been of interest especially in the fields of chemistry and biochemistry [Barron 1982]. In recent decades, it has become of increasing interest in the electromagnetic theory for chirally composite materials [Lidell et al. 1994, Lakhtakia 1990, Lakhtakia et al. 1990]. Polarized discrimination in underwater imagery for chiral spheroids medium was studied by Ablitt et al. (1999).

However, most of researches are limited to discussion of the effective permittivity of an unbounded medium of completely-random, small particles or continuous medium with a linearly polarized wave incidence. Actually, polarized scattering from layering chiral particulate media has not been studied. During recent decades the polarimetry technology such as the advance of SAR (synthetic aperture radar) imagery has been able to measure full 2×2-D complex scattering amplitude matrix and 4×4-D Mueller matrix of four scattering Stokes parameters [Ulaby and Elachi 1990]. Jin (1992,1994) developed a theoretical model and the Mueller matrix solution for fully polarimetric scattering from a layer of non-uniformly oriented, random non-spherical particles. Numerical results of co-polarized, cross-polarized scattering and polarization degree had been calculated.

Based upon Jin's approach (1992,1994), this session presents the derivations of the scattering amplitude matrix, $f_{pq}\,(p,q=v,h)$, of spatially-oriented, weakly chiral, and electrically small spheroids. All functions of the ensemble averages $<f_{pq}f_{st}^*>\;(p,q,s,t=v,h)$ are explicitly formulated. They are used to construct the 4×4-D Mueller matrix solution of the scattered Stokes parameters [Jin 1992,1994]. Co-polarized and cross-polarized backscattering σ_c, σ_x , the polarization degree m_s, which describes partial polarization of scattered fields through random media, and transmitting of four Stokes parameters through the medium are calculated. Numerical results can show how the chirality modulates Polarimetric scattering pattern and functional dependence on the physical parameters.

It has been known that the co-polarized σ_c maximum is largely located at the polarization angles ($\chi=0^\circ,\psi=0^\circ,180^\circ$) or ($\chi=0^\circ,\psi=90^\circ$) depending on spatial orientation and particle shape. Our result first demonstrates that the chirality can asymmetrically shift this location, due to the chiral parameter appearing in the depolarized scattering amplitude functions. This shift can be evaluated by the measurement of the scattered Stokes parameters, and can be used to invert the chiral parameter in some simple cases [Jin and Chang 2003].

8-1.1 Scattering Amplitude Functions of Spatially Oriented, Chiral Spheroids

Consider an electromagnetic wave

$$\begin{aligned}\overline{E}_{inc} &= \hat{e}_i E_0 = (\hat{x}_{b1}\hat{x}_{b1} + \hat{x}_{b2}\hat{x}_{b2} + \hat{x}_{b3}\hat{x}_{b3})\cdot\hat{e}_i E_0 \\ &= (\hat{x}_{b1}\hat{x}_{b1} + \hat{x}_{b2}\hat{x}_{b2} + \hat{x}_{b3}\hat{x}_{b3})\cdot(\hat{v}_i E_{v0} + \hat{h}_i E_{h0})\end{aligned} \tag{8-1.1}$$

is incident upon a chiral spheroid. The local coordinates of a spheroid (with the axial radii a, b and c) are defined as $\hat{x}_{bj}(j=1,2,3)$. It can be transformed into the principle coordinates $\hat{x}_j(j=1,2,3)$, i. e. $\hat{x}_1=\hat{x},\hat{x}_2=\hat{y},\hat{x}_3=\hat{z}$, by operating rotation of the Euler angles (α,β,γ) of the spheroid [Jin 1994]. $\hat{v}_i, \hat{h}_i$ are the vertically and horizontally polarized vectors of the incident field in the principle coordinates, respectively.

The scattered field can be written as

$$\overline{E}_s(\overline{r}) = \begin{bmatrix} E_{vs} \\ E_{hs} \end{bmatrix} = \frac{e^{ik_0 r}}{r} \begin{bmatrix} f_{vv} & f_{vh} \\ f_{hv} & f_{hh} \end{bmatrix} \begin{bmatrix} E_{vi} \\ E_{hi} \end{bmatrix} = \frac{e^{ik_0 r}}{r} \overline{\overline{f}} \cdot \overline{E}_{inc} \tag{8-1.2}$$

where $k_0 = \omega\sqrt{\mu_0 \varepsilon_0}$ is the free space wavenumber; $f_{pq}(\theta_s, \varphi_s; \theta_i, \varphi_i)$, $p,q = v,h$ are the scattering amplitude functions of a single scatterer. θ_s, φ_s and θ_i, φ_i are the scattering and incident angles, respectively. The harmonic $e^{-i\omega t}$ time dependence is assumed and suppressed throughout.

The ellipsoid is made of isotropic chiral material, the constitutive relationship of which is described by the Drude-Born-Fedorov constitutive relations [Lidell et al. 1994]

$$\overline{D} = \varepsilon(\overline{E} + \xi \nabla \times \overline{E}), \quad \overline{B} = \mu(\overline{H} + \xi \nabla \times \overline{H}) \tag{8-1.3}$$

where ξ is the chiral parameter.

Assuming that the size of the chiral ellipsoid is small at the Rayleigh approximation of low frequency limit (i.e. the size parameter $k_0 a << 1$), the scattered field can be written as [Lakhtakia 1990, Lidell et al. 1994]

$$\overline{E}_s(\overline{r}) = \frac{e^{ik_0 r}}{4\pi r} [\omega^2 \mu_0 (\overline{\overline{I}} - \hat{k}_s \hat{k}_s) \cdot \overline{p} - \omega k_0 \hat{k}_s \times \overline{m}] \tag{8-1.4}$$

where $\hat{k}_s$ is the unit scattering vector; $\overline{p}, \overline{m}$ are the electric and magnetic dipoles, respectivelly. $\overline{p}, \overline{m}$ can be expressed by the incident waves as [Lakhtakia 1990].

$$\begin{bmatrix} \overline{p} \\ \overline{m} \end{bmatrix} = \begin{bmatrix} \overline{\overline{\alpha}}_{ee} & i\overline{\overline{\alpha}}_{em} \\ -i\overline{\overline{\alpha}}_{em} & \overline{\overline{\alpha}}_{mm} \end{bmatrix} \begin{bmatrix} \overline{E}_{inc} \\ \overline{H}_{inc} \end{bmatrix} \tag{8-1.5}$$

where

$$\overline{\overline{\alpha}}_{ee} = \sum_{j=1}^{3} \alpha_{eej} \hat{x}_{bj} \hat{x}_{bj} \tag{8-1.6a}$$

$$\overline{\overline{\alpha}}_{em} = \sum_{j=1}^{3} \alpha_{emj} \hat{x}_{bj} \hat{x}_{bj} \tag{8-1.6b}$$

$$\overline{\overline{\alpha}}_{mm} = \sum_{j=1}^{3} \alpha_{mmj} \hat{x}_{bj} \hat{x}_{bj} \tag{8-1.6c}$$

here

$$\alpha_{eej} = \frac{4}{3} \pi abc (\varepsilon_0 / L_j \Delta_j) [(\varepsilon_r \zeta^2 / k^2 - 1)(\mu_r \zeta^2 / k^2 - 1 + L_j^{-1}) - -\zeta^4 \xi^2 / k_0^2] \tag{8-1.7a}$$

$$\alpha_{emj} = (\frac{4}{3} \pi abc) \zeta^2 \xi / \omega L_j^2 \Delta_j \tag{8-1.7b}$$

$$\alpha_{mmj} = \frac{4}{3} \pi abc (\mu_0 / L_j \Delta_j) [(\mu_r \zeta^2 / k^2 - 1)(\varepsilon_r \zeta^2 / k^2 - 1 + L_j^{-1}) - -\zeta^4 \xi^2 / k_0^2] \tag{8-1.7c}$$

with $k = \omega\sqrt{\mu\varepsilon}$, $\zeta^2 = k^2 / (1 - k^2 \xi^2)$, and it is assumed that $k^2 \xi^2 << 1$, and

$$\Delta_j = (\varepsilon_r \zeta^2 / k^2 - 1 + L_j^{-1})(\mu_r \zeta^2 / k^2 - 1 + L_j^{-1}) - \zeta^4 \xi^2 / k_0^2, \ j = 1,2,3 \tag{8-1.8a}$$

$$L_1 = \frac{1}{2} abc \int_0^{\infty} ds (s + a^2)^{-3/2} (s + b^2)^{-1/2} (s + c^2)^{-1/2} \tag{8-1.8b}$$

$$L_2 = \frac{1}{2} abc \int_0^{\infty} ds (s + a^2)^{-1/2} (s + b^2)^{-3/2} (s + c^2)^{-1/2} \tag{8-1.8c}$$

$$L_3 = 1 - L_1 - L_2 \tag{8-1.8d}$$

The integrations of L_1, L_2 and L_3 can be analytically done for small spheroids, for example, needles $(L_1 = L_2 = 0.5, L_3 = 0)$ or disks $(L_1 = L_2 = 0, L_3 = 1)$ [Jin 1994].

Substituting Eqs. (8-1.1) and (8-1.5) into Eq. (8-1.4), we derived the scattering amplitude functions of a small chiral spheroid as follows:

$$f_{vv}(\theta_s,\varphi_s;\theta_i,\varphi_i) = \frac{1}{4\pi}\left[\omega^2\mu_0 \sum_{j=1}^{3} \alpha_{eej} C_{vvj} + \omega^2\varepsilon_0 \sum_{j=1}^{3} \alpha_{mmj} C_{hhj} + i\omega k_0 \sum_{j=1}^{3} \alpha_{emj}(C_{vhj} - C_{hvj})\right] \tag{8-1.9a}$$

$$f_{vh}(\theta_s,\varphi_s;\theta_i,\varphi_i) = \frac{1}{4\pi}\left[\omega^2\mu_0 \sum_{j=1}^{3} \alpha_{eej} C_{vhj} - \omega^2\varepsilon_0 \sum_{j=1}^{3} \alpha_{mmj} C_{hvj} - i\omega k_0 \sum_{j=1}^{3} \alpha_{emj}(C_{vvj} + C_{hhj})\right] \tag{8-1.9b}$$

$$f_{hv}(\theta_s,\varphi_s;\theta_i,\varphi_i) = \frac{1}{4\pi}\left[\omega^2\mu_0 \sum_{j=1}^{3} \alpha_{eej} C_{hvj} - \omega^2\varepsilon_0 \sum_{j=1}^{3} \alpha_{mmj} C_{vhj} + i\omega k_0 \sum_{j=1}^{3} \alpha_{emj}(C_{vvj} + C_{hhj})\right] \tag{8-1.9c}$$

$$f_{hh}(\theta_s,\varphi_s;\theta_i,\varphi_i) = \frac{1}{4\pi}\left[\omega^2\mu_0 \sum_{j=1}^{3} \alpha_{eej} C_{hhj} + \omega^2\varepsilon_0 \sum_{j=1}^{3} \alpha_{mmj} C_{vvj} + i\omega k_0 \sum_{j=1}^{3} \alpha_{emj}(C_{vhj} - C_{hvj})\right] \tag{8-1.9d}$$

where

$$C_{pqj} = (\hat{p}_s \cdot \hat{x}_{bj})(\hat{x}_{bj} \cdot \hat{q}_i), \quad p,q = v,h,\ j = 1,2,3 \tag{8-1.10}$$

$\hat{v}_s, \hat{h}_s$ are the vertically and horizontally polarized vectors of the scattered field, respectively.

Let the scatterer have $\varepsilon/\varepsilon_0 = \varepsilon_r, \mu_r = \mu/\mu_0 = 1$ and $a=b\neq c$ as spheroid. Considering the case of weak chirality $k^2\xi^2 << 1$, Eq. (8-1.8a) yields

$$\Delta_j = [1 + (\varepsilon_r - 1)L_j]/L_j^2 \tag{8-1.10a}$$

Eqs. (8-1.7a~d) can be simplified to keep the first order term $k\xi$ as follows:

$$\begin{aligned} f_{vv}(\theta_s,\varphi_s;\theta_i,\varphi_i) = \frac{3}{2k_0}\{&[t_1(\hat{v}_s\cdot\hat{v}_i) + (t_0 - t_1)(\hat{v}_s\cdot\hat{z}_b)(\hat{z}_b\cdot\hat{v}_i)] \\ &+ i\kappa[t_1(\hat{v}_s\cdot\hat{h}_i - \hat{h}_s\cdot\hat{v}_i) + (t_0 - t_1)((\hat{v}_s\cdot\hat{z}_b)(\hat{z}_b\cdot\hat{h}_i) - (\hat{h}_s\cdot\hat{z}_b)(\hat{z}_b\cdot\hat{v}_i))]\} \end{aligned} \tag{8-1.11a}$$

$$\begin{aligned} f_{vh}(\theta_s,\varphi_s;\theta_i,\varphi_i) = \frac{3}{2k_0}\{&[t_1(\hat{v}_s\cdot\hat{h}_i) + (t_0 - t_1)(\hat{v}_s\cdot\hat{z}_b)(\hat{z}_b\cdot\hat{h}_i)] \\ &- i\kappa[t_1(\hat{v}_s\cdot\hat{v}_i + \hat{h}_s\cdot\hat{h}_i) + (t_0 - t_1)((\hat{v}_s\cdot\hat{z}_b)(\hat{z}_b\cdot\hat{v}_i) + (\hat{h}_s\cdot\hat{z}_b)(\hat{z}_b\cdot\hat{h}_i))]\} \end{aligned} \tag{8-1.11b}$$

$$\begin{aligned} f_{hv}(\theta_s,\varphi_s;\theta_i,\varphi_i) = \frac{3}{2k_0}\{&[t_1(\hat{h}_s\cdot\hat{v}_i) + (t_0 - t_1)(\hat{h}_s\cdot\hat{z}_b)(\hat{z}_b\cdot\hat{v}_i)] \\ &+ i\kappa[t_1(\hat{v}_s\cdot\hat{v}_i + \hat{h}_s\cdot\hat{h}_i) + (t_0 - t_1)((\hat{v}_s\cdot\hat{z}_b)(\hat{z}_b\cdot\hat{v}_i) + (\hat{h}_s\cdot\hat{z}_b)(\hat{z}_b\cdot\hat{h}_i))]\} \end{aligned} \tag{8-1.11c}$$

$$\begin{aligned} f_{hh}(\theta_s,\varphi_s;\theta_i,\varphi_i) = \frac{3}{2k_0}\{&[t_1(\hat{h}_s\cdot\hat{h}_i) + (t_0 - t_1)(\hat{h}_s\cdot\hat{z}_b)(\hat{z}_b\cdot\hat{h}_i)] \\ &+ i\kappa[t_1(\hat{v}_s\cdot\hat{h}_i - \hat{h}_s\cdot\hat{v}_i) + (t_0 - t_1)((\hat{v}_s\cdot\hat{z}_b)(\hat{z}_b\cdot\hat{h}_i) - (\hat{h}_s\cdot\hat{z}_b)(\hat{z}_b\cdot\hat{v}_i))]\} \end{aligned} \tag{8-1.11d}$$

where $\hat{z}_b \equiv \hat{x}_{b3}$, and

$$\kappa = \varepsilon_r k_0 \xi / (\varepsilon_r - 1) \tag{8-1.12a}$$

$$t_0 = \frac{2}{9}\frac{k_0^3 a^2 c(\varepsilon_r - 1)}{1 + (\varepsilon_r - 1)L_3} \quad \text{and} \quad t_1 = \frac{2}{9}\frac{k_0^3 a^2 c(\varepsilon_r - 1)}{1 + (\varepsilon_r - 1)L_1} \tag{8-1.12b}$$

$$t_d = t_0 - t_1 \tag{8-1.12c}$$

Note that the chiral parameter κ now appears in Eqs. (8-1.11a~d). These equations are the basis for the following derivations and discussions.

To ensure that the scattering amplitude functions are sufficiently accurate so that the optical theorem can be used for extinction calculation, t_0, t_1 are replaced by more accurate T_0, T_1 in Eqs. (8-1.11a~d) [Jin 1994]

$$T_0 = t_0(1+it_0) \quad \text{and} \quad T_1 = t_1(1+it_1) \tag{8-1.13a}$$

$$T_d = T_0 - T_1 = t_d[1+i(t_0+t_1)] \tag{8-1.13b}$$

The scattering amplitude functions of randomly oriented spheroids are obtained by averaging spatial orientation of the Euler angle (α, β, γ). Employing the optical theorem, the extinction matrix $\overline{\overline{\kappa}}_e$ is expressed by the scattering amplitude function in the forward direction ($\theta_s = \theta_i, \varphi_s = \varphi_i \equiv 0$), which are

$$< f_{vv}^o > = \frac{3}{2k_0}[T_1 + (T_0 - T_1)\cdot c_{vv}^o] \tag{8-1.14a}$$

$$< f_{vh}^o > = \frac{3}{2k_0}\{(T_0 - T_1)\cdot c_{vh}^o - i\kappa[2T_1 + (T_0 - T_1)\cdot(c_{vv}^o + c_{hh}^o)]\} \tag{8-1.14b}$$

$$< f_{hv}^o > = \frac{3}{2k_0}\{(T_0 - T_1)\cdot c_{hv}^o + i\kappa[2T_1 + (T_0 - T_1)\cdot(c_{vv}^o + c_{hh}^o)]\} \tag{8-1.14c}$$

$$< f_{hh}^o > = \frac{3}{2k_0}[T_1 + (T_0 - T_1)\cdot c_{hh}^o] \tag{8-1.14d}$$

where the superscript o denotes the forward scattering, and

$$c_{vv}^o = <\cos^2\gamma><\sin^2\beta>\cos^2\theta_i + 2<\cos\gamma><\sin\beta\cos\beta>\sin\theta_i\cos\theta_i + <\cos^2\beta>\sin^2\theta_i \tag{8-1.15a}$$

$$c_{vh}^o = -<\sin\gamma\cos\gamma><\sin^2\beta>\cos\theta_i - <\sin\gamma><\sin\beta\cos\beta>\sin\theta_i \tag{8-1.15b}$$

$$c_{hv}^o = c_{vh}^o \tag{8-1.15c}$$

$$c_{hh}^o = -<\sin^2\gamma><\sin^2\beta> \tag{8-1.15d}$$

The angular brackets <•> denote the orientation averaging, for example

$$< f(\beta) > = \frac{1}{\cos\beta_1 - \cos\beta_2}\int_{\beta_1}^{\beta_2} p(\beta) f(\beta)\sin\beta d\beta \tag{8-1.16a}$$

$$< f(\gamma) > = \frac{1}{\gamma_2 - \gamma_1}\int_{\gamma_1}^{\gamma_2} p(\gamma) f(\gamma) d\gamma \tag{8-1.16b}$$

where $p(\beta)$ and $p(\gamma)$ are, respectively, the probablity density functions of the Euler angles β and γ, which are spatially distributed whithin the range (β_1, β_2) and (γ_1, γ_2), respectively. Note that for the spheroids (semi-axes $a = b \neq c$), the Euler angle α would not appear in the averaging. If (β_1, β_2) does not uniformly cover the whole range (0°, 180°), or (γ_1, γ_2) not cover the whole range (0°, 360°), the orientation is usually described as non-uniform. In the following calculations, $p(\beta)$, $p(\gamma) = 1$ over the range (β_1, β_2), (γ_1, γ_2), respectively.

It can be seen that besides the non-uniform orientation, the chirality κ of Eq. (8-1.12a) will also yield $< f_{vh}^o >, < f_{hv}^o > \neq 0$ which cause non-diagonal $\overline{\overline{\kappa}}_e$. This fact was unfortunately lost in many literatures.

In the backward ($\varphi_i \equiv 0, \varphi_s = \pi - \varphi_i$) direction, Eqs. (8-11a~d) become

$$f_{vv}^b = \frac{3}{2k_0}[-T_1\cdot\cos(\theta_s + \theta_i) + (T_0 - T_1)\cdot c_{vv}^b + i\kappa(T_0 - T_1)\cdot(c_{vh}^b - c_{hv}^b)] \tag{8-1.17a}$$

$$f_{vh}^b = \frac{3}{2k_0}\{(T_0 - T_1)\cdot c_{vh}^b - i\kappa[-T_1\cdot(1+\cos(\theta_s + \theta_i)) + (T_0 - T_1)\cdot(c_{vv}^b + c_{hh}^b)]\} \tag{8-1.17b}$$

$$f_{hv}^b = \frac{3}{2k_0}\{(T_0 - T_1)\cdot c_{hv}^b + i\kappa[-T_1\cdot(1+\cos(\theta_s+\theta_i)) + (T_0 - T_1)\cdot(c_{vv}^b + c_{hh}^b)]\} \quad (8\text{-}1.17c)$$

$$f_{hh}^b = \frac{3}{2k_0}[-T_1 + (T_0 - T_1)\cdot c_{hh}^b + i\kappa(T_0 - T_1)\cdot(c_{vh}^b - c_{hv}^b)] \quad (8\text{-}1.17d)$$

where

$$c_{vv}^b = -\cos^2\gamma\sin^2\beta\cos\theta_s\cos\theta_i + \cos\gamma\sin\beta\cos\beta\sin(\theta_s-\theta_i) + \cos^2\beta\sin\theta_s\sin\theta_i \quad (8\text{-}1.18a)$$

$$c_{vh}^b = \sin\gamma\cos\gamma\sin^2\beta\cos\theta_s - \sin\gamma\sin\beta\cos\beta\sin\theta_s \quad (8\text{-}1.18b)$$

$$c_{hv}^b = \sin\gamma\cos\gamma\sin^2\beta\cos\theta_i + \sin\gamma\sin\beta\cos\beta\sin\theta_i \quad (8\text{-}1.18c)$$

$$c_{hh}^b = -\sin^2\gamma\sin^2\beta \quad (8\text{-}1.18d)$$

The explicit formulations of $< f_{pq} f_{st}^* >$ in backward direction are listed in Appendix 8-1.A1. They are used to construct the phase matrix in the Mueller matrix solution for the backscattering calculation.

8-1.2 The Mueller Matrix of Random Chiral Spheroids

We now consider the polarimetric scattering from a layer of non-uniformly oriented, randomly distributed, weakly chiral and electrically small spheroids over an underlying medium. By solving the VRT equation and boundary conditions [Jin 1992,1994], the Mueller matrix solution has been obtained as

$$\begin{bmatrix} I_{vs} \\ I_{hs} \\ U_s \\ V_s \end{bmatrix} = [M_{ij}(\theta_s,\phi_s;\theta_i,\phi_i)]\cdot\begin{bmatrix} 0.5(1-\cos 2\chi\cos 2\psi) \\ 0.5(1+\cos 2\chi\cos 2\psi) \\ -\cos 2\chi\sin 2\psi \\ \sin 2\chi \end{bmatrix} \quad (8\text{-}1.19)$$

The elements of the Mueller matrix, $M_{ij}, i,j = 1,\cdots,4$, are expressed by the phase matrix of the random scatters and reflectivity matrix of the boundary surface. It can be found in Jin (1994).

As a simple case of an unbounded layer, the Mueller matrix is simply reduced to

$$\overline{\overline{M}} = n_0\begin{bmatrix} <|f_{vv}|^2> & <|f_{vh}|^2> & \mathrm{Re}< f_{vv}f_{vh}^* > & -\mathrm{Im}< f_{vv}f_{vh}^* > \\ <|f_{hv}|^2> & <|f_{hh}|^2> & \mathrm{Re}< f_{hv}f_{hh}^* > & -\mathrm{Im}< f_{hv}f_{hh}^* > \\ 2\mathrm{Re}< f_{vv}f_{hv}^* > & 2\mathrm{Re}< f_{vh}f_{hh}^* > & \mathrm{Re}< f_{vv}f_{hh}^* + f_{vh}f_{hv}^* > & -\mathrm{Im}< f_{vv}f_{hh}^* - f_{vh}f_{hv}^* > \\ 2\mathrm{Im}< f_{vv}f_{hv}^* > & 2\mathrm{Im}< f_{vh}f_{hh}^* > & \mathrm{Im}< f_{vv}f_{hh}^* + f_{vh}f_{hv}^* > & \mathrm{Re}< f_{vv}f_{hh}^* - f_{vh}f_{hv}^* > \end{bmatrix} \quad (8\text{-}1.20)$$

where n_0 is the number of the particles per unit volume (the fractional volume can be defined as $n_0 v_0$).

Once the Mueller matrix $\overline{\overline{M}}$ is obtained, the co- and cross-polarized backscattering coefficients σ_c, σ_x, and the polarization degree $m_s = [(I_{vs} - I_{hs})^2 + U_s^2 + V_s^2]^{1/2}/(I_{vs} + I_{hs})$ are calculated.

Fig. 8-1.1(a) describes a layer of vertically aligned small needles with all $\beta = 0°$ over an underlying conducting medium. Figs. 8-1.1(b~d) show σ_c, σ_x and m_s versus the elliptic angle χ when the azimuth angle $\psi = 0$ of the incidence polarization. The solid lines represent the left-handed chirality $k\xi = 0.2$, the dashed lines are the right-handed $k\xi = -0.2$, and dotted lines are for achiral case, i.e. $k\xi = 0$.

It can be seen that the chirality $k\xi$ decreases σ_c and m_s and significantly enhances σ_x especially at linearly polarized incidence (LP, χ=0). In this sense, the chirality enhances the medium randomness. It also shows that the left and right-handed chirality ($\pm k\xi$) shift the maximum of σ_c , σ_x and m_s variations and yield a significant difference of σ_c and m_s, especially at the left-handed and right-handed circularly polarized (LCP and RCP) incidences.

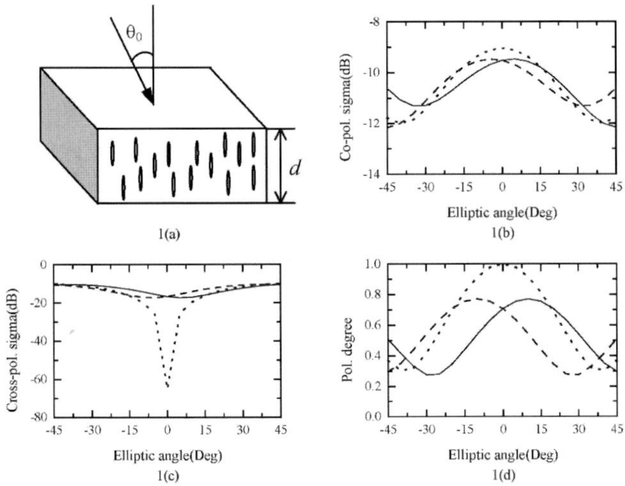

Fig. 8-1.1. σ_c, σ_x and m_s vs. χ for a layer of aligned needles with $\beta = 0°$, $\gamma \in (0°,360°)$
f=5GHz; d=20cm; θ_0=30°; a=b=0.1cm, c=1.5cm; ε_r =10.0; fractional volume = 0.1

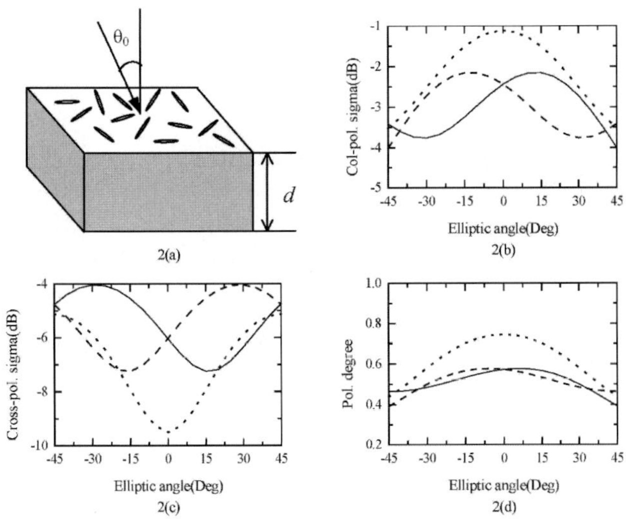

Fig. 8-1.2. σ_c, σ_x and m_s vs. χ for a layer of random needles with $\beta = 90°$, $\gamma \in (0°,360°)$, the other parameters are the same as in Fig. 8-1.1.

As a comparison, Fig. 8-2a shows σ_c , σ_x and m_s versus (χ, ψ) of a layer of horizontally-laid chiral needles with all $\beta = 90°$, $\gamma \in (0°,360°)$.

Figs. 8-1.3(a~c) and Figs. 8-1.4(a~c) show σ_c , σ_x and m_s versus polarization angles (χ, ψ), respectively, of a layer of aligned left-handed chiral ($k\xi$ = 0.2) and achiral ($k\xi$ = 0) spheroids as Fig. 8-1.1(a) describes, respectively. Comparing Figs. 8-1.3(a~c) with Figs. 8-1.4(a~c), it shows that the chirality causes the asymmetric shift of polarimetric scattering maximum on the graph plane (χ, ψ).

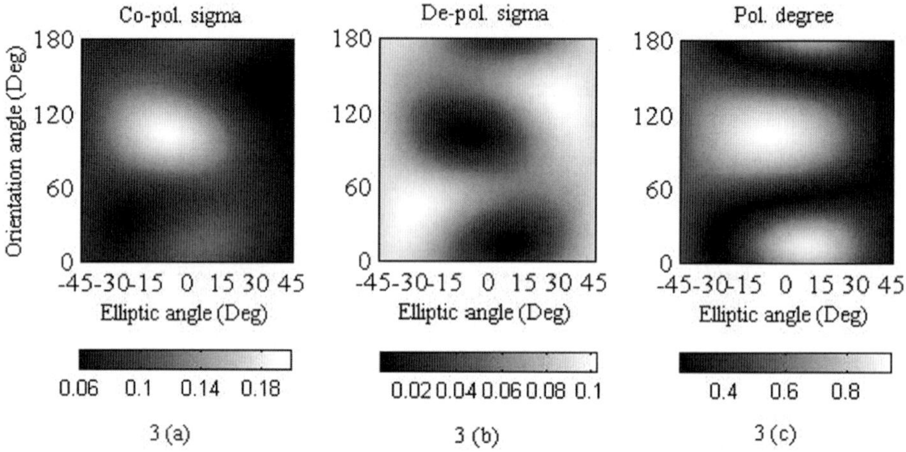

Fig. 8-1.3 σ_c , σ_x and m_s vs. (χ, ψ) for aligned chiral needles ($k\xi$=0.2) of Fig. 8-1.1(a).

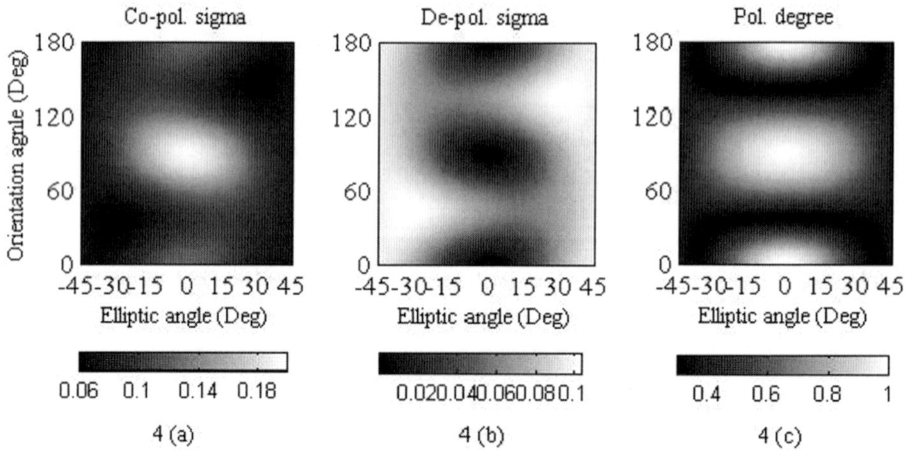

Fig. 8-1.4. σ_c, σ_x and m_s vs. (χ, ψ) for aligned achiral needles ($k\xi$=0) of Fig. 8-1.1(a)

The σ_c maximum is usually at the location of $(\chi = 0°, \psi = 0°,180°)$ or $(\chi = 0°, \psi = 90°)$ depending on spatial orientation and shape of the particles. Thus, it requires, respectively, that

$$\partial P_n(\chi,\psi)/\partial\chi\,|_{\chi=0,\psi=0} = 0 \tag{8-1.21a}$$

$$\text{or } \partial P_n(\chi,\psi)/\partial\chi\,|_{\chi=0,\psi=90^\circ}=0 \tag{8-1.21b}$$

As a simple case of a layer of unbounded medium, calculations of Eqs. (8-1.19,20) show that Eqs. (8-1.21a,b), respectively, require

$$0 = M_{42} + M_{24} = 2\,\mathrm{Im}\,(<f_{vh}f_{hh}^*>) - \mathrm{Im}(<f_{hv}f_{hh}^*>) \tag{8-1.22a}$$

$$0 = M_{41} + M_{14} = 2\,\mathrm{Im}\,(<f_{vv}f_{hv}^*>) - \mathrm{Im}(<f_{vv}f_{vh}^*>) \tag{8-1.22b}$$

From Eqs. (8-1.17a-d), it can be seen that Eqs. (8-1.22a,b) cannot be hold if the chiral parameter κ is not equal to zero. On other words, the chirality can shift the σ_c maximum location on (χ, ψ).

Further, let $\partial P_n(\chi,\psi)/\partial\chi\,|_{\psi=90^\circ}=0$ and see what the χ is. It yields

$$\begin{aligned}0 &= -(I_{vs}-I_{hs})\sin 2\chi + V_s\cos 2\chi \\ &\quad + 0.5(I_{vs}+I_{hs})' + 0.5V_s'\sin 2\chi + 0.5(I_{hs}-I_{vs})'\cos 2\chi\end{aligned} \tag{8-1.23}$$

where the superscript of prime denote $\partial/\partial\chi$. Calculations of Eq.(19) can show that

$0.5(I_{vs}+I_{hs})' \sim (M_{14}+M_{24})\sin 2\chi$ and $0.5V_s' \sim M_{44}\cos 2\chi$

are much less than $(I_{vs}-I_{hs}) \sim 0.5(M_{11}+M_{22})\cos 2\chi$,

and $0.5(I_{vs}-I_{hs})' \sim 0.5(M_{14}-M_{24})\cos 2\chi$ is much less than $V_s \sim 0.5(M_{41}+\mathrm{M}_{42})$.

The last three terms on RHS of Eq. (8-1.23) are now neglected. Thus, it yields the χ shift at the σ_c maximum expressed by the scattered Stokes parameters as follows

$$\tan 2\chi = \frac{V_s}{I_{vs}-I_{hs}} \tag{8-1.24}$$

It can be seen that the fourth Stokes parameter $V_s \neq 0$ causes the χ shift. It is produced by the chirality. If both V_s and $I_{vs}-I_{hs}$ approach zero, e.g. scattering from uniformly oriented scatterers or isotropic scatter media, χ cannot be well defined by Eq.(8-1.24).

Consider a simple case of vertically aligned needles $(\hat{z}_b=\hat{z})$ as Fig. 8-1.1a. No averaging of (β,γ) is needed for uniformly aligned particles, and reflection from underlying surface is simply neglected. In backscattering direction, it yields $\hat{v}_s=\hat{v}_i, \hat{h}_s=-\hat{h}_i$ and the functions of Eqs.(11a-d) can be easily calculated. It can be found that

$$V_s \approx M_{41} = 2\,\mathrm{Im}(f_{vv}f_{hv}^*) = 2\,\mathrm{Im}(f_{hv}^*)f_{vv} \tag{8-1.25}$$

and

$$I_{vs}-I_{hs} \approx M_{11} = f_{vv}^{\ 2} \tag{8-1.26}$$

Thus, it yields

$$\tan 2\chi \approx \frac{2\,\mathrm{Im}(f_{hv}^*)}{f_{vv}} = -2\kappa\frac{t_d\sin^2\theta_i}{t_1+t_d\sin^2\theta_i} \tag{8-1.27}$$

The chirality parameter κ or $k_0\xi$ of Eq. (8-1.12a) can be inverted from Eq.(8-1.27). From the result of Fig. 8-1.2a, the shift $\chi=-8^\circ$ yields $k_0\xi=0.24$. Comparing with the true value of $k_0\xi=0.2$, the retrieval of chirality parameter is fairly good.

8-1.3 Transmitting through the Chiral Particulate Medium

Solving the VRT equation, it has been obtained [Jin 1994] that the down-going Stokes vector $\bar{I}(\pi-\theta,\varphi,z)$ is as

$$\bar{I}(\pi-\theta,\varphi,z) = \overline{\overline{E}}(\pi-\theta,\varphi)\cdot\overline{\overline{D}}[\beta(\pi-\theta,\varphi)z\sec\theta]\cdot\overline{\overline{E}}^{-1}(\pi-\theta,\varphi)\cdot\bar{I}_0\delta(\cos\theta-\cos\theta_0)\delta(\varphi-\varphi_0)$$
$$+\int_z^0 dz'\left\{\overline{\overline{E}}(\pi-\theta,\varphi)\cdot\overline{\overline{D}}[\beta(\pi-\theta,\varphi)(z-z')\sec\theta]\cdot\overline{\overline{E}}^{-1}(\pi-\theta,\varphi)\cdot\overline{W}(\theta,\varphi,z)\right\} \tag{8-1.28}$$

where $\bar{I}_0\delta(\cos\theta-\cos\theta_0)\delta(\varphi-\varphi_0)$ is the incident Stokes vector at incident angles $(\pi-\theta_0,\varphi_0)$. Assuming that multiple scattering is negligible at low frequency limit, the zero-th order solution of Eq. (8-1.28) can be obtained as

$$\bar{I}(\pi-\theta,\varphi,z) \approx \overline{\overline{E}}(\pi-\theta,\varphi)\cdot\overline{\overline{D}}[\beta(\pi-\theta,\varphi)z\sec\theta]\cdot\overline{\overline{E}}^{-1}(\pi-\theta,\varphi)$$
$$\cdot\bar{I}_0\delta(\cos\theta-\cos\theta_0)\delta(\varphi-\varphi_0) \tag{8-1.29}$$

It should be interesting to see how much different Eq. (8-1.29) is compared with simple exponential attenuation in isotropic medium.

As the electromagnetic wave penetrates into the chiral medium, transmitting of four Stokes parameters $I\ (=(I_v+I_h)/2)$, $Q\ (=(I_v-I_h)/2)$, U and V are calculated as the penetrated depth $z=-d$ by using Eq. (8-1.29).

Fig. 8-1.5(a) shows the first Stokes parameter I versus transmitting depth for v- and h-polarized incidence in the case described by Fig. 8-1.1(a). The solid lines indicate the case of chiral needles $k\xi=\pm 0.2$. The dotted lines are the case of achiral needles $k\xi=0$. Variations of I vs. d are monotonically decreasing. The difference of chiral and achiral needles for *LP* incidence is trivial.

As a comparison, Fig. 8-1.5(b) gives I vs. d for right-handed circular (*RCP*) and left-handed circular (*LCP*) polarized incidences. It can be seen that the chirality significantly affects the difference for *RCP* and *LCP* incidences.

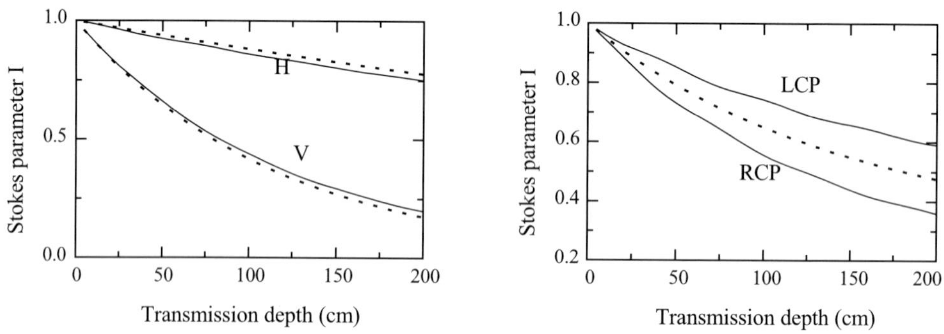

5(a) 5(b)

Fig. 8-1.5. First Stokes parameter I vs. transmission depth

7-4.1 Iterative Inversion for Canopy Biomass Parameters

Figs. 8-1.6(a, b) show the second Stokes parameter Q versus transmitting depth d for a layer of chiral and achiral needles at v, h-*LP* and *RCP*, *LCP* incidences, respectively. It is interesting to see oscillating attenuation of Q as a function of d. Oscillation might be due to phase variation of the electric fields E_v and E_h modulated by the chirality $k\xi$ as through the layer.

Figs. 8-1.7(a, b) show the third Stokes parameter U versus d.

Either non-uniform orientation or chirality can yield non-diagonal $\overline{\overline{\kappa}}_e$ and full matrix $\overline{\overline{E}}$

due to the fact of $< f_{vh}^{o} >, < f_{hv}^{o} > \neq 0$. It makes coupling of four Stokes parameters and yields oscillating variation of Q and U through the chiral medium. It is derivable that uniformly oriented chiral spheroids yields oscillation, but achiral spheroids can not.

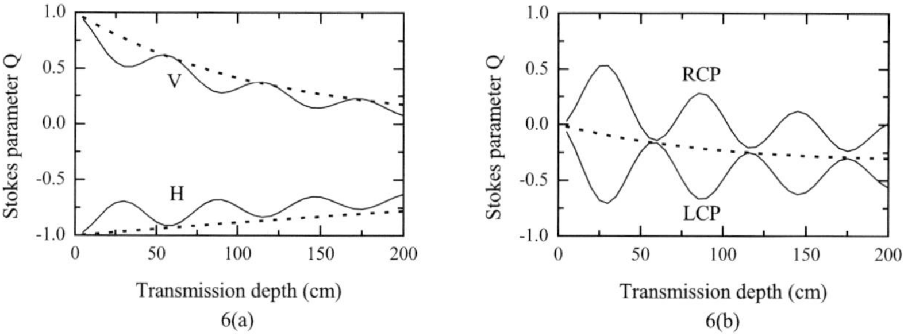

Fig.8-1.6. Second Stokes parameter Q vs. transmission depth.

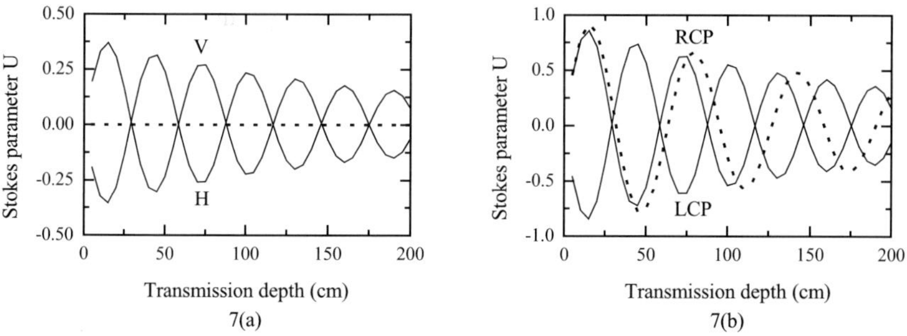

Fig. 8-1.7. Third Stokes parameter U vs. transmission depth

Appendix 8-1A

The formulations of $< f_{pq}^{b} f_{st}^{b*} >$ are listed as follows,

$$\left\langle \left|f_{vv}^{b}\right|^{2} \right\rangle = \frac{9}{4k_0^2} \left\{ |T_1|^2 \cos^2(\theta_s + \theta_i) + |T_0 - T_1|^2 \cdot \left[VV2 - 2\,\mathrm{Im}(k) \cdot (VVVH - VVHV) + |\kappa|^2 \cdot VMH2 \right] \right.$$
$$\left. - 2\,\mathrm{Re}\left[T_1 (T_0 - T_1)^* \right] \cdot \cos(\theta_s + \theta_i) \cdot VV - 2\,\mathrm{Im}\left[\kappa^* T_1 (T_0 - T_1)^* \right] \cdot \cos(\theta_s + \theta_i) \cdot VMH \right\}$$

$$\left\langle \left|f_{vh}^{b}\right|^{2} \right\rangle = \frac{9}{4k_0^2} \left\{ |\kappa|^2 |T_1|^2 \left[1 + \cos(\theta_s + \theta_i) \right]^2 - 2\,\mathrm{Im}\left[\kappa T_1 (T_0 - T_1)^* \right] \cdot \left[1 + \cos(\theta_s + \theta_i) \right] \cdot VH \right.$$
$$- 2|\kappa|^2\, \mathrm{Re}\left[T_1 (T_0 - T_1)^* \right] \cdot \left[1 + \cos(\theta_s + \theta_i) \right] \cdot VPH$$
$$\left. + |T_0 - T_1|^2 \cdot \left[VH2 + 2\,\mathrm{Im}(k) \cdot (VVVH + HHVH) + |\kappa|^2 \cdot VPH2 \right] \right\}$$

$$\left\langle \left|f_{hv}^b\right|^2 \right\rangle = \frac{9}{4k_0^2}\Big\{|\kappa|^2|T_1|^2 \cdot [1+\cos(\theta_s+\theta_i)]^2 + 2\,\mathrm{Im}\big[\kappa T_1 (T_0 - T_1)^*\big]\cdot[1+\cos(\theta_s+\theta_i)]\cdot HV$$

$$-2|\kappa|^2\,\mathrm{Re}\big[T_1(T_0-T_1)^*\big]\cdot[1+\cos(\theta_s+\theta_i)]\cdot VPH$$

$$+|T_0-T_1|^2\cdot\big[HV2 - 2\,\mathrm{Im}(k)\cdot(VVHV+HHHV)+|\kappa|^2\cdot VPH2\big]\Big\}$$

$$\left\langle \left|f_{hh}^b\right|^2 \right\rangle = \frac{9}{4k_0^2}\Big\{|T_1|^2 + |T_0-T_1|^2\cdot\big[HH2 - 2\,\mathrm{Im}(k)\cdot(HHVH - HHHV)+|\kappa|^2\cdot VMH2\big]$$

$$-2\,\mathrm{Re}\big[T_1(T_0-T_1)^*\big]\cdot HH - 2\,\mathrm{Im}\big[\kappa^* T_1(T_0-T_1)^*\big]\cdot VMH\Big\}$$

$$\left\langle f_{vv}^b f_{vh}^{b*} \right\rangle = \frac{9}{4k_0^2}\Big\{i\kappa^*|T_1|^2\cos(\theta_s+\theta_i)\cdot[1+\cos(\theta_s+\theta_i)]$$

$$-T_1(T_0-T_1)^*\cos(\theta_s+\theta_i)\cdot\big[VH + i\kappa^*\cdot VPH\big]$$

$$+T_1^*(T_0-T_1)\cdot[1+\cos(\theta_s+\theta_i)]\cdot\big[|\kappa|^2\cdot VMH - i\kappa^*\cdot VV\big]$$

$$+|T_0-T_1|^2\cdot\big[VVVH + i\kappa^*\cdot(VV2+VVHH)+i\kappa\cdot(VH2-VHHV)\big]-|\kappa|^2\cdot VPHM\Big\}$$

$$\left\langle f_{vv}^b f_{hv}^{b*} \right\rangle = \frac{9}{4k_0^2}\Big\{-i\kappa^*|T_1|^2\cos(\theta_s+\theta_i)\cdot[1+\cos(\theta_s+\theta_i)]$$

$$-T_1(T_0-T_1)^*\cos(\theta_s+\theta_i)\cdot\big[HV - i\kappa^*\cdot VPH\big]$$

$$-T_1^*(T_0-T_1)\cdot[1+\cos(\theta_s+\theta_i)]\cdot\big[|\kappa|^2\cdot VMH - i\kappa^*\cdot VV\big]$$

$$+|T_0-T_1|^2\cdot\big[VVHV - i\kappa^*\cdot(VV2+VVHH)-i\kappa\cdot(HV2-VHHV)\big]+|\kappa|^2\cdot VPHM\Big\}$$

$$\left\langle f_{hv}^b f_{hh}^{b*} \right\rangle = \frac{9}{4k_0^2}\Big\{i\kappa|T_1|^2\cdot[1+\cos(\theta_s+\theta_i)] - T_1^*(T_0-T_1)\cdot[HV - i\kappa\cdot VPH]$$

$$-T_1(T_0-T_1)^*\cdot[1+\cos(\theta_s+\theta_i)]\cdot\big[|\kappa|^2\cdot VMH + i\kappa\cdot HH\big]$$

$$+|T_0-T_1|^2\cdot\big[HHHV + i\kappa\cdot(HH2+VVHH)+i\kappa^*\cdot(HV2-VHHV)\big]+|\kappa|^2\cdot VPHM\Big\}$$

$$\left\langle f_{vh}^b f_{hh}^{b*} \right\rangle = \frac{9}{4k_0^2}\Big\{-i\kappa|T_1|^2\cdot[1+\cos(\theta_s+\theta_i)] - T_1^*(T_0-T_1)\cdot[VH + i\kappa\cdot VPH]$$

$$+T_1(T_0-T_1)^*\cdot[1+\cos(\theta_s+\theta_i)]\cdot\big[|\kappa|^2\cdot VMH + i\kappa\cdot HH\big]$$

$$+|T_0-T_1|^2\cdot\big[HHVH - i\kappa\cdot(HH2+VVHH)-i\kappa^*\cdot(VH2-VHHV)\big]-|\kappa|^2\cdot VPHM\Big\}$$

$$\left\langle f_{vv}^b f_{hh}^{b*} \right\rangle = \frac{9}{4k_0^2}\Big\{|T_1|^2\cos(\theta_s+\theta_i) - T_1^*(T_0-T_1)\cdot[VV + i\kappa\cdot VMH]$$

$$+T_1(T_0-T_1)^*\cos(\theta_s+\theta_i)\cdot\big[i\kappa^*\cdot VMH - HH\big]$$

$$+|T_0-T_1|^2\cdot\big[VVHH + i\kappa\cdot(HHVH-HHHV)-i\kappa^*\cdot(VVVH-VVHV)\big]+|\kappa|^2\cdot VMH2\Big\}$$

$$\left\langle f_{vh}^b f_{hv}^{b*} \right\rangle = \frac{9}{4k_0^2}\Big\{-|\kappa|^2|T_1|^2\cdot[1+\cos(\theta_s+\theta_i)]^2$$

$$+T_1^*(T_0-T_1)\cdot[1+\cos(\theta_s+\theta_i)]\cdot\big[i\kappa^*\cdot VH + |\kappa|^2\cdot VPH\big]$$

$$+T_1(T_0-T_1)^*\cdot[1+\cos(\theta_s+\theta_i)]\cdot\big[i\kappa\cdot HV + |\kappa|^2\cdot VPH\big]$$

$$+|T_0-T_1|^2\cdot\big[VHHV - i\kappa\cdot(VVHV+HHHV)-i\kappa^*\cdot(VVVH+HHVH)\big]-|\kappa|^2\cdot VPH2\Big\}$$

where

$$VV = -\langle\cos^2\gamma\rangle\langle\sin^2\beta\rangle\cos\theta_s\cos\theta_i + \langle\cos\gamma\rangle\langle\sin\beta\cos\beta\rangle\sin(\theta_s-\theta_i)+\langle\cos^2\beta\rangle\sin\theta_s\sin\theta_i$$

$$VH = \langle \sin\gamma \cos\gamma \rangle \langle \sin^2\beta \rangle \cos\theta_s - \langle \sin\gamma \rangle \langle \sin\beta\cos\beta \rangle \sin\theta_s$$

$$HV = \langle \sin\gamma \cos\gamma \rangle \langle \sin^2\beta \rangle \cos\theta_i + \langle \sin\gamma \rangle \langle \sin\beta\cos\beta \rangle \sin\theta_i$$

$$HH = -\langle \sin^2\gamma \rangle \langle \sin^2\beta \rangle$$

$$\begin{aligned} VV2 = {} & \langle \cos^4\gamma \rangle \langle \sin^4\beta \rangle \cos^2\theta_s \cos^2\theta_i + \langle \cos^2\gamma \rangle \langle \sin^2\beta\cos^2\beta \rangle \sin^2(\theta_s - \theta_i) + \langle \cos^4\beta \rangle \sin^2\theta_s \sin^2\theta_i \\ & -2\langle \cos^3\gamma \rangle \langle \sin^3\beta\cos\beta \rangle \cos\theta_s \cos\theta_i \sin(\theta_s - \theta_i) - 2\langle \cos^2\gamma \rangle \langle \sin^2\beta\cos^2\beta \rangle \cos\theta_s \cos\theta_i \sin\theta_s \sin\theta_i \\ & +2\langle \cos\gamma \rangle \langle \sin\beta\cos^3\beta \rangle \sin\theta_s \sin\theta_i \sin(\theta_s - \theta_i) \end{aligned}$$

$$\begin{aligned} VH2 = {} & \langle \sin^2\gamma \cos^2\gamma \rangle \langle \sin^4\beta \rangle \cos^2\theta_s + \langle \sin^2\gamma \rangle \langle \sin^2\beta\cos^2\beta \rangle \sin^2\theta_s \\ & -2\langle \sin^2\gamma \cos\gamma \rangle \langle \sin^3\beta\cos\beta \rangle \cos\theta_s \sin\theta_s \end{aligned}$$

$$\begin{aligned} HV2 = {} & \langle \sin^2\gamma \cos^2\gamma \rangle \langle \sin^4\beta \rangle \cos^2\theta_i + \langle \sin^2\gamma \rangle \langle \sin^2\beta\cos^2\beta \rangle \sin^2\theta_i \\ & +2\langle \sin^2\gamma \cos\gamma \rangle \langle \sin^3\beta\cos\beta \rangle \cos\theta_i \sin\theta_i \end{aligned}$$

$$HH2 = \langle \sin^4\gamma \rangle \langle \sin^4\beta \rangle$$

$$\begin{aligned} VVHH = {} & \langle \sin^2\gamma \cos^2\gamma \rangle \langle \sin^4\beta \rangle \cos\theta_s \cos\theta_i - \langle \sin^2\gamma \rangle \langle \sin^2\beta\cos^2\beta \rangle \sin\theta_s \sin\theta_i \\ & -\langle \sin^2\gamma \cos\gamma \rangle \langle \sin^3\beta\cos\beta \rangle \sin(\theta_s - \theta_i) \end{aligned}$$

VHHV = VVHH

$$\begin{aligned} VVVH = {} & -\langle \sin\gamma \cos^3\gamma \rangle \langle \sin^4\beta \rangle \cos^2\theta_s \cos\theta_i - \langle \sin\gamma \rangle \langle \sin\beta\cos^3\beta \rangle \sin^2\theta_s \sin\theta_i \\ & + \langle \sin\gamma \cos^2\gamma \rangle \langle \sin^3\beta\cos\beta \rangle \cos\theta_s \cdot [\cos\theta_i \sin\theta_s + \sin(\theta_s - \theta_i)] \\ & + \langle \sin\gamma \cos\gamma \rangle \langle \sin^2\beta\cos^2\beta \rangle \sin\theta_s \cdot [\cos\theta_s \sin\theta_i - \sin(\theta_s - \theta_i)] \end{aligned}$$

$$\begin{aligned} VVHV = {} & -\langle \sin\gamma \cos^3\gamma \rangle \langle \sin^4\beta \rangle \cos\theta_s \cos^2\theta_i + \langle \sin\gamma \rangle \langle \sin\beta\cos^3\beta \rangle \sin\theta_s \sin^2\theta_i \\ & - \langle \sin\gamma \cos^2\gamma \rangle \langle \sin^3\beta\cos\beta \rangle \cos\theta_i \cdot [\cos\theta_s \sin\theta_i - \sin(\theta_s - \theta_i)] \\ & + \langle \sin\gamma \cos\gamma \rangle \langle \sin^2\beta\cos^2\beta \rangle \sin\theta_i \cdot [\cos\theta_i \sin\theta_s + \sin(\theta_s - \theta_i)] \end{aligned}$$

$$HHVH = -\langle \sin^3\gamma \cos\gamma \rangle \langle \sin^4\beta \rangle \cos\theta_s + \langle \sin^3\gamma \rangle \langle \sin^3\beta\cos\beta \rangle \sin\theta_s$$

$$HHHV = -\langle \sin^3\gamma \cos\gamma \rangle \langle \sin^4\beta \rangle \cos\theta_i - \langle \sin^3\gamma \rangle \langle \sin^3\beta\cos\beta \rangle \sin\theta_i$$

VPH = VV + HH

VMH = VH – HV

VPH2 = VV2 + HH2 + 2 • VVHH

VMH2 = VH2 + HV2 – 2 • VHHV

VPHM = VVVH – VVHV + HHVH – HHHV

8-2 Polarimetric Scattering from a Layer of Spatially Oriented Metamaterial Small Spheroids

In early 1968, Vesalago proposed a medium with both the permittivity ε and permeability μ as negative, and studied electromagnetic (EM) wave propagation with abnormal phenomena, e.g. opposite directivity of the phase velocity and Poynting vector, reversal of Doppler shift, and anomalous refraction at the boundary of the media. These phenomena have been remained for verification, even though the ionosphere has been found with negative $\varepsilon_{eff}(\omega)$ in case of low frequency below the plasma frequency [Rotman 1962]. In late 1996 and 1999, Pendry et al. suggested that an array of metal wires and split ring resonators (SRRs), respectively, exhibit negative effective permittivity $\varepsilon < 0$ and permeability $\mu < 0$ when frequency is close to the resonance frequency of these cells. In 2001, Smith et al. constructed a metamaterial of $\varepsilon < 0$ and $\mu < 0$ with a periodical array of SRRs and thin wires, and experimentally demonstrated the negative refraction phenomena. From then on, metamaterial has attracted great attention in many fields, such as electromagnetics, materials science, electronics, and potential applications are also studied, for example, such as high directivity antenna, perfect lens [Pendry 2000], phase compensator [Engheta 2002] etc.

This kind of material was named by Vesalago as Left-Handed Media (LHM), because the vectors, $\overline{H}$ and $\hat{k}$ form a left-handed triplet instead of a right-handed triplet [Veselago 1968] as in the conventional Right-Handed Media (RHM). Sometimes, this material is also called as Negative Refractive Index Material, Backward Wave Medium, Double Negative Material and some others.

However most researches are limited to study of the effective permittivity and propagation through an infinite methamaterial medium [Pendry 2000, Engheta 2002]. No discussion has been studied for polarimetric scattering from a layer of metamaterial particulate media. In polarimetric scattering and radiative transfer from a layer of random medium, the 2×2-dimensional complex scattering amplitude functions and the 4×4-D Mueller matrix for Stokes scattered vector can be measured by the polarimetry technique such as synthetic aperture radar (SAR) imagery technology [Jin and Chang 2003]. Jin (1992,1994,2003) presented a Mueller matrix solution for numerical simulation of polarimetric scattering from a layer of random and non-uniformly oriented non-spherical particles. Co-polarized and cross-polarized bistatic scattering coefficients can be numerically calculated.

Following the last session of chiral spheroids, this session derives the complex scattering amplitude functions of random metamaterial small spheroids under the Rayleigh approximation, and then constructs the Mueller matrix. The bistatic scattering from a layer of random metamaterial small spheroids over an underlying medium are numerically simulated. Variations of horizontally $(\hat{h})$ and vertically $(\hat{v})$ co-polarized backscattering coefficients σ_{hh}, σ_{vv} and their difference $\sigma_{hh}-\sigma_{vv}$ upon frequency are obtained. Scattering patterns of co-polarized backscattering σ_c, cross-polarized σ_x and polarizability degree m_s of a layer of random metamaterial small spheroids under illumination of an elliptically polarized wave are demonstrated [Ye and Jin 2005].

8-2.1 Polarizability matrix of bianisotropic spheroids

The macroscopic constitutive relations of bianisotropic media are generally written as

$$\begin{cases} \overline{D} = \overline{\overline{\varepsilon}} \cdot \overline{E} + \overline{\overline{\xi}} \cdot \overline{H} = \varepsilon_0 \overline{E} + \overline{P}_e \\ \overline{B} = \overline{\overline{\varsigma}} \cdot \overline{E} + \overline{\overline{\mu}} \cdot \overline{H} = \mu_0 \overline{H} + \overline{P}_m \end{cases} \tag{8-2.1}$$

where $\overline{\overline{\varepsilon}}$, $\overline{\overline{\xi}}$, $\overline{\overline{\varsigma}}$ and $\overline{\overline{\mu}}$ are 3×3-D matrices, $\overline{E}$ and $\overline{H}$ are the electric and magnetic fields, $\overline{D}$ and $\overline{B}$ are the electric and magnetic displacements, $\overline{P}_e$ and $\overline{P}_m$ are polarization and magnetization dyadics, respectively. Eq. (8-2.1) can yield

$$\begin{cases} \overline{P}_e = (\overline{\overline{\varepsilon}} - \varepsilon_0 \overline{\overline{I}}) \cdot \overline{E} + \overline{\overline{\xi}} \cdot \overline{H} \\ \overline{P}_m = \overline{\overline{\varsigma}} \cdot \overline{E} + (\overline{\overline{\mu}} - \mu_0 \overline{\overline{I}}) \cdot \overline{H} \end{cases} \tag{8-2.2}$$

where $\overline{\overline{I}}$ is unit diagonal matrix.

Under the Rayleigh approximation for small particle, the internal fields inside a small spheroid (its semi-axes are, respectively, a, b and c) can be obtained as [Lindell and Shivola 1995]

$$\begin{cases} \overline{E} = \overline{E}_i - \dfrac{\overline{\overline{L}}}{\varepsilon_0} \cdot \overline{P}_e \\ \overline{H} = \overline{H}_i - \dfrac{\overline{\overline{L}}}{\mu_0} \cdot \overline{P}_m \end{cases} \tag{8-2.3}$$

where $\overline{E}_i$ and $\overline{H}_i$ are incident fields, the depolarization dyadic $\overline{\overline{L}}$ is written as

$$\overline{\overline{L}} = g_1 \hat{x}_b \hat{x}_b + g_2 \hat{y}_b \hat{y}_b + g_3 \hat{z}_b \hat{z}_b \tag{8-2.4}$$

here $g_i (i = 1,2,3)$ is related to the geometrical dimension of the spheroid, and its expression can be found in Jin (1994). The local coordinates $(\hat{x}_b, \hat{y}_b, \hat{z}_b)$ defined by the semi-axes of the spheroid can be transformed into the principal coordinates $(\hat{x}, \hat{y}, \hat{z})$ by rotations of the Euler angles (α, β, γ).

From Eqs. (8-2.2, 8-2.3), it is derived to obtain

$$\begin{aligned} &\left[\overline{\overline{I}} + (\overline{\overline{\varepsilon}}_r - \overline{\overline{I}}) \cdot \overline{\overline{L}} - \overline{\overline{\xi}}_r \cdot \overline{\overline{L}} \cdot (\overline{\overline{I}} + \overline{\overline{\mu}}_r \cdot \overline{\overline{L}} - \overline{\overline{L}})^{-1} \cdot \overline{\overline{\varsigma}}_r \cdot \overline{\overline{L}}\right] \cdot \overline{P}_e \\ &\quad = \varepsilon_0 \left[\overline{\overline{\varepsilon}}_r - \overline{\overline{I}} - \overline{\overline{\xi}} \cdot \overline{\overline{L}} \cdot (\overline{\overline{I}} + \overline{\overline{\mu}}_r \cdot \overline{\overline{L}} - \overline{\overline{L}})^{-1} \cdot \overline{\overline{\varsigma}}_r \cdot \overline{\overline{L}}\right] \cdot \overline{E}_i \\ &\quad + \sqrt{\varepsilon_0 \mu_0} \overline{\overline{\xi}}_r \cdot \left[\overline{\overline{I}} - \overline{\overline{L}} \cdot (\overline{\overline{I}} + \overline{\overline{\mu}}_r \cdot \overline{\overline{L}} - \overline{\overline{L}})^{-1} \cdot (\overline{\overline{\mu}}_r - \overline{\overline{I}})\right] \cdot \overline{H}_i \end{aligned} \tag{8-2.5}$$

$$\begin{aligned} &\left[\overline{\overline{I}} + (\overline{\overline{\mu}}_r - \overline{\overline{I}}) \cdot \overline{\overline{L}} - \overline{\overline{\varsigma}}_r \cdot \overline{\overline{L}} \cdot (\overline{\overline{I}} + \overline{\overline{\varepsilon}}_r \cdot \overline{\overline{L}} - \overline{\overline{L}})^{-1} \cdot \overline{\overline{\xi}}_r \cdot \overline{\overline{L}}\right] \cdot \overline{P}_m \\ &\quad = \mu_0 \left[\overline{\overline{\mu}}_r - \overline{\overline{I}} - \overline{\overline{\varsigma}}_r \cdot \overline{\overline{L}} \cdot (\overline{\overline{I}} + \overline{\overline{\varepsilon}}_r \cdot \overline{\overline{L}} - \overline{\overline{L}})^{-1} \cdot \overline{\overline{\xi}}_r \cdot \overline{\overline{L}}\right] \cdot \overline{H}_i \\ &\quad + \sqrt{\varepsilon_0 \mu_0} \overline{\overline{\varsigma}}_r \cdot \left[\overline{\overline{I}} - \overline{\overline{L}} \cdot (\overline{\overline{I}} + \overline{\overline{\varepsilon}}_r \cdot \overline{\overline{L}} - \overline{\overline{L}})^{-1} \cdot (\overline{\overline{\varepsilon}}_r - \overline{\overline{I}})\right] \cdot \overline{E}_i \end{aligned} \tag{8-2.6}$$

where $\overline{\overline{\varepsilon}}_r = \overline{\overline{\varepsilon}} / \varepsilon_0$, $\overline{\overline{\mu}}_r = \overline{\overline{\mu}} / \mu_0$, $\overline{\overline{\xi}}_r = \overline{\overline{\xi}} / \sqrt{\varepsilon_0 \mu_0}$, $\overline{\overline{\varsigma}}_r = \overline{\overline{\varsigma}} / \sqrt{\varepsilon_0 \mu_0}$.

Supposing the uniform distribution of $\overline{P}_e$ and $\overline{P}_m$ inside the spheroid, the electric and magnetic dipole moments $\overline{p}$ and $\overline{m}$ are expressed as follows

$$\begin{cases} \overline{p} = V\overline{P}_e = \overline{\overline{\alpha}}_{ee} \cdot \overline{E}_i + \overline{\overline{\alpha}}_{em} \cdot \overline{H}_i \\ \overline{m} = V\overline{P}_m = \overline{\overline{\alpha}}_{me} \cdot \overline{E}_i + \overline{\overline{\alpha}}_{mm} \cdot \overline{H}_i \end{cases} \tag{8-2.7}$$

where $\overline{\overline{\alpha}}_{ee}$, $\overline{\overline{\alpha}}_{em}$, $\overline{\overline{\alpha}}_{me}$, $\overline{\overline{\alpha}}_{mm}$ are the polarizability matrices, $V = (4\pi/3)abc$ is the volume of a spheroid. Because of the reciprocity of artificial synthetic metamaterial $\overline{\overline{\varepsilon}}_r = \overline{\overline{\varepsilon}}_r^t$, $\overline{\overline{\mu}}_r = \overline{\overline{\mu}}_r^t$, $\overline{\overline{\xi}}_r = -\overline{\overline{\varsigma}}_r^t$ [Lindell et al. 1995, Ishimaru et al. 2003] and shape symmetry of the spheroid, the polarizability matrices also satisfy the symmetric relations such as $\overline{\overline{\alpha}}_{ee} = \overline{\overline{\alpha}}_{ee}^t$, $\overline{\overline{\alpha}}_{mm} = \overline{\overline{\alpha}}_{mm}^t$ and $\overline{\overline{\alpha}}_{em} = -\overline{\overline{\alpha}}_{me}^t$, where the superscript t denotes the transpose.

The problem of chirality generated from the coupling factors $\bar{\bar{\xi}}_r$ and $\bar{\bar{\varsigma}}_r$ had been discussed in the last session. In this session, $\bar{\bar{\varepsilon}}_r$ and $\bar{\bar{\mu}}_r$ are assumed as diagonal and the coupling $\bar{\bar{\xi}}_r$ and $\bar{\bar{\varsigma}}_r$ are simply neglected. Thus, the polarizability matrices are reduced to diagonal and the diagonal elements are written as

$$\alpha_{een} = \varepsilon_0 \frac{\varepsilon_{rn} - 1}{1 + (\varepsilon_{rn} - 1) g_n} \frac{4\pi}{3} abc \tag{8-2.8}$$

$$\alpha_{mmn} = \mu_0 \frac{\mu_{rn} - 1}{1 + (\mu_{rn} - 1) g_n} \frac{4\pi}{3} abc \tag{8-2.9}$$

$$\alpha_{emn} = \alpha_{men} = 0 \tag{8-2.10}$$

where n=1,2,3.

The scattering amplitude functions of a scatter spheroid, such as oblate $(a = b >> c)$ or prolate $(a = b << c)$ spheroid, are derived from Eqs. (8-2.7) and (8-2.10)

$$\begin{aligned} f_{vv} = \frac{3}{2k_0} \{ & [t_{e1}(\hat{v}_s \cdot \hat{v}_i) + (t_{e0} - t_{e1})(\hat{v}_s \cdot \hat{z}_b)(\hat{z}_b \cdot \hat{v}_i)] \\ & + [t_{m1}(\hat{h}_s \cdot \hat{h}_i) + (t_{m0} - t_{m1})(\hat{h}_s \cdot \hat{z}_b)(\hat{z}_b \cdot \hat{h}_i)] \} \end{aligned} \tag{8-2.11}$$

$$\begin{aligned} f_{vh} = \frac{3}{2k_0} \{ & [t_{e1}(\hat{v}_s \cdot \hat{h}_i) + (t_{e0} - t_{e1})(\hat{v}_s \cdot \hat{z}_b)(\hat{z}_b \cdot \hat{h}_i)] \\ & - [t_{m1}(\hat{h}_s \cdot \hat{v}_i) + (t_{m0} - t_{m1})(\hat{h}_s \cdot \hat{z}_b)(\hat{z}_b \cdot \hat{v}_i)] \} \end{aligned} \tag{8-2.12}$$

$$\begin{aligned} f_{hv} = \frac{3}{2k_0} \{ & [t_{e1}(\hat{h}_s \cdot \hat{v}_i) + (t_{e0} - t_{e1})(\hat{h}_s \cdot \hat{z}_b)(\hat{z}_b \cdot \hat{v}_i)] \\ & - [t_{m1}(\hat{v}_s \cdot \hat{h}_i) + (t_{m0} - t_{m1})(\hat{v}_s \cdot \hat{z}_b)(\hat{z}_b \cdot \hat{h}_i)] \} \end{aligned} \tag{8-2.13}$$

$$\begin{aligned} f_{hh} = \frac{3}{2k_0} \{ & [t_{e1}(\hat{h}_s \cdot \hat{h}_i) + (t_{e0} - t_{e1})(\hat{h}_s \cdot \hat{z}_b)(\hat{z}_b \cdot \hat{h}_i)] \\ & + [t_{m1}(\hat{v}_s \cdot \hat{v}_i) + (t_{m0} - t_{m1})(\hat{v}_s \cdot \hat{z}_b)(\hat{z}_b \cdot \hat{v}_i)] \} \end{aligned} \tag{8-2.14}$$

$$t_{e0} = \frac{2k_0^3 a^2 c}{9} \frac{\varepsilon_{r3} - 1}{1 + (\varepsilon_{r3} - 1) g_3}, \quad t_{e1} = \frac{2k_0^3 a^2 c}{9} \frac{\varepsilon_{r1} - 1}{1 + (\varepsilon_{r1} - 1) g_1} \tag{8-2.15}$$

$$t_{m0} = \frac{2k_0^3 a^2 c}{9} \frac{\mu_{r3} - 1}{1 + (\mu_{r3} - 1) g_3}, \quad t_{m1} = \frac{2k_0^3 a^2 c}{9} \frac{\mu_{r1} - 1}{1 + (\mu_{r1} - 1) g_1} \tag{8-2.16}$$

Here the horizontal and vertical polarization vectors are defined as

$$\hat{h}_d = \frac{\hat{z} \times \hat{k}_d}{|\hat{z} \times \hat{k}_d|}, \quad \hat{v}_d = \hat{h}_d \times \hat{k}_d \qquad (d \equiv s, i) \tag{8-2.17}$$

As mentioned in Eq. (8-1.13), t of Eqs. (8-2.11~14) is replaced by T as

$$T = t(1 + it) \tag{8-2.18}$$

where t represents t_{e0}, t_{e1}, t_{m0} and t_{m1}, respectively [Jin 1994, Tsang et al. 1985].

8-2.2 Numerical Results

The scattering amplitude functions of both the metamaterial and dielectric prolate spheroids $(a = 0.1cm, c = 1.5cm)$ are calculated at $\theta_i = 30°$, $f = 7GHz$. The constructive parameters are taken as

$$\begin{cases} \varepsilon_{xx} = \varepsilon_{yy} = -1.4 - j4.8 \\ \mu_{xx} = \mu_{yy} = -1.8 - j2.5 \end{cases} \text{(metamaterial)}, \text{ and } \begin{cases} \varepsilon_{xx} = \varepsilon_{yy} = 1.45 \\ \mu_{xx} = \mu_{yy} = 1.0 \end{cases} \text{(dielectric)}$$

Fig. 8-2.1 compares the scattering amplitude functions $|f_{vv}|$ and $|f_{hh}|$ on the incident plane (i.e. $\varphi_s - \varphi_i = 0, \pi$) for metamaterial (solid circle line) and dielectric (no-filled circle line) spheroids. At this case, f_{vh} and f_{hv} are almost zero. It can be seen that scattering patterns of both $|f_{vv}|$ and $|f_{hh}|$ of the dielectric spheroid are symmetric along the incident direction. And $|f_{vv}|$ is strong in forward and backward direction (i.e. $\theta_s = 30°, \theta_s = 210°$, respectively), $|f_{hh}|$ is almost uniform in all directions. But, as a comparison, scattering patterns of both $|f_{vv}|$ and $|f_{hh}|$ of the metamaterial spheroid are symmetric along the spheroid axis $\hat{z}$. And $|f_{vv}|$ and $|f_{hh}|$ are strong around $\theta_s = 0°$ and become weaker around $\theta_s = 180°$ (one order less than $\theta_s = 0°$).

Conventional dielectric spheroid excited by incident EM wave generates polarization and its polarization direction is along the incident electric field $\overline{E}_i$. But the metamaterial spheroid excited by the incidence generates both polarization and magnetization, which are, respectively, opposite to the direction of the incident $\overline{E}_i$ and $\overline{H}_i$ due to negative ε' and μ'. It is a reason why the electric and magnetic dipole moments inside the spheroid are strong around $\theta_s = 0°$ and weak around $\theta_s = 180°$.

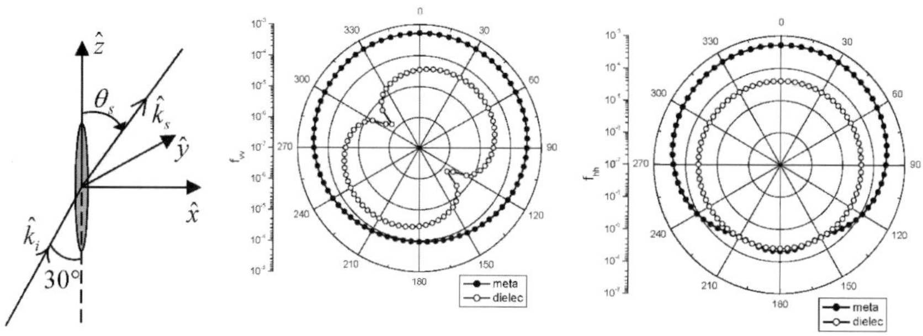

Fig. 8-2.1. Scattering amplitude functions of metamaterial and dielectric spheroids

Bistatic scattering from a layer of horizontally $(\beta = 90°)$ and vertically $(\beta = 0°)$ aligned prolate spheroids over an underlying perfectly conducting surface is calculated. The models are shown in Fig. 8-2.2. The following parameters are taken: uniform $\gamma \in (0°, 360°)$, particle fractional volume $f_s = 0.1$, layer thickness $d = 30cm$, frequency of incident wave $f = 7GHz$, and incident angle $\theta_0 = 30°$ $(\theta_i = \pi - \theta_0)$. The bistatic scattering coefficients $\sigma_c(\theta_s, \theta_0 = 30°)$ of metamaterial and dielectric spheroids on the incident plane are presented in Fig. 8-2.3.

It can be seen that co-polarized scattering of metamaterial spheroids (solid circle line) are apparently stronger than dielectric spheroids (no-filled circle line). Especially, in the case of horizontal orientation $(\beta = 90°)$, scattering from metamaterial spheroids is much stronger at $\theta_s > 0$ than at $\theta_s < 0$ (in Figs. 8-2.3 (a) and (c)).

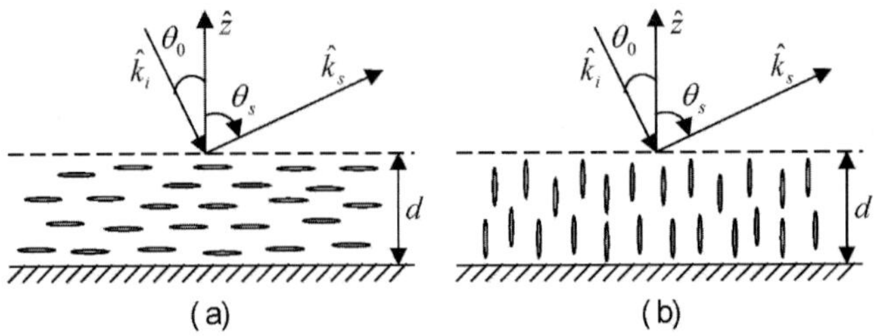

Fig. 8-2.2. Geometric model of a layer of random and horizontally or vertically aligned spheroids

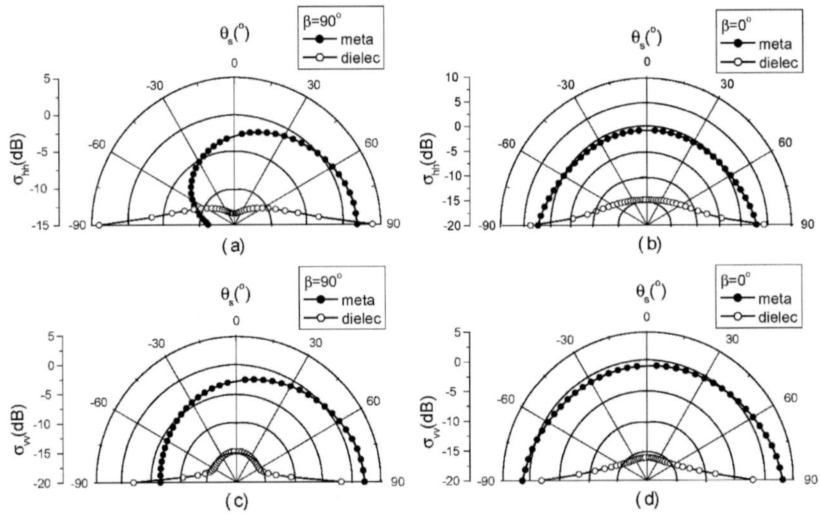

Fig. 8-2.3. σ_{vv} and σ_{hh} vs. scattering angles θ_s

It might be interpreted by Fig. 8-2.4(a) that as an EM wave is incident upon, two main rays of scattering wave come from forward direction due to scattering of the particle and multi-reflections of particle-underlying surface (see the solid line and dashed line in Fig. 8-2.4(a)). All of them contribute to the forward direction $(\theta_s > 0)$.

In another case of vertical orientation indicated by $\beta = 0°$, main rays as shown in Fig. 8-2.4(b) contribute to forward and backward directions in couple, and yield equal scattering at θ_s and $-\theta_s$ as shown in Figs. 8-2.3(b) and (d).

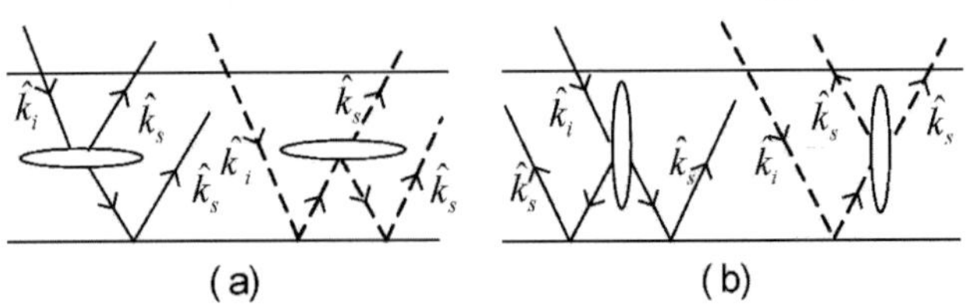

Fig. 8-2.4. Main rays of scattering from spheroids

Due to dispersion of artificially synthesized metamaterial, the permittivity ε and permeability μ become simultaneously negative at specific frequencies. Suppose to have

$\varepsilon_{xx} = \varepsilon_{yy} = \varepsilon_1$, $\mu_{xx} = \mu_{yy} = \mu_1$ of metamaterial as cited in Yao and Li (2004) shown in Fig. 8-2.5, and $\varepsilon_{zz} = 1.5$, $\mu_{zz} = 1.0$, which are simply invariant with frequency. As a contrast, the spheroid scatters with isotropic permittivity of Lorentzian dielectric FR-4 [Koledintseva et al. 2002] and dispersive relation of

$$\varepsilon_r(\omega) = \varepsilon_\infty + \frac{(\varepsilon_s - \varepsilon_\infty) f_p^2}{f_0^2 - f^2 + j \cdot f \cdot \Delta f} \tag{8-2.27}$$

are also compared. Here $\varepsilon_\infty = 4.181$, $\varepsilon_s = 4.307$, $f_0 = f_p = 17GHz$, $\Delta f = 150GHz$, and $\mu_r = 1$.

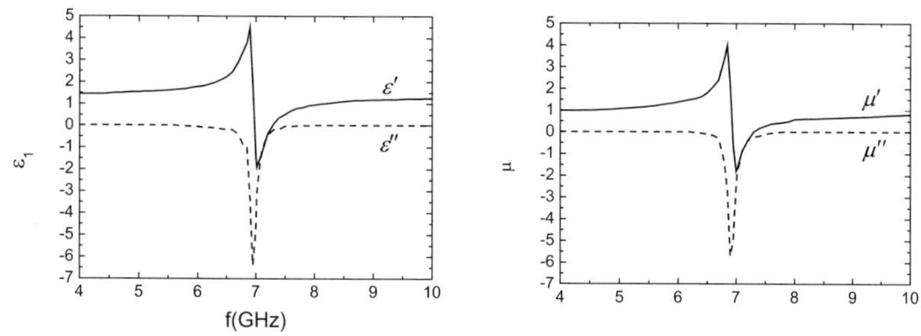

Fig. 8-2.5. Constructive parameters ε_1 and μ_1 of metamaterial

Fig. 8-2.6 shows variation of co-polarized backscattering coefficients σ_{hh} and σ_{vv} of a layer of random, horizontally $(\beta = 90^\circ)$ and vertically $(\beta = 0^\circ)$ aligned metamaterial (solid points) and dielectric (hollow points) spheroids versus frequency. At the frequencies to make ε_r and μ_r as negative, σ_{hh} and σ_{vv} of metamaterial spheroids increase quickly, especially for the case of $\beta = 0^\circ$, and take peaks around $f = 7GHz$ because of the dispersion of ε and μ indicated in Fig. 8-2.5. But the σ_{hh} and σ_{vv} of dielectric FR-4 change slightly with frequency.

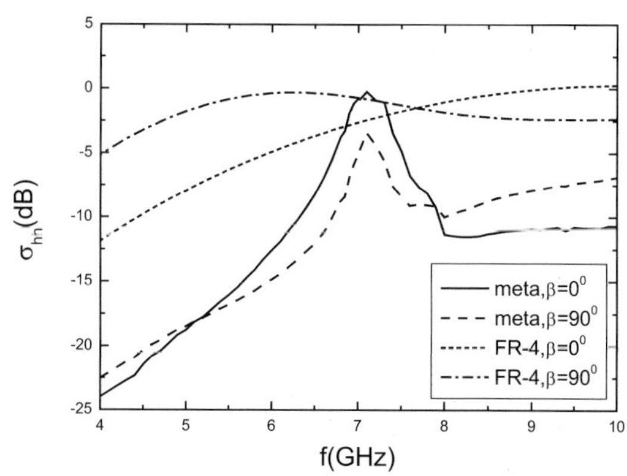

(a)

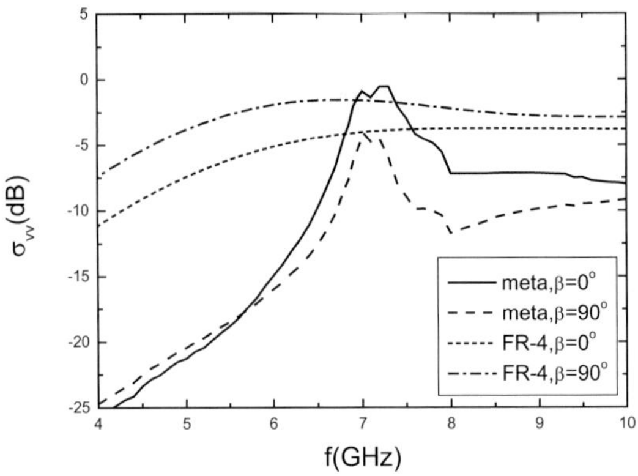

(b)

Fig. 8-2.6. Variation of backscattering coefficients σ_{hh} (a) and σ_{vv} (b) vs. frequency

Fig. 8-2.7 compares the difference $\sigma_{hh} - \sigma_{vv}$ of metamaterial and dielectric FR-4 spheroids layer with $\beta = 0°$ and $\beta = 90°$ orientation distributions. The results indicate that the $\sigma_{hh} - \sigma_{vv}$ of metamaterial spheroids varies remarkably with frequency, and takes a peak around $f = 7.1GHz$, which happens to be the resonance frequency of metamaterial as shown in Fig. 8-2.5. But $\sigma_{hh} - \sigma_{vv}$ of dielectric spheroids varies slightly with frequency, although it is dispersive.

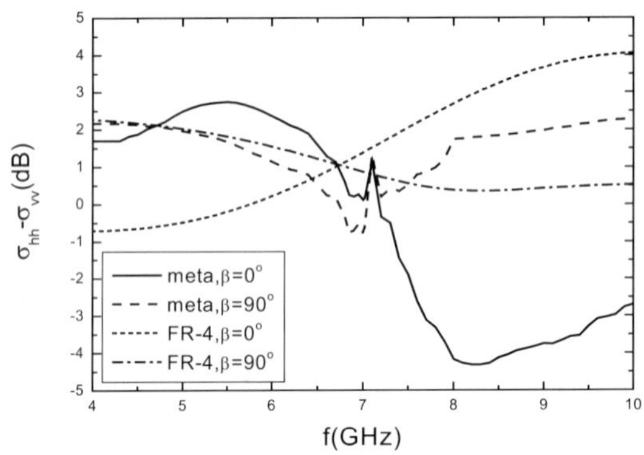

Fig. 8-2.7. Variation of $\sigma_{hh} - \sigma_{vv}$ of a layer of spheroids vs. frequency

Figs. 8-2.8 and 8-2.9 show backscattering σ_c, σ_x and m_s vs. polarization angles (χ, ψ) when the metamaterial spheroids, respectively, take the parameters ($\varepsilon_r = -1.4$, $\mu_r = -1.8$) and $(\varepsilon_r = -1.4 - j4.8, \mu_r = -1.8 - j2.5)$. Orientation of random spheroids are assumed as non-uniform $\gamma \in [0°, 90°]$ and uniform $\beta \in [0°, 180°]$. Compared with the case of conventional dielectric spheroids [Jin and Chang 2003] or Fig. 8-1.4, the backscattering σ_c, σ_x and m_s of metamaterial spheroids exhibit significantly asymmetry modes.

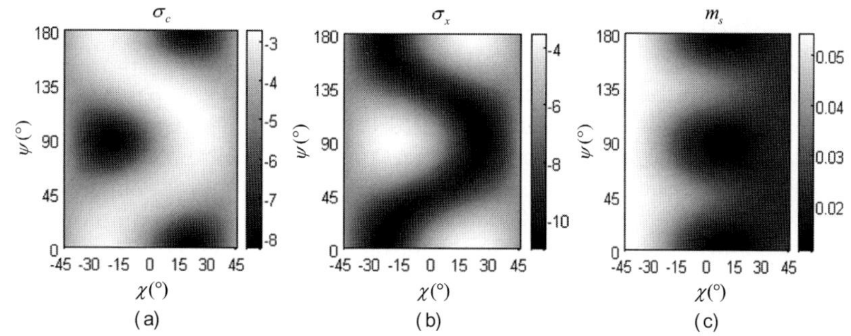

Fig. 8-2.8. σ_c, σ_X and m_S vs (χ,ψ) of a layer of lossless metamaterial spheroids with non-uniform orientation distribution γ

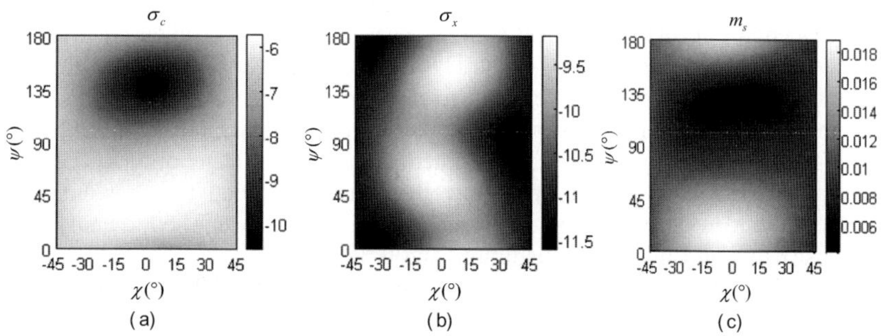

Fig. 8-2.9. σ_c, σ_X and m_S vs. (χ,ψ) of a layer of lossy metamaterial spheroids with non-uniform orientation distribution γ

In Fig. 8-2.8 of the scattering pattern of lossless metamaterial spheroids, the maximum of σ_c shifts to the elliptic polarization ($\chi \neq 0$), the maximum of σ_x shifts in the opposite direction, and m_s of right-handed circular polarized wave ($\chi = -45°$) is the highest. These variations are related to non-uniform γ and constitutive parameters of metamaterial.

In Fig. 8-2.9 of scattering pattern of lossy metamaterial spheroids, all maxims of σ_c, σ_X and m_s also demonstrate asymmetric shifts.

8-3 Time-Frequency Analysis of EM Pulse Response from a Spherical Target

Computation of the radar cross-section (RCS) of complex scatterer has been extensively studied [Aden and Kerker 1951, Gaunaurd et al. 1981]. The incident wave is usually a plane wave to illuminate the target. During recent years, the ultra-wideband (UWB) radar system employs the pulse with short duration as the interrogating signal. Since the impulse echo from the target includes all the resonance features of the RCS within the spectrum of the incident pulse, the scattering mechanism retrieved from the backscattering data can be applied to analysis and detection of the target characteristics [Strifors et al. 1994].

Under a pulse incidence, scattering from a complex target is contributed by several scattering mechanisms. To clarify scattering mechanism from backscattering data in either the time or frequency domains is of greatly helpful to scattering analysis and detection of complex targets. A novel algorithm called "joint time-frequency analysis" (JTFA) to retrieve both the time and frequency information in local area has been studied [Moghaddar and Walton 1993, Cohen 1989]. A most popular method of the JTFA is the short time Fourier-transform (STFT). However, the STFT is restrictive due to its fixed resolution in both the time and frequency domains. In recent years, the wavelet theory [Kaiser and Friendly 1994] has attracted great attention in various electromagnetic (EM) problems. The wavelet transform (WT) of the JTFA, which has good frequency resolution at low frequency and good time resolution at high frequency as called "multi-resolution characteristics", is currently applied to analysis of radar target echoes [Kim and Ling 1993, Nishimoto and Ikuno 1997].

In this session, the WT is applied to analysis of pulse scattering from a spherical target and is compared with the STFT. Numerical examples of the WT and STFT for analysis of backscattering signatures from a perfectly conducting sphere, a dielectric sphere or a dielectric spherical shell are discussed [Chen and Jin 2003].

8-3.1 The Short Time Fourier-Transform (STFT) and Wavelet Transform (WT) Methods

Consider a plane wave incident upon a radially-stratified two-layers sphere. The outer-sphere has the radius a_0 and refractive index m_1, and the inner-core has b_0, m_2, respectively. The calibrated RCS of a Mie sphere has been derived as follows [Strifors et al. 1994]:

$$\frac{\sigma}{\pi a_0^2} = \lim_{r\to\infty}\left(\frac{2r}{a_0}\frac{|\overline{E}_{sc}|}{|\overline{E}_{inc}|}\right)^2 = \left|\frac{1}{jx}\sum_{n=1}^{\infty}(-1)^n(2n+1)[A_n(x)+B_n(x)]\right|^2 \tag{8-3.1}$$

where $x = ka_0$, and k is the wave-number in free space.

When $r \to \infty$, the form-function $f_\infty(\theta, x)$ is defined as

$$f_\infty(\theta = 0, x) = \frac{1}{jx}\sum_{n=1}^{\infty}(-1)^n(2n+1)[A_n(x)+B_n(x)] \tag{8-3.2}$$

Thus, the backscattering electric field can be written as

$$\overline{E}_{sc} = E_0\frac{a_0}{2r}[\hat{\theta}\cos\phi - \hat{\phi}\sin\phi]e^{j(\omega t - kr)}f_\infty(0,x) \tag{8-3.3}$$

where E_0 is the amplitude of incident wave $\overline{E}_{inc}$, ω is the circular frequency in free-space. θ

and ϕ are the co-latitude and azimuth angles, respectively, $\hat{\theta}$ and $\hat{\phi}$ are the unit vectors in those directions. The Mie scattering coefficients A_n and B_n are well known [see Strifors et al. 1994].

We consider a pulse of short duration $g_0(t)$ incident upon the sphere. The Fourier transform pair yields:

$$G(x) = \int_{-\infty}^{\infty} g(\tau)e^{-jx\tau}d\tau, \quad g(\tau) = \frac{1}{2\pi}\int_{-\infty}^{\infty} G(x)e^{jx\tau}dx \tag{8-3.4}$$

where $\tau = ct/a_0$, which gives $\omega t = \tau x$, then $g_0(t)$ can be rewritten as $g(\tau)$. $G(x)$ is the spectrum of incident pulse $g(\tau)$. At any frequency within the spectrum, the transient response $\overline{E}_{sc}(0,\tau)$ is obtained by Eqs. (8-3.3, 8-3.4) [Gaunaurd et al. 1991]:

$$E_{sc}(0,\tau) = E_0 \frac{a_0}{2r}\int_{-\infty}^{\infty} G(x) f_{\infty}(0,x)e^{jx(\tau - r/a_0)}dx \tag{8-3.5}$$

where $(r/a_0) > 2$ corresponds to the observer location outside the sphere.

In the case of a perfectly conducting sphere $m_1 = m_2 \to \infty$, the scattering coefficients A_n, B_n in Eq. (8-3.2) can reduce to:

$$A_n(x) = -\frac{\psi_n(x)}{\zeta_n(x)}, \quad B_n(x) = \frac{\partial\psi_n(x)/\partial x}{\partial\zeta_n(x)/\partial x} \tag{8-3.6}$$

To demonstrate how the scattering spectrum varies with time, the STFT was proposed as a new Fourier analysis [Cohen 1989]:

$$F_g(\omega,t) = \int_{-\infty}^{\infty} f(t)w_a(t-t)e^{-j\omega t}\,dt \tag{8-3.7}$$

where $w_a(\cdot)$ is a window function. The Hamming window or Blackman window are usually applied. Here the Blackman window function is employed as

$$w_a(t) = 0.42 - 0.5\cos(\frac{2\pi t}{T_0}) + 0.08\cos(\frac{4\pi t}{T_0}), \quad 0 \le t < T_0 \tag{8-3.8}$$

where T_0 is the time width of the window.

The time and frequency resolutions of the STFT are, respectively, proportional to the effective duration and bandwidth of the window $w_a(t)$. Consequently, the STFT has the trade-off between time and frequency resolutions: on one hand, a good time resolution requires a short window; on the other hand, a good frequency resolution requires a narrow-band filter i.e. a long window. However, the STFT cannot satisfy two wishes simultaneously because of its fixed window width.

Because the signal frequency is inversely proportional to its period length, the width of time-window should become automatically wide for low frequency to present the full spectrum and become narrow for high frequency to present high accuracy. Contrary to the fixed resolution of the STFT, the wavelet transform (WT) can achieve changeable resolution in one domain (either time or frequency) and multi-resolution in another domain. The continuous wavelet transform of time domain signal $f(t)$ is defined as follows:

$$W_t(a,b) = \frac{1}{\sqrt{a}}\int_{-\infty}^{\infty} f(t)\Psi(\frac{t-b}{a})dt \tag{8-3.9}$$

where $\Psi(t)$ is the mother wavelet. The scale parameter a reflects the flexibility scale of the wavelet $\Psi((t-b)/a)$ and theshift parameter b representstheshiftpositionofwavelet on the time axisAccording to Eq. (8-3.9), the basicdifference between the WT and STFT is: when the scale parameter a is changed, both the duration and band width of the wavelet are changed but its shape remains the same. In contrast to the STFT, the WT employs a short window at high frequency and a long window at low frequency. Suppose to have the Gaussian wavelet:

$$\Psi(t) = \frac{1}{\sqrt{2\pi}\sigma_t} e^{-t^2/2\sigma_t^2} e^{-j\omega_0 t} \tag{8-3.10}$$

where the constants σ_t and ω_0 determine the width and center frequency of the wavelet $\Psi(t)$, respectively.

Then, Eq. (8-3.9) can be rewritten as

$$W_t(T,\Omega) = \sqrt{\frac{\Omega}{2\pi\omega_0\sigma_t^2}} \int_{-\infty}^{\infty} f(t) e^{-\frac{\Omega^2}{2\sigma_t^2\omega_0^2}(t-T)^2} e^{-j\Omega(t-T)} dt \tag{8-3.11}$$

where $\Omega = \omega_0 / a$, $T = b$. Descritizing time and defining

$$F_n(0,\Omega) = f(n\Delta t) e^{-\frac{\Omega^2}{2\sigma_t^2\omega_0^2}(n\Delta t)^2} e^{-j\Omega(n\Delta t)} \tag{8-3.12}$$

Eq. (8-3.11) at $T = m\Delta T$ can be rewritten as

$$W_t(m\Delta T,\Omega) = \sqrt{\frac{\Omega}{2\pi\omega_0\sigma_t^2}} e^{j\Omega m\Delta T} \sum_{n=-\infty}^{\infty} F_n(0,\Omega) e^{-\frac{\Omega^2}{2\sigma_t^2\omega_0^2} m\Delta T(m\Delta T - 2n\Delta t)} \Delta t \tag{8-3.13}$$

8-3.2 Numerical Example

Fig. 8-3.1 shows the once differentiated Gaussian pulse as the incident pulse at frequency 5.64*GHz*.

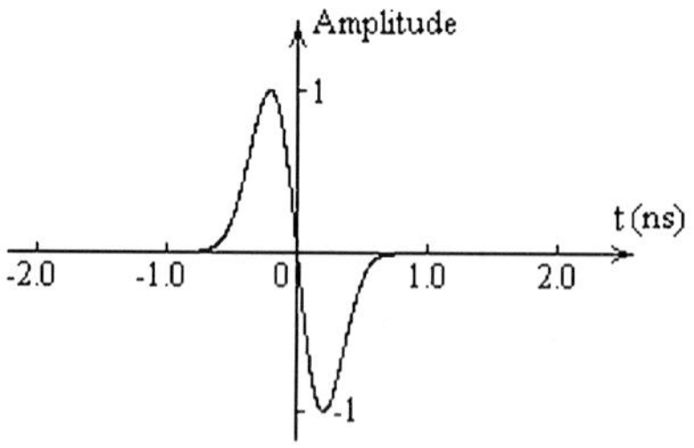

Fig. 8-3.1. Incident pulse

Fig. 8-3.2 shows backscattering data from a perfectly conducting sphere within 30*ns* and the spectrogram of STFT (upper plot) and WT (lower plot) representation. The sphere radius is $a_0 = 25.0cm$. The data is the backscattering intensity normalized by setting its maximum value equal to unity. Because the linear scale in each case is not of interest for our discussion, we employ this normalization in all examples. The width of time window in the STFT is chosen to be $T_0 = 3.75ns$.

The spots in the STFT spectrogram, as well as the vertical lines in lower spectrogram of the WT, correspond to the response of different scattering processes. The perfectly conducting sphere does not have the internal reflections but only two scattering mechanisms: specular reflection (S) and creeping (C). Two scattering processes in both the STFT and WT spectrogram can be clearly identified.

Fig. 8-3.3 shows the case of a dielectric sphere. The dielectric sphere has $a_0 = 25.0cm$ and $m = 3.0$. The scattering processes, such as the specular reflection (S), *j*-times internal reflection $(I_j, j = 1,2,3,\cdots)$, creeping (C) and combination of internal reflection and creeping $(IC_j, j = 1,2,3,\cdots)$ are presented in these two spectrograms. It seems not direct to classify these mechanisms from the complicated impulse response either in the time or frequency domain. Because of the superior multi-resolution characteristics, the WT is more suitable than STFT for identifying weaker or more concentrated scattering processes.

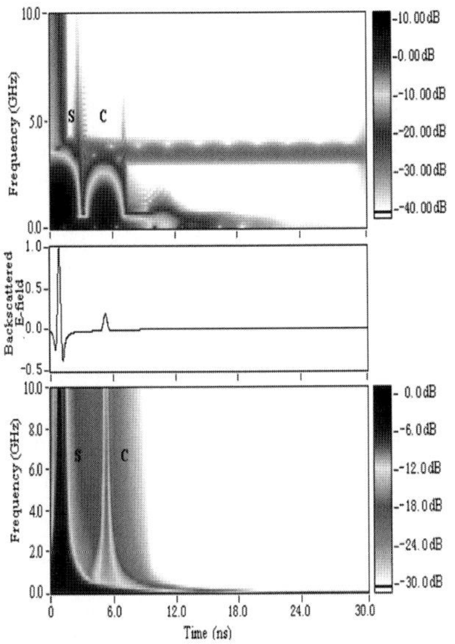

Fig. 8-3.2. Backscattering of a perfect conducting sphere and its spectrograms

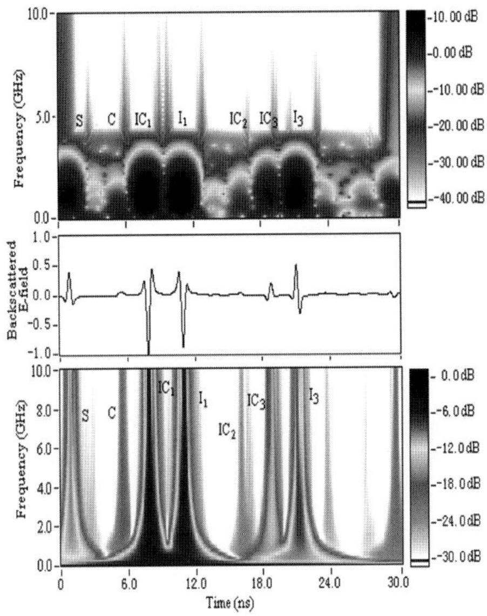

Fig. 8-3.3. Backscattering of a dielectric sphere and its spectrograms

Fig. 8-3.4 and Fig. 8-3.5 give an example to illustrate the better capability of the WT over STFT. Fig. 8-3.4 shows an air-filled dielectric spherical shell. The outer radius of the shell is $a_0 = 25.0cm$, the shell thickness is $d = 1.0cm$ and the refractive index is m=2.0.

Fig. 8-3.5 shows the same dielectric shell but with the shell thickness $d = 4.0cm$. Because the WT has good time resolution at high frequency, it is easy to obtain the time location of certain scattering mechanism in the WT spectrogram. Whereas the STFT presents the frequency information inside the window, viz. the spectrum of one time point is correlated to the time point around it. This characteristic prevents the STFT from providing accurate time location of scattering mechanism, as shown in Fig. 8-3.4.

When the shell thickness increases to 4.0*cm* of Fig. 8-3.5, less scattering mechanisms have overlapped, so the WT spectrogram can distinguish more of them even on the late-time. Whereas the STFT cannot.

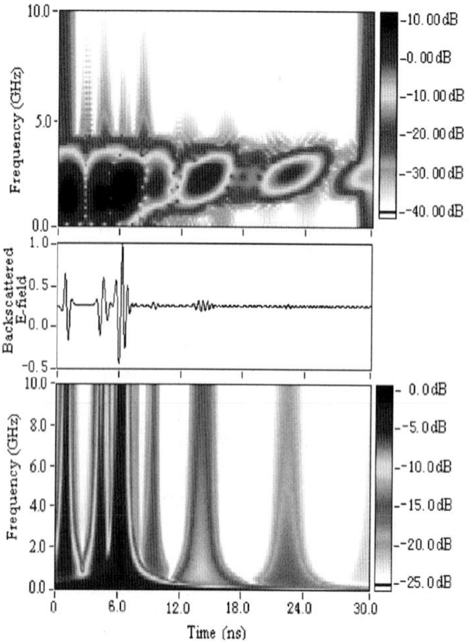

Fig. 8-3.4. Backscattering of a thin dielectric spherical shell and its spectrograms

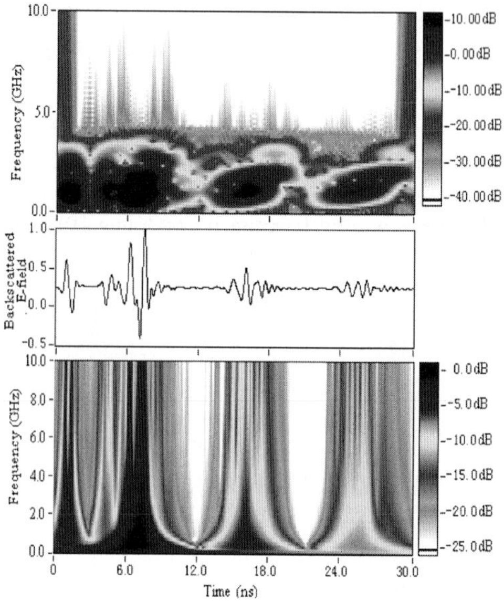

Fig. 8-3.5. Backscattering of a thick spherical shell and

To classify the radar target from transient backscattered responses, it is important to clarify scattering mechanism from the backscattering data. Complexity of the scattering processes, however, makes it difficult to retrieve them directly in either the time or frequency domain. Our comparison of the STFT and WT methods shows that the WT with the multi-resolution characteristics is more feasible than STFT to clarifying scattering mechanism and analyzing impulse response from complex single or multiple scatterers.

8-4 Monte Carlo Simulation for Correlated Scattering of Very Densely Random Spherical Particles

An important feature of the scattering of dense particles is that the correlation of particles' positions affects the scattering physics. Due to the correlated scattering and coherent wave interaction among the dense particles, the classical approximation of independent scattering becomes invalid [Ishimaru and Kuga 1980, Koh 1992, Tsang et al. 1985, 2001, Jin 1994]. To account for the correlation between densely random particles, the effective field approximation (EFA), quasi-crystalline approximation (QCA), quasi-crystalline approximation with coherent potential (QCA-CP) derived from the first moment of the scattered field are studied. The effective propagation constant, scattering and extinction coefficients of dense small spherical particles are analytically calculated. The Percus-Yevick pair distribution function, which was initially used in the study of liquid molecules, is employed for study of dense ice grains of snowpack (fractional volume less than 0.4) [Tsang et al 1985, Jin 1994]. The QCA-CP is also applied to derive the dense media radiative transfer (DMRT) equation [Tsang et al. 1985, 2001, Jin 1994] from the second moment of the scattered fields.

During recent years, Monte Carlo (MC) numerical simulation with the direct solution of Maxwell's equations has been applied to calculation of the extinction coefficient and effective propagation constant of dense particles, and extended to study of scattering of dense and sticky particles. These discussions are based on the case for ice grains of snowpack with the fractional volume less than 0.4. However, correlated scattering from very dense random particles (i.e. fractional volume more than 0.4. Theoretically, the fractional volume of spherical particles with single size can reach 0.74) has not been studied. Some techniques, such as microwave interference of special foam clouds and darkening of hollow particles smokes, need to understand the scattering mechanism of very dense particles.

In this session, two MC methods (i.e. the sequential addition method and random shuffling method) are adopted to generate randomly spatial positions of dense spherical particles. Using the electric field volumetric integral equation and expanding the internal field of each particle with basis functions, the internal and scattered fields of densely distributed spherical particles are derived. The scattering, absorption and extinction coefficients are numerically calculated from the first moment of the incoherent field. Effective permittivity and propagation constant are obtained by comparing the coherent scattered fields of discrete particle media and a homogeneous media. As validation, our results are compared with those from EFA, QCA, QCA-CP and Maxwell-Garnett (MG) mixing formula for lower particle fractional volume. Finally, backscattering and forward-scattering from a layer of very dense and random spherical particles versus some parameters such as the layer depth, particle dielectric property etc. are calculated.

The simulation results show that very dense particles demonstrate particles clustering and enhance scattering as bigger or clustered particles. The scattering and extinction coefficients versus fractional volume (very dense, i.e. 0.7) and other parameters are also presented.

8-4.1 Simulation of Dense and Random Distribution of Particles

A cubic volume is selected as a test volume V, where the random positions of particles embedded are generated by two MC methods.

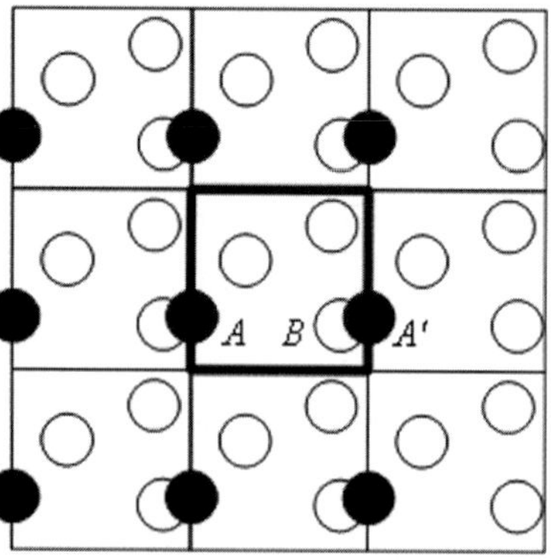

Fig. 8-4.1. Particle media in two-dimensional geometry. The central square represents the test volume, one of the particles in the test volume and its eight images are shown as solid circles.

A. Sequential addition method (SAM)

In this method, the particles are randomly put into the test volume one by one until all the particles are filled in [Tsang and Kong 2001]. The distribution of dense and random particles is viewed as the repetitions of the test volume in three-dimensional space. Overlapping of the particles is forbidden. If a particle overlaps the image of another particle, it is also seen as overlapped. For example, the particles A and B in Fig. 8-4.1 do not overlap in the test volume, but B overlaps A', one of the images of the particle A, so the particle B must be re-entered.

B. Random shuffling method (RSM)

In the RSM [Metropolis et al. 1953], the particles are initially placed in the test volume periodically, each particle is then shuffled and displaced randomly as follows:

$$x = x + \Delta\eta_1, y = y + \Delta\eta_2, z = z + \Delta\eta_3 \tag{8-4.1}$$

where x, y, z are the particle coordinates, Δ is the maximum displacement in each movement, η_1, η_2, η_3 are independent random numbers uniformly distributed in the range [-1,1]. If the new position overlaps one of the other particles, return the particle to its original position. Otherwise, update the coordinates of the displaced particle. Similarly, overlapping particle, including overlapping its images, is forbidden. To make random and uniform distribution of particles, we adjust Δ so that $A_c \approx 50\%$, where A_c is the accepted rate. Besides, the number of movement must be large enough (i.e. > 1000).

In this session, we firstly place all the particles in the test volume according to the face centered cubic structure (see Appendix 8-4A) so that the distribution of the particles is very dense. Taking the test volume V for a given fractional volume $f_v (= Nv_0 / V, v_0$ is a particle volume, N is the particles number in V), random shuffling process is applied.

8-4.2 Solution of Electric Field Volumetric Integral Equation of a Coated Particle

Consider a plane wave $\overline{E}^e(\overline{r})$ incident on N randomly distributed coated small spherical particles (Fig. 8-4.2). The coated particle j is centered at $\overline{r}_j$ and has the permittivity ε_j and the volume $V_j, j = 1,2,\cdots,N$. The background media has the permittivity ε .

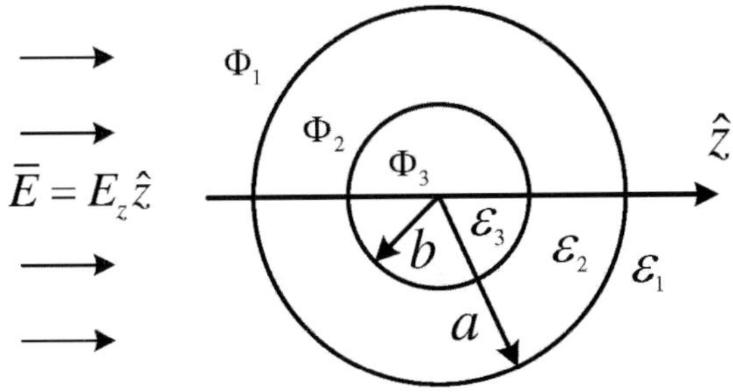

Fig. 8-4.2. Geometry of a coated particle

The volumetric integral equation, derived from Maxwell's equations for a collection of the particles, is

$$\overline{E}(\overline{r}) = \overline{E}^e(\overline{r}) + k^2 \sum_{j=1}^{N} \int_{V_j} d\overline{r}' g(\overline{r},\overline{r}')(\varepsilon_{rj}-1)\overline{E}_j(\overline{r}') - \sum_{j=1}^{N} \nabla \int_{V_j} d\overline{r}' \nabla' g(\overline{r},\overline{r}')(\varepsilon_{rj}-1)\overline{E}_j(\overline{r}') \quad (8\text{-}4.2)$$

where $\overline{r} \in V_j$, $g(\overline{r},\overline{r}')$ is the free-space scalar Green's function, $\varepsilon_{rj} = \varepsilon_j / \varepsilon$.

Let the internal field of the j-th particle be

$$\overline{E}_j(\overline{r}) = \overline{E}(\overline{r}) \quad (8\text{-}4.3)$$

The internal field $\overline{E}_j(\overline{r})$ is expanded into three basis functions. The basis functions are chosen following the electrostatic solution of the coated spherical particle (see Appendix 8-4A).

$$\overline{E}_j(\overline{r}) = \sum_{\alpha=1}^{3} c_{j\alpha} \overline{f}_{j\alpha}(\overline{r}) \quad (8\text{-}4.4)$$

For $b \le r \le a$,

$$\overline{f}_{j1}(\overline{r}) = \frac{\varepsilon_{rj3}+2\varepsilon_{rj2}}{3\varepsilon_{rj2}}\hat{x} + \frac{\varepsilon_{rj3}-\varepsilon_{rj2}}{3\varepsilon_{rj2}} b_j^3[\hat{r}\frac{2}{r^3}\sin\theta\cos\phi - \frac{1}{r^3}(\hat{\theta}\cos\theta\cos\phi - \hat{\phi}\sin\phi)] \quad (8\text{-}4.5a)$$

$$\overline{f}_{j2}(\overline{r}) = \frac{\varepsilon_{rj3}+2\varepsilon_{rj2}}{3\varepsilon_{rj2}}\hat{y} + \frac{\varepsilon_{rj3}-\varepsilon_{rj2}}{3\varepsilon_{rj2}} b_j^3[\hat{r}\frac{2}{r^3}\sin\theta\sin\phi - \frac{1}{r^3}(\hat{\theta}\cos\theta\sin\phi + \hat{\phi}\cos\phi)] \quad (8\text{-}4.5b)$$

$$\overline{f}_{j3}(\overline{r}) = \frac{\varepsilon_{rj3}+2\varepsilon_{rj2}}{3\varepsilon_{rj2}}\hat{z} + \frac{\varepsilon_{rj3}-\varepsilon_{rj2}}{3\varepsilon_{rj2}} b_j^3[\hat{r}\frac{2}{r^3}\cos\theta + \frac{1}{r^3}\hat{\theta}\sin\theta] \quad (8\text{-}4.5c)$$

For $0 \le r \le b$,

$$\overline{f}_{j1}(\overline{r}) = \hat{x},\ \overline{f}_{j2}(\overline{r}) = \hat{y},\ \overline{f}_{j3}(\overline{r}) = \hat{z}, \quad (8\text{-}4.6)$$

From Eqs. (8-4.2~4), it yields

$$\sum_{\alpha=1}^{3} c_{i\alpha}\overline{f}_{i\alpha}(\overline{r}) = \overline{E}^e(\overline{r}) + \sum_{\substack{j=1 \\ j\neq i}}^{N} \sum_{\alpha=1}^{3} c_{j\alpha}\overline{q}_{j\alpha}(\overline{r}) + \sum_{\alpha=1}^{3} c_{i\alpha}\overline{q}_{i\alpha}(\overline{r}) \quad (8\text{-}4.7)$$

where

$$\overline{q}_{j\alpha}(\overline{r}) = k^2 \int_{V_j} d\overline{r}' g(\overline{r},\overline{r}')(\varepsilon_{rj}-1)\overline{f}_{j\alpha}(\overline{r}') - \nabla \int_{V_j} d\overline{r}' \nabla' g(\overline{r},\overline{r}')(\varepsilon_{rj}-1)\overline{f}_{j\alpha}(\overline{r}') \quad (8\text{-}4.8)$$

Dot-multiplying the test function $(\varepsilon_{ri}-1)\overline{f}_{i\beta}(\overline{r})$ to Eq. (8-4.7), the integration over the region V_j yields

$$\sum_{\alpha=1}^{3} c_{i\alpha} \int_{V_i} d\bar{r}(\varepsilon_{ri}-1)\bar{f}_{i\beta}(\bar{r})[\bar{f}_{i\alpha}(\bar{r})-\bar{q}_{i\alpha}(\bar{r})]$$
$$= \int_{V_i} d\bar{r}(\varepsilon_{ri}-1)\bar{f}_{i\beta}(\bar{r})\cdot\bar{E}^{e}(\bar{r}) + \sum_{\substack{j=1\\ j\neq i}}^{N}\sum_{\alpha=1}^{3} c_{j\alpha}\int_{V_i} d\bar{r}(\varepsilon_{ri}-1)\bar{f}_{i\beta}(\bar{r})\cdot\bar{q}_{j\alpha}(\bar{r}) \tag{8-4.9}$$

with $i=1,2,\cdots,N;\ \beta=1,2,3$.

Based on the Rayleigh approximation of small particle, Eq. (8-4.9) becomes

$$\sum_{\alpha=1}^{3} c_{i\alpha} \int_{V_i} d\bar{r}(\varepsilon_{ri}-1)\bar{f}_{i\beta}(\bar{r})[\bar{f}_{i\alpha}(\bar{r})-\bar{q}_{i\alpha}(\bar{r})] = \int_{V_i} d\bar{r}(\varepsilon_{ri}-1)\bar{f}_{i\beta}(\bar{r})\cdot\bar{E}^{e}(\bar{r})$$
$$+\sum_{\substack{j=1\\ j\neq i}}^{N}\sum_{\alpha=1}^{3} k^2 c_{j\alpha}\int_{V_i} d\bar{r}(\varepsilon_{ri}-1)\bar{f}_{i\beta}(\bar{r})\cdot\bar{\bar{G}}(\bar{r}_i,\bar{r}_j)\int_{V_j} d\bar{r}'(\varepsilon_{rj}-1)\bar{f}_{j\alpha}(\bar{r}') \tag{8-4.10}$$

It yields

$$c_{i\beta}K_i = \bar{s}_{i\beta}\cdot\bar{E}^{e}(\bar{r}) + \sum_{\substack{j=1\\ j\neq i}}^{N}\sum_{\alpha=1}^{3} k^2 c_{j\alpha}\bar{s}_{i\beta}\cdot\bar{\bar{G}}(\bar{r}_i,\bar{r}_j)\cdot\bar{s}_{j\alpha} \tag{8-4.11}$$

where

$$K_i = \frac{(\varepsilon_{ri3}+2\varepsilon_{ri2})((\varepsilon_{ri2}+2)+2b_i^3(\varepsilon_{ri2}-\varepsilon_{ri3})(1-\varepsilon_{ri2})/a_i^3}{9\varepsilon_{ri2}}L_i \tag{8-4.12}$$

$$\bar{s}_{i1} = \int_{V_i} d\bar{r}(\varepsilon_{ri}-1)\bar{f}_{i1}(\bar{r}) = L_i\hat{x} \tag{8-4.13a}$$

$$\bar{s}_{i2} = \int_{V_i} d\bar{r}(\varepsilon_{ri}-1)\bar{f}_{i2}(\bar{r}) = L_i\hat{y} \tag{8-4.13b}$$

$$\bar{s}_{i3} = \int_{V_i} d\bar{r}(\varepsilon_{ri}-1)\bar{f}_{i3}(\bar{r}) = L_i\hat{z} \tag{8-4.13c}$$

$$L_i = \frac{4\pi}{3}\left[\frac{2\varepsilon_{ri2}+\varepsilon_{ri3}}{3\varepsilon_{ri2}}(a_i^3-b_i^3)(\varepsilon_{ri2}-1)+b_i^3(\varepsilon_{ri3}-1)\right] \tag{8-4.13d}$$

For the case of $\varepsilon_2=\varepsilon_3=\varepsilon_s$, the coated particle returns to a homogeneous particle and Eq. (8-4.11) reduce to the result of Tsang et al. (2001). For the case of $\varepsilon_3=\varepsilon_0$, the core of the coated particle is air and Eq. (8-4.11) reduce to the result of Chen et al. (2003).

After solving the matrix equations of (8-4.11), all coefficients and the internal field of each particle $\bar{E}_j(\bar{r}),\ j=1,2,\cdots,N,\ \bar{r}\in V_j$ are obtained.

The power absorbed by N particles can be calculated by

$$P_{abs} = \frac{1}{2}\omega\sum_{j=1}^{N}\int_{V_j} d\bar{r}\varepsilon_j''(\bar{r})|\bar{E}_j(\bar{r})|^2 \tag{8-4.14}$$

where ε_j'' is the imaginary part of ε_j. The absorption coefficient is

$$\kappa_{abs} = P_{abs}/[|\bar{E}^{e}|^2 V/(2\eta)] \tag{8-4.15}$$

In the far-field, the scattered field can be written as

$$\bar{E}_s(\bar{r}) = \frac{k^2 e^{ikr}}{4\pi r}(\hat{v}_s\hat{v}_s+\hat{h}_s\hat{h}_s)\sum_{j=1}^{N}\int_{V_j} d\bar{r}(\varepsilon_{rj}-1)\bar{E}_j(\bar{r})e^{-i\bar{k}_s\cdot\bar{r}_j} \equiv (E_{vs}\hat{v}_s+E_{hs}\hat{h}_s)\frac{e^{ikr}}{r} \tag{8-4.16}$$

The scattered power is

$$P_s = \frac{1}{2\eta}\int_0^{2\pi} d\phi_s\int_0^{\pi} d\theta_s \sin\theta_s(|E_{vs}|^2+|E_{hs}|^2) \tag{8-4.17}$$

The scattered power can be decomposed into coherent and incoherent parts. The coherent field is obtained by averaging the scattered field over the MC realizations. Thus, the scattering coefficient is derived as

$$\kappa_s = \frac{P_s^{inco}}{|\overline{E}^e|^2 V/(2\eta)} = \int_0^{2\pi} d\phi_s \int_0^{\pi} d\theta_s \sin\theta_s [<|E_{vs} - <E_{vs}>|^2> + <|E_{hs} - <E_{hs}>|^2>]/(|\overline{E}^e|^2 V) \quad (8\text{-}4.18)$$

where the angular brackets represent averaging over realizations.

The extinction coefficient is $\kappa_e = \kappa_s + \kappa_{abs}$ and the albedo is $\tilde{\omega} = \kappa_s / \kappa_e$.

8-4.3 Calculation of the Effective Permittivity

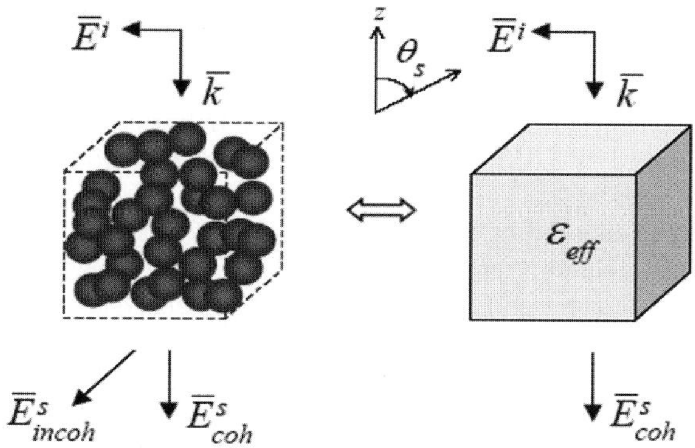

Fig. 8-4.3. To find the effective permittivity

As shown in Fig. 8-4.3, a homogeneous dielectric body V_0 with the same geometrical shape and size as the test volume V is constructed. To find when the scattered fields of the test volume V and the homogeneous V_0 are (approximately) equal as a same incident wave is upon, the effective permittivity of the test volume V, ε_{eff}, is defined as identical to that of V_0 [Zurk et al. 1995,1996, Siqueira and Saranbandi 2000]. In this session, the finite-element method [Morgem and Mei 1979] is utilized to calculate the scattered field of the homogenous dielectric body V_0. The effective propagation can be obtained from the effective permittivity.

$$K = k_0\sqrt{\varepsilon_{eff}} \quad (8\text{-}4.19)$$

where k_0 is the wave number of the vacuum.

8-4.4 Numerical Results

Several cases of the lossless, low loss and high loss spherical particles and water bubbles are studied. The parameters in the simulations are as follows: the particle size factor is $ka = 0.2$, the radius is $a = 0.6mm$, the outer and inner radii of water bubble are $0.6mm$ and $0.4mm$, respectively. The water permittivity is taken as $49.6 + i34.6$ (at 15.9 GHz) in this session.

A. Convergence of κ_e / k_0

Table 8-4.1 shows κ_e / k_0 for different test volume V with the fractional volume $f_v = 15\%$. It can be seen that when the test volume V and the number of particles N are large enough, κ_e / k_0 well converges to correct value. So it must satisfy $V > \lambda_0^3$ and N large enough [Zurk et al. 1996].

Table 8-4.1. Relationship between convergence of κ_e / k_0 and V, N

N	V (mm^3)	λ_0^3 (mm^3)	V / λ_0^3	$\kappa_e / k_0 (\times 10^{-4})$
500	14.0		0.45	1.36
1000	27.9		0.90	1.45
1200	33.5	31.0	1.08	1.62
2000	55.9		1.80	1.60
3000	83.8		2.70	1.61

B. Random distribution of dense particle

A truly random distribution of particles must be uniform. In lower fractional volume, the realization generated by SAM is more uniform than that by RSM. In higher fractions, the situation reverses.

We can check the distribution of particles as follows:

Dividing the test volume V equally into $n \times n \times n$ small cubes, calculate the number of particles in each small cube and compare the number difference from the truly random situation. Thc smaller the deviation is, the more uniform is the distribution of particles.

In Table 8-4.2, the number of particles is 2048 and the test volume is divided into $2 \times 2 \times 2$ small cubes. If the distribution of particles is completely uniform, the number of particles in each small cube is 256. We can see that the realization generated by SAM is more uniform than that by RSM when the fractional volume is 25%. For the case of the fractional volume 60%, the realization by RSM is very uniform while SAM does not work.

Table 8-4.2. The distribution of particles

f_v	25%								60%				
	SAM				RSM				SAM	RSM			
particles in each cube	251	260	254	258	253	239	260	254	N/A	256	256	256	256
	253	258	256	258	264	253	259	266		256	256	256	256
variance	8.25				62.5					0			

Thus, we adopt SAM for $f_v \leq 35\%$ and RSM for $f_v > 35\%$, as SAM does not work with high rate of overlapping.

C. Comparison of κ_e / k_0 with other approaches

Comparison of the extinction coefficients κ_e between our method and other approaches such as QCA, QCA-CP and Zurk et al. (1995) is shown in Table 8-4.3 and Fig. 8-4.4.

In Fig. 8-4.4, it can be seen that κ_e / k_0 of QCA and QCA-CP increases as a function of fractional volume f_v to reach the maximum around f_v=0.15 and then swings a little during f_v within 0.2 and 0.55. When f_v<0.3, our result is between QCA and QCA-CP. When f_v>0.3, our result is bigger than QCA and less than QCA-CP. Very dense distribution f_v~ (0.2-0.6) demonstrates particles clustering and enhance scattering as bigger or clustered particles. When f_v keeps increasing and reaches very dense ($f_v \geq 0.6$), κ_e / k_0 decreases and eventually becomes 0 of a homogeneous medium together with QCA and QCA-CP. Our results at lower fractional volume are exactly matched with numerical result of Zurk et al. (1995).

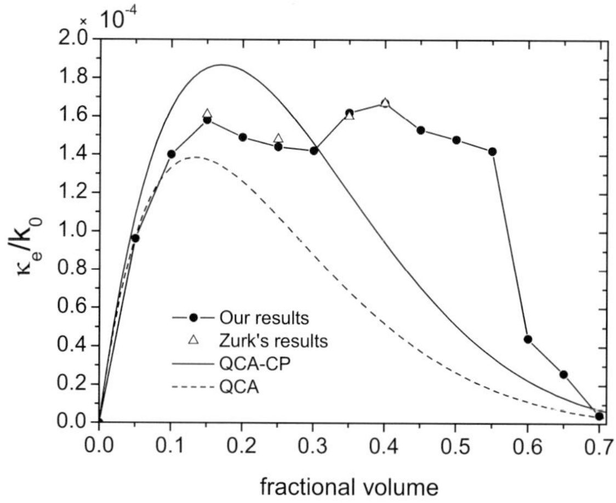

Fig. 8-4.4. Comparison of normalized extinction coefficients ($\varepsilon_s = 3.2\varepsilon_0$)

Table 8-4.3. The extinction coefficient κ_e

f_v (%)	$\kappa_e / k_0 \times 10^{-4}$			N	realizations
	QCA	QCA-CP	Our method		
5	0.97	1.08	0.962	2000	20
10	1.33	1.64	1.40		
15	1.38	1.85	1.58		
20	1.26	1.84	1.49		
25	1.08	1.69	1.44	3000	
30	0.88	1.46	1.42		
35	0.69	1.20	1.62		
40	0.52	0.95	1.67	4000	
45	0.38	0.71	1.53		
50	0.27	0.51	1.48		
55	0.18	0.35	1.42		
60	0.11	0.23	0.443		
65	0.069	0.14	0.260		
70	0.038	0.075	0.0444		

D. Effective permittivity

Scattering fields between a cubic body of densely distributed particles V and a homogeneous medium V_0 are compared in Figs. 8-4.5(a, b). It can be seen in Fig. 8-4.5(a) that the scattered field of random particles of V with f_v=20% is approximately identical to that of homogeneous V_0 with relative permittivity of 1.3. Thus the relative effective permittivity of the particle medium V with f_v=20% is offered as $\varepsilon_{r,eff} = 1.3$. Fig. 8-4.5(b) shows that the imaginary part of the particle permittivity has little effect on the scattered field when its value is actually very small.

We can calculate the imaginary part of the effective permittivity as follows

$$\varepsilon''_{r,eff} = \kappa_e = 2k_0 \mathrm{Imag}(\sqrt{\varepsilon'_{r,eff} + i\varepsilon''_{r,eff}}) \tag{8-4.20}$$

Assuming that $\varepsilon''_{r,eff} << \varepsilon'_{r,eff}$, it yields

$$\varepsilon''_{r,eff} = \kappa_e \sqrt{\varepsilon'_{r,eff}} / k_0 \tag{8-4.21}$$

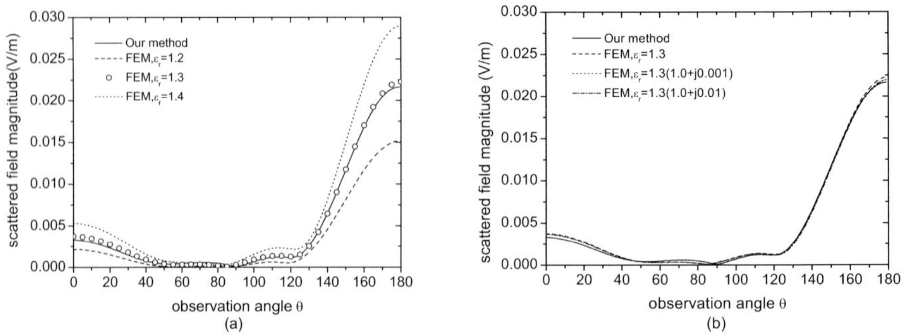

Fig. 8-4.5. Comparison of scattering fields between densely distributed particles ($f_v = 20\%$, ε_s=(3.2+i0.0032)ε_0) and a homogeneous body

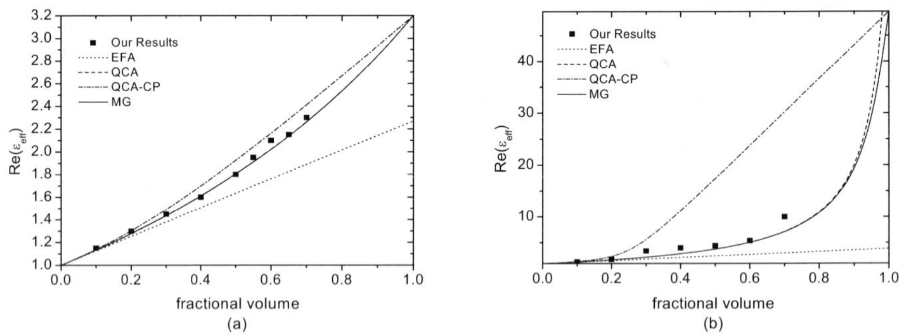

Fig. 8-4.6. Comparison of real part of effective permittivity (a) $\varepsilon_s = (3.2 + i0.0032)\varepsilon_0$ (b) $\varepsilon_s = (49.587 + i34.595)\varepsilon_0$

Figs. 8-4.6(a, b) shows ε'_{eff} of ice and water particles as a function of f_v. With the coherent potential introduced into QCA-CP, the effective propagation constant K is calculated. When the particle permittivity is small and the difference between K and k_0 is not remarkable and QCA-CP is almost equivalent to other methods (Fig. 8-4.6(a)). When the particle permittivity is large and the difference between K and k_0 is remarkable, the result of QCA-CP becomes higher (Fig. 8-4.6(b)) and not suitable. The propagation constant of our method, QCA and MG mixing formula is k_0, so all of these results are in good agreement. Since EFA is good only for small fractional volume, it becomes deviated from others when the fractional volume is more than 10%.

Figs. 8-4.7(a, b) shows ε''_{eff} of ice and water particles as a function of f_v. It can be seen that our results are in good agreement with that of MG mixing formula. As QCA and

QCA-CP assume $K' >> K''$ (K', K'' are the real and imaginary parts of K, respectively), it yields $\varepsilon_{eff} = K^2 / k_0^2 \approx (K'^2 + 2K'K'') / k_0^2$. When the condition $K' >> K''$ is not satisfied, the difference between QCA and QCA-CP is remarkable (Fig. 8-4.7b).

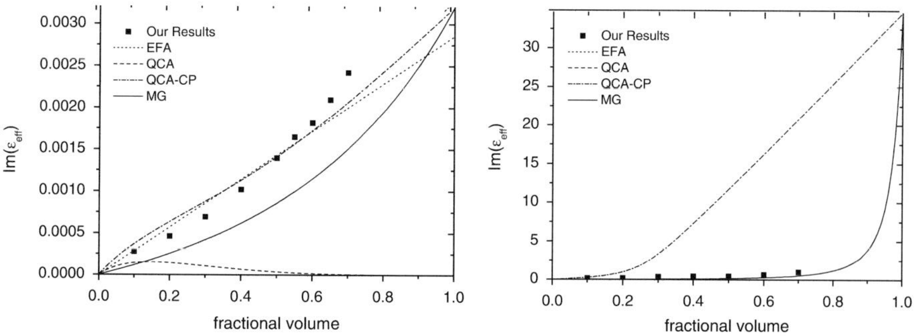

(a) (b)

Fig. 8-4.7. Comparison of imaginary part of effective permittivity
(a) $\varepsilon_s = (3.2 + i0.0032)\varepsilon_0$ (b) $\varepsilon_s = (49.587 + i34.595)\varepsilon_0$

8-4.5 Forward-Scattering and Backscattering from a Layer of Dense Particles

Consider a plane wave incident at the direction $(\pi - \theta_0, \phi_0 = 0)$ upon a layer of random particles with the layer thickness d (Fig. 8-4.8). The first-order scattering intensities iteratively derived from the VRT are obtained as Jin (1994)

$$\overline{I}^{(1)}(\theta, \phi, z = 0) = \overline{\overline{P}}(\theta, \phi; \pi - \theta_0, 0) \cdot \overline{I}_0 \frac{\sec\theta}{\kappa_e(\sec\theta_0 + \sec\theta)}[1 - e^{-\kappa_e d(\sec\theta_0 + \sec\theta)}] \tag{8-4.22a}$$

$$\overline{I}^{(1)}(\pi - \theta, \phi, z = -d) = \overline{\overline{P}}(\pi - \theta, \phi; \pi - \theta_0, 0) \cdot \overline{I}_0 \frac{\sec\theta}{\kappa_e(\sec\theta_0 - \sec\theta)}[e^{-\kappa_e d \sec\theta} - e^{-\kappa_e d \sec\theta_0}] \tag{8-4.22b}$$

where $\overline{I}, \overline{I}_0$ are the scattered and incident intensities, respectively, and $\overline{\overline{P}}$ is the scattering phase matrix.

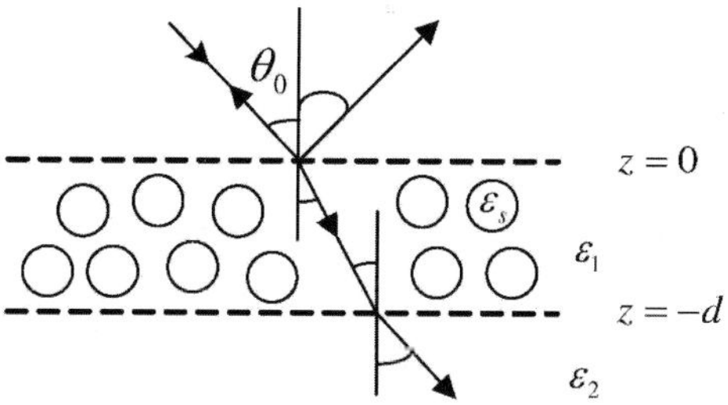

Fig. 8-4.8. Geometry of the problem

From Eqs. (8-4.22a,b), we can obtain the first-order forward-scattering coefficients

$$\gamma_{vv}^{(1)}(\theta_0) = 4\pi P_{11}(\pi - \theta_0, 0; \pi - \theta_0, 0) d \sec\theta_0 e^{-\kappa_e d \sec\theta_0} \tag{8-4.23a}$$

$$\gamma_{hh}^{(1)}(\theta_0) = 4\pi P_{22}(\pi - \theta_0, 0; \pi - \theta_0, 0) d \sec\theta_0 e^{-\kappa_e d \sec\theta_0} \tag{8-4.23b}$$

and the first-order backscattering coefficients

$$\sigma_{vv}^{(1)}(\theta_0) = 4\pi P_{11}(\theta_0, \pi; \pi - \theta_0, 0) \frac{\cos\theta_0}{2\kappa_e}(1 - e^{-2\kappa_e d \sec\theta_0}) \tag{8-4.24a}$$

$$\sigma_{hh}^{(1)}(\theta_0) = 4\pi P_{22}(\theta_0, \pi; \pi - \theta_0, 0)\frac{\cos\theta_0}{2\kappa_e}(1 - e^{-2\kappa_e d \sec\theta_0}) \tag{8-4.24b}$$

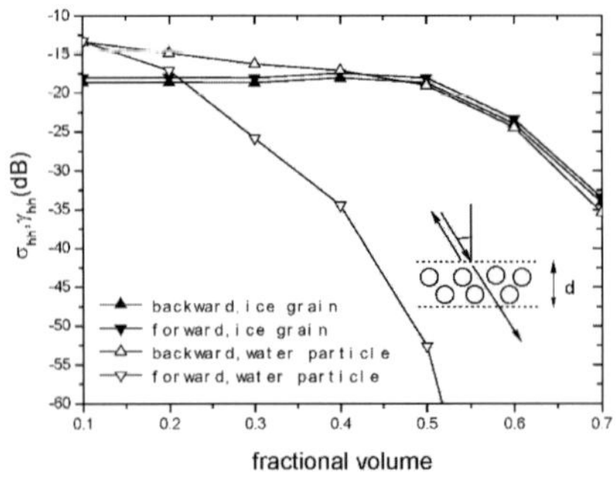

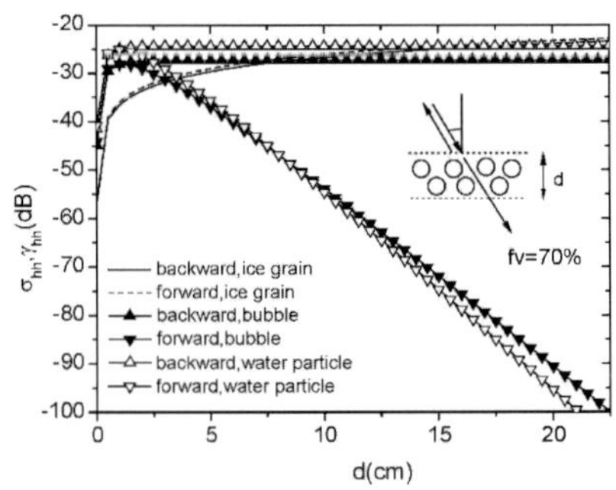

Fig.8-4.9. γ_{hh}, σ_{hh} vs. f_v $(d = 20cm, \theta_0 = 30^\circ)$ Fig. 8-4.10. γ_{hh}, σ_{hh} vs. d $(f_v = 70\%, \theta_0 = 30^\circ)$.

Fig. 8-4.9 shows the forward scattering and backscattering coefficients of a layer of dense particles (ice or water) as a function of f_v. It can be seen that forward scattering of water particles is significantly reduced due to high loss of the water particles.

Fig. 8-4.10 shows the forward scattering and backscattering of a layer of dense particles (ice, water or water bubble) versus the layer thickness.

Appendix 8-4A

Face Centered Cubic Structure

Spherical particles are arranged at the corners and center of each cube face. Let the side length of cube is a.

Assuming that the particles touch along face diagonals, the radius of the spherical particle is $\sqrt{2}a/4$. The spheres contained in the cube are 1/8 sphere whose center is the corner and 1/2 sphere whose center is the face of cube.

The total volume of the spheres contained in the cube is

$$(8\times\frac{1}{8} + 6\times\frac{1}{2})\times\frac{4}{3}\pi(\frac{\sqrt{2}}{4}a)^3 = \frac{\sqrt{2}}{6}\pi a^3 \tag{8-4A.1}$$

The fractional volume is

$$\frac{\sqrt{2}}{4}\pi a^3 / a^3 \approx 74\% \tag{8-4A.2}$$

which is the maximum packing for a system in which all spheres have equal diameter

Arranging many small cubes shown in Fig. 8-4.A1 to construct a big cube V, the fractional volume of the big cube V is also 74%, as shown in Fig. 8-4.A2.

The electrostatic field of a small coated particle is derived by using the Laplace equation. Consider an electromagnetic wave $\overline{E}^e = E_z\hat{z}$ incident upon a coated spherical particle with a shell of permittivity ε_2 and a core of permittivity ε_3 (Fig. 8-4.2). The permittivity of the background medium is ε_1. The coated particle is of outer radius a and inner radius b.

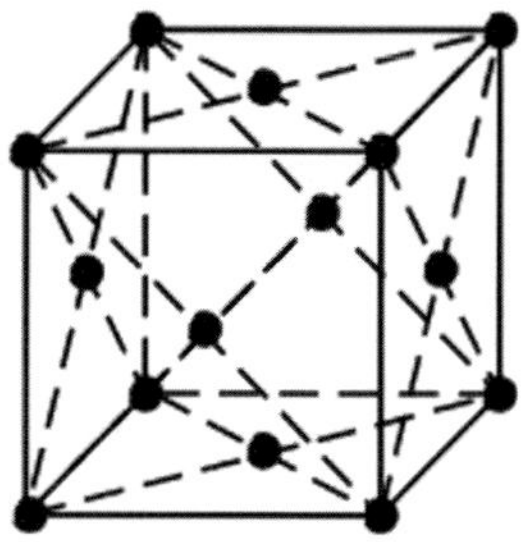

Fig. 8-4.A1. A face centered cubic structure

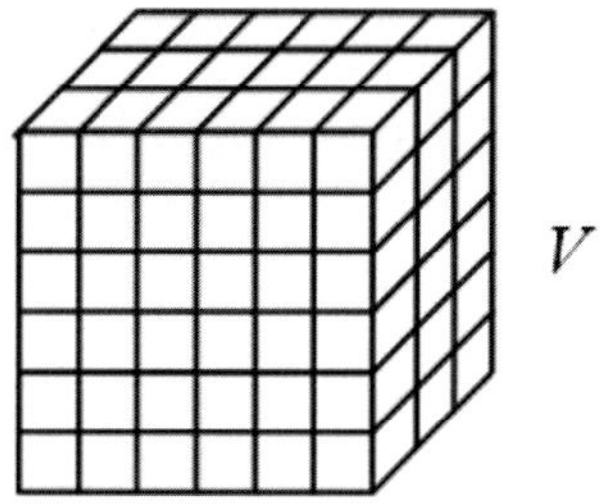

Fig. 8-4.A2. The fractional volume of a cube V

Let Φ_1, Φ_2, Φ_3 be the scalar potentials outside the particle, within the shell and in the core of the coated particle, respectively. Solving the Laplace equation, it yields

$$\Phi_1 = A_1 r\cos\theta + B_1\cos\theta / r^2, \quad \Phi_2 = A_2 r\cos\theta + B_2\cos\theta / r^2, \quad \Phi_3 = A_3 r\cos\theta \tag{8-4A.3}$$

The boundary conditions are

$$\Phi_1 = -E_z r\cos\theta \quad \text{as} \quad r \to \infty \tag{8-4A.4a}$$

$$\Phi_1 = \Phi_2 \quad \text{and} \quad \varepsilon_1 \frac{\partial \Phi_1}{\partial r} = \varepsilon_2 \frac{\partial \Phi_2}{\partial r} \quad \text{as} \quad r = a \tag{8-4A.4b}$$

$$\Phi_2 = \Phi_3 \quad \text{and} \quad \varepsilon_2 \frac{\partial \Phi_2}{\partial r} = \varepsilon_3 \frac{\partial \Phi_3}{\partial r} \quad \text{as} \quad r = b \tag{8-4A.4c}$$

From (8-4A.3) and (8-4A.4), the coefficients A_2, B_2 and A_3 can be easily obtained as follows

$$A_2 = \frac{-3(\varepsilon_{r3} + 2\varepsilon_{r2})E_z}{(\varepsilon_{r3} + 2\varepsilon_{r2})(\varepsilon_{r2} + 2) + 2b^3(\varepsilon_{r2} - \varepsilon_{r3})(1 - \varepsilon_{r2}) / a^3} \tag{8-4A.5a}$$

$$B_2 = \frac{-3(\varepsilon_{r2} - \varepsilon_{r3})b^3 E_z}{(\varepsilon_{r3} + 2\varepsilon_{r2})(\varepsilon_{r2} + 2) + 2b^3(\varepsilon_{r2} - \varepsilon_{r3})(1 - \varepsilon_{r2}) / a^3} \tag{8-4A.5b}$$

$$A_3 = \frac{-9\varepsilon_{r2} E_z}{(\varepsilon_{r3} + 2\varepsilon_{r2})(\varepsilon_{r2} + 2) + 2b^3(\varepsilon_{r2} - \varepsilon_{r3})(1 - \varepsilon_{r2}) / a^3} \tag{8-4A.5c}$$

where $\varepsilon_{r2} = \varepsilon_2 / \varepsilon_1$、$\varepsilon_{r3} = \varepsilon_3 / \varepsilon_1$.

The electric field of the shell is

$$\overline{E}_2 = -\nabla\Phi_2 = -\hat{z}A_2 + \frac{B_2}{r^3}(\hat{r}2\cos\theta + \hat{\theta}\sin\theta)$$

$$= \frac{9\varepsilon_{r2}}{(\varepsilon_{r3} + 2\varepsilon_{r2})(\varepsilon_{r2} + 2) + 2b^3(\varepsilon_{r2} - \varepsilon_{r3})(1 - \varepsilon_{r2}) / a^3} \tag{8-4A.6a}$$

$$\times[\frac{2\varepsilon_{r2} + \varepsilon_{r3}}{3\varepsilon_{r2}}\hat{z} - \frac{(\varepsilon_{r2} - \varepsilon_{r3})b^3}{3\varepsilon_{r2}}\frac{1}{r^3}(\hat{r}2\cos\theta + \hat{\theta}\sin\theta)]E_z$$

The electric field of the core is

$$\overline{E}_3 = -\nabla\Phi_3 = -\hat{z}A_3 = \frac{9\varepsilon_{r2}}{(\varepsilon_{r3} + 2\varepsilon_{r2})(\varepsilon_{r2} + 2) + 2b^3(\varepsilon_{r2} - \varepsilon_{r3})(1 - \varepsilon_{r2}) / a^3}\hat{z}E_z \tag{8-4A.6b}$$

Similarly, for the case $\overline{E}^e = E_x\hat{x}$,

$$\overline{E}_2(\overline{r}) = -\hat{x}A_2 + \frac{B_2}{r^3}[\hat{r}2\sin\theta\cos\phi - (\hat{\theta}\cos\theta\cos\phi - \hat{\phi}\sin\phi)] \tag{8-4A.7a}$$

$$\overline{E}_3 = -\hat{x}A_3 \tag{8-4A.7b}$$

For the case $\overline{E}^e = E_y \hat{y}$,

$$\overline{E}_2(\overline{r}) = -\hat{y}A_2 + \frac{B_2}{r^3}[\hat{r}2\sin\theta\cos\phi - (\hat{\theta}\cos\theta\cos\phi + \hat{\phi}\sin\phi)] \quad (8\text{-}4A.8a)$$

$$\overline{E}_3 = -\hat{y}A_3 \quad (8\text{-}4A.8b)$$

8-5 References

Aden, A. and M. Kerker (1951), "Scattering of electromagnetic waves from two concentric spheres", *Journal of Applied Physics*, 22 1242~1246

Ablitt, B.P. *et al.* (1999), "Imaging and multiple scattering through media containing optically active particles", *Waves in Random Media*, 9: 561-572.

Bohren C.F. (1974), "Light scattering by an optically active sphere", *Chemical Physics Letters*, 29: 458-462.

Barron L.D. (1982), *Molecular Light Scattering and Optical Activity*. Cambridge Univ. Press, UK.

Cohen, L. (1989), *Proceeding IEEE*, 77: 941

Chen, D. and L. Tsang (2003), "Microwave emission and scattering of foam based on Monte Carlo simulations of dense media", *IEEE Transaction Geoscience Remote Sensing*, 41(4): 782-790.

Chen, D. and Y.Q. Jin (2003), "Time-Frequency Analysis of Electromagnetic Pulse Response from a Spherical Target", *Chinese Physics Letters*, 20(5): 660-663.

Engheta, N. (2002), "An idea for thin subwavelength cavity resonators using metamaterials with negative permittivity and permeability", *IEEE Antennas and Wireless Propagation Letters*, 1(1): 10-13.

Gaunaurd, G., H. Uberall and P. Moser (1981), *Journal of Applied Physics*, 52 :35.

Gaunaurd, G., H. Strifors H, W. Wertman (1991), *Journal of Electromagnetic Waves and Applications*, 5: 75.

Ishimaru, A. and Kuga Y., (1982), "Attenuation constant of a coherent field in a dense distribution of particles", *Journal of the Optical Society of America*, 72:1317-1320.

Ishimaru, A. et al. (2003), "Generalized constitutive relations for metamaterials based on the quasi-static lorentz theory", *IEEE Transaction on Antennas and Propagation*, 51(10): 2550-2557.

Jin, Y.Q. (1992), "A Mueller matrix approach to complete polarimetric scattering from a layer of non-uniformly oriented, non-spherical scatters", *Journal of Quantitative Spectroscopy and Radiative Transfer,* 48(3): 295-306.

Jin, Y.Q. (1994), *Electromagnetic Scattering Modeling for Quantitative Remote Sensing,* Singapore: World Scientific 32-64.

Jin, Y.Q. and M. Chang (2003), "Polarimetric Back-scattering and Shift of Polarization Angle from Random Chiral Spheroids", *Electromagnetics,* 23(3): 237-252.

Jin, Y.Q., Y. Chen and Z. Wu (2004), "Retrievals and Applications of Remote Sensing Information from Multi- disciplinary Researches", *IEEE GRSS Newsletters*, September 2004: 7-11.

Koh, G. (1992), "Experimental study of electromagnetic wave propagation in dense random media", *Waves in Random Media*, 2:39-48.

Kim H. and H. Ling (1993), *IEEE Transaction Antennas Propagation* 41: 200

Kaiser G., A. Friendly (1994), *Guide to Wavelets*, (New York: Birkhauser)

Koledintseva, M.Y., D.J. Pommerenke, J.L. Drewniak, (2002), "FDTD analysis of printed circuit boards containing wideband Lorentzian dielectric dispersive meida", *IEEE International Symposium on Electromagnetic Compatibility*, 2: 830-833.

Lakhtakia, A. (1990), "Palarizability dyadics of small chiral ellipsoids", *Chemical Physics Letters ,* 174(6): 583-586.

Lakhtakia, A., V.K. Varadan and V.V. Varadan (1990), "Dilute random distribution of small chiral spheres", Applied Optics 29(25): 3627-3632.

Lindell, I.V., *et al.* (1994), *Electromagnetic Waves in Chiral and Bi-Isotropic Media,* Artech House.

Lindell, I.V., A.H. Sihvola *et al.* (1990), *Electromagnetic Waves in Chiral and Bi-Isotropic Media*, Boston: Artech House.

Lu, C.C. and W.C. Chew (1995), "The application of recursive aggregate T-matrix algorithm in the Monte Carlo simulations of the extinction rate of random distribution of particles", *Radio Science*, 30:25-28.

Metropolis, N., A.W. Rosenbluth *et al.* (1953), "Equation of State Calculations by Fast Computing Machines", *Journal of Chemical Physics,* 21(6): 1087-1092.

Morgan M.A. and K.K. Mei (1979), "Finite-element computation of scattering by inhomogeneous penetrable bodies of revolution", *IEEE Transaction Antennas Propagation*, 27(2): 202-214.

Moghaddar, A. and E. Walton (1993), *IEEE Transaction Antennas Propagation*, 41: 677

Nishimoto, M. and H. Ikuno (1997), *IEICE Transaction Electron.* E80-C: 1440

Pendry, J.B. and F.J. Garcia (1996), "Computational studies of photonic band gaps in metals", *IEE Colloquium on Semiconductor Optical Microcavity Devices and Photonic Bandgaps* (Digest 267): 5/1-5/6.

Pendry, J.B. et al (1999), "Magnetism from conductors and enhanced nonlinear phenomena", *IEEE Transaction on Microwave Theory and Techniques*, 47(11): 2075-2084.

Pendry, J.B. (2000), "Negative refraction makes a perfect lens", *Physical Review Letters*, 85: 3966-3969.

Rotman, W. (1962), "Plasma simulation by artificial dielectrics and parallel-plate media", *IEEE Transaction on Antennas and Propagation*, 10(1): 82 – 95.

Strifors, H., G. Gaunaurd, B. Brusark and S. Abrahamson (1994), *IEEE Transaction Antennas Propagation*, 42: 453.

Siqueira, P.R. and K. Sarabandi (2000), "T-matrix determination of effective permittivity for three-dimensional dense random media", *IEEE Transaction Antennas Propagation*, 48(2): 317-327.

Smith, D.R. and W.J. Padilla (2000), "Composite medium with simultaneously negative permeability and permittivity", *Physical Review Letters*, 84(18): 4184–4187.

Tsang, L., J.A. Kong and R. Shin (1985), *Theory of Microwave Remote Sensing*, New York: Wiley.

Tsang, L., C.E. Mandt and K.H. Ding (1992), "Monte Carlo simulations of extinction rate of dense media with randomly distributed dielectric spheres based on solution of Maxwell's equations", *Optics Letters,* 17:314-316.

Tsang, L. and J.A. Kong (2001), *Scattering of Electromagnetic Waves: Numerical Simulations,* New York: John Wiley.

Tsang, L. and J.A. Kong (2001), *Scattering of Electromagnetic Waves: Advanced Topics*, New York: John Wiley.

Ulaby, F.T. and C. Elachi *ed.* (1990), *Radar Polarimetry for Geoscience Applications*, Mass: Artech House.

Veselago, V.G. (1968), "The electrodynamics of substances with simultaneously negative values of ε and μ", *Soviet Physics USPEKI,* 10(4): 509-514.

Yang, Q. and Y.Q. Jin (2005), "Numerical Monte Carlo Simulation of Correlated Scattering from Very Densely Random Spherical Particles", *Journal of Applied Physics* in press.

Yao, H.Y. and L.W. Li (2004), "Performance analysis of metamaterials with two-dimensional isotropy", *Proceedings of Annual Symposium of Singapore-MIT*, 58: 19-20.

Ye, H. and Y.Q. Jin (2005), "Polarimetric Scattering from a Layer of Random Metamaterial Small Spheroids", *Applied Physics*, 31: 3~9.

Zurk, L.M., L. Tsang *et al.* (1995), "Monte Carlo simulation of the extinction rate of densely packed sphere with clustered and non-clustered geometries", *Journal of the Optical Society of America*, 12:1772-1781.

Zurk, L.M., L. Tsang and D.P. Winebrenner (1996), "Scattering properties of dense media from Monte Carlo simulations with application to active remote sensing of snow", *Radio Science*, 31: 803-819.

Index